Linear Algebra by Example

This book offers a modular, concept-driven introduction to linear algebra designed for undergraduate students across mathematics, engineering, computer science, and the sciences. Emphasizing core ideas such as symmetry, transformation, and structure, it supports flexible pacing and multiple instructional pathways while remaining accessible to students with diverse mathematical backgrounds.

What sets this text apart is its intentional structure and student-centered design. Each chapter begins with a visual flowchart that makes dependencies explicit, allowing instructors to customize coverage. Proofs are introduced gradually through examples and activities rather than formal presentation, helping students develop mathematical reasoning with confidence.

Key features include:

- Modular chapter design with dependency flowcharts
- Scaffolded examples and classroom-tested activities
- Integrated SageMath and Python exploration
- Built-in AI prompts for responsible, productive use
- Applications ranging from classic topics to ray tracing, molecular symmetry, and structural engineering

This book is intended for first undergraduate courses in linear algebra, including service courses and active-learning classrooms, and is suitable for both instruction and self-study.

Erik Wallace is an Assistant Professor of Instruction in Mathematics at Temple University. He earned a BA in Mathematics from Hartwick College and a PhD in Mathematics from Indiana University, specializing in number theory. He has extensive experience teaching linear algebra and related courses to non-majors and focuses on active learning, accessibility, and meaningful applications, integrating computation, SageMath, Python, and AI-assisted tools into his teaching.

Linear Algebra by Example
An Active Approach

Erik Wallace

CRC Press is an imprint of the
Taylor & Francis Group, an **informa** business

First edition published 2027
by CRC Press
2385 NW Executive Center Drive, Suite 320, Boca Raton FL 33431

and by CRC Press
4 Park Square, Milton Park, Abingdon, Oxon, OX14 4RN

CRC Press is an imprint of Taylor & Francis Group, LLC

ISBN: 978-1-041-23504-0 (hbk)
ISBN: 978-1-041-23508-8 (pbk)
ISBN: 978-1-003-73749-0 (ebk)

DOI: 10.1201/9781003737490

Typeset in CMR10
by KnowledgeWorks Global Ltd.

Publisher's note: This book has been prepared from camera-ready copy provided by the authors.

This book is dedicated to Jeff Rients and Emtinan Alqurashi, who each provided me with game changing insights into teaching.

Contents

Preface to the Instructor

What does a well-functioning classroom look like? I would say it is one in which there is a high level of student engagement while meeting course goals. There is a model of student engagement based on three variables[16]:

- **Value.** The extent to which a student finds the class to be useful, interesting, or important.
- **Expectation.** The extent to which they believe they can succeed in the class.
- **Environment.** The extent to which the classroom atmosphere is perceived as being positive or negative.

I have achieved 90% engagement with this book, and I have done so in a way that is fair and equitable.

I do not believe that this book on its own is sufficient for producing these results, however, it does contain certain aspects of content and style that I believe to be contributing factors. I will try to address some of the other factors below in *How I use this book*, *Assessments*, and *Value*.

As for course goals, there are many options to consider including:

1. Gaining an understanding of how linear algebra is applied in other subject areas.
2. Developing competence with computations by hand.
3. Developing problem solving skills.
4. Learning numerical methods in linear algebra.
5. Learning how to program with Python and Sage.
6. Learning how to interact with AI.

All of these goals are facilitated by various features of this book. There are AI prompts to teach interaction with AI. There is a section introducing students to Sage and Python. The Sage and Python code is sometimes requested from the AI to write algorithms or prepare data and examples for applications. When the algorithm corresponds to built-in features of Sage, then code is also used to check the correctness of what the AI produces. Other examples and activities are meant to build computational skills, which then can be used as a tool in problem solving. There are also various parts of this book that lay groundwork for proofs and other content in higher level math classes.

Your choice of goals can influence the path that you choose to take through the book, which is facilitated by a modular structure. Each chapter starts with a diagram indicating the dependence between subsections within that chapter. For example, if you want to focus on numerical methods, you can choose a path that leads to the SVD as efficiently as possible. That choice has two consequences: first, the material covered will use Sage and Python code more heavily and, second, at least single variable calculus will have to be assumed because limits provide an essential viewpoint to many numerical methods. On the other hand, it is also possible to take a path that does not assume calculus at all. Even if a subsection has some activities on calculus, you may still be able to cover it if there are enough other activities in that section that don't depend on calculus. See the subsection on Sample Course Layouts below for more detail. No matter what path you choose, there are two things you should do to obtain high engagement and equity:

1. Course goals should be assessed with a grade.

2. Course content must be self-contained.

The importance of the course being self-contained cannot be under estimated. Chapter 1 provides the computational tools for all subsequent activities. It is unwise to go through any part of Chapter 1 faster than the students can handle.

The rest of this preface is divided into separate subsections, focusing on different issues. These subsections are in the following order: How I Use This Book, Sample course Layouts, The Efficacy of Examples, Assessments, Value, and Thoughts on AI.

How I Use This Book

You can use this book however you want, but I am providing a description of my own technique as an indication of at least one way that high engagement and equity can be achieved. Students have a lot going on outside of our classrooms. I try to deal with this in two ways:

1. I try to create a classroom environment where students can feel at home.

2. I try to avoid expecting too much work done outside of class.

The first day is critical for laying the foundation for a positive environment. I do not do any new material on the first day. I hand out a "visual syllabus" that is easier to digest even if a classic syllabus is still posted, along with a one page schedule. I have slides prepared to introduce them to the class. The first slide is an icebreaker: "Your name, your major, your favorite number, and something you are good at and how you became good at it." The last prompt gets them thinking about what learning is. The next two slides show the front and back side of the visual syllabus so that it is on the screen as we are discussing it. The remaining slides go into more detail about course features.

I always make sure to include two slides indicating what a typical class day looks like. Here is what a non-quiz day looks like for me

- Intro questions

- New material with slides

- Activities

A day with a quiz is the same except that the quiz comes before the intro questions. If students will be using Sage on their laptops, then I help them install Sage at the end of the first day. Most students use either Windows or Mac, so getting them to move with the Windows users on one side of the room and Mac users on the other side is helpful.

The main goal of the intro questions to get prior knowledge into active memory, which is comparable to what is described in chapter 14 of [1]. This book includes introductory questions at the beginning of most subsections. Some questions may be easy, others may require a few hints; but ideally, they should dovetail with the new material for that day so that the students can connect the new concepts with their prior knowledge.

Unlike some other active learning approaches, I do not assume that the students have looked at any materials prior to class. They have the schedule so they could look at the book in advance, but the purpose of the slides is to get everyone on the same page. Often it is fine to put all of the new material before the activities. However, there are times when the new material is so new that it is necessary to stop and do an activity to ensure that the students understand it before continuing. For example, in the first section on representation theory, I would only go through the new material up to Activity 7.1.5, then stop and let the students do that activity before continuing. Subsection 4.2.1 is another case where it is important to alternate between new material and activities. Some students like having the slides in advance. But when I am alternating between new material and activities, I like to have solutions on the slides to recap the results. This approach makes it necessary to have two versions of the slides: one without the solutions that can be posted before class, and one with the solutions to be used during class. Either way, I try to have relevant examples on the screen as students are working on the activities, and when it is not possible I switch between slides as needed.

The activities are strongly scaffolded, meaning that they are broken down into steps that guide the students. Although I am trying to get the activities in a digital form, I still usually distribute the activities on paper printouts. As the students work on activities I walk around, monitor progress, answer questions, and assist as necessary. Inevitably, because groups work at different paces, some groups will get finished before others. This happens both when alternating between new material and activities, and when putting all of the activities at the end of class. If a group finishes early, I let them work on other things. When alternating between new material and activities, I try to avoid starting an activity unless there is enough time remaining for most students to complete it. I always collect activity sheets at the end of class, then I can hand the activity sheets back out in the next class. Most students are good about documenting their work by taking images of the activities with their phone, or using a separate notebook, but it is good to remind them.

The sort of interaction that I just described is much harder to pull off remotely. I have been forced to hold certain class days remotely, but the level of success depends on the topic. Section 4.1 is one that I would never do remotely. Once, when classes were being held by zoom due to bad weather, I actually had to reorganize the schedule so that Section 4.1 could be done in person. The problem with Section 4.1 is that most students are not comfortable using definitions. Even with the hint in Activity 4.1.3 it is necessary to go around group by group and make sure that everyone understands what they are doing. Ideally, it might be nice to end class with a discussion or reflection that ties everything together, but with students working at different paces it is not always practical. An alternative is to assign brief reflections or comprehension checks outside of class.

What if a student is absent? I tell them the examples and activities that we went over. They can then look at those in the book and get caught up. What happens if a student comes in late? First, I welcome them unconditionally. They join their group and can get caught up with the activities. If too many people from a group are missing on account of students being late or absent, I may encourage some groups to merge.

There are many different approaches to active learning, as indicated by [3]. The approach that I have just described does not fit neatly into any of the categories the authors of that article mention, but is probably the closest to a flipped classroom except that I do not assume preparation outside of class. Other approaches to using this book may be possible. I think that many of the activities could be adapted to a think-pair-share format. The exercises contain problems that go beyond the examples and activities, which could be discussed in the classroom, and there are also comprehension and reflection questions which take a deeper dive.

Sample Course Layouts

The modular structure of the book is meant to provide a high level of flexibility for course design, but Table 0.1 at the end of this preface provides three examples of how I might lay out a course. These sample course layouts are based on the following assumptions:

- A term roughly 40 class days in length with 50 minutes per class.
- Division of the material into small units, with one test per unit.

Unit tests take a full class period. With 7 units test days and the first day removed from the total, this leaves roughly 32 class days available for new material. An additional 2 or 3 days should be removed at the end if you are planning group presentations for projects.

Each line of the table is roughly one class day. Of course, a particular class might get through certain sections faster or slower than expected. It is always a good idea to come prepared with more material than needed and delay or omit material based on what actually happens. Say that you want to cover block matrices in Unit 1, but run out of time. You can still introduce them in a later unit on the fly when you actually need them. Or, you could change the schedule itself by increasing the length of Unit 1 and decreasing the length of a later unit.

If an institution decided to make Linear algebra a 4 credit class, that would really open up new possibilities about how much can be covered. But to use my version of competency grading for a 4 credit class it would be better to keep the class length the same, and increase the total number of class periods. Otherwise, Unit test days would become awkward.

Chapter 7 does not have a flow chart at the beginning but basically needs to be covered in order. Three days is enough if you omit Theorem 7.3.7, but you would need a fourth if you want to include it. Calculus is not assumed, but Theorem 7.3.7 is hard without Sage. Chapter 7 is useful for at least two reasons: as background for chemical symmetries or as warm up for an abstract algebra class. A chemistry department may or may not care about chemical symmetries, and a community college may not offer abstract algebra. But if it is known that a certain number of students will eventually need the material, there could still be an incentive to include it. Communication between departments and institutions, and communication with students about what they actually need, will go a long way toward making an integrated and meaningful curriculum.

No Calculus/low tech. This layout is traditional but more applied rather than proof based. By low tech, I mean that Sage activities would be omitted, but not necessarily Desmos activities. Desmos does work on smart phones, although it is better on a laptop or tablet.

If calculus is not assumed you would omit calculus based example activity. For instance, even though Example 4.3.6 is useful connecting the concepts of Null Space and Column Space with prior knowledge for students who already know calculus, Example 4.3.6 should be sufficient for students who don't.

Chapter 7 is both calculus free and low tech, but could also be swapped out for another topic. The section on Markov Chains in Chapter 8 can also be covered in about 3 days. With only 28 days of new material in the list, there is room for another topic or two, or for extra time on certain sections if needed. Other topics could include LU factorization, or One-to-One and onto (if preparation for proof is a goal).

Traditional with the invertible matrix theorem (IMT). This layout could still be done without assuming calculus but puts a much stronger emphasis on the IMT. At 33 days of new material, it may be hard to include both Unit 7 and a project depending on the exact length of the term. The schedule becomes even more packed if you want to include Checking Linearity or the Choice of Scalars, which would add one or two more days.

Modern with Sage. While the previous two layouts follow a pattern of one unit per chapter, the modern layout is radically different. By removing row reduction, the remaining key concepts from Chapters 2 and 3 can be fit into a single unit. Chapter 6 can then be expanded into two units, giving more time to cover the SVD.

Calculus shows up in several ways. The QR algorithm is a limit, and the introduction to numerical stability and condition numbers uses differentiation. Therefore, single variable calculus should be a prerequisite. It also helps if Taylor series or partial derivatives are familiar, but in both cases it is possible to reason by analogy. Example 5.6.9 and Activity 5.6.10 technically provide a way of doing the QR algorithm completely by hand! I found that example by hand, but it is very demanding on algebra skills and uses knowledge of continued fractions and Fibonacci numbers. So, I am not sure I would spend class time on Example 5.6.9 and Activity 5.6.10. Maybe they would be fine for an out-of-class assignment. Activity 5.6.7 and Activity 5.6.8 are fine in class.

Some practical notes. I often find it practical to combine several similar activities together. For example, once students have been provided with the formulas for several different cases of the characteristic polynomial, they can usually do several of the characteristic polynomial activities in one shot. Any activities or homework with Sage are setup in Jupyter notebooks. In particular, I have the entire subsection introducing students to programming with Python and Sage in a Jupyter notebook. But I also emphasize to the students that they can use AI to help them write code, and sometimes I make Sage assignments that intentionally include AI in the workflow. For example, I have a Jupyter notebook that explores the partial fraction decomposition with Sage and AI.

For maximum engagement it is best for out-of-class assignments (except a project) to be worth no more than the 20% of the final grade. The reason is that certain students will not turn in any homework. Such students often see very low value in the class initially. Competency grading has a natural effect of increasing value, but homework may still not be submitted. If homework is worth 30%, then suddenly their maximum grade is 70%, and that will immediately cause them to have low expectation. Expectation drives value, so then their sense of value plummets when they realize that they have no hope for a good grade. For this reason, I have come to favor the following percentages: Activities 20%, Homework 20%, Project 10%, Quiz/Exam 50%. When using Sage in class, I also include Homework with Sage. Thus the homework category might be further subdivided as Exercises 10%, Sage 5%, AI 5%. Written reflections or reading comprehension might count in the Homework category if done outside of class, or in the Activities category if done inside class.

The next section on assessments goes into more detail on how I design problems for quizzes and tests. The principle of avoiding double jeopardy is very important for quiz and test problems. But the exercises do include multi-step problems. If these were done online, it would be an easy matter to allow for multiple attempts. But for paper based homework resubmitting and regrading assignments could be too complicated. I have actually asked for resubmissions of homework as well, but I have not had to do it very often. I select the exercises or design assessments based on what is covered in class. I also try to be mindful of the total number of pages that the students are required to do outside of class, since many students have jobs or commute.

The Efficacy of Examples

Part of the idea of my approach with examples is that they can be used to shed light on abstract concepts. When reading a book like [23], I often write down examples to make sure that I understand all of the details. But is that really what the students get out of it? I have used True/False questions on quizzes and have seen students compute examples to figure them out, so apparently the answer is yes.

Examples have many other uses, including:

1. as a problem solving tool,
2. as landmarks for what is true or false, or
3. as a way to test the correctness of computer code.

If one of your goals is to ease students into understanding abstract mathematics, such as proofs and theorems, then both of the first two points are useful. A theorem may have built-in assumptions. Why are they needed? What goes wrong if one of the assumptions is removed? Is there an example showing that the theorem is no longer true if the assumption is removed? The proof itself might be written abstractly. How can the student be confident that they have understood each step? We want students to know that one or two examples of something isn't proof, but if a student can't think of any examples then it could be that they don't understand what is going on. Furthermore, a sufficient number of well chosen examples can be suggestive of how to prove something. Here is where using examples as a problem solving tool comes in. How do you prove something new? Finding a proof is a type of problem. A problem might be too big to tackle in general at first glance. We might try a version with small numbers, and work out some simple cases as suggested by Polya [20]. Those computations might reveal certain assumptions that have to be made in order for a proof even to be possible. They might also be suggestive of a big picture giving a hint as to how to do the proof. True/False questions can encourage this way of thinking about examples, if you teach students to try to check computationally rather than just guessing.

This book shies away from formal proof, but aims instead to give students a taste of concepts which can then serve as prior knowledge in subsequent classes. For example, the material on one-to-one and onto in subsection 4.3.2 is intended to be a gentle introduction to the way these concepts show up in proof. The material on choice of scalars in subsection 4.2.4 technically involves proof by contradiction. In subsection 2.3.1, row equivalence is explained as an equivalence relation. Checking that something is a linear transformation in 4.2.3 introduces the standard style of proof seen with homomorphisms. When I taught an introduction to proofs course immediately after linear algebra, I was able to refer directly to things that I knew they had seen in this book. Similarly, even though Chapter 7 was added mainly to support the section on chemical symmetries, it could also serve as prior knowledge in an Abstract Algebra class if it was covered.

Assessments

So far as I can tell, the design and grading of assessments plays a key role in both engagement and equity. In [4], the authors say that assessment is more helpful for retention of concepts than repetition. So perhaps the first thing to consider when choosing assessments is what you want the students to remember.

Assessments are connected to engagement because of the correlation between value and expectation. Expectation is strongly correlated with grades, and assessments have a direct impact on grades. Good grades promote positive feelings toward a class and even toward a subject as a whole, which then can increase the sense of value that subject has for them. However, learning is almost never instantaneous. It may take several tries before a student gets a concept, which is why it is often recommended to separate *formative assessments* with *summative assessments.* A summative assessment should be reflected in the final grade, while formative assessments generally should not be.

I use a the following approach to competency based grading, which provides a smooth transition between formative and summative assessment. Quizzes and exams count as one category. Students get 3 tries per problem type: 1st on a quiz, 2nd on a unit test, 3rd on the final. The grade on a problem type is the maximum score over the tries for that problem type, and their overall grade in the quiz/test category is the average over all problem types. In this way, formative grades are replaced by summative grades. There are other strategies, but a strategy of giving students multiple tries will directly impact students' expectation of success, and indirectly impact their sense of value. With both expectation and value increasing, you can expect engagement to increase.

The content and design of computational assessments is also critical for equitable grading. In-class assessments such as quiz and test questions should be broken into independent pieces to avoid Double Jeopardy. For example, the following question would lead to a Double Jeopardy effect:

Given the matrix

$$\mathbf{A} = \begin{bmatrix} 0 & 3 \\ 1 & 2 \end{bmatrix}$$

compute the eigenvalues and eigenvectors, then diagonalize the matrix.

If a student remembers how to compute the eigenvectors but forgets how to compute the eigenvalues, they will not even be able to start and their actual knowledge will not be accurately reflected. Instead, each piece should be a separate problem:

1. One problem for computing the characteristic polynomial of a matrix.
2. One problem for computing the eigenvalues given the characteristic polynomial.
3. One problem for computing an eigenvector given an eigenvalue.
4. One problem for computing the diagonalization given the eigenvalues and corresponding eigenvectors of a matrix.

Now each detail is tested and is its own problem type, and the assessment more accurately reflects what the student knows. Granted, it is also important to see that students can put things together, but I tend to assign such problems on homework. If you look at the exercises, you will see many multi-part problems among them. While students have a lot going on outside of class, I have found that a one week grace period is usually enough for homework to be submitted. Since homework is reflected in the final grade, and is therefore summative, I also let students resubmit any written homework. Online homework systems usually allow for multiple tries as well.

Value

Among the variables influencing engagement, value is probably the trickiest to control. I have already explained the connection between grades and value in *Assessment.* But another very important factor is the aspect of usefulness. Many students taking linear algebra are in other majors. It is helpful for them to see applications to their majors. There is a trend of using real life applications to introduce topics such as in the eventmath project, to which I have actually contributed. However, I do see a conflict with another aspect of value, which is the extent to which the student understands what is going on in class. Typically, the more realistic the problem, the harder the math. Quite often, what students like about my problems is that the clutter is cut out: they can handle the calculations and can see what is going on in class. So I tend to use the applications as something to build up to, which is why most of the applications are in the last chapter. But given the modular structure of the book, you could easily bring those applications in at any time. You could also have students ask an AI to suggest applications within their subject area.

Simply including applications in a class does not seem to be enough to improve the sense of value. Ideally there should be integration between classes. Having linear algebra as a requirement or an elective within a major is not as meaningful as having it as a prerequisite for a class, and deliberately including material in support of the class for which it is a prerequiste. For example, if a very large number of business majors need linear algebra for the simplex method, it could make sense to include that in the class for them. At a big university there could be enough students for a separate class, but in case there are many different majors, I have found it practical to assign projects, which makes it possible for the students to personalize the class without forcing material on other students.

Thoughts on AI

Recently there have been major advances in AI. What can we expect a future with AI to look like? I believe that chess gives the best precedent. Prior to the match in which Deep Blue beat Gary Kasparov, the best humans were better than computers at chess. By now, Stockfish and Leela Zero are the current best computer programs at Chess. They are both free, can run on a laptop, and would beat not only the best humans but even Deep Blue. Essentially anyone with an internet connection and decent hardware now has expert level chess analysis at their fingertips. Meanwhile, humans still play chess but they now use computers to explore ideas.

My own experiments with ChatGPT indicate that it is below the expert level as of this writing. But depending on when you read this, that may no longer be the case; and there are many other options including Claude, Deepseek, Gemini, Grok, Kimi, and Qwen. AI is developing very fast. The chess analogy shows that competing against AI is hopeless, which means that we should not be teaching our students to compete against AI. Instead we should be teaching them to interact with AI in a way analogous to how Chess players have used Stockfish to explore ideas. I have included AI prompts in an attempt to start teaching this kind of interaction. Just as people have not stopped playing chess, I do not see massive job loss happening due to AI. Instead, I see the nature of the jobs changing. I see people being freed from tasks that are beyond human capabilities, allowing for their energies to be reallocated in fulfilling ways better suited to the human condition.

TABLE 0.1: Three sample course layouts

No Calc/low tech	Traditional with IMT	Modern with Sage
Unit 1		
1.1.1, 1.1.2, 1.1.4 1.1.5, 1.1.6 1.2.1-3 1.3.1-3	1.1.1, 1.1.2, 1.1.4 1.1.5, 1.1.6 1.2.1-3 1.3.1-3 1.3.5, 1.3.6	1.1.1, 1.1.2, 1.1.4 1.1.5, 1.1.6 1.2.1-3, 1.4.1 1.3.1-3, 1.4.2 1.1.3
Unit 2		
2.1 2.2 2.1.1, 2.3.1 2.3.2	2.1 2.2 2.1.1, 2.3.1 2.3.2 2.3.3	2.1 (except 2.1.1) 2.2.1-2.2.3 3.1.1, 3.1.3 3.1.5, 3.2 3.3.1, 3.3.3
Unit 3		
3.1.1, 3.1.2 3.1.3, 3.1.5 3.2, 3.3.1 3.3.2, 3.3.3	3.1.1, 3.1.2 3.1.3, 3.1.5 3.2, 3.3.1 3.3.2, 3.3.3 3.3.4	4.1, 4.2.1 4.2.1 4.3.1 1.5, 3.1.4, 4.7.1 4.8 4.4.1, 4.5
Unit 4		
4.1, 4.2.1 4.2.1 4.3.1 4.4.1, 4.5	4.1, 4.2.1 4.2.1 4.3.1 4.3.2, 4.4.1 4.4.2 4.2.2, 4.2.4	5.1.1 5.2.1-5.2.3, 5.2.5 5.4 5.3.1 5.6
Unit 5		
5.1.1 5.2.1-5.2.3 5.2.8 5.2.9 5.3.1	5.1.1 5.2.1-5.2.3 5.2.4 5.2.8 5.3.1	6.1.1, 6.1.2 6.1.3, 6.1.4 6.2.1, 6.2.2 6.4.1, 6.4.2, 6.6
Unit 6		
6.1.1, 6.1.2 6.1.3, 6.1.4 6.2.1, 6.2.2 6.4.1	6.1.1, 6.1.2 6.1.3, 6.1.4 6.2.1, 6.2.2 6.2.3	6.5.1 6.5.2 6.5.3
Unit 7		
7.1 7.2, 7.3 7.3	7.1 7.2, 7.3 7.3	7.1 7.2, 7.3 7.3

Preface to the Student

When am I ever going to use this? What is this good for? Linear Algebra is a very general and a very powerful subject. It is used all over mathematics and in other disciplines as well. But don't take my word for it. Ask an AI:

AI Prompt 0.0.1. What is an application of Linear Algebra to _________?

I encourage everyone to practice interacting with AI. If you don't already have an account, you should set one up. This book includes AI prompts that explore how to use AI as part of your workflow. In general, it helps to be as specific as possible. But beware! As of this writing, AIs still may give incorrect results. For this reason, you should not just trust the AI, but always verify.

Here are some other features of the book that I hope you will find useful:

- A section on how to read a math book.
- TL;DR boxes that provide references to key information.
- A section of hints for the exercises.

I also recommend reading "A short guide to mathematics" after this preface.

Examples are very useful, but it is important to approach them with the right mindset. Here are some ways that examples are used in this book:

- To illustrate abstract concepts.
- As landmarks of what is true or false.
- To build computational proficiency.

Mathematicians do general calculations algebraically, but in the back of our minds we understand the context of the algebra. Examples provide us with an understanding of that context. If we are solving a problem, we also might use examples to explore. Therefore, we need to be able to compute examples without being told. If we get a computer to do the example we still need to know that the computer is doing the right thing. We might test the computer on a small example or with a basic reality check. Does the answer make sense? If you approach this book with this attitude toward examples, I believe that you will benefit greatly and hopefully have fun in the process.

Acknowledgments

This book began its life as a set of lecture notes with examples for my Finite Math students at Indian University, which were inspired by a similar set of notes by Christian Hoffland. Had I not seen Hoffland's notes at all, I probably never would have started writing this book. I was exposed to the grid method of multiplying matrices at Indiana by Tristan Tager. At UConn I became disillusioned with lecture. Adam Giambrone pointed me in the direction of active learning.

Many people at Hartwick College assisted in the development of this book. James Cochran and Joelle Ocheltree in the writing center helped polish the prose. Joelle was also one of my linear algebra students and was able to provide some feedback from the student perspective. Cody Webb in Chemistry and Karl Seeley in Economics provided feedback on subject specific content. W. Donald Cotter of Mount Holyoke introduced me to the use of representation theory in chemical symmetries over a zoom call while I was still at Hartwick. The section on the structural engineering of trusses was heavily influenced by my memory of Ken Carper's class at Washington State University when I was a student.

At Temple University, the biggest influences came from Daniel Szyld, Jeff Rients, and Emtinan Alqurashi. Szyld's feedback led, among other things, to the modular structure of the book and the addition of more material on modern algorithms. I had seen a book with a modular structure before, specifically *Quantum Field Theory* by Szrednicki. Jeff Rients introduced me to concept maps and the importance of prior knowledge, among many other concepts. Emtinan Alqurashi introduced me to *Discovering Statistics with IBM SPSS Statistics* by Andy Field. His book includes many comic-book-style characters that appear in different types of boxes for different purposes. I wasn't about to adopt the comic book style, but the TL;DR boxes and warnings in this book have some of the same goals. Emtinan also led me to start thinking about how to use AI directly in the pedagogical process. There are countless other influences, and I am probably not fully aware of many of them. I express my deepest thanks to all of the people who influenced this book, whether directly named here or not.

A Short Guide to Mathematics

Introduction

Students often want to know how to do well in a math class. There is no secret, and yet it is not obvious or not always easy to carry out. Sometimes events outside of class get in the way. But there are also preventable issues, and sometimes there are misconceptions. What is math? What is learning? These are not easy questions, but it is my belief that everyone is capable of becoming proficient in math. So, how?

What is math?

I would say that math is one aspect of what it means to be human. Math is fundamentally about thinking, logic, learning, and solving problems. People had practical real life problems. Someone may have asked "How many cattle do I have?" To answer this question, they had to arrive at the concept of counting numbers. How did they do it? We are taught how to count by our parents, but a very long time ago someone had to come up with the idea without ever being told. Later someone may have asked "How do I get from Alexandria to Athens?" The Mediterranean sea is between Alexandria and Athens. Someone with a boat would need to know how to navigate. They would need to look at the stars, and plot their course. Star charts and maps may have been at least part of the motivation behind geometry (although buildings also played a role). In ancient times, people did not have Google Maps or GPS. In fact, Google Maps and GPS would not exist without the insights about geometry that have been known since ancient times. The study of numbers and geometry quickly led to other questions. Among these questions is a very central one: how do we know? It is the justification of knowledge that led to logic and proof.

Today we have the benefit of several millennia of accumulated knowledge of mathematics. However, many things remain unknown. There are still questions and unsolved problems, and there is currently no hint that we will ever run out of problems or questions. When we encounter a problem that we don't know the solution to, what do we do? Google? What if Google doesn't turn up the solution? What if nobody knows the solution? Then we are in the same situation as the first person in history to ask "How many cattle do I have?" We have a problem to solve, and we need to think and analyze the problem, and try to find the solution somehow. In trying to find the solution, it is unlikely that our first guess will be correct. So, we try something else. We may try many wrong ideas until we finally get something that works. We are not automatically born being able to do something, since at first we cannot even walk, nor are we immediately teleported to the finish line. Instead there is a gradual process. We crawl first. We try to stand. We fall down, and maybe even cry. So there is a struggle, but by perseverance we eventually succeed. Another way of saying this is that we should expect many failures in the process of attaining success. That is to say, failure should not be looked at as a bad thing, but as a natural part of being human, and as a natural part of problem solving and learning. This is the viewpoint that we mathematicians embrace: being confused and searching until the light bulb goes off; the aha moment when we realize the answer. Therefore, you should be patient with yourself and not expect immediate results, but keep trying.

The several millennia of accumulated knowledge does come with benefits, depending on the approach we take. For instance, the solutions which are already known give us tools, which make the search for answers more efficient. These tools come in many forms. First, there are the known results to previously solved problems. If we use certain basic facts often enough, like the single digit multiplication table or the quadratic formula, then it becomes useful to have those things memorized. Second, there are known methods. Being familiar with known methods is often even more useful than memorizing basic facts. A method known to work for one type of problem might be adapted to work in a new context, to solve a problem whose solution is currently unknown. Third, it is certainly true that modern technology is the result of the collective development in math and science over time, and that it remains useful in the further exploration of these fields.

Problem solving

"Where do I start?" If this question occurs to you, then the follow up should be immediate: *start by trying to understand the problem.* Ask yourself. What is given? What are we asked to find? Are there units? Is anything extra assumed that is not written in the problem (like a formula for area, or a trig identity)? Read the problem carefully to make sure that you don't miss any important details. Definitions are also important. A very slight change in a definition can completely change a problem.

The next step is to *come up with a plan.* If the problem has units, then keeping the units in the calculations can be useful for working through the problem and catching mistakes. A well labeled diagram can also help lead through a problem. In some cases, if the numbers given are too big, it can be helpful to try a similar problem with smaller numbers. Most of the problems I give for activities in this book already are designed to have numbers that are smaller and as simple as possible, but it is important to see that those problems generalize.

Collaboration

It is common in many professions for teams of people to collaborate on solving a problem. When you work in a team, keep several things in mind

- Be respectful of each other,
- Don't let one person do all the work,
- Don't be afraid to make guesses. Even if the first guess is wrong, that is how progress is made.
- Don't rely too much on technology.

Using technology

While it is true that technology is useful for further exploration and problem solving in math and science, that depends entirely on the way it is used. The burden is on us to use the technology correctly, responsibly, and ethically. There is a classic phrase "garbage in garbage out" (GIGO for short). It means that, if you enter the wrong information into a computer or calculator, then you should not expect the answer to be correct. But there are even more reasons that the answer entered in a computer or calculator might not be correct. Maybe the correct information was entered, but not in the correct way. For example parentheses might be needed for a calculator to do the intended order of operations correctly. If units are involved, maybe a unit conversion was not done correctly or not done at all. Even NASA had an accident that ultimately was caused by not doing a conversion between metric and inches (see [18]). A human designing the hardware or software could have made a mistake in the design. Even Intel made a mistake along these lines, resulting in incorrect division on some of its processors (see [7] for details). Meanwhile, bugs are routine in software. There could be data corruption. Computers generally make errors less commonly than humans, but that doesn't mean that they never make errors. Hardware is physical and can deteriorate, causing more errors as it ages. A processor that overheats will also make more errors than a processor with a sufficient amount of cooling. It is standard to use codes for error correction, but they only work if there are not too many errors. As such, while the probability of a computer making errors may be small, it is never zero. The section on achieving success will go into more detail about the use of technology.

Achieving success

Math builds on itself more than any other subject. Not only do higher level classes depend on lower level classes, but even within a class, material builds on previous material on a day to day basis. If you miss a day of class, then you miss the material for that day. You cannot expect to understand material on the next day, which may assume familiarity with material that you missed. The more days of class that are missed, the harder it gets to keep pace. There are, at times, legitimate reasons for missing class, but that does not make it any easier. Missing class makes it hard to keep up regardless of the reason.

The importance of attendance cannot be overstated, but it is also part of a bigger picture about attitude and lifestyle. Poor attendance leads to the need to catch up. The need to catch up leads to cramming and lack of sleep. Lack of sleep in turn makes it difficult to focus and retain information. Good attendance should therefore be paired with a regular sleep schedule, healthy eating habits, and spreading work out over time to minimize the need for cramming. Being well rested will generally make it easier to focus on the material.

Take homework for example. Let's say that there are 10 problems on an assignment. You should start as soon as possible. Spend no more than a minute or two on each problem, you may get a few of the easier ones, and you will at least have started thinking about the harder problems. Stop working on those problems. Sleep on it. While you are asleep, or walking around doing other things, the brain may be working on them subconsciously. When you come back, you may find the problems easier to do. Something may jump out that you didn't notice before. If you are really confused, you can ask a friend, or go to office hours. On the other hand, if you put off your homework until the day before it is due, you will not have the opportunity to sleep on it. Moreover, you will be cramming on every assignment, and unlikely to remember the content.

The use of computers, calculators, and other aspects of modern technology can be helpful or it can be a hindrance: it all depends on how they are used. While "learning" is hard to define, it seems very closely connected to memory. Even when calculators are allowed, or if it is allowed to look up things in books and other resources, then you usually are expected to demonstrate the memory of some method. Computers and calculators are very effective at doing large problems, or tedious calculations, quickly and without complaining. But if they are used in a way that avoids the exercise of memory, then I would say the technology is being overused and misused. It is particularly unhelpful to try to google solutions to problems, or look up answers in the back of a book and write it down without understanding what is written down. On the other hand, the best use of computers is to support and extend human skills, not to replace human skills.

0

General Skills

0.1 Reading a Math Book

TL;DR

This section discusses some common misunderstandings with mathematical vocabulary and notation, and the way that we use them in sentences. It also takes a deeper look at where the vocabulary and notation come from.

Sometimes math is described as being like a language, except that it is universal. I agree up to a point. There are things about math that seem universal, like logic, but there are also regional and cultural differences in the way that math is expressed. Learning how to read math is very much like learning to read a language, but it is also written within a language, and extends that language with notation and vocabulary. To learn how to read and write in a language like English, we first begin with the alphabet. We then have to learn how to spell words. Once we have words, we can combine them into sentences. Understanding a sentence requires us to learn grammar. The alphabet, words, and grammar do not go away in math, rather they are extended. Therefore, learning how to read a math book requires the same steps.

Step 1: Understanding notation and vocabulary.

True or False? $-5 > 2$

The correct answer is False, but that is not the point. The point is that this question is phrased in a way that requires knowledge of several things:

1. That the symbol $>$ means "greater than".
2. Positive and negative numbers and their order.
3. What "greater than" means.

If a person thinks that $-5 > 2$ is true, they might still answer the question correctly if it is rephrased as

True or False? -5 is greater than 2,

in which case they may understand the concept but not the symbol. Perhaps they are getting $>$ and $<$ backwards. On the other hand, if they give an incorrect answer to both the original version and the rephrased version, maybe they would give a correct answer to

True or False? $5 > 2$

DOI: 10.1201/9781003737490-0

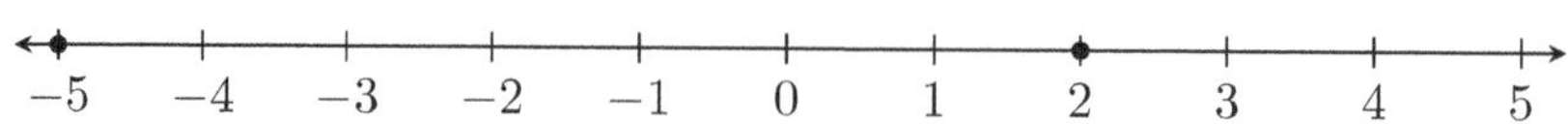

FIGURE 0.1: The number line with −5 and 2 marked

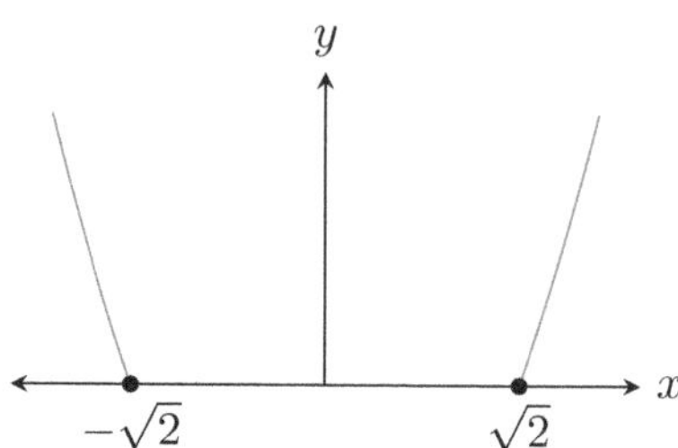

FIGURE 0.2: Roots

Then they are understanding the symbol and what greater than means in the context of positive numbers, but do not fully understand how the concept extends to negative numbers. They may know all too well that 5 is bigger than 2 and think of $>$ as a mouth eating the bigger number, which is a common way of introducing the symbol for positive numbers. But with negatives involved, they need to know the number line. They need to know that -5 is plotted to the left of 2 on a number line as shown in figure 0.1.

But some people may still get the order of -5 and 2 backwards. And why do we draw the number line left to right in the first place? Because we read left to right in the European languages, and the number line was developed in those countries. For a person used to reading right to left, as Hebrew and Arabic are, the number line itself may seem backwards. Even a Chinese person who is used to reading vertically and then right to left may feel that the number line is sideways. At least when it comes to two-dimensional graphs, it is easier to agree that positive numbers should be up because we stand up, unless we are doing gymnastics perhaps. The point is that when you are reading $-5 > 2$, you need to understand all of these details in order to understand why $-5 > 2$ is false.[1]

You will need to learn the meaning not only of symbols in math but also of words. How do you usually learn the meaning of words? You can memorize words, but it helps if you know how to use a word correctly in a sentence, and it also can help if you know the etymology of a word. What does etymology mean? It is the history of a word: where it came from and how its use developed over time.

You can look up the etymology of etymology if you want to, but consider the word *radical.* The word radical comes from the Latin "radix," which is also where we get the word radish. A radish is a root. The square root of 2 is sometimes called radical 2 and is written as $\sqrt{2}$. The cube root of 2, $\sqrt[3]{2}$, is also called a radical. In this sense, radical and root are nearly interchangeable.

But why should it make sense to think of $\sqrt{2}$ as a root? Roots are in the ground. In math, zero is kind of like the ground. We get $\sqrt{2}$ when asking for the roots of $x^2 - 2$; that is the values of x where $x^2 - 2 = 0$. If you graph $x^2 - 2$ from the x-axis up, then it almost looks like $x^2 - 2$ is growing out of $\pm\sqrt{2}$:

You also need to know that $\sqrt{2}$ is the positive root, and that when asking for the roots of $x^2 - 2$, we could also ask for:

[1] At advanced levels of math we say that $a > b$ is true if $a - b$ is positive. Technically this gives a different way to understand inequalities so long as you can subtract with negative numbers. But subtraction with negative numbers is also commonly explained with the number line.

1. the x-intercepts of $x^2 - 2$, in other words the places where $x^2 - 2$ intercepts or intersects with the x; or

2. the zeros of $x^2 - 2$, that is the places where $x^2 - 2$ is equal to zero.

These are all synonyms, and they all follow the usual meaning of words in the English language. For example, in English grammar, the word radical is also used in the sense of a root. I do not know if a radical person is somehow getting to the root of something, but that is a modern usage of the word.

We also use the word radix in math. English speaking countries use a dot as the decimal separator so, for example, 3.14 is an approximation of π. However, many people don't realize that the same notation is used in other bases. For example, 11.001001 gives an approximation of π in binary. This notation is sometimes called radix-base notation, and the dot or comma is called the radix, in other words the root of the number. Here we no longer mean it in the sense of a square root, but we still seem to be referring to zero: the fewer symbols you have to the left or right of the radix, the closer you are to zero. So, once again zero is the ground.

It is possible to be confused between the spelling, pronunciation, or meaning between common words. There are classic examples like:

1. their, there, or they're;

2. dessert and desert; or

3. right (which can mean either "correct" or the direction as in right and left)

"Do I turn left here? Right." Why should we be surprised when there is similar confusion in mathematical vocabulary or symbols? The symbol $(1, 2)$ is used for

1. the point with x coordinate 1 and y coordinate 2, and

2. the open interval $1 < x < 2$.

The only way to tell the difference is from context and, believe me, there are even worse examples of reused notation.

Comprehension Check 0.1.1. How would you graph the point $(-4, 3)$ as a point in two-dimensions? What is the x-coordinate? What is the y coordinate? What quadrant is it in? How would you graph $(-4, 3)$ as an open interval?

Comprehension Check 0.1.2. Is it possible for $(3, -4)$ to be an open interval? Why or why not?

Different words such as *interval* and *integral* can also be confused. Even the word *integral* itself can be used in more than one way:

1. as a noun in the sense of an integral in calculus,

2. as an adjective, such as in the value of x being an integer (an "integral value").

This problem with the word integral is specific to English, but other languages may have analogous problems with other words.

It is common to use Greek letters, in honor of the ancient Greek mathematicians, but the choice of letter is language specific. We use a capital sigma, $\sum$, from the Greek alphabet for a summation, because sigma makes the sound of an S, and S is the first letter in sum.[2] So,

$$\sum_{n=1}^{4} n = 1 + 2 + 3 + 4.$$

As another example, the Greek letter theta, θ, is often used for angles. So 2θ is two times theta, not twenty. Greek letters are used so commonly in math that it may be worth your time to learn the Greek alphabet.

Step 2: Reading and writing full sentences in math.

When we include mathematical words or symbols into an equation, we do not abruptly leave the sentence and go into math land. Instead, the sentence continues and the math becomes an organic part of the grammar of the sentence. If you have gotten this far, then you have already encountered sentences with mathematical words and symbols in them. Let's have a look at some of these sentences. Consider the sentence

> The cube root of 2, $\sqrt[3]{2}$, is also called a radical.

The commas are being used parenthetically just as they are in the sentence

> My best friend, Joe, is going to a baseball game.

Of course, the point of including $\sqrt[3]{2}$ in the sentence is to emphasize how to read it. It may not make sense to literally read the sentence as

> The cube root of 2, the cube root of 2, is also called a radical.

But as written on paper "the cube root of 2" and $\sqrt[3]{2}$ are visually different, and the grammar is meant to express a structural relationship between them.

There are sometimes reasons that we may put a formula on a line by itself, or even on several lines:

1. We don't like to read long paragraphs filled with formulas.
2. We sometimes want to emphasize a formula.
3. We want to provide a reference number to a formula.
4. We need to do a calculation that requires several steps that may be best to put on more than one line.

For example, the difference of squares identity is important enough to put on one line with a reference number:

$$a^2 - b^2 = (a + b)(a - b). \tag{0.1}$$

Because of the colon after number, the sentence does not end until after the equation. The equation itself might be read as

> a squared minus b squared is equal to a plus b times a minus b.

Numbering equations can be useful to organize complicated calculations or when a certain equation is needed frequently. It is much better to use an equation number than to say the equation above or below, unless it is immediately above or immediately below. How do you know that you found the equation being referred to? The equation number gives it

[2] Ironically, sum comes from Latin, not Greek.

an address, and in the pdf version of this book, they are clickable. So I can now refer to equation (0.1) in the special case of $a = x$ and $b = 2$, to show that

$$x^2 - 4 = (x + 2)(x - 2).$$

I can also show you how to derive the formula by breaking it down on several lines, which do not need to be numbered:

$$\begin{aligned}(a + b)(a - b) &= a^2 - ab + ab - b^2, \\ &= a^2 - b^2.\end{aligned}$$

I would generally read $=$ as "is equal to." In the case of this multi-line equation, there is also an implied "which" after the comma:

> a plus b times a minus b is equal to a squared minus ab plus ab minus b^2, which is then equal to a squared minus b squared.

In class, I might give even more detail by referring to the English acronym FOIL (First Outside Inside Last), and to the fact that $-ab$ and $+ab$ cancel. As equations get more complicated you may not want to literally read them in one sentence. You need to be thinking about each part of the equation, which may involve remembering things like FOIL.

Comprehension Check 0.1.3. How would you write

$$x^2 + 3x - 2 = 0$$

as a sentence?

As an example of how relationships between words can change the meaning, consider the following problem:

Factor the largest power of x out of the numerator and denominator of

$$\frac{x - 3x^2}{x^2 - 4}.$$

I have seen answers like this:

$$\frac{x(1 - 3x)}{(x + 2)(x - 2)}.$$

This is factoring – equation (0.1) was used in the denominator – but it is not what the instructions say to do. I have seen this answer even with a step by step example on the exact same page of how to factor out the largest power of x. The word "factor" is a very important mathematical word. It is being used as a verb here. But how is the verb being used? What is it acting on? Is it modified in any way? We are told not simply to factor, but to factor *the largest power of x out*. So, to solve the problem, we first need to figure out what the largest power of x is. It is x^2. We then need to know exponent rules and the distributive property. So the result should have been

$$\frac{x^2}{x^2} \cdot \frac{\frac{1}{x} - 3}{1 - \frac{4}{x^2}}.$$

Key words, like factor, are important, but the entire sentence needs to be read for an understanding of what those key words are doing.

AI Prompt 0.1.1. Can you explain to me how the word "factor" is used as a noun in math? How is it different when factor is used as a verb? Can you give some clear examples of each?

We often try to write in a way so that the word order matches the order that the symbols are written on the page. But grammar sometimes allows that to be changed. We can write $4 - 3$ as

Four minus three

or

Three subtracted from four

The fact that one sentence starts with four while the other starts with three can be confusing. Both are correct grammatically, but the word "from" plays a critical role in our ability to switch the order. You should not think of "Four minus three," as $3-4$. You may understand what you are doing, but when you have to communicate that to someone else, you need them to understand. Grammatically, "Four minus three" means $4 - 3$. Particularly when you learn new concepts, you should try to pay attention to the exact way in which the words and symbols are being used.

Comprehension Check 0.1.4. Convert each sentence below into math symbols:

1. "two x minus three"
2. "two x subtracted from three."

Step 3: Understanding the math behind what is written.

Many books and articles in math are written in very abstract way. A definition or theorem may be stated without examples and used in proofs. How do you know for sure that you understand a definition or theorem? You may think you understand it, but even professional mathematicians have managed to fool themselves.

One approach that is recommended in this book is to use examples to make the abstract more concrete and show the issues that can occur in full detail. For some people the extra detail can be overwhelming. But if you are not aware of all of the details, do you understand it? Consider the following definition

Definition 0.1.1. A *rational number* is a number that can be written in the form $\frac{a}{b}$, where a and b are integers and $b \neq 0$.

Q: So, if I write $\frac{\pi}{2}$, does that mean it is rational?
A: No, because π is not an integer.
Q: What are the integers?
A: The whole numbers, their negatives, and zero: $\ldots, -2, -1, 0, 1, 2, \ldots$.
Q: Why is π not an integer?
A: Well, earlier in this section, we said that π is approximately 3.14, which means that it is between 3 and 4. That is why we draw the number line without gaps. In Figure 0.1 the location of the integers is only where the vertical tick marks are. The space in between is filled up with other decimals like 3.14.
Q: So, what then would be rational?
A: 3.14 is rational. It is $\frac{314}{100}$, so $a = 314$ and $b = 100$, and by the way it is not exactly equal to π.
Q: So, why can't b be zero?
A: Because it is in the denominator and we can't divide by zero.

You should observe that the answers contained many examples. I did not list all of the integers, which is not possible. On the other hand I was not satisfied with saying just "whole numbers, their negatives, and zero," because that could easily lead to more questions. What are negatives? What are whole numbers? Where do the questions stop? The questions may not always stop with an example, because an example could lead to more questions. But the questions that are raised by examples are usually good: they can lead to deeper understanding.

Comprehension Check 0.1.5. $\frac{1}{2}$ is rational because we can take $a = 1$ and $b = 2$. Can you give another choice of a and b that works?

Comprehension Check 0.1.6. Are integers rational numbers? For example, can you write 5 in the form $\frac{a}{b}$ where a and b are integers and $b \neq 0$? If so what a and b work?

There is a lot more that can be said about π and division by zero, but let's consider another situation.

Property 0.1.2. If $ab = 0$, then $a = 0$ or $b = 0$.

I have stated this property without telling you what a and b are. If a and b are integers, real numbers, or complex numbers, this property is true. You use this property when factoring a quadratic equation, and set each factor individually to zero:

$$\text{if} \quad (x-1)(x+3) = 0, \quad \text{or} \quad x - 1 = 0 \quad \text{then} \quad x + 3 = 0.$$

Then we would solve each of those equations individually and get $x = 1$ and $x = -3$ as the roots. But as you will see in this book, property 0.1.2 is not true if a and b are matrices.

Property 0.1.2 also does not generalize to inequalities. It is not true that

"$ab > 0$ implies, $a > 0$ or $b > 0$."

Why? Because both a and b could be negative. It then only takes one example to show that it is not true: Take $a = -2$ and $b = -1$. Then $ab = (-2)(-1) = 2$ and $2 > 0$ is true, but both $-2 > 0$ and $-1 > 0$ are false.

Comprehension Check 0.1.7. Find a value of x so that $x-1$ and $x+3$ are both negative. For that value of x is $(x-1)(x+3)$ positive or negative? So dose $(x-1)(x+3) > 0$ mean that $x - 1 > 0$ or $x + 3 > 0$?

Being able to compute examples is a useful skill to have. If you are reading a math article or a book, and you are not sure if something is true, or if you are not sure you understand, you might want to try scribbling a calculation in the margin and see what happens. One effect that the concrete examples in this book have is that they show you how to do such calculations by hand. Certain calculations may be too hard by hand. In that case, then it is also helpful to know how to give the calculation to a computer. The next section is intended to open the door to calculations on the computer.

0.2 Getting Up and Running with Python and Sage

TL;DR

If you don't want any trouble installing, just setup an account on cocalc.com and work through subsections 0.2.4 and 0.2.5.

This book contains activities and exercises that can be used with Sage and Python. Many students may come into a linear algebra class with some programming skills already, but for those who do not this section should provide enough of an introduction to get started.

The easiest way to get access to Sage is to setup an account on cocalc.com. You can then go straight to subsection 0.2.4, which will show you how to create a CoCalc Project and then create a Jupyter notebook. But if you want to setup Sage on your computer, these are the options that I would recommend:

1. Installing with conda-forge (subsection 0.2.1).
2. Installing with docker (subsection 0.2.1). If you have a Mac with an ARM processor, docker is not an option.

Keep in mind that Sage is technically a Linux application. When you install Sage on Windows or Mac, you are actually installing it in a Linux like environment along with all of the Python files that it needs. The current documentation can be found on the Sage website, www.sagemath.org.

Once you are satisfied with the installation, you should read subsection 0.2.3, showing how to create a Jupyter Notebook on your own system.

Finally, you can read subsection 0.2.5, which provides an introduction programming in Sage with a Jupyter Notebook.

0.2.1 Installing with Docker.

First, you need to install Docker. Go to docs.docker.com and follow the instructions for setting up Docker Desktop on your system.

On Windows, you should use the WSL2 backend for Docker. WSL stands for Windows Subsystem Linux. Here is a brief description, but more detail can be found on the Microsoft website. To install WSL, open PowerShell as follows: type "PowerShell" into the windows search, right click on it and select "run as administrator." Then run the following code:

```
wsl --install
```

You may need to restart. When the installation is complete, "Ubuntu" should be listed in the start menu. Open it and create a user name and password. The WSL2 backend for docker can be selected during installation or in settings.

Once you have Docker, you will need the sagemath docker image, which is located at hub.docker.com/r/sagemath/sagemath. Docker Desktop provides a graphical interface (GUI) and a command line interface (CLI).

Using the GUI, you can just search for sagemath and pull the image.

Using the CLI, you must run the code

```
docker pull sagemath/sagemath
```

to pull the docker image for sagemath.

To start the JupyterNotebook interface run

```
docker run -p 8888:8888 sagemath/sagemath:latest sage-jupyter
```

Then open a browser and enter http://localhost:8888 in the address bar, or use the link from the terminal. You can now skip to subsection 0.2.3 on Using Jupyter.

0.2.2 Installing with Conda-Forge

You will need a Conda installation. The Sage website recommends miniforge.

On Windows, you will need to setup WSL (Windows Subsystem Linux). First open a power shell, by typing "power shell" into the windows search bar and select it. Then in the power shell, run the following code:

```
wsl --install
```

To download miniforge on Mac or Linux, open a terminal. Copy and paste the code from https://doc.sagemath.org/html/en/installation/conda.html into the terminal, and hit enter.

On Windows, you run the same code within the Linux subsystem.

Once you have conda, you can run

```
mamba create -n sage sage python=3.11
```

or

```
conda create -n sage sage python=3.11
```

to install Sage. Sage will install in a virtual environment called sage.

If the environment is active, your command line prompt should say (sage). If not, then you should run the following code to activate the environment

```
conda activate sage
```

Then to start a Jupyter notebook with the Sage kernel, you can run

```
sage -n
```

On the other hand, you can run

```
jupyter notebook
```

or

```
jupyter lab
```

and select the Sage kernel, depending on whether you have jupyter notebook or jupyter lab installed. You can now skip to subsection 0.2.3 on Using Jupyter.

0.2.3 Using Jupyter

Once Jupyter has been launched in your browser, you should create a folder called Jupyter-Notebooks:

- In the Jupyter Notebook interface, click New and select New Folder from the dropdown menu.
- In the Jupyter Lab interface, click on the new folder icon in the sidebar.

Inside the new folder, you can create a new Jupyter notebook and select the SageMath kernel:

- In the Jupyter Notebook interface, click New and select SageMath.
- In the Jupyter Lab interface, click on the SageMath icon.

All of the code in this book has been tested with the SageMath kernel, but some code may work with the python3 kernel instead. You can now skip to subsection 0.2.5 on programming in Sage.

0.2.4 Using CoCalc

1. Login, then click on *Your Projects.*
2. Select *Create a project* and name it HelloWorld.
3. A project may have multiple files. Create a new file, and select Jupyter Notebook, then select SageMath as the kernel.

Once you are in the new notebook, you should notice that you can enter code or ask ChatGPT to generate code.

0.2.5 Programming in Sage

You should now be in a Jupyter notebook with the SageMath kernel running. The rest of this section is intended to give a brief overview of some of the coding concepts in this book. It is not meant to be a full introduction to python.

Data types and operations

Python has several built-in data types, including:

- `int`, which is used to represent integers in python.
- `float`, which is used to represent floating point numbers in python. As the name suggests, the decimal point can move.
- `bool`, to store True/False values. These correspond to single bits.
- `set`, like a finite set in math (order of elements does not matter and the elements are unique).
- `tuple`, elements can be repeated and order matters. Tuples are "immutable."
- `list`, elements can be repeated and order matters. Lists are "mutable."

If the words "mutable" and "immuatable" are unfamiliar, do not worry; the difference will be illustrated shortly. If you are not sure what type something is, you can check using the type function:

```
type(True)
```

Enter this in a cell, and click Run. The output should be `<class 'bool'>`.

Python also has some built-in operations for each type, and sometimes the same symbol can mean different things for different types.

For `int` and `float` types + is just addition. In another cell, enter

```
1+1
```

and click Run. The output should be 2, as expected since $1 + 1 = 2$. But in the case of a list, + is concatenation, a word meaning "to join together." In another cell, enter

```
[1,3,5]+[2,4]
```

and click Run. The output should be `[1,3,5,2,4]`. So, we joined two lists into a larger list.

For `int` and `float` types `*` is multiplication. In another cell, enter

```
2*3
```

and click Run. The output should be 6. But for list you can use * to repeat elements. In another cell, enter

```
2*[1,2,3]
```

and click Run. The output should be `[1,2,3,1,2,3]`.

In Python, exponentiation can be done by `**`. So, for example 2^3, which is $2 \times 2 \times 2$, can be computed by `2**3`. But for calculators the standard notation is `2^3`. Sage recognizes both `2**3` and `2^3`. Try each of them in separate cells. In both cases, the result should be 8.

Now let's get a deeper understanding of tuples and lists. In Python, you can define a variable with `=`. For example in another cell, enter

```
L = [4,5,8]
```

and click run. The output will be blank, but you have just defined a list called `L`. To see it, you can run

```
L
```

in another cell. The numbers separated by the commas are called **elements**. The list has 3 elements. You can check this by using the `len` function (an abbreviation for length). In another cell, enter

```
len(L)
```

The output should be 3. Since order matters, each elements has a location that is specified by an index. Python is 0 indexed, meaning that the first index is 0. Check this by running

```
L[1]
```

The output should be 5, not 4, because the first index is 0, so `L[0]` is 4.

Comprehension Check 0.2.1. What is `L[2]`?

You can also use `=` to redefine variables. You can always redefine the whole variable, but for a mutable object, like a list, you can also redefine parts of a variable specified by index. If we want to change the value of `L[0]` to 3, we can do that with

```
L[0] = 3
```

Try running this code, and then check that `L` was actually changed, by running

```
L
```

There should be no errors. But if you try the same thing with a Tuple you will get an error. First, define a tuple in one cell

```
T = (4,5,8)
```

then in another cell, run

```
T[0] = 3
```

You will get an error telling you that elements of tuples cannot be changed. This is what it means when we say that a tuple is immutable. We cannot change individual elements, nor can we append elements to a tuple. Why do we need tuples if we have lists? Because tuples are more efficient. It takes extra work behind the scenes to make it possible to append elements and change their values. It is possible to change a tuple into a list. For example, try running

```
list(T)
```

But we can also work with tuples directly. Even though we cannot change a tuple, we can create new tuples from old tuples. For example,

```
T = T+(0,)
```

works just fine, but it works by looking up `T`, using it in a calculation, then redefining `T`. The calculation in this case involves constructing the tuple `(0,)`, which has one element, then combining it with `T` into a new tuple with 4 elements. You can check the result by running

```
T
```

You can learn more about data types and functions from www.w3schools.com

Logic and loops

First, run

```
a = 2
```

There should be no output, but we defined `a`. What is `a`? Run

```
a
```

You should get 2 as an output. Is `a`, equal to 2? We can check by running

```
a == 2
```

The output should be `True`. So `=` assigns values, and `==` checks if things are equal.

Comprehension Check 0.2.2. What will happen if you run `a == 3` next? What if you run `a = 3`? What if you run `a == 3` after `a = 3`?

We can also check if things are not equal. Try running

```
a != 5
```

Regardless of whether $a = 2$ or $a = 3$, it should be true that $a \neq 5$, which is what `a != 5` is checking, so the output should be `True` in both cases.

We can check if $a > 0$ by running

```
a > 0
```

The output should be `True` so long as `a` is positive, which both 2 and 3 are. We can negate a statement with `not`. For example, if we run

```
not True
```

the output is `False`. A number that is not greater than 0 is less than or equal to zero, thus

```
not a > 0
```

and

```
a <= 0
```

should give the same output.

Comprehension Check 0.2.3. Is zero positive? If you are not sure, try redefining `a` to be zero and running `a > 0` again.

We can also use `a < 0` to check if `a` is less than 0, and `a<=0` to check if `a` is less than or equal to 0, which we would usually write as $a \leq 0$.

We can use `and`, `or`,and `not` exactly as they are used in logic:

P	Q	P and Q
True	True	True
True	False	False
False	True	False
False	False	False

P	Q	P or Q
True	True	True
True	False	True
False	True	True
False	False	False

P	not P
True	False
False	True

For example, "less than or equal to" is an or statement, so even though you could use `a<=0`, you should still get the same result with

```
a < 0 or a == 0
```

Comprehension Check 0.2.4. What can be said about `a < 0 and a == 0`? If you are not sure, try it on three cases: $a = -2$, $a = 0$, and $a = 2$.

Warning 0.2.1. Warning When using, `or`, `and` and `not` together, you need to be very careful with order of operations, just as you would be with addition and multiplication:

1. Compare the output of

   ```
   (True or True) and False
   ```

   with

   ```
   True or (True and False)
   ```

2. Compare the output of

   ```
   not (True or False)
   ```

   with

   ```
   (not True) or (not False)
   ```

Suppose you have the list

```
L = [3,5,8]
```

and you want to double each number in the list. You can't just do `2*L`, because what you will get is a list with each number repeated twice. Instead, you should run

```
[2*n for n in L]
```

If you want the output to be a set instead of a list, you can just change `[` and `]` to `{` and `}` like this: [3]

```
{2*n for n in L}
```

Maybe you don't want everything in `L`. Maybe you only want to double the numbers greater than 4. You can do that by including an if statement:

```
[2*n for n in L if n>4]
```

[3]Using (and) instead technically produces a generator object.

There is another python type called range, which is commonly used together with for loops. For example

```
[n for n in range(10)]
```

If you want only the odd numbers, you could use an if statement, but range already has the built in flexibility to do this as follows:

```
[n for n in range(1,10,2)]
```

In general `range(a,b,m)` gives you $a, a+m, a+2m, \cdots$ up to but not including b.

Comprehension Check 0.2.5. What values of a, b, m in `range(a,b,m)` gives positive integers with 3 in the ones place but less than 100?

This is a good time to point out that an integer in Sage is not the same as an `int`. If you run

```
[type(n) for n in range(10)]
```

you will see that range contains ints, but if you are running the SageMath kernel, then

```
type(2)
```

will not give you int. Integers have many properties, like whether it is prime. Sage provides the `is_prime()` function to check if a number is prime, but this function does not apply to ints. You will get an error if you run

```
[n for n in range(10) if n.is_prime()]
```

But you can turn `n` into Sage's integer type by `ZZ(n)`, thus

```
[n for n in range(10) if ZZ(n).is_prime()]
```

will give you the primes less than 10.

There is another way of writing a for loop that we will use as well, particularly when we want to see how a result changes from one step to the next. Enter the following code in one cell and run it:

```
L = list()             #Define an empty list L
print(L)               #Show L on the screen
for n in range(10):    #Start a for loop over the values 0...9
    L += [n^2,]        #Square n, then append it to L
    print(n,L)         #Show the current value of both n and L
print(n,L)             #Print the final values of n and L
```

The # gives us the ability to make comments. The comments explain what each line of code is doing, but there are two key details:

1. the colon signals a change in the level of indentation
2. the indentation must be consistent, since it signals to python what is inside and what is outside the for loop.

We can prevent code from running just by sticking a # in front of it. For example, if you replace the `print(n,L)` inside the for loop with `#print(n,L)`, and run the code again, only the print statements outside the for loop will execute, and we will see only the initial and final value of `L`.

Functions

Definition 0.2.1. Fibonacci numbers can be defined recursively as follows

$$\begin{aligned} F_0 &= 0 \\ F_1 &= 1 \\ F_n &= F_{n-1} + F_{n-2} \quad \text{for } n > 1 \end{aligned} \tag{0.2}$$

We want to construct them using a function. One strategy is to construct all of the Fibonacci numbers up to n as a list:

```
def FList(n):
    F = [0,1]
    print(F[0])
    print(F[1])
    for k in range(2,n+1):
        print(k,len(F))
        F += [F[-1]+F[-2],]
        print(k,len(F))
        print(F[k])
    return F[:n+1]
```

If you copy this code into a cell, and click Run the output should be blank if the function is successfully defined.

Just as with a for loop, the colon indicates a change in the indentation level. We use `def` to signal that a function is defined. That function is named `FList`. It takes one argument, `n`. That argument is supposed to be an integer, but we have not bothered to check. The two print statements after `F=[0,1]` show that `F[0]` is 0 and `F[1]` is 1, which agrees with the first two lines of (0.2) in definition 0.2.1.

If $n \leq 1$, then `range(2,n+1)` will be empty and will not execute at all. Within the for loop, the print statement `print(k,len(F))` shows that k is initially equal to the length of `F`, which means that the last index in `F` is $k-1$. `F[-1]` gets the last element of `F`, which is F_{k-1} and `F[-2]` gets the second to last element, which is F_{k-2}. They are added to get F_k, which uses the third equation of (0.2) in definition 0.2.1, and this is new value is appended to `F`. Now `F` has length $k+1$, while k remains the same. It is now the last index, so `print(F[k])` does not give an error, rather it gives the value of F_k.

By the indentation, you can see that the return statement is outside the for loop. It tells us to leave the function and return `F[:n+1]`. Why not just `F`? Because if $n = 0$, then `F` is still `[0,1]`, but we want `[0]`. The syntax `F[:n+1]`, specifies that the range of indices must be less than $n+1$. If $n = 0$, then $n+1 = 1$ so `F[:n+1]` gives us `[0]`.

Try running

```
FList(7)
```

The function `FList` is not very space efficient. If we only need F_7, the computation can be done more efficiently by keeping only two at a time. It may seem fast for small values of n, but we can do better: we can compute F_n directly from n by the Binet formula

$$F_n = \frac{1}{\sqrt{5}}(\phi^n - (-\phi)^{-n}) \quad \text{where} \quad \phi = \frac{1+\sqrt{5}}{2} \tag{0.3}$$

so we do not need a for loop at all. Run the code

```
def F(n):
    K.<phi> = NumberField(x^2-x-1)
    return (phi^n-(-phi)^(-n))/(phi+phi^-1)
```

The first line constructs ϕ (`phi`) as a root of $x^2 - x - 1$. Then the return statement is essentially equation (0.3) because $\phi + \phi^{-1} = \sqrt{5}$.

Comprehension Check 0.2.6. Use the quadratic formula to check to find the roots of $x^2 - x - 1$; one of them should be

$$\phi = \frac{1+\sqrt{5}}{2}.$$

Show that

$$\left(\frac{\sqrt{5}-1}{2}\right)\left(\frac{\sqrt{5}+1}{2}\right) = 1$$

so

$$\phi^{-1} = \frac{\sqrt{5}-1}{2}.$$

Is it now clear that $\phi + \phi^{-1} = \sqrt{5}$? If not, check by computing $\phi + \phi^{-1}$.

In the long run, `F(n)` should be faster, but for small n it is slower because the computations with ϕ are not as easy as adding integers. The `timeit` function makes it possible to compare the speeds of these two functions for different values of n. For individual values of n we can get times like this

```
timeit('F(100)')
```

and

```
timeit('FList(100)')
```

Times may be listed in:

- ns (nanoseconds) 1 ns = 10^{-9} s
- µs(microseconds) 1 µs = 10^{-6} s
- ms (milliseconds) 1 ms = 10^{-3} s

etc. My times were 73.8 µsfor `F(100)` and 15.6 µsfor `FList(100)`. Your times will probably be different, but it looks like $n = 100$ isn't big enough. You can keep changing the numbers, but if you will be collecting a lot of times, it is worthwhile to get an efficient approach.

```
def timeitrange(S1,S2,n,b=10):
    L = [b^k for k in range(n)]
    for x in L:
        print(x)
        print(timeit(S1+str(x)+S2))
    return None
```

The argument `b=10` specifies a default value of 10 for the base `b`, so b optional. We start by creating a list `L` of powers. We then do a for loop over the powers. Within the loop, we print the current power, and the result of `timeit` for that power. The `timeit` function takes a string as an argument. So each power is converted to a string by `str(x)`, then strings `S1` and `S2` are attached at the front and the back. If we want to get times for the function `F`, then we use `"F("` for `S1` and `")"` for `S2`. So if x is 10, then `S1+str(x)+S2` is `"F(10)"`. Try running

```
timeitrange("F(",")",5)
timeitrange("FList(",")",5)
timeitrange("FList(",")",5,2)
```

There are many other ways of writing a function that generates Fibonacci numbers. If you have access to ChatGPT, Claude, or another AI, you might try asking it something like this:

AI Prompt 0.2.1. Can you write a function in Sage for computing Fibonacci numbers?

The function ChatGPT gave me was this:[4]

```
def fibonacci(n):
    print("n="+str(n)+" Open")
    if n == 0:
        return 0
    elif n == 1:
        return 1
    else:
        Fn = fibonacci(n-1) + fibonacci(n-2)
        print("F"+str(n)+" = "+str(Fn))
        print("n="+str(n)+" To be Closed")
        return Fn
```

This is technically a correct method using a recursion. However, there are two types of recursion: "tail recursions" and "non-tail recursions."

A non-tail recursion occurs if you need the result of the next step of the recursion before you can finish the computation of the result in the current step. Here we need to know both `fibonacci(n-1)` and `fibonacci(n-2)` before the return statement involving them is executed, so ChatGPT gave me a non-tail recursion. New instances of fibonacci keep having to be called all the way down to fibonnci(0), which returns 1. Try running the following code for yourself

```
fibonacci(5)
```

You will observe that `n=5` is the first to open and the last to close. This is a problem, because all partial results have to be kept in memory until the function closes and returns a value.

A tail recursion is essentially the opposite: it is not necessary to go all the way to the end of the recursion to finish the computation of earlier steps.

AI Prompt 0.2.2. Can you give an implementation of the Fibonacci numbers using a tail recursion?

The function ChatGPT gave me was this:[5]

```
def fibonacci_tail_recursive(n, a=0, b=1):
    print("n="+str(n)+" Open")
    print("a="+str(a)+" and b="+str(b))
    if n == 0:
        return a
    if n == 1:
        return b
    print("n="+str(n)+" Closed")
    return fibonacci_tail_recursive(n-1, b, a+b)
```

[4]The print statements were added by me for clarity.

[5]Once again, the print statements were added by me.

If you run

```
fibonacci_tail_recursive(5)
```

you will observe that `n=5` closes, giving up its memory, before `n=4` opens. The trick is that the Fibonnaci numbers are computed from the bottom up. The values of `a` and `b` are initialized as F_0 and F_1 respectively. If $n \geq 2$, then it computes F_2 and passes $n-1$, $a = F_1$, and $b = F_2$ to a new instance of the function. Then we get F_2 and F_3 etc., passing Fibonnaci numbers in pairs until we get the final correct result.

In any case, it should now be clear that a tail recursion uses less memory, which generally means that tail recursions are faster. You can check the times as follows:

```
timeitrange("fibonacci(",")",5,2)
timeitrange("fibonacci_tail_recursive(",")",5,2)
```

Exercises

Problem 0.1. Rewrite each statement in words (read it aloud as you would to a friend):

(a) $7 \leq 10$

(b) $2x - 3y = 6$

(c) $(x + a)^2 = x^2 + 2ax + a^2$

Problem 0.2. Rewrite each English sentence using mathematical notation

(a) Negative two x minus four is greater than eight.

(b) The absolute value of negative four is equal to four.

(c) a squared plus b squared is equal to c squared.

(d) Sine two theta equals two times sine theta times cosine theta.

(e) Negative b plus or minus the square root of b squared minus four a c all over $2a$.

Problem 0.3 (With the Python 3 kernel)**.** This problem explores automatic conversion between types. Run each line below in a separate cell

```
type(1)
type(1.0)
1+1.0
type(1+1.0)
1==1.0
```

Look carefully at the results, and try to answer each question below. Then ask an AI to explain.

(a) Why is there no error in the calculation of 1+1.0?

(b) Why is 1==1.0 True even though 1 and 1.0 do not have the same type?

Redo the problem with 1 or 1.0 replaced by True.

int, float, and bool are all compatible because they are all different types of numbers. But a list is not a number. Redo the problem with 1 replaced by [1]. There should now be an error.

Problem 0.4 (With the Python 3 kernel)**.** This problem explores the relationship between /, //, and % in Python 3 (Python 2 works differently). $123 \div 10$ can be expressed as a decimal or as quotient and remainder. Run each line below in a separate cell

```
123/10
123//10
123%10
123//10+(123%10)/10
```

Look carefully at the results, and try to answer each question below. Ask an AI to explain if necessary.

(a) Which one of /, //, and % gives the result as a decimal?

(b) Which one of /, //, and % is the quotient?

(c) Which one of /, //, and % is the remainder?

(d) Why are `123/10` and `123//10+(123%10)/10` equal?

Run each line below in a separate cell

```
3/1
3.14/1
3//1
3.14//1
-3//1
-3.14//1
```

Plot 3, 3.14, 4, -3.14, -3, and -4 on a number line. Look carefully at the results and your plot, then try to answer each question below. Ask an AI to explain if necessary.

(e) Why are `3//1` and `3.14//1` the same but not `3/1` and `3.14/1`?

(f) Is `3.14//1` to the left or right of 3.14?

(g) Is -3 to the left or right of -3.14?

(h) `3.14//1` is the "whole part" of 3.14. Why is `3.14//1` not -3?

In math, `//1` is called the floor function.

Problem 0.5. Use list comprehension to create a list for each of the following

(a) Powers of 2 from 2^0 up to 2^8.

(b) The squares of 1,2,...,10.

(c) The fractions $1/1, 1/3, 1/5 \ldots 1/101$.

(d) The decimals 0.0, 0.1, 0.2, $\ldots$ 1.0.

Problem 0.6. Construct the same lists as in problem 0.5 using the following method instead:

- Define L as an empty list.
- Use a for loop to append each item to L.
- Print L within the for loop so that you can see L grow at each step.

Problem 0.7 (with AI). The Sieve of Eratosthenes is an ancient method to find numbers. This problem explores different ways to implement the Sieve of Eratosthenes in python.

1. Ask an AI to write a function in Python 3 to find the prime numbers up to N using the Sieve of Eratosthenes and return a list. Also ask for it to use print statements to show the intermediate steps of the calculation.
2. Use the function to find the prime numbers under 100.
3. Ask an AI to write a function for the Sieve of Eratosthenes using numpy arrays instead that removes numbers from the array by a bit mask.
4. Use the new function to find the prime numbers under 100.
5. Remove or comment out the print statements and compare the speed for different sizes of N. The speed should be asymptotically the same, but which implementation is faster? Ask the AI if it can come up with additional speed improvements without switching to a more advanced algorithm.

1

Matrix Arithmetic

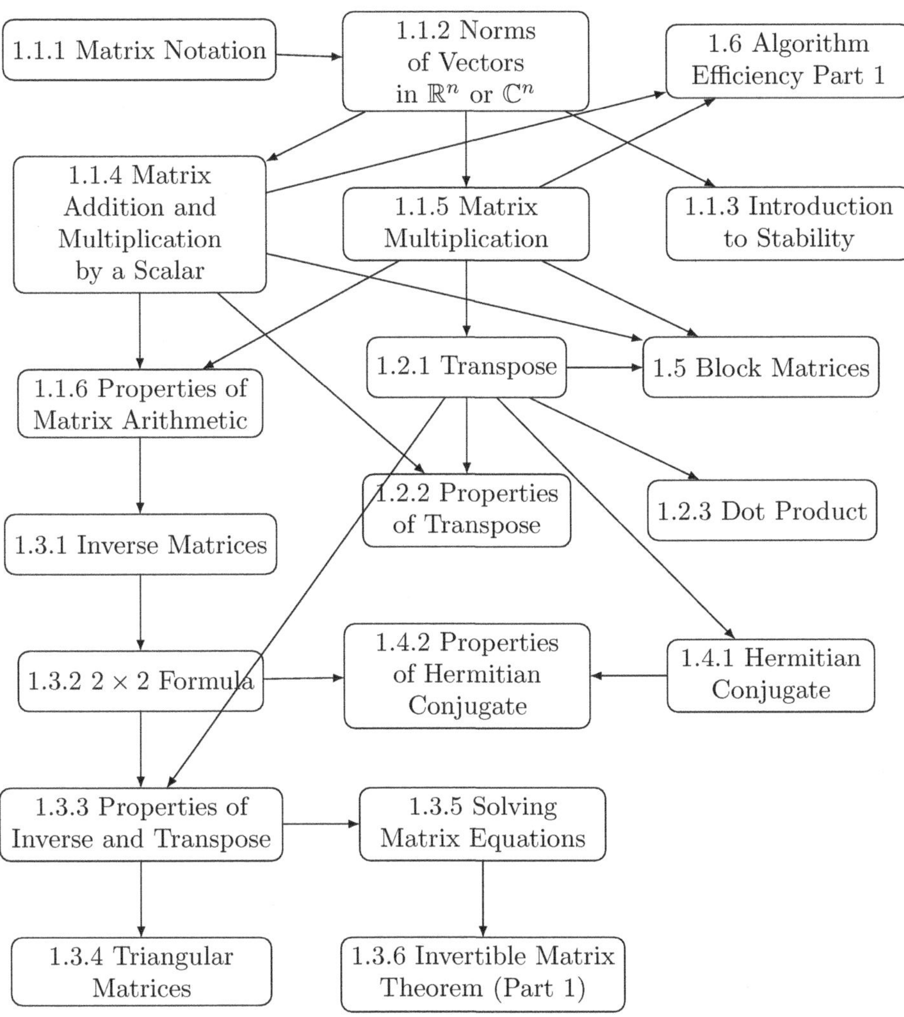

DOI: 10.1201/9781003737490-1

1.1 Addition and Multiplication

TL;DR

Subsection 1.1.1 is foundational and should not be skipped unless you already know what vectors and matrices are, as well as terminology such as row, column and dimension. Key concepts:

1. In Definition 1.1.12, matrix multiplication **AB** is defined as the dot product of the row vectors in **A** with the column vectors in **B**. I have seen some students write down the outer product of **A** and **B**, which is not correct; the outer product is covered in Subsection 4.7.1. Multiplication of a matrix by a scalar (a number) is also possible, but it usually does not lead to confusion since only one matrix is involved. Example 1.1.15 covers all of the basics.

2. Sage and NumPy do not use the same syntax for matrix multiplication:

   (a) In Sage, matrix multiplication is `A*B`.

   (b) In NumPy, matrix multiplication is `np.dot(A,B)`, whereas `A*B` is componentwise multiplication.

   Pay close attention to the data types of `A` and `B`: if they are NumPy arrays, then `A*B` will be interpreted following the NumPy syntax, not the Sage syntax.

3. Matrices must have compatible dimensions for addition or multiplication to be defined, as illustrated by Example 1.1.15. When addition and multiplication are defined, most of the familiar properties of arithmetic apply to matrices (see Propositions 1.1.14, 1.1.15, 1.1.16, and 1.1.17) except that generally $\mathbf{AB} \neq \mathbf{BA}$ (see Example 1.1.28).

1.1.1 Matrix Notation

Definition 1.1.1. A **vector** is a an ordered list of numbers that can be written horizontally as a **row vector**,

$$\mathbf{x} = \begin{bmatrix} x_1 & x_2 & \cdots & x_n \end{bmatrix},$$

or vertically vector as a **column vector**,

$$\mathbf{x} = \begin{bmatrix} x_1 \\ x_2 \\ \vdots \\ x_n \end{bmatrix},$$

The individual numbers x_i are called the **components**.

Definition 1.1.2. A **matrix** is a rectangular array of m rows and n columns:

$$\mathbf{A} = \begin{bmatrix} a_{11} & a_{12} & \cdots & a_{1n} \\ a_{21} & a_{22} & \cdots & a_{2n} \\ \vdots & \vdots & & \vdots \\ a_{m1} & a_{m2} & \cdots & a_{mn} \end{bmatrix},$$

and is said to have **dimension** $m \times n$ (read m by n). The number a_{ij} is in the i-th row and j-th column and is called the i, j **entry** of the matrix. For example, the 2nd row of the above matrix is emphasized in gray. Each coefficient in that row has a 2 in the first subscript.

Remark 1.1.1. Note that rows are consistently listed before columns:

- a_{ij} is in row i and column j
- $\mathbf{A}$ is $m \times n$ because it m rows and n columns.

I believe that this convention comes from reading left to right and then down. Imagine reading along the first row. The column number increases going left to right, then resets when you go to the next row.

Comprehension Check 1.1.1. What is the 2nd row of $\left[\begin{smallmatrix}1 & 2 & 3\\4 & 5 & 6\end{smallmatrix}\right]$?

Example 1.1.1. Write down the dimensions of the matrices

$$\mathbf{A} = \begin{bmatrix} 2 & 1 & 0 \\ 0 & -1 & -2 \end{bmatrix} \text{ and } \mathbf{B} = \begin{bmatrix} 2 & 0 \\ 1 & -1 \\ 0 & -2 \end{bmatrix}.$$

Solution

$\mathbf{A}$ has 2 rows and 3 columns, so it is 2×3.
$\mathbf{B}$ has 3 rows and 2 columns, so it is 3×2.

AI Prompt 1.1.1. Explain the history (etymology) of the mathematical terms 'matrix' and 'vector'. Which languages do they come from, and what did the original words mean?

In Sage, a matrix can be defined by

```
A = matrix([[2, 1, 0],
            [0,-1,-2]])
```

and a vector can be defined by

```
v = vector([3,4])
```

The definition of `A` indeed gives us a 2×3 matrix, but v works as either a row or a column vector: its dimension is just 2, not 1×2 (for a row vector) or 2×1 (for a column vector).

```
v_row = matrix([[3,4]])
v_col = matrix([[3],
                [4]])
```

In NumPy, all matrices and vectors are arrays. For example, the following code defines `v` and `A` directly in NumPy:

```
import numpy as np

v = np.array([3,4])
A = np.array([[2,1,0],
              [0,-1,-2]])
```

NumPy also provides a function called shape that matches the concept of dimension in Definition 1.1.2

```
np.shape(A)
np.shape(v)
```

You should see that the shape of v is just 2, which agrees with Sage. In NumPy, you can change the shape, so long as the total number of entries does not change. So for example,

```
v_row = v.reshape((1,2))
v_col = v.reshape((2,1))
```

will turn v into a row or a column vector. However,

```
A.reshape((5,2,3))
```

will not work. If you are hoping for 5 copies of the original matrix `A`, then you need

```
np.broadcast_to(A,(5,2,3))
```

Note that the output array now has 3 indices, versus 2 indices for a matrix, or 1 index for a vector. If `A` is a matrix (or vector) in Sage, then

```
np.array(A)
```

will convert it into a NumPy array. Going the other direction,

```
vector(v)
```

will work if v has one index, and

```
matrix(A)
```

will work if `A` has two indices. But the number of indices must match! Neither

```
matrix(v)
vector(A)
```

will work, and there is no clear option for more than two indices. Finally, we should point out that Python is zero indexed as discussed in Section 0.2.

1.1.2 Norms of Vectors in $\mathbb{R}^n$ or $\mathbb{C}^n$

Q 1.1.1. What is the Pythagorean theorem?

Q 1.1.2. What is the absolute value of -4?

Q 1.1.3. What is the maximum of 4 and 3?

Q 1.1.4. What is the complex conjugate of $4 + 3i$?

When we speak of a vector in $\mathbb{R}^n$, we mean that the vector has n-components, all of which are real (numbers in $\mathbb{R}$). Vectors in $\mathbb{C}^n$ are discussed at the end of this subsection. A vector

$$\mathbf{v} = \begin{bmatrix} x \\ y \end{bmatrix}$$

in $\mathbb{R}^2$ can be represented graphically by an arrow starting from the origin, $(0, 0)$, and ending at (x, y). In this way we can think of vectors as having both a length and a direction (indicated by the arrow). The length can be determined by the Pythagorean theorem:

$$\| \mathbf{v} \| = \sqrt{x^2 + y^2}.$$

We read $\| \mathbf{v} \|$ as the "length of $\mathbf{v}$" or the "norm of $\mathbf{v}$."

Comprehension Check 1.1.2. How would you read $\| \mathbf{x} \| = 3$?

Example 1.1.2. Graph the vector

$$\mathbf{v} = \begin{bmatrix} 4 \\ 3 \end{bmatrix}$$

and find its length.

Solution

It is usual to take the 1st component as the x-component and the 2nd component as the y-component, since x comes before y in the alphabet.
The vector $\mathbf{v}$ lies along the hypotenuse of a right triangle as shown in Figure 1.1a. The x and y components of $\mathbf{v}$ are the legs of the right triangle, so

$$\| \mathbf{v} \| = \sqrt{4^2 + 3^2} = \sqrt{5^2} = 5.$$

The fact that the length is an integer means that $3, 4, 5$ is an example of what we call a Pythagorean triple.

In higher dimensions the Pythagorean theorem generalizes as follows

Definition 1.1.3. Let $\mathbf{x}$ be a vector in $\mathbb{R}^n$, with components

$$x_1, x_2 \dots x_n.$$

Then the length of $\mathbf{x}$ is

$$\| \mathbf{x} \| = \sqrt{x_1^2 + x_2^2 + \cdots x_n^2}.$$

This generalization to higher dimensions can be built up recursively as illustrated in the next example.

Example 1.1.3. As illustrated in Figure 1.1b, the vector

$$\mathbf{w} = \begin{bmatrix} 4 \\ 3 \\ 2 \end{bmatrix}$$

lies along the diagonal of a box. If we ignore the z-component of $\mathbf{w}$, then we get the vector $\mathbf{v}$ from Example 1.1.2. As shown in Figure 1.1b, the vector $\mathbf{v}$ is the diagonal of the bottom of the box, which lies in the xy-plane.

The shaded triangle in Figure 1.1b is a right triangle with

- base $\| \mathbf{v} \|$ (which is 5 as found in Example 1.1.2), and
- height equal to the z-component of $\mathbf{w}$ (which is 2).

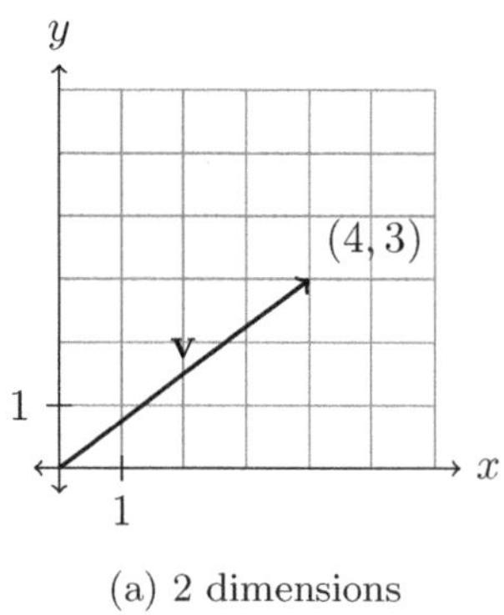

(a) 2 dimensions

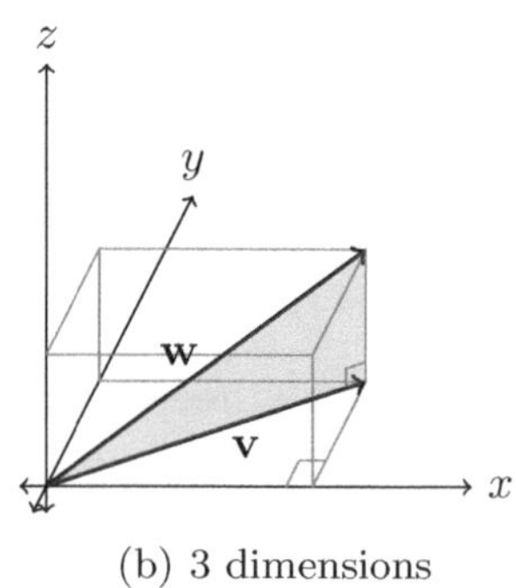

(b) 3 dimensions

FIGURE 1.1: Graphing vectors

So, we can get the length of **w** by applying the Pythagorean theorem to the shaded triangle.

$$\| \mathbf{w} \| = \sqrt{5^2 + 2^2} = \sqrt{25 + 4} = \sqrt{29}$$

On the other hand, if we use Definition 1.1.3, then

$$\| \mathbf{w} \| = \sqrt{4^2 + 3^2 + 2^2} = \sqrt{16 + 9 + 4} = \sqrt{29}.$$

Remark 1.1.2. The method of Example 1.1.2 allows the length in higher dimensions to be obtained from the Pythagorean theorem by induction.

Definition 1.1.3 is a special case of a more general concept.

Definition 1.1.4. Let $1 \le p < \infty$. If $\mathbf{x}$ is a vector in $\mathbb{R}^n$ or $\mathbb{C}^n$ with components

$$x_1, x_2, \ldots x_n,$$

then the p-norm of $\mathbf{x}$ is

$$\| \mathbf{x} \|_p = (|x_1|^p + |x_2|^p + \ldots + |x_n|^p)^{\frac{1}{p}}.$$

The ∞-norm or max norm of $\mathbf{x}$ is

$$\| \mathbf{x} \|_\infty = \max(|x_1|, |x_2|, \ldots |x_n|).$$

We would read

$$\| \mathbf{x} \|_2 < 5$$

as "the 2-norm of $\mathbf{x}$ is less than 5."

Comprehension Check 1.1.3. How would you read $\| \mathbf{x} \|_\infty < 1$? How would you read $\| \mathbf{x} \|_\infty + \| \mathbf{y} \|_\infty < 2$?

You should observe that, for a vector in $\mathbb{R}^n$, the 2-norm is just the length of the vector as stated in Definition 1.1.3 because $\sqrt{x} = x^{\frac{1}{2}}$. So, in Example 1.1.2, we could have computed the length as

$$\| \mathbf{v} \|_2 = (|4|^2 + |3|^2)^{\frac{1}{2}} = (5^2)^{\frac{1}{2}} = 5.$$

The squares may make the absolute values look redundant. But in general, the absolute values are critical because if $p = \frac{5}{4}$, then $(-1)^{\frac{5}{4}}$ is not a real number but

$$|-1|^{\frac{5}{4}} = 1.$$

Why bother with anything other than $p = 2$? There are many reasons, but some values of p are seen more commonly than others. For $p = 1$, we have

$$\| \mathbf{x} \|_1 = |x_1| + |x_2| + \cdots + |x_n|.$$

In probability and statistics, it is common to consider vectors of probabilities, which may be defined as follows.

Definition 1.1.5. A vector $\mathbf{x}$ in $\mathbb{R}^n$ is a probability vector if:

1. the components x_i are all non-negative, and
2. the sum of the components is equal to 1.

   So, if $\mathbf{x}$ is a probability vector, then

$$\| \mathbf{x} \|_1 = |x_1| + |x_2| + \cdot + |x_n| = x_1 + x_2 + \cdots x_n = 1,$$

which makes the 1-norm a particularly natural thing to consider in the context of probability.

Activity 1.1.4. (With Desmos)
Go to https://www.desmos.com/ in a browser and click on Graphing Calculator.
Most things can be entered in desmos by just typing:

- You can get x^2 by typing `x^2`
- You can get $\sqrt{\ }$ by typing `sqrt`
- Most keyboards have `|`, but on a phone it could hidden on a different screen.

Consider the vector

$$\mathbf{v} = \begin{bmatrix} x \\ y \end{bmatrix}$$

We will use desmos to study the effect that changing p has on the equality

$$\| \mathbf{v} \|_p = 1.$$

1. ($\| \mathbf{v} \|_2 = 1$) In the first cell, enter

$$\sqrt{x^2 + y^2} = 1$$

   You should see a circle with radius 1. Any vector $\mathbf{v}$ with the arrowhead landing on this circle will have length 1.

2. ($\| \mathbf{v} \|_2 = 1$) In the second cell, enter

$$|x| + |y| = 1$$

   You should see a square with corners at $(1, 0), (0, 1), (-1, 0), (0, -1)$. The square is completely inside the circle. Any vector $\mathbf{v}$ with the arrowhead landing on this square will have $\| \mathbf{v} \|_1 = 1$.

3. In a third cell, enter

$$\max(|x|, |y|) = 1$$

   You should see a square with horizontal and vertical edges. The edges are tangent to the circle at the points $(1, 0), (0, 1), (-1, 0), (0, -1)$, and the circle is completely inside the square. Any vector $\mathbf{v}$ with the arrowhead landing on this square will have $\| \mathbf{v} \|_\infty = 1$.

4. In a fourth cell, enter

$$(|x|^p + |y|^p)^{\frac{1}{p}} = 1$$

and make a slider. When the slider is created, the value of p will be 1, and the curve will match the inner square. As you move the slider to the right toward 2, the curve will remain inside the circle and approach it. Once you pass 2 and continue moving toward 10, the curve will be outside the circle approaching the outer square. This illustrates the fact that

$$\lim_{p \to \infty} \| \mathbf{v} \|_p = \| \mathbf{v} \|_\infty$$

Theorem 1.1.6. *Let* $\mathbf{v}$ *be in* $\mathbb{R}^n$ *or* $\mathbb{C}^n$ *and let* $p < q$. *Then*

$$\| \mathbf{v} \|_p \geq \| \mathbf{v} \|_q \tag{1.1}$$

$$\| \mathbf{v} \|_p \leq n^{\frac{1}{p}} \| \mathbf{v} \|_\infty \tag{1.2}$$

Although the first inequality in Theorem 1.1.6 can be understood in terms of Activity 1.1.4, it may be easier to consider the computational example in Activity 1.1.5. The second inequality follows from considering replacing the components of $\mathbf{v}$ by $\| \mathbf{v} \|_\infty$, which amounts to rounding up in absolute value.

Activity 1.1.5. Let

$$\mathbf{v} = \begin{bmatrix} -1 \\ 2 \end{bmatrix}$$

Compute

$$\begin{aligned} \| \mathbf{v} \|_1 &= |-1| + |2| \\ \| \mathbf{v} \|_2 &= \sqrt{|-1|^2 + |2|^2} \\ \| \mathbf{v} \|_\infty &= \max(|-1|, |2|) \end{aligned}$$

then verify each inequality below:

$$\| \mathbf{v} \|_1 > \| \mathbf{v} \|_2 \qquad \| \mathbf{v} \|_2 > \| \mathbf{v} \|_\infty$$

$$\| \mathbf{v} \|_1 < 2 \| \mathbf{v} \|_\infty \qquad \| \mathbf{v} \|_2 < \sqrt{2} \| \mathbf{v} \|_\infty$$

A complex number is a number in the form

$$a + bi$$

where a and b are real and $i = \sqrt{-1}$. It is common to graph complex numbers by thinking of a and b as components of a 2 dimensional vector. As a 2-dimensional vector, the length would be given by the Pythagorean theorem, and that is exactly how the absolute value of a complex number is defined:

$$|a + bi| = \sqrt{a^2 + b^2}. \tag{1.3}$$

Example 1.1.6. The complex number $4 + 3i$ can be thought of as a vector with horizontal component 4 and vertical component 3 as drawn in Figure 1.1a. The horizontal axis is the real axis, and the vertical axis is the imaginary axis.

In Example 1.1.2 we saw that the length was 5, so

$$|4 + 3i| = \sqrt{4^2 + 3^2} = \sqrt{5^2} = 5.$$

If $z = a + bi$ is a complex number, its complex conjugate is $\overline{z} = a - bi$. Multiplying a complex number by its conjugate is one way of getting the square of the absolute value, because

$$\begin{aligned} z\overline{z} = (a+bi)(a-bi) &= a^2 - (bi)^2 \\ &= a^2 - b^2(i)^2 = a^2 - b^2(-1) = a^2 + b^2 = |z|^2. \end{aligned}$$

Example 1.1.7. If $z = 4 + 3i$, then the complex conjugate is $\overline{z}$ and

$$\begin{aligned} z\overline{z} = (4+3i)(4-3i) &= 4^2 - (3i)^2 \\ &= 16 - 9(i)^2 = 16 - 9(-1) = 16 + 9 = 25 = 5^2. \end{aligned}$$

A vector $\mathbf{z}$ in $\mathbb{C}^n$ is a vector with n components

$$z_1, z_2 \ldots z_n$$

that are complex. Definition 1.1.4 and Theorem 1.1.6 remain valid with the absolute value of a complex number defined by equation (1.3).

Example 1.1.8. Given

$$\mathbf{v} = \begin{bmatrix} 4+3i \\ -2 \end{bmatrix}$$

compute $\|\mathbf{v}\|_1$, $\|\mathbf{v}\|_2$, and $\|\mathbf{z}\|_\infty$. Then check that

$$\|\mathbf{v}\|_1 > \|\mathbf{v}\|_2 \qquad \|\mathbf{v}\|_2 > \|\mathbf{v}\|_\infty$$
$$\|\mathbf{v}\|_1 < 2\|\mathbf{v}\|_\infty \qquad \|\mathbf{v}\|_2 < \sqrt{2}\|\mathbf{v}\|_\infty$$

Solution

Example 1.1.6 shows that $|4 + 3i| = 5$, so

$$\begin{aligned} \|\mathbf{v}\|_1 &= |4+3i| + |-2| = 5 + 2 = 7 \\ \|\mathbf{v}\|_2 &= \sqrt{|4+3i|^2 + |-2|^2} = \sqrt{5^2 + 2^2} = \sqrt{25+4} = \sqrt{29} \\ \|\mathbf{v}\|_\infty &= \max(|4+3i|, |-2|) = \max(5, 2) = 5. \end{aligned}$$

As for the inequalities, we have:

$$\|\mathbf{v}\|_1 = 7 > \sqrt{29} = \|\mathbf{v}\|_2 \qquad \|\mathbf{v}\|_2 = \sqrt{29} > 5 = \|\mathbf{v}\|_\infty$$
$$\|\mathbf{v}\|_1 = 7 < 10 = 2\|\mathbf{v}\|_\infty \qquad \|\mathbf{v}\|_2 = \sqrt{29} < 5\sqrt{2} = 2\|\mathbf{v}\|_\infty .$$

To see that the inequalities with the square roots are true:

1. you can approximate them (e.g. with your calculator), or

2. you can square them.

Taking the second approach, squaring 7 and $\sqrt{29}$ gives 49 and 29. Since $49 > 29$, then $7 > \sqrt{29}$.
You can even get rough approximations for square roots without a calculator. Maybe you have a few square numbers memorized, like

$$1,\ 4,\ 9,\ 25,\ 36,\ 49,\ 64,\ 81,\ 100,\ \ldots$$

Since 29 is between 25 and 36, then $\sqrt{29}$ is between 5 and 6.

Activity 1.1.9. Given $z = 1 + 2i$, and

$$\mathbf{v} = \begin{bmatrix} 1+2i \\ 3 \end{bmatrix}$$

1. Compute the complex conjugate $\overline{z}$.
2. Compute $|z|$ and show that $|z|^2 = z\overline{z}$.
3. Compute $\| \mathbf{v} \|_1$, $\| \mathbf{v} \|_2$, and $\| \mathbf{z} \|_\infty$. Then check that

$$\| \mathbf{v} \|_1 > \| \mathbf{v} \|_2 \qquad \| \mathbf{v} \|_2 > \| \mathbf{v} \|_\infty$$
$$\| \mathbf{v} \|_1 < 2 \| \mathbf{v} \|_\infty \qquad \| \mathbf{v} \|_2 < \sqrt{2} \| \mathbf{v} \|_\infty$$

1.1.3 Introduction to Numerical Stability

Q 1.1.5. What is $\ln(1)$?

Q 1.1.6. What is the derivative of $\ln(x)$?

Q 1.1.7. What is an equation for the tangent line to $\ln(x)$ at $x = 1$?

In some calculations it is possible for a very slight change in the input to result in a very large change in the result. Intuitively, such unpredictable behavior is what we mean by numerical instability. To quantify numerical stability or instability we must pay attention to how small changes to an input will impact the final result of a calculation. Unless otherwise stated, the content on numerical stability in this book will follow [11].

If x is a number or $\mathbf{x}$ is a vector, we will use $\widehat{x}$ or $\widehat{\mathbf{x}}$ for an approximation.

We will use Δ to indicate a difference, such as

$$\Delta x = \widehat{x} - x \quad \text{or} \quad \Delta \mathbf{x} = \widehat{\mathbf{x}} - \mathbf{x}.$$

This means that Δx could be positive or negative and that $\Delta \mathbf{x}$ is a vector. Hopefully both are small. We can measure how small they are in two ways:

1. By the **Absolute error**

$$\mathrm{E}_{\mathrm{abs}}(\widehat{x}) = |\Delta x| = |x - \widehat{x}| \quad \text{or} \quad \mathrm{E}_{\mathrm{abs}}(\widehat{\mathbf{x}}) = \| \Delta \mathbf{x} \| = \| \widehat{\mathbf{x}} - \mathbf{x} \| . \tag{1.4}$$

2. By the **Relative error**

$$\mathrm{E}_{\mathrm{rel}}(\widehat{x}) = \frac{|\Delta x|}{|x|} = \frac{|x - \widehat{x}|}{|x|} \quad \text{or} \quad \mathrm{E}_{\mathrm{rel}}(\widehat{\mathbf{x}}) = \frac{\| \Delta \mathbf{x} \|}{\| \mathbf{x} \|} = \frac{\| \widehat{\mathbf{x}} - \mathbf{x} \|}{\| \mathbf{x} \|}. \tag{1.5}$$

Many science classes include a discussion of significant figures (or digits). How can we be sure how many significant digits we have? Higham suggests the following definition:

"$\widehat{x}$ agrees with x to p significant digits if the absolute error is less than half a unit in the p-th significant digit of x."

By this definition we would say that $\widehat{x} = 0.1451$ agrees with $x = 0.1449$ to three significant digits because $|x - \widehat{x}| = 0.0002$ which is less than half of 0.001. However, just by looking at the numbers, it seems that only two digits match: 0.14. Part of the problem is that the very concept of significant figures assumes a base, and carrying can make the correct number of digits unpredictable. In this case,

$$0.1449 + 0.0001$$

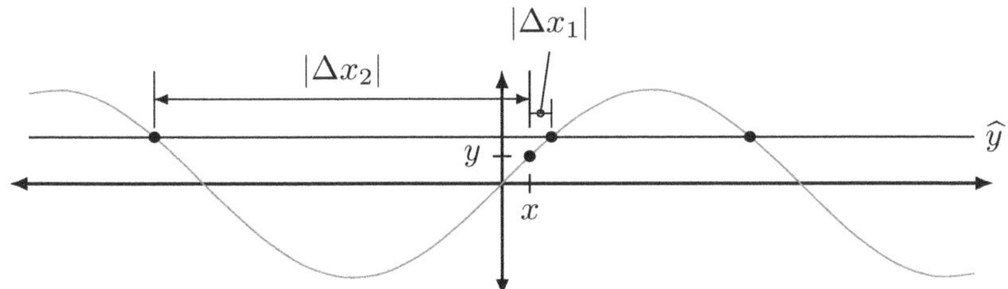

FIGURE 1.2: Backward error illustrated with $f(x) = \sin(x)$

becomes 0.1450 by carrying, and three digits of .1451 and .1450 do match. One advantage of the relative error is that it does not depend on the base. Still using $\widehat{x} = 0.1451$ and $x = 0.1449$, we have

$$\mathrm{E}_{\mathrm{rel}}(\widehat{x}) = \frac{|.1449 - .1451|}{.1449} \approx 0.00139.$$

A disadvantage of the relative error is that it is undefined when $x = 0$.

Suppose that we have a function f of one variable, and that $\widehat{y}$ is an approximation to $y = f(x)$, then $\mathrm{E}_{\mathrm{rel}}(\widehat{y})$ is a **forward error**. So a forward error quantifies the error of the output.

On the other hand, even though $\widehat{y}$ may not be equal to y, we may consider all $\widehat{x}$ such that

$$f(\widehat{x}) = \widehat{y}.$$

Then, we can consider how far each $\widehat{x}$ is from x. Which one is the closest? If $\Delta x = \widehat{x} - x$, we must find

$$\min |\Delta x|,$$

which is the minimum absolute error. We can then divide by $|x|$ to get the relative error. This error is the **backward error**.

How can we be sure that we have understood the concept of backward error? As with any definition or theorem it is important to understand it in context. What is actually capable of happening? We have a minimum, which suggests that there could be more than one Δx. How many are we capable of having? How do we know that we have any? Is it possible for the minimum to be large? To answer these questions it is helpful to understand function concepts.

How do we know that we have any Δx? Since $\Delta x = \widehat{x} - x$, we need a solution of $f(\widehat{x}) = \widehat{y}$ to exist. Since $\widehat{y}$ is a given approximation, it must be in the range of f for a solution to exist. Then you think "what functions do I know that don't have all real numbers as the range?" There are many. But take $f(x) = \sin(x)$. The range of $\sin(x)$ is $[-1, 1]$. Now you pick a number that is not in the range, like $\widehat{y} = 1.1$. Then

$$\sin(\widehat{x}) = 1.1$$

has no solutions. $\widehat{y} = 1.1$ could make sense as an approximation to $y = 1$, which occurs when $x = \frac{\pi}{2}$.

How many Δx are we capable of having? Again, since $\Delta x = \widehat{x} - x$, we are asking how many solutions can exist to $f(\widehat{x}) = \widehat{y}$. Each input of a function must have one output, otherwise it is not a function. But it is possible that several different inputs could have the same output. Again if $f(x) = \sin(x)$, and if we take $\hat{y} = 0.497$, then there are infinitely many $\hat{x}$ that work. A few of them are illustrated in Figure 1.2.

Is it possible for the minimum $|\Delta x|$ to be large? If the minimum $|\Delta x|$ is large, it means that none of the $\widehat{x}$ are close to x. In the case of $f(x) = \sin(x)$, the minimum

$|\Delta x|$ is small, but $\sin(x)$ is an example of a continuous function. The formal definition of a continuous function tells us that if $y = f(x)$ and $\hat{y}$ is in a small interval around y, then $f(\hat{x}) = \hat{y}$ has a solution in a small interval around x. So if we want an example with a large minimum $|\Delta x|$, we need a function that is not continuous. What discontinuities do we know of? Asymptotes, jump discontinuities, or removable discontinuities come to mind. But it is best not to over complicate things. Pick the simplest functions you can think of. Making a function out of lines is pretty easy. We only need one discontinuity, and we can place it anywhere. Why not at zero?

Activity 1.1.10 (Desmos optional)**.** Consider the function

$$f(x) = \begin{cases} x & \text{if } x \neq 0 \\ 1 & \text{if } x = 0 \end{cases}$$

(a) Graph the function. In Desmos this can be done as follows

(i) Enter $\frac{x^2}{x}$ in the 1st cell

(ii) Enter $(0, 1)$ in the 2nd cell.

(iii) Enter $(0, 0)$ in the 3rd cell. Then click on the solid circle in the side bar. A box should pop up with some options, including one to change the circle type to an open circle.

(b) Take $x = 0$, so that $y = 1$. Label the point (x, y) on the graph. In Desmos, you can add a label by clicking the checkbox and typing the label.

(c) Suppose $\widehat{y} = 1.1$. What is the value of $\widehat{x}$?

(d) Plot and label the point $(\widehat{x}, \widehat{y})$ on the graph.

(e) Compute $\Delta y = \widehat{y} - y$ and $\Delta x = \widehat{x} - x$.

Suppose we have an algorithm for approximating values of a function f. We say that the algorithm is **backward stable** if for all x it it produces a $\widehat{y}$ with small backward error, that is if there exists Δx small such that

$$\widehat{y} = f(x + \Delta x)$$

If f has two derivatives, then

$$\widehat{y} - y = f(x + \Delta x) - f(x) = f'(x)\Delta x + \frac{f''(x + t\Delta x)}{2}(\Delta x)^2 \tag{1.6}$$

where $0 < t < 1$. If the f'' term is dropped, we are left with

$$f(x + \Delta x) - f(x) \approx f'(x)\Delta x,$$

which is the tangent line approximation near x. So one way to look at the f'' term is as a correction to the tangent line approximation. Where does the correction come from? It is one way of expressing the remainder of a Taylor polynomial. If you are unfamiliar with Taylor series, equation (1.6) can be seen as a generalization of the mean value theorem (MVT) to higher derivatives. The condition that t is between 0 and 1 means that $x + t\Delta x$ is between x and $\widehat{x} = x + \Delta x$.

Since f has two derivatives, it is continuous. So the situation studied in Activity 1.1.10 cannot occur. If we divide by $y = f(x)$, then

$$\begin{aligned}\frac{\widehat{y}-y}{y} &= \frac{f'(x)}{f(x)}\cdot\Delta x + \frac{f''(x+t\Delta x)}{2f(x)}(\Delta x)^2\\ &= \frac{xf'(x)}{f(x)}\cdot\frac{\Delta x}{x} + \frac{f''(x+t\Delta x)}{2f(x)}(\Delta x)^2.\end{aligned} \tag{1.7}$$

You should recognize that

$$\frac{|\widehat{y}-y|}{|y|} = \mathrm{E}_{\mathrm{rel}}(\widehat{y}) \quad\text{and}\quad \frac{|\Delta x|}{|x|} = \mathrm{E}_{\mathrm{rel}}(\widehat{x}).$$

The **condition number** is

$$\kappa(x) = \frac{|f'(x)x|}{|f(x)|}. \tag{1.8}$$

So equation (1.7) tells us that

$$\mathrm{E}_{\mathrm{rel}}(\widehat{y}) \approx \kappa(x)\,\mathrm{E}_{\mathrm{rel}}(\widehat{x}).$$

If $\kappa(x)$ is very small, it is possible for $\mathrm{E}_{\mathrm{rel}}(\widehat{y})$ to be small while $\mathrm{E}_{\mathrm{rel}}(\widehat{x})$ is big. On the other hand, if $\kappa(x)$ is very big, it is possible for $\mathrm{E}_{\mathrm{rel}}(\widehat{y})$ to be big, while $\mathrm{E}_{\mathrm{rel}}(\widehat{x})$ is small. Thus $\kappa(x)$ gives us a way to quantify the relationship between forwards and backward errors.

Example 1.1.11. Suppose $f(x) = \ln(x)$, then $f'(x) = \frac{1}{x}$. So the condition number is

$$\kappa(x) = \frac{|f'(x)x|}{|f(x)|} = \frac{|\frac{1}{x}x|}{|\ln(x)|} = \frac{1}{|\ln(x)|}.$$

Since $\ln(1) = 0$, $\kappa(x)$ is very big when $x \approx 1$.

At first, this may seem a bit odd, because $x - 1$ is the tangent line to $\ln(x)$ at $x = 1$. "What is the problem with a tangent line approximation" you ask. In this case, the problem is that $\ln(x)$ has a zero that is not at zero: $\ln(1) = 0$. If we consider $f(x) = \ln(x+1)$ instead, now $f(0) = 0$. The problem with the condition number goes away because $f'(x) = \frac{1}{x+1}$, so

$$\kappa(x) = \frac{|f'(x)x|}{|f(x)|} = \frac{|x|}{|\ln(x+1)|}\cdot\frac{1}{|x+1|}.$$

What happens near 0? First, observe that

$$\lim_{x\to 0}\frac{1}{x+1} = 1,$$

then by L'Hôspital's rule

$$\lim_{x\to 0}\frac{x}{\ln(x+1)} = \lim_{x\to 0}\frac{1}{\frac{1}{x+1}} = \frac{1}{1} = 1.$$

By multiplying these together, we find $\kappa(0) = 1$. There is a problem near -1, because $\ln(x+1)$ has a vertical asymptote at $x = -1$. But the Taylor series for $\ln(x+1)$ diverges there anyway.

Activity 1.1.12. Given $f(x) = x^n$

(a) Compute $f'(x)$.

(b) Compute $\kappa(x) = \frac{|f'(x)x|}{|f(x)|}$. Simplify completely.

Repeat with $f(x) = e^x$.

Activity 1.1.13. Given $f(x) = \sin(x)$

(a) Compute $f'(x)$.

(b) Compute $\kappa(x) = \frac{|f'(x)x|}{|f(x)|}$

(c) Use L'Hôspital's to evaluate

$$\lim_{x \to 0} \frac{x}{\sin(x)}$$

(d) Use the result of the previous limit to compute $\kappa(0)$.

The concept of a condition number can be generalized to functions from $\mathbb{R}^n$ to $\mathbb{R}^m$ by using norms. Right now we will consider the case where $n = 1$. In this case we have a vector of functions

$$\mathbf{f}(t) = \begin{bmatrix} f_1(t) \\ f_2(t) \\ \vdots \\ f_m(t) \end{bmatrix},$$

which is often called a **parametric curve**. The variable t is called the parameter. If $\mathbf{f}'(t)$ is the component-wise derivative, then the condition number is

$$\kappa(t) = \frac{\| \mathbf{f}'(t) \|_2 \cdot |t|}{\| \mathbf{f}(t) \|_2} \tag{1.9}$$

Example 1.1.14. We can describe the points on the unit circle by

$$\mathbf{f}(t) = \begin{bmatrix} \sin(t) \\ \cos(t) \end{bmatrix}$$

with t in $[-\pi, \pi]$. Then

$$\mathbf{f}'(t) = \begin{bmatrix} \cos(t) \\ -\sin(t) \end{bmatrix}.$$

Using the Pythagorean trig identity, the norms are

$$\| \mathbf{f}(t) \|_2 = \sqrt{\sin^2(t) + \cos^2(t)} = \sqrt{1} = 1,$$
$$\| \mathbf{f}'(t) \|_2 = \sqrt{\cos^2(t) + \sin^2(t)} = \sqrt{1} = 1.$$

So the condition number is

$$\kappa(t) = \frac{\| \mathbf{f}'(t) \|_2 \cdot |t|}{\| \mathbf{f}(t) \|_2} = \frac{1 \cdot |t|}{1} = |t| \leq \pi$$

for all t in $[-\pi, \pi]$.

1.1.4 Matrix Addition and Multiplication by a Scalar

Definition 1.1.7. Addition for matrices. If $\mathbf{A} = [a_{ij}]$ and $\mathbf{B} = [b_{ij}]$ are matrices with the same dimension, then their sum is defined to be $\mathbf{A} + \mathbf{B} = [a_{ij} + b_{ij}]$.

Definition 1.1.8. Multiplication of a matrix by a scalar. If $\mathbf{A} = [a_{ij}]$ is a matrix, and c is an arbitrary real number (scalar), then $c\mathbf{A} = [ca_{ij}]$.

AI Prompt 1.1.2. Explain the history (etymology) of the mathematical term 'scalar.' What language does it come from, and what did the original word mean?

Warning 1.1.3. You should always check the dimensions of the matrix to make sure that they are compatible before launching into a calculation. Addition is only defined for two matrices if they have the same dimension. If a calculation has a step where you would be adding two matrices with different dimensions, then you should not even start the calculation. In that case, the answer does not exist.

Example 1.1.15. Given the matrices

$$\mathbf{A} = \begin{bmatrix} 2 & 1 & 0 \\ 0 & -1 & -2 \end{bmatrix}, \quad \mathbf{B} = \begin{bmatrix} 1 & 2 & 3 \\ 4 & 5 & 6 \end{bmatrix}, \quad \mathbf{C} = \begin{bmatrix} 0 & -1 \\ 1 & 0 \end{bmatrix}, \quad \mathbf{D} = \begin{bmatrix} 2 & 0 \\ 1 & -1 \\ 0 & -2 \end{bmatrix},$$

1. find the sum of each pair of different matrices, if the sum exists, and
2. compute $3\mathbf{A}$.

Solution

1. First. check the dimensions. **A** and **B** have dimension 2×3, **C** has dimension 2×2, and **D** has dimension 3×2. The only compatible pair is **A** and **B**. We obtain

$$\begin{aligned} \mathbf{A} + \mathbf{B} &= \begin{bmatrix} 2 & 1 & 0 \\ 0 & -1 & -2 \end{bmatrix} + \begin{bmatrix} 1 & 2 & 3 \\ 4 & 5 & 6 \end{bmatrix} \\ &= \begin{bmatrix} 2+1 & 1+2 & 0+3 \\ 0+4 & -1+5 & -2+6 \end{bmatrix} = \begin{bmatrix} 3 & 3 & 3 \\ 4 & 4 & 4 \end{bmatrix} \end{aligned}$$

2. Multiplication by a scalar always works, so there is no need to check dimensions:

$$3\mathbf{A} = 3\begin{bmatrix} 2 & 1 & 0 \\ 0 & -1 & -2 \end{bmatrix} = \begin{bmatrix} 6 & 3 & 0 \\ 0 & -3 & -6 \end{bmatrix}.$$

Whether the `A`, `B`, `C`, and `D` are defined as matrices in Sage or as NumPy Arrays, both

```
A+B
3*A
```

will produce the correct output. You will, of course, get an unfriendly error if you try to add two matrices of incompatible sizes

```
A+D
```

...but they are the wrong size, so this should be expected!

Activity 1.1.16. Using the matrices

$$\mathbf{A} = \begin{bmatrix} 2 & 1 & 0 \\ 0 & -1 & -2 \end{bmatrix}, \quad \mathbf{B} = \begin{bmatrix} 1 & 2 & 3 \\ 4 & 5 & 6 \end{bmatrix}$$

calculate $\mathbf{B} - 2\mathbf{A}$.

Given two vectors $\mathbf{v}, \mathbf{w}$ and a scalar c, then :

1. $c\mathbf{v}$ simply stretches the vector, and it may change the direction it points if c is a negative.
2. The sum $\mathbf{v} + \mathbf{w}$ and difference $\mathbf{v} - \mathbf{w}$ correspond to the diagonals of a parallelogram.

Activity 1.1.17. Let $\mathbf{v} = \begin{bmatrix} 2 \\ 1 \end{bmatrix}$ and $\mathbf{w} = \begin{bmatrix} 1 \\ 3 \end{bmatrix}$. Use the rules for matrix addition and multiplication by a scalar to compute the following vectors:

$$2\mathbf{v}, \quad -\mathbf{v}, \quad \mathbf{v} + \mathbf{w}, \quad \mathbf{v} - \mathbf{w}, \quad \mathbf{w} - \mathbf{v}$$

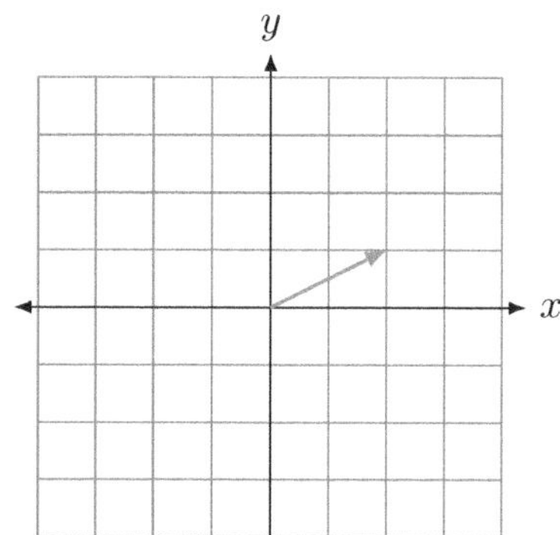

then graph them. $\mathbf{v}$ has been drawn for you.

Your final graph should also illustrate "tip to tail addition," in the sense that if $\mathbf{w}$ is moved so that it starts where $\mathbf{v}$ ends, the arrows describe a path along the perimeter of the parallelogram ending at $\mathbf{v} + \mathbf{w}$.

Proposition 1.1.9. *Let* $\mathbf{v}$ *and* $\mathbf{w}$ *be vectors in* $\mathbb{R}^n$ *or* $\mathbb{C}^n$ *and let* c *be a scalar. Then*

$$\| c\mathbf{v} \|_p = |c| \, \| \mathbf{v} \|_p, \tag{1.10}$$

$$\| \mathbf{v} + \mathbf{w} \|_p \leq \| \mathbf{v} \|_p + \| \mathbf{w} \|_p. \tag{1.11}$$

Note that $\mathbf{v} - \mathbf{w} = \mathbf{v} + (-1)\mathbf{w}$, so

$$\| \mathbf{v} - \mathbf{w} \|_p = \| \mathbf{v} + (-1)\mathbf{w} \|_p \leq \| \mathbf{v} \|_p + |-1| \, \| \mathbf{w} \|_p = \| \mathbf{v} \|_p + \| \mathbf{w} \|_p.$$

The inequality (1.11) is called the **triangle inequality**, because in the case where $p = 2$ it shows that the sum of the lengths of two sides of a triangle is greater than or equal to the length of the third side.

Activity 1.1.18. Use your results from Activity 1.1.17 to compute the following norms:

$$\| \mathbf{v} \|_1, \; \| \mathbf{w} \|_1, \; \| -\mathbf{v} \|_1, \; \| 2\mathbf{v} \|_1, \; \| \mathbf{v} + \mathbf{w} \|_1, \; \| \mathbf{v} - \mathbf{w} \|_1.$$

Then verify that

$$\| 2\mathbf{v} \|_1 = 2 \, \| \mathbf{v} \|_1 \qquad \| -\mathbf{v} \|_1 = \| \mathbf{v} \|_1$$

$$\| \mathbf{v} + \mathbf{w} \|_1 \leq \| \mathbf{v} \|_1 + \| \mathbf{w} \|_1 \qquad \| \mathbf{v} - \mathbf{w} \|_1 \leq \| \mathbf{v} \|_1 + \| \mathbf{w} \|_1$$

Definition 1.1.10. A vector $\mathbf{u}$ in $\mathbb{R}^n$ or $\mathbb{C}^n$ is called a **unit vector** with respect to the p-norm if

$$\| \mathbf{u} \|_p = 1.$$

By Proposition 1.1.9, we can rescale a vector to a unit vector if we divide it by its norm:

$$\mathbf{u} = \frac{1}{\| \mathbf{v} \|_p} \mathbf{v}. \tag{1.12}$$

The process of dividing a vector by its norm is called **normalization**.

Example 1.1.19.

Activity 1.1.20. Given

$$\mathbf{v} = \begin{bmatrix} 4 \\ 3 \end{bmatrix}$$

compute the normalization using equation (1.12) for each norm below, and then check that the new vector is a unit vector with respect to that norm.

1. $\| \mathbf{v} \|_1 = 7$
2. $\| \mathbf{v} \|_2 = 5$
3. $\| \mathbf{v} \|_\infty = 4$

Since the components of $\mathbf{v}$ are non-negative, you should observe that normalizing with respect to the 1-norm gives a probability vector.

1.1.5 Matrix Multiplication

Q 1.1.8. Is $4 \times (2 + 3) = 4 \times 2 + 3$?

Q 1.1.9. What is the dimension of

$$\mathbf{A} = \begin{bmatrix} 1 & 2 & 3 \\ 4 & 5 & 6 \end{bmatrix}?$$

Aside from multiplication by a scalar, there is also the concept of matrix multiplication. We begin with the simplest case: a row vector times a column vector. If you have seen the dot product before, this should look familiar.

Definition 1.1.11. Multiplication of a row vector by a column vector. Given a $1 \times n$ row vector $\mathbf{A} = [a_i]$ and an $n \times 1$ column vector $\mathbf{B} = [b_i]$, the product $\mathbf{AB}$ is

$$a_1 b_1 + a_2 b_2 + \cdots + a_n b_n.$$

Remark 1.1.4. Pay close attention to the fact that the row vector is on the left and the column vector is on the right.

Example 1.1.21. Find the products

$$\begin{bmatrix} 1 & 2 & 3 \end{bmatrix} \begin{bmatrix} 2 \\ 1 \\ 0 \end{bmatrix} \text{ and } \begin{bmatrix} 1 & 2 \end{bmatrix} \begin{bmatrix} -2 \\ 1 \end{bmatrix}$$

Solution

$$\begin{bmatrix}1 & 2 & 3\end{bmatrix}\begin{bmatrix}2\\1\\0\end{bmatrix} = 1\cdot 2 + 2\cdot 1 + 3\cdot 0 = 2+2+0 = 4$$

$$\begin{bmatrix}1 & 2\end{bmatrix}\begin{bmatrix}-2\\1\end{bmatrix} = 1(-2) + 2\cdot 1 = -2+2 = 0$$

Definition 1.1.12. Multiplication of two matrices. If $\mathbf{A} = [a_{ij}]$ is an $m \times n$ matrix and $\mathbf{B} = [b_{jk}]$ is an $n \times l$ matrix, then the product is defined by $\mathbf{AB} = [c_{ik}]$, where

$$c_{ik} = \sum_{j=1}^{n} a_{ij}b_{jk} = a_{i1}b_{1k} + a_{i2}b_{2k} + \cdots + a_{in}b_{nk}.$$

The rule basically amounts to taking the dot product of the i-th row of $\mathbf{A}$ with the k-th column of $\mathbf{B}$ to get the component c_{ik} in the product (the sum is over j). For this reason, it is computationally expedient to arrange the problem in a grid:

$$\begin{array}{cc} & \begin{bmatrix} & b_{1k} & \\ & b_{2k} & \\ & \vdots & \\ & b_{nk} & \end{bmatrix} \\ \begin{bmatrix} \\ a_{i1} & a_{i2} & \cdots & a_{in} \\ \\ \end{bmatrix} & \begin{bmatrix} & \downarrow & \\ \rightarrow & \boxed{c_{ik}} & \\ & & \end{bmatrix} \end{array}$$

Warning 1.1.5. Once again, you should always check the dimensions before launching into a calculation. Matrix multiplication is defined by taking the dot product of rows of $\mathbf{A}$ with columns of $\mathbf{B}$. In order for that to work, the rows of $\mathbf{A}$ must have the same length as the columns of $\mathbf{B}$. This means that the number of columns of $\mathbf{A}$ must be equal to the number of rows of $\mathbf{B}$. It is easy to check by writing the dimensions below the matrices:

$$\begin{array}{cc} \mathbf{A} & \mathbf{B} \\ m \times n & n \times l \end{array}$$

Since the n's are next to each other, then $\mathbf{AB}$ is defined. Starting a matrix computation when the result does not exist is a waste of time.

Example 1.1.22. Given the matrices

$$\mathbf{A} = \begin{bmatrix}1 & 1\\1 & 0\\1 & -1\end{bmatrix}, \quad \mathbf{B} = \begin{bmatrix}1 & 2 & 3\\4 & 5 & 6\end{bmatrix}, \quad \mathbf{C} = \begin{bmatrix}1 & 1\\0 & 1\end{bmatrix},$$

compute each of the following, if it exists:

$$\mathbf{AB}, \quad \mathbf{BA}, \quad (\mathbf{AB})\mathbf{C}, \quad \text{and} \quad (\mathbf{BA})\mathbf{C}$$

Solution

First, check the dimensions!

$$\begin{array}{ccccc} \mathbf{A} & \mathbf{B} & & \mathbf{AB} \\ 3\times \not{2} & \not{2}\times 3 & \Longrightarrow & 3\times 3 \\ \mathbf{B} & \mathbf{A} & & \mathbf{BA} \\ 2\times \not{3} & \not{3}\times 2 & \Longrightarrow & 2\times 2 \end{array}$$

Thus **AB** and **BA** both exist. Again, 3×2 and 2×3 are not interpreted as products in this context. They are not equal to 6. The numbers 2 and 3 are not factors, and we are not canceling them. 3×2 is an expression for the dimension of the matrix, and we are crossing out the axis that we sum along. The numbers crossed out must be equal.
It should be clear that $(\mathbf{AB})\mathbf{C}$ does not exist because **AB** has 3 columns while **C** has 2 rows and $3 \neq 2$.

$$\begin{array}{ccccc} \mathbf{AB} & \mathbf{C} & & (\mathbf{AB})\mathbf{C} \\ 3\times 3 & 2\times 2 & \Longrightarrow & \text{DNE} \\ \mathbf{BA} & \mathbf{C} & & (\mathbf{BA})\mathbf{C} \\ 2\times \not{2} & \not{2}\times 2 & \Longrightarrow & 2\times 2 \end{array}$$

Therefore, we do not even begin the computation of $(\mathbf{AB})\mathbf{C}$, and since **BA** shows up in the computation of $(\mathbf{BA})\mathbf{C}$, using the result for **BA** saves time.

$$\begin{array}{cc} & \begin{bmatrix} 1 & 2 & 3 \\ 4 & 5 & 6 \end{bmatrix} \\ \begin{bmatrix} 1 & 1 \\ 1 & 0 \\ 1 & -1 \end{bmatrix} & \begin{bmatrix} 5 & 7 & 9 \\ 1 & 2 & 3 \\ -3 & -3 & -3 \end{bmatrix} = \mathbf{AB} \end{array} \qquad \begin{array}{cc} & \begin{bmatrix} 1 & 1 \\ 1 & 0 \\ 1 & -1 \end{bmatrix} \\ \begin{bmatrix} 1 & 2 & 3 \\ 4 & 5 & 6 \end{bmatrix} & \begin{bmatrix} 6 & -2 \\ 15 & -2 \end{bmatrix} = \mathbf{BA} \end{array}$$

$$\begin{array}{cc} & \begin{bmatrix} 1 & 1 \\ 0 & 1 \end{bmatrix} \\ \begin{bmatrix} 6 & -2 \\ 15 & -2 \end{bmatrix} & \begin{bmatrix} 6 & 4 \\ 15 & 13 \end{bmatrix} = (\mathbf{BA})\mathbf{C} \end{array}$$

Remark 1.1.6. With numbers, we are used to being able to change the order of multiplication such as $2 \cdot 3 = 3 \cdot 2 = 6$. Given two matrices **A** and **B**, it is possible that **AB** may exist while **BA** does not. Example 1.1.22 shows a situation where both exist but have different dimension. Two matrices cannot be equal if they have different dimension. To be equal, two matrices must have the same dimension and the same components. Later we will see that even if both **AB** and **BA** exist and have the same dimension, then they still may not be equal. Thus, unlike multiplication with numbers, the order of multiplication matters for matrices.

There is a huge difference between the syntax for matrix multiplication in Sage versus that in NumPy. If `A`, `B`, and `C` from Example 1.1.22 are Sage matrices then

```
A*B
```

is matrix multiplication. But if `A`, `B`, and `C` are NumPy arrays, then

```
np.dot(A,B)
```

is matrix multiplication, while

```
A*B
```

will give an error. That is because in NumPy

```
A*C
```

is component-wise multiplication of matrices.

Activity 1.1.23. Using the matrices

$$\mathbf{A} = \begin{bmatrix} 1 & -1 & 0 \\ 0 & -1 & 1 \end{bmatrix} \quad \mathbf{B} = \begin{bmatrix} 3 & -1 \\ 2 & 0 \\ 1 & 1 \end{bmatrix} \quad \mathbf{C} = \begin{bmatrix} 0 & 1 \\ 1 & 0 \end{bmatrix}$$

find each of the following if they exist:

$$(\mathbf{AB})\mathbf{C}, \quad \mathbf{A}(\mathbf{B}+\mathbf{C}), \quad \text{and} \quad \mathbf{AB}+\mathbf{C}$$

1.1.6 Properties of Matrix Arithmetic part 1

Q 1.1.10. What are some of the arithmetic properties of ordinary numbers? If you do not know the names of the properties, then just name some examples.

Q 1.1.11. If I rotate 90 degrees clockwise in front of a mirror, then what will the rotation look like in the mirror?

Definition 1.1.13. The $n \times n$ **identity matrix** $\mathbf{I}_n$, or sometimes just $\mathbf{I}$, is the $n \times n$ matrix with 1's down the diagonal and 0's everywhere else.

Example 1.1.24. For $n = 2$ we have

$$\mathbf{I}_2 = \begin{bmatrix} 1 & 0 \\ 0 & 1 \end{bmatrix}$$

Proposition 1.1.14. *If* $\mathbf{A}$ *is an* $n \times m$ *matrix, then* $\mathbf{I}_n\mathbf{A} = \mathbf{A}$ *and* $\mathbf{A} = \mathbf{A}\mathbf{I}_m$.

Remark 1.1.7. So just multiplying a number 1 doesn't change it, multiplying a matrix $\mathbf{A}$ by $\mathbf{I}_m$ doesn't change $\mathbf{A}$, except that there are different sizes of $\mathbf{I}_m$. The size needs to be correct for multiplication to be defined. Even so, the subscript m is sometimes omitted if it can be understood from context.

The identity matrix of any size can be constructed in Sage or NumPy. The syntax is similar but slightly different. In Sage $\mathbf{I}_2$ is:

```
matrix.identity(2)
```

In NumPy there are two options $\mathbf{I}_2$:

```
import numpy as np

np.identity(2)
np.eye(2)
```

If $\mathbf{A}$ is an $m \times n$ matrix, then multiplying on the right by one column of the $n \times n$ identity matrix has the effect of picking out that column of $\mathbf{A}$. Similarly, multiplying on the left by one row of the $m \times m$ identity matrix has the effect of picking out that row of $\mathbf{A}$.

Activity 1.1.25. Using the matrices

$$\mathbf{A} = \begin{bmatrix} 1 & 2 & 3 \\ -3 & -2 & -1 \end{bmatrix}, \quad \mathbf{u} = \begin{bmatrix} 1 & 0 \end{bmatrix}, \quad \text{and} \quad \mathbf{v} = \begin{bmatrix} 0 \\ 0 \\ 1 \end{bmatrix}$$

compute

1. $\mathbf{uA}$, and
2. $\mathbf{Av}$.

Observe that $\mathbf{u}$ is the 1st row of $\mathbf{I}_2$ and $\mathbf{v}$ is the 3rd column of $\mathbf{I}_3$.

Proposition 1.1.15. *Given two $m \times n$ matrices $\mathbf{A}$ and $\mathbf{B}$ and arbitrary scalars c and d, the following rules are true in general.*

1. $\mathbf{A} + \mathbf{B} = \mathbf{B} + \mathbf{A}$
2. $(c + d)\mathbf{A} = c\mathbf{A} + d\mathbf{A}$
3. $c(\mathbf{A} + \mathbf{B}) = c\mathbf{A} + c\mathbf{B}$
4. $(cd)\mathbf{A} = c(d\mathbf{A})$

Activity 1.1.26. Verify each part of the previous proposition using the matrices

$$\mathbf{A} = \begin{bmatrix} 2 & 1 & 0 \\ 0 & -1 & -2 \end{bmatrix}, \quad \mathbf{B} = \begin{bmatrix} 1 & 2 & 3 \\ 4 & 5 & 6 \end{bmatrix}$$

and the constants $c = -1$ and $d = 2$.

Proposition 1.1.16. *The following rules are true in general for matrices $\mathbf{A}$, $\mathbf{B}$, and $\mathbf{C}$ whose sizes are appropriate for the sums and products to be defined:*

1. *Associative:* $\mathbf{A}(\mathbf{BC}) = (\mathbf{AB})\mathbf{C}$
2. *Left Distributive:* $\mathbf{C}(\mathbf{A} + \mathbf{B}) = \mathbf{CA} + \mathbf{CB}$
3. *Right Distributive:* $(\mathbf{A} + \mathbf{B})\mathbf{C} = \mathbf{AC} + \mathbf{BC}$

Activity 1.1.27. Verify that each part of Proposition 1.1.16 is true for the matrices

$$\mathbf{A} = \begin{bmatrix} 1 & 2 \\ 3 & 4 \end{bmatrix}, \quad \mathbf{B} = \begin{bmatrix} 1 & 0 \\ 0 & -1 \end{bmatrix}, \quad \mathbf{C} = \begin{bmatrix} 0 & 1 \\ 1 & 0 \end{bmatrix}$$

Although we have been consistently multiplying scalars on the left, it is acceptable to multiply them on the right as well. It is defined in the same way – component-wise – and so the result is the same.

Proposition 1.1.17. *For an arbitrary matrix $\mathbf{A}$ and arbitrary scalar c*

$$c\mathbf{A} = \mathbf{A}c.$$

However, the definition of the product of two matrices $\mathbf{A}$ and $\mathbf{B}$ is quite different, and so generally $\mathbf{AB} \neq \mathbf{BA}$.

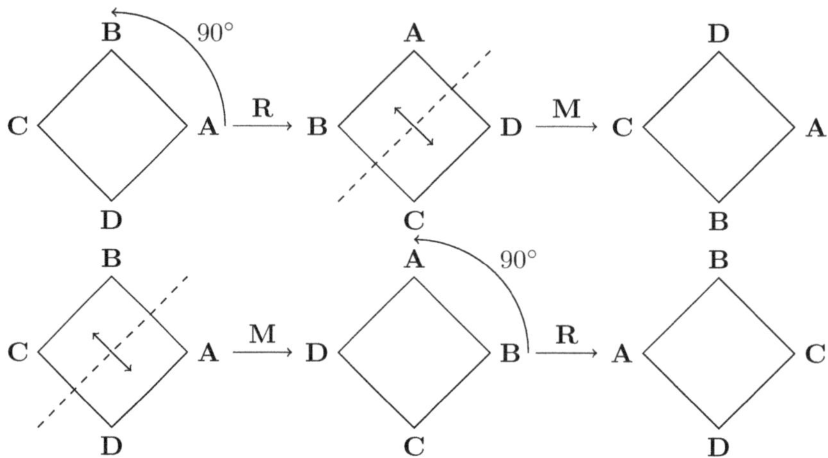

FIGURE 1.3: Lack of commutativity of $\mathbf{R}$ and $\mathbf{M}$

Example 1.1.28. First, suppose $\mathbf{A}$ is 2×2 and $\mathbf{B}$ is 2×3. Then

$$\begin{array}{cccc} \mathbf{A} & \mathbf{B} & & \mathbf{AB} \\ 2\times \not{2} & \not{2}\times 3 & \Longrightarrow & 2\times 3 \\ \mathbf{B} & \mathbf{A} & & \mathbf{BA} \\ 2\times 3 & 2\times 2 & \Longrightarrow & \text{DNE} \end{array}$$

How can $\mathbf{AB} = \mathbf{BA}$ if one of them is not even defined? It is not possible.

Even if they are both defined, though, they may not have the same shape. We have already seen this in Example 1.1.22 where $\mathbf{A}$ was a 3×2 matrix and $\mathbf{B}$ was a 2×3 matrix. Both $\mathbf{AB}$ and $\mathbf{BA}$ were defined, but $\mathbf{AB}$ had dimension 3×3 while $\mathbf{BA}$ had dimension 2×2. For two matrices to be equal, every component must match, but comparing components is only possible if the shape is the same. So if $\mathbf{AB}$ is 3×3 and $\mathbf{BA}$ is 2×2, it is not possible to say that they are equal.

But even if both $\mathbf{AB}$ and $\mathbf{BA}$ exist and have the same shape, it is still typical that $\mathbf{AB} \neq \mathbf{BA}$. One of the most natural examples that I know of comes from the symmetries of the square.

Figure 1.3 shows what happens to a square if 90 degrees rotation counterclockwise, $\mathbf{R}$, is done first followed by a reflection across a mirror, $\mathbf{M}$ (indicated by the dotted line), and what happens when those operations are reversed. This is something you might try doing with a square cut out of paper or cardboard. The corners are labeled **A**, **B**, **C**, and **D** to keep track of what happens. It should be clear from the final position of the labels that the result is different depending on the order of $\mathbf{R}$ and $\mathbf{M}$. As we will see in a later section, these operations can be described with the matrices

$$\mathbf{R} = \begin{bmatrix} 0 & -1 \\ 1 & 0 \end{bmatrix} \quad \text{and} \quad \mathbf{M} = \begin{bmatrix} 0 & 1 \\ 1 & 0 \end{bmatrix}.$$

Doing one operation and then the other amounts to matrix multiplication. The fact that the result is not the same means that $\mathbf{MR} \neq \mathbf{RM}$.

Activity 1.1.29. Compute $\mathbf{RM}$ and $\mathbf{MR}$. Observe that they are not equal.

1.2 Transpose

1.2.1 Introduction to Transpose

Q 1.2.1. What is the first row of

$$\mathbf{A} = \begin{bmatrix} 1 & 2 & 3 \\ 4 & 5 & 6 \end{bmatrix}?$$

Definition 1.2.1. Given an $m \times n$ matrix $\mathbf{A}$, the **transpose** of $\mathbf{A}$, written $\mathbf{A}^T$, is the $n \times m$ whose columns are formed from the rows of $\mathbf{A}$, keeping their order the same. In terms of indices, if $\mathbf{A} = [a_{ij}]$, then $\mathbf{A}^T = [a_{ji}]$.

Example 1.2.1. What is the transpose of $\mathbf{A} = \begin{bmatrix} 1 & 2 & 3 \\ 4 & 5 & 6 \end{bmatrix}$?

Solution

$$\begin{bmatrix} 1 & 2 & 3 \\ 4 & 5 & 6 \end{bmatrix} \longrightarrow \mathbf{A}^T = \begin{bmatrix} 1 & 4 \\ 2 & 5 \\ 3 & 6 \end{bmatrix}.$$

Notice that while $\mathbf{A}$ has dimension 2×3, $\mathbf{A}^T$ has dimension 3×2.

If `A` is a matrix in Sage or a NumPy array with two indices, then

```
A.T
```

is the transpose. If `v` is a vector in Sage or a NumPy array with one index, then

```
v.T
```

does nothing. This should not be surprising. The transpose reverses the indices, but `v` has only one index so there is nothing to reverse. When we defined `v_row` earlier, it had two indices, and so

```
v_row.T == v_col
```

should be True. If `A` is a NumPy array has more than 2 indices, then

```
A.T
```

will still reverse the indices. For example if the indices of `A` are i, j, k, then the indices of `A.T` will be k, j, i. That may not be what you want. If the array is a stack of matrices, then we may want to switch only the last two indices. This can be done with the following code:

```
np.Transpose(A,(0,2,1))
```

Definition 1.2.2. A **symmetric matrix** is a matrix $\mathbf{A}$ satisfying the property $\mathbf{A}^T = \mathbf{A}$.

The equation $\mathbf{A}^T = \mathbf{A}$ implies that $\mathbf{A}$ must be a square matrix, since otherwise $\mathbf{A}$ and $\mathbf{A}^T$ do not even have the same shape and so cannot possibly be equal.

Example 1.2.2. The matrix

$$\mathbf{A} = \begin{bmatrix} 1 & 2 \\ 2 & 3 \end{bmatrix}$$

is a symmetric matrix.

Comprehension Check 1.2.1. For what values of t is $\begin{bmatrix} 1-t^2 & -2t \\ 2t & 1-t^2 \end{bmatrix}$ symmetric?

Proposition 1.2.3. *Given any matrix* $\mathbf{A}$*, the matrix* $\mathbf{S} = \mathbf{A}^T\mathbf{A}$ *is symmetric (meaning* $\mathbf{S}^T = \mathbf{S}$*).*

Activity 1.2.3. Illustrate Proposition 1.2.3 using the matrix

$$\mathbf{A} = \begin{bmatrix} 2 & 1 & 0 \\ 0 & -1 & -2 \end{bmatrix},$$

by computing both $\mathbf{A}^T\mathbf{A}$ and $\mathbf{A}(\mathbf{A}^T)$. The resulting matrices will be different, but both should be symmetric.

Remark 1.2.1. While multiplying a matrix $\mathbf{A}$ by its transpose will give a symmetric matrix, multiplying two symmetric matrices $\mathbf{A}$ and $\mathbf{B}$ by each other, may not result in a symmetric matrix.

Activity 1.2.4. The matrices

$$\mathbf{A} = \begin{bmatrix} 1 & 1 \\ 1 & 0 \end{bmatrix} \quad \text{and} \quad \mathbf{B} = \begin{bmatrix} 1 & 2 \\ 2 & 3 \end{bmatrix}$$

are clearly symmetric. Compute $\mathbf{AB}$ and $\mathbf{BA}$. You will see that $\mathbf{BA} = (\mathbf{AB})^T$, but that neither is symmetric so $\mathbf{AB} \neq \mathbf{BA}$.

1.2.2 Properties of Transpose

Theorem 1.2.4. *Let* $\mathbf{A}$ *and* $\mathbf{B}$ *denote matrices whose sizes are appropriate for the following sums and products.*

1. $(\mathbf{A}^T)^T = A$
2. $(\mathbf{A} + \mathbf{B})^T = \mathbf{A}^T + \mathbf{B}^T$
3. *For any scalar* c, $(c\mathbf{A})^T = c\mathbf{A}^T$
4. $(\mathbf{AB})^T = \mathbf{B}^T\mathbf{A}^T$

Activity 1.2.5. Verify that all parts of the previous theorem are true for

$$\mathbf{A} = \begin{bmatrix} 1 & 2 \\ 3 & 4 \end{bmatrix}, \quad \mathbf{B} = \begin{bmatrix} 1 & 0 \\ 0 & -1 \end{bmatrix}, \quad \text{and} \quad c = 2.$$

1.2.3 Dot Product

Q 1.2.2. What does it mean for two lines to be perpendicular?

Q 1.2.3. What is $\cos(0)$?

Q 1.2.4. When is $\cos(\theta) = 0$?

Definition 1.2.5. Given two column vectors $\mathbf{u}$ and $\mathbf{v}$ in $\mathbb{R}^n$, their dot product is

$$\mathbf{u} \cdot \mathbf{v} = \mathbf{u}^T \mathbf{v}$$

Activity 1.2.6. Given

$$\mathbf{u} = \begin{bmatrix} 4 \\ 3 \end{bmatrix} \quad \text{and} \quad \mathbf{v} = \begin{bmatrix} -3 \\ 4 \end{bmatrix}$$

compute $\mathbf{v} \cdot \mathbf{v}$ and $\mathbf{u} \cdot \mathbf{v}$, then graph $\mathbf{u}$ and $\mathbf{v}$.

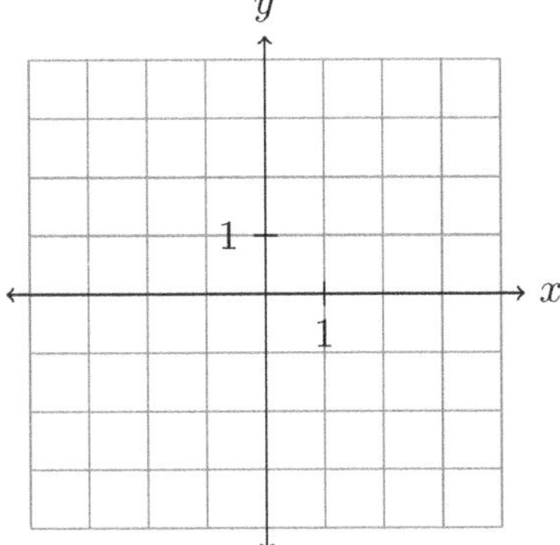

You should observe several things. First,

$$\mathbf{v} \cdot \mathbf{v} = \| \mathbf{v} \|_2^2 . \tag{1.13}$$

Second, the vectors $\mathbf{u}$ and $\mathbf{v}$ are perpendicular: the slope of the line that $\mathbf{u}$ lies along is $\frac{3}{4}$, the negative reciprocal of $\frac{3}{4}$ is $-\frac{4}{3}$, and that is the slope of the line that $\mathbf{v}$ lies along. Meanwhile, the dot product of $\mathbf{u}$ and $\mathbf{v}$ is zero. All of these facts are special cases of the general formula

$$\mathbf{u} \cdot \mathbf{v} = \| u \|_2 \| v \|_2 \cos(\theta) \tag{1.14}$$

where θ is the angle between $\mathbf{u}$ and $\mathbf{v}$.

If $\mathbf{u} = \mathbf{v}$, then $\theta = 0$.

If $\mathbf{v}$ is perpendicular to $\mathbf{u}$, then $\theta = \pm\frac{\pi}{2}$ (in radians) or ± 90 degrees.

We will see shortly how the dot product can be used to derive the Law of Cosines, but to do so, we need the following theorem.

Theorem 1.2.6. *Let $\mathbf{u}, \mathbf{v}, \mathbf{w}$ be real column vectors of the same dimension and let c be a scalar. Then the following are true:*

1. $(\mathbf{u} + \mathbf{v}) \cdot \mathbf{w} = \mathbf{u} \cdot \mathbf{w} + \mathbf{v} \cdot \mathbf{w}$
2. $(c\mathbf{u}) \cdot \mathbf{v} = c(\mathbf{u} \cdot \mathbf{v})$
3. $\mathbf{u} \cdot \mathbf{v} = \mathbf{v} \cdot \mathbf{u}$

Theorem 1.2.6 follows from the properties of transpose, together with the properties of matrix arithmetic. The first two properties give us linearity in the first component of the dot product, that is in the blank: $\_\_ \cdot \mathbf{v}$. two properties give us linearity in t. The third property gives us symmetry between $\mathbf{u}$ and $\mathbf{v}$, and so we also have linearity in the second component of the dot product: $\mathbf{u} \cdot \_\_$.

Activity 1.2.7. Given

$$\mathbf{u} = \begin{bmatrix} 1 \\ 0 \end{bmatrix}, \quad \mathbf{v} = \begin{bmatrix} 3 \\ 2 \end{bmatrix}, \quad \mathbf{w} = \begin{bmatrix} 1 \\ 1 \end{bmatrix},$$

and $c = -1$, verify each part of Theorem 1.2.6.

In Figure 1.4, the direction of the arrow for $\mathbf{u}-\mathbf{v}$ can be understood in terms of tip to tail addition: When we add $\mathbf{u}-\mathbf{v}$ to $\mathbf{v}$, $-\mathbf{v}$ and $\mathbf{v}$ cancel, giving us $\mathbf{u}$. We will derive the law of cosines,

$$\| \mathbf{w} \|^2 = \| \mathbf{u} \|^2 + \| \mathbf{v} \|^2 - 2 \| u \| \cdot \| v \| \cos(\theta), \tag{1.15}$$

by simplifying $\| \mathbf{u}-\mathbf{v} \|_2^2$:

$$\begin{aligned} \| \mathbf{u}-\mathbf{v} \|_2^2 = (\mathbf{u}-\mathbf{v}) \cdot (\mathbf{u}-\mathbf{v}) &= \mathbf{u} \cdot (\mathbf{u}-\mathbf{v}) - \mathbf{v} \cdot (\mathbf{u}-\mathbf{v}) \\ &= \mathbf{u} \cdot \mathbf{u} - \mathbf{u} \cdot \mathbf{v} - \mathbf{v} \cdot \mathbf{u} + \mathbf{v} \cdot \mathbf{v} \\ &= \mathbf{u} \cdot \mathbf{u} + \mathbf{v} \cdot \mathbf{v} - 2\mathbf{u} \cdot \mathbf{v} \\ &= \| \mathbf{u} \|_2^2 + \| \mathbf{v} \|_2^2 - 2 \| u \|_2 \| v \|_2 \cos(\theta) \end{aligned}$$

In the first two lines we used linearity, in the third line we used the symmetric property, end in the fourth, we used equations (1.13) and (1.14).

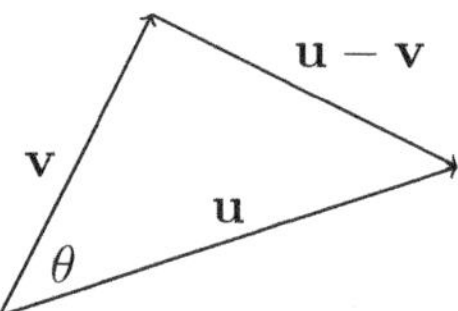

FIGURE 1.4: The law of cosines

Activity 1.2.8. Given

$$\mathbf{u} = \begin{bmatrix} 1 \\ 1 \end{bmatrix} \quad \text{and} \quad \mathbf{v} = \begin{bmatrix} 2 \\ 0 \end{bmatrix},$$

1. graph the vectors

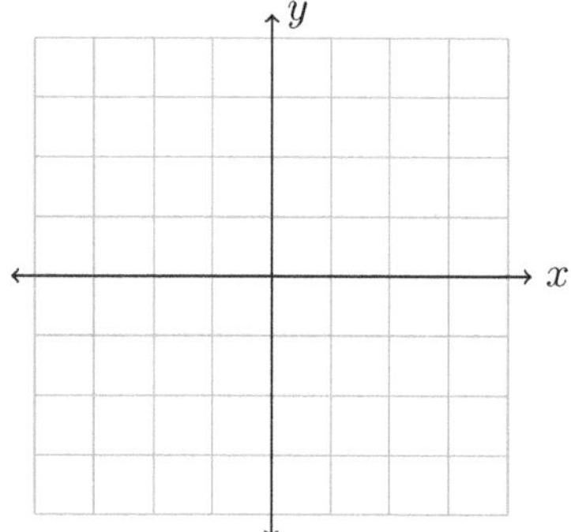

2. compute $\mathbf{u} \cdot \mathbf{v}$ using components,
3. compute $\mathbf{u} \cdot \mathbf{v}$ as $\| u \| \cdot \| v \| \cos(\theta)$,
4. compute $\mathbf{w} = \mathbf{u} - \mathbf{v}$ and verify the law of cosines.

Proposition 1.2.7. *Given two vectors* $\mathbf{u}$ *and* $\mathbf{v}$*, the projection of* $\mathbf{u}$ *onto* $\mathbf{v}$ *is given by*

$$\frac{\mathbf{u} \cdot \mathbf{v}}{\mathbf{v} \cdot \mathbf{v}} \mathbf{v} \quad \textit{or equivalently} \quad \| \mathbf{u} \| \cos(\theta) \frac{\mathbf{v}}{\| \mathbf{v} \|}$$

where θ *is the angle between* $\mathbf{u}$ *and* $\mathbf{v}$.

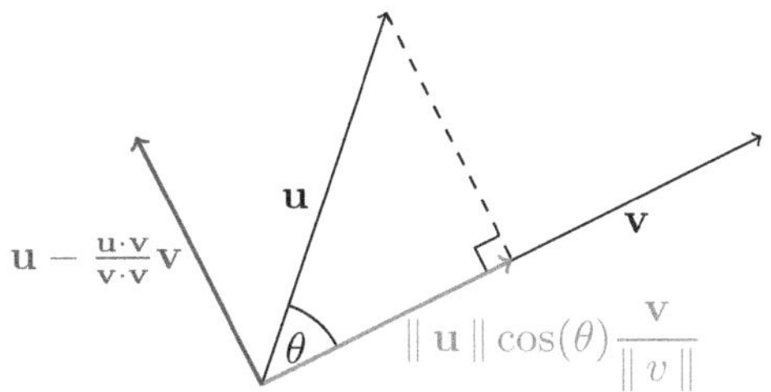

FIGURE 1.5: Vector projection

If $\mathbf{u}$ is being projected onto $\mathbf{v}$, then the hypotenuse is $\| \mathbf{u} \|$, the length of $\mathbf{u}$ (see Figure 1.5). The side along $\mathbf{v}$ has length $\| u \| \cos(\theta)$, where θ is the angle between $\mathbf{u}$ and $\mathbf{v}$. To get a vector with that length in the direction of $\mathbf{v}$, we multiply by $\frac{\mathbf{v}}{\| \mathbf{v} \|}$, which has length 1.

If we divide equation (1.14) by equation(1.13), then

$$\frac{\mathbf{u} \cdot \mathbf{v}}{\mathbf{v} \cdot \mathbf{v}} = \frac{\| \mathbf{u} \|}{\| \mathbf{v} \|} \cos(\theta).$$

Multiplying both sides by $\mathbf{v}$ shows that the two formulas given in Proposition 1.2.7 for the vector projection are equivalent.

The projection of $\mathbf{u}$ onto $\mathbf{v}$ can be thought of as the component of $\mathbf{u}$ in the direction of $\mathbf{v}$. Therefore, by subtracting it from $\mathbf{u}$, we obtain a vector orthogonal to $\mathbf{v}$, as illustrated in Figure 1.5.

Proposition 1.2.8 (Cauchy-Schwarz for the dot product). *Let* $\mathbf{u}$ *and* $\mathbf{v}$ *be vectors in* $\mathbb{R}$. *Then*

$$|\mathbf{u} \cdot \mathbf{v}| \leq \| \mathbf{u} \|_2 \cdot \| \mathbf{v} \|_2 \tag{1.16}$$

This proposition is a special case of Proposition 6.2.2.

Activity 1.2.9. Given

$$\mathbf{u} = \begin{bmatrix} 1 \\ -1 \end{bmatrix} \quad \text{and} \quad \mathbf{v} = \begin{bmatrix} 1 \\ 2 \end{bmatrix},$$

1. graph the vectors,

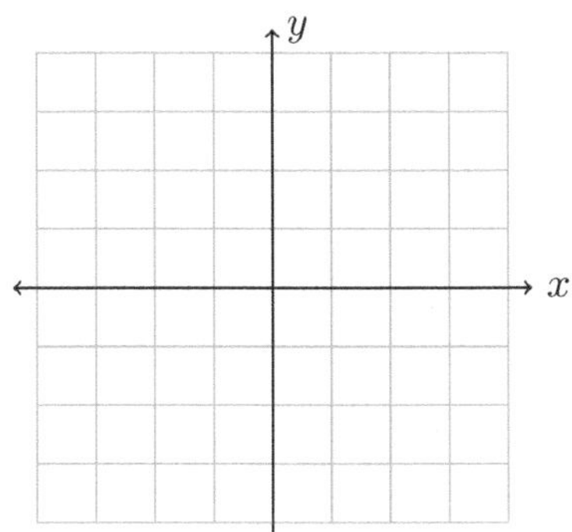

2. compute $\mathbf{u} \cdot \mathbf{v}$, $\| \mathbf{u} \|$, and $\| \mathbf{v} \|$
3. verify the inequality

$$|\mathbf{u} \cdot \mathbf{v}| \leq \| \mathbf{u} \|_2 \cdot \| \mathbf{v} \|_2 .$$

1.3 Inverse Matrices

1.3.1 Intro to Inverse Matrices

Q 1.3.1. What is the solution to $2x = 1$?

Definition 1.3.1. An $n \times n$ matrix $\mathbf{A}$ is **invertible** if there exists a matrix $\mathbf{B}$ such that $\mathbf{AB} = \mathbf{BA} = \mathbf{I}_n$. In this case the matrix $\mathbf{B}$ is its inverse and is written $\mathbf{A}^{-1}$.

Example 1.3.1. If $\mathbf{A} = \begin{bmatrix} 0 & -1 \\ 1 & 0 \end{bmatrix}$, then

$$\mathbf{A}^{-1} = \begin{bmatrix} 0 & 1 \\ -1 & 0 \end{bmatrix}$$

because

$$\begin{array}{cc} & \begin{bmatrix} 0 & -1 \\ 1 & 0 \end{bmatrix} \\ \begin{bmatrix} 0 & 1 \\ -1 & 0 \end{bmatrix} & \begin{bmatrix} 1 & 0 \\ 0 & 1 \end{bmatrix} \end{array} \quad \text{and} \quad \begin{array}{cc} & \begin{bmatrix} 0 & 1 \\ -1 & 0 \end{bmatrix} \\ \begin{bmatrix} 0 & -1 \\ 1 & 0 \end{bmatrix} & \begin{bmatrix} 1 & 0 \\ 0 & 1 \end{bmatrix} \end{array}$$

Activity 1.3.2. If $\mathbf{A} = \begin{bmatrix} 1 & 1 \\ 0 & 1 \end{bmatrix}$, verify that

$$\mathbf{A}^{-1} = \begin{bmatrix} 1 & -1 \\ 0 & 1 \end{bmatrix}.$$

by checking $\mathbf{A}^{-1}\mathbf{A} = \mathbf{I}_2$ and $\mathbf{I}_2 = \mathbf{AA}^{-1}$.

1.3.2 The 2×2 Case

Q 1.3.2. If a vector is rotated 90 degrees counter-clockwise, then what rotation will return the vector to its original direction?

Q 1.3.3. What is the effect of reflecting across a mirror twice?

There is a general result that can be quickly computed in the 2×2 case, but first we want to build up some intuition for it.

Example 1.3.3. Show that if $\mathbf{A} = \begin{bmatrix} 0 & b \\ c & 0 \end{bmatrix}$ and $bc \neq 0$, then

$$\mathbf{A}^{-1} = \begin{bmatrix} 0 & \frac{1}{c} \\ \frac{1}{b} & 0 \end{bmatrix}.$$

Solution

$$\begin{array}{cc} & \begin{bmatrix} 0 & b \\ c & 0 \end{bmatrix} \\ \begin{bmatrix} 0 & \frac{1}{c} \\ \frac{1}{b} & 0 \end{bmatrix} & \begin{bmatrix} 1 & 0 \\ 0 & 1 \end{bmatrix} \end{array}$$

Activity 1.3.4. Show that if $\mathbf{A} = \begin{bmatrix} a & 0 \\ 0 & d \end{bmatrix}$ and $ad \neq 0$, then

$$\mathbf{A}^{-1} = \begin{bmatrix} \frac{1}{a} & 0 \\ 0 & \frac{1}{d} \end{bmatrix}.$$

Proposition 1.3.2. *A* 2×2 *matrix*

$$\mathbf{A} = \begin{bmatrix} a & b \\ c & d \end{bmatrix}$$

is invertible if $ad - bc \neq 0$, *and in this case the inverse is*

$$\mathbf{A}^{-1} = \frac{1}{ad - bc} \begin{bmatrix} d & -b \\ -c & a \end{bmatrix}.$$

If $ad - bc = 0$, *then* $\mathbf{A}$ *is not invertible.*

This proposition includes both Example 1.3.3 and Activity 1.3.4 as special cases. If $b = c = 0$, then

$$\frac{1}{ad} \begin{bmatrix} a & 0 \\ 0 & d \end{bmatrix} = \begin{bmatrix} \frac{1}{a} & 0 \\ 0 & \frac{1}{d} \end{bmatrix},$$

and if $a = d = 0$, then

$$\frac{1}{-bc} \begin{bmatrix} 0 & -b \\ -c & 0 \end{bmatrix} = \begin{bmatrix} 0 & \frac{1}{c} \\ \frac{1}{b} & 0 \end{bmatrix}.$$

Example 1.3.5. Given the matrix

$$\mathbf{A} = \begin{bmatrix} 3 & 5 \\ 1 & 2 \end{bmatrix}$$

compute $\mathbf{A}^{-1}$ using Proposition 1.3.2.

Solution

We first check $3 \cdot 2 - 5 \cdot 1 = 6 - 5 = 1$, which is not zero, so the inverse exists. Then

$$\mathbf{A}^{-1} = \frac{1}{1} \cdot \begin{bmatrix} 2 & -5 \\ -1 & 3 \end{bmatrix}.$$

Activity 1.3.6. Check that the solution in the previous example works by computing $\mathbf{A}^{-1}\mathbf{A}$.

Activity 1.3.7. Given the matrix

$$\mathbf{A} = \begin{bmatrix} 1 & 1 \\ 2 & 0 \end{bmatrix}$$

compute $\mathbf{A}^{-1}$ using Proposition 1.3.2.

Example 1.3.8. We have seen in Example 1.1.28 and Activity 1.1.29 that matrices can be used to represent the symmetries of the square. The matrix

$$\mathbf{R} = \begin{bmatrix} 0 & -1 \\ 1 & 0 \end{bmatrix}$$

represents counterclockwise rotation by 90 degrees. The inverse matrix should represent clockwise rotation by 90 degrees. From the 2×2 formula, we have $ad-bc = 0\cdot 0-(1)(-1) = 1$, so

$$\mathbf{R}^{-1} = \begin{bmatrix} 0 & 1 \\ -1 & 0 \end{bmatrix}.$$

Why is this clockwise rotation by 90 degrees? Later we will see how to get these matrices from their action on vectors. But in terms of what we know now, we can use the fact that clockwise rotation 90 degrees is the same as rotating counterclockwise by 90 degrees three times. So, it should be true that $\mathbf{R}^{-1} = \mathbf{R}^3$. Let's check:

$$\begin{bmatrix} 0 & -1 \\ 1 & 0 \end{bmatrix} \underbrace{\begin{bmatrix} 0 & -1 \\ 1 & 0 \end{bmatrix}\begin{bmatrix} 0 & -1 \\ 1 & 0 \end{bmatrix}}_{\begin{bmatrix} -1 & 0 \\ 0 & -1 \end{bmatrix}} \; \begin{bmatrix} 0 & 1 \\ -1 & 0 \end{bmatrix} = \mathbf{R}^3.$$

So $\mathbf{R}^{-1} = \mathbf{R}^3$, which is what we expected from the geometry.

Activity 1.3.9. The matrix

$$\mathbf{M} = \begin{bmatrix} 0 & 1 \\ 1 & 0 \end{bmatrix}$$

represents a mirror reflection. Show with the 2×2 formula that $\mathbf{M}^{-1} = \mathbf{M}$. It hopefully makes sense that doing a mirror reflection twice gets you back where you started.

1.3.3 Properties of Inverse and Transpose

Q 1.3.4. What of the following is $(\mathbf{AB})^T$ equal to?
(a) $\mathbf{A}^T\mathbf{B}^T$ (b) $\mathbf{B}^T\mathbf{A}^T$ (c) Neither of these

Theorem 1.3.3.

1. *If* $\mathbf{A}$ *is an invertible matrix, then so is* $\mathbf{A}^{-1}$*, and*
$$(\mathbf{A}^{-1})^{-1} = \mathbf{A}.$$
2. *If* $\mathbf{A}$ *is an invertible matrix, then so is* $\mathbf{A}^T$*, and*
$$(\mathbf{A}^T)^{-1} = (\mathbf{A}^{-1})^T.$$
3. *If* $\mathbf{A}$ *and* $\mathbf{B}$ *are* $n \times n$ *invertible matrices, then so is* $\mathbf{AB}$*, and*
$$(\mathbf{AB})^{-1} = \mathbf{B}^{-1}\mathbf{A}^{-1}.$$

The first part of this theorem is actually a restatement of Corollary 1.3.9. The second part does contain the new statement that

$$(\mathbf{A}^T)^{-1} = (\mathbf{A}^{-1})^T,$$

although the equivalence of the invertibility of $\mathbf{A}$ and $\mathbf{A}^T$ is known from Theorem 1.3.8.

Activity 1.3.10. Let

$$\mathbf{A} = \begin{bmatrix} 1 & 1 \\ 0 & 1 \end{bmatrix} \quad \text{and} \quad \mathbf{B} = \mathbf{A}^T = \begin{bmatrix} 1 & 0 \\ 1 & 1 \end{bmatrix}.$$

Use Proposition 1.3.2 to compute $\mathbf{A}^{-1}$ and $\mathbf{B}^{-1}$, then verify:

1. $(\mathbf{A}^{-1})^{-1} = \mathbf{A}$,
2. $(\mathbf{A}^T)^{-1} = (\mathbf{A}^{-1})^T$,
3. $(\mathbf{AB})^{-1} = \mathbf{B}^{-1}\mathbf{A}^{-1}$,

thus illustrating the previous theorem.

1.3.4 Triangular Matrices

Definition 1.3.4. An $n \times n$ matrix $\mathbf{A}$ is:

1. **upper triangular** if $a_{ij} = 0$ for all $i > j$,
2. **lower triangular** if $a_{ij} = 0$ for all $i < j$,
3. **diagonal** if $a_{ij} = 0$ for all $i \neq j$

Comprehension Check 1.3.1. What row and column is a_{12}? Is it below the diagonal or above the diagonal?

Example 1.3.11. Given

$$\mathbf{U} = \begin{bmatrix} 1 & 2 & 3 \\ 0 & 5 & 6 \\ 0 & 0 & 9 \end{bmatrix}, \quad \mathbf{L} = \begin{bmatrix} 1 & 0 & 0 \\ 4 & 5 & 0 \\ 7 & 8 & 9 \end{bmatrix}, \quad \text{and} \quad \mathbf{D} = \begin{bmatrix} 1 & 0 & 0 \\ 0 & 5 & 0 \\ 0 & 0 & 9 \end{bmatrix},$$

$\mathbf{U}$ is upper triangular, $\mathbf{L}$ is lower triangular, and $\mathbf{D}$ is diagonal. Often the places where the zeros are above or below the diagonal will be left blank. There may be zeros on the diagonal, which are not usually left blank. The case where all of the entries on the diagonal are zero is also interesting.

Example 1.3.12. Let

$$\mathbf{N} = \begin{bmatrix} 0 & 1 \\ 0 & 0 \end{bmatrix}$$

Show that $\mathbf{N}^2 = \mathbf{0}$.

Solution

$$\begin{array}{cc} & \begin{bmatrix} 0 & 1 \\ 0 & 0 \end{bmatrix} \\ \begin{bmatrix} 0 & 1 \\ 0 & 0 \end{bmatrix} & \begin{bmatrix} 0 & 0 \\ 0 & 0 \end{bmatrix} \end{array}$$

Activity 1.3.13. Given

$$\mathbf{N} = \begin{bmatrix} 0 & 1 & 0 \\ 0 & 0 & 1 \\ 0 & 0 & 0 \end{bmatrix},$$

show that $\mathbf{N}^3 = \mathbf{0}$

Definition 1.3.5. An $n \times n$ matrix $\mathbf{N}$ is **nilpotent** if there exists a positive integer k, such that $\mathbf{N}^k = \mathbf{0}$.

Theorem 1.3.6. *Let $\mathbf{A}$ and $\mathbf{B}$ be two upper triangular $n \times n$ matrices. Then:*

1. $\mathbf{A}+\mathbf{B}$ *is upper triangular.*
2. $\mathbf{AB}$ *is upper triangular.*
3. $\mathbf{A}$ *is invertible if and only if* $a_{ii} \neq 0$ *for all* i.
4. *If* $\mathbf{A}$ *is invertible, then* $\mathbf{A}^{-1}$ *is upper triangular.*
5. *If* $\mathbf{A}$ *is invertible and* $\mathbf{B}$ *is nilpotent, then* $\mathbf{A}+\mathbf{B}$ *is invertible.*
6. $\mathbf{A}^T$ *is lower triangular.*

The analogous facts are true if the words upper and lower are reversed.

Activity 1.3.14. Given

$$\mathbf{A} = \begin{bmatrix} 1 & 1 \\ 0 & 1 \end{bmatrix} \quad \text{and} \quad \mathbf{B} = \begin{bmatrix} 0 & 1 \\ 0 & 0 \end{bmatrix}$$

verify each part of Theorem 1.3.6. Apply Proposition 1.3.2 for determining when a matrix is invertible, and for computing the inverse when it exists. For part 3 check that $\mathbf{A}$ is invertible, while $\mathbf{B}$ is not.

Theorem 1.3.7. *Let*

$$\mathbf{D} = \begin{bmatrix} d_{11} & & \\ & \ddots & \\ & & d_{nn} \end{bmatrix} \quad \textit{and} \quad \mathbf{L}_i = \begin{bmatrix} 1 & & & & & \\ \vdots & \ddots & & & & \\ 0 & & 1 & & & \\ 0 & & a_{i+1,i} & \ddots & & \\ \vdots & & \vdots & & 1 & \\ 0 & & a_{ni} & & 0 & 1 \end{bmatrix}$$

where $d_{11}, \ldots, d_{nn} \neq 0$ *where* i *is a specific column. Then*

$$\mathbf{D}^{-1} = \begin{bmatrix} \frac{1}{d_{11}} & & \\ & \ddots & \\ & & \frac{1}{d_{nn}} \end{bmatrix} \quad \textit{and} \quad \mathbf{L}_i^{-1} = \begin{bmatrix} 1 & & & & & \\ \vdots & \ddots & & & & \\ 0 & & 1 & & & \\ 0 & & -a_{i+1,i} & \ddots & & \\ \vdots & & \vdots & & 1 & \\ 0 & & -a_{ni} & & 0 & 1 \end{bmatrix} \tag{1.17}$$

Theorem 1.3.7 may be proved either by row reduction, or simply by multiplying the matrices. The following activity illustrates the second approach. To use row reduction instead, see Activity 2.3.12

Activity 1.3.15. Given

$$\mathbf{D} = \begin{bmatrix} 1 & & \\ & 2 & \\ & & 3 \end{bmatrix} \quad \text{and} \quad \mathbf{L}_1 = \begin{bmatrix} 1 & & \\ 2 & 1 & \\ 3 & 0 & 1 \end{bmatrix}$$

show that

$$\mathbf{D}^{-1} = \begin{bmatrix} 1 & & \\ & \frac{1}{2} & \\ & & \frac{1}{3} \end{bmatrix} \quad \text{and} \quad \mathbf{L}_1^{-1} = \begin{bmatrix} 1 & & \\ -2 & 1 & \\ -3 & 0 & 1 \end{bmatrix}$$

by doing the matrix multiplication $\mathbf{D}^{-1}\mathbf{D} = \mathbf{I}$ and $\mathbf{L}_1^{-1}\mathbf{L}_1 = \mathbf{I}$.

Theorems 1.3.6, 1.3.7, and 1.3.3 together can be used to efficiently compute the inverse of any invertible triangular matrix, as demonstrated by the next activity.

Activity 1.3.16. Given

$$\mathbf{L}_1 = \begin{bmatrix} 1 & & \\ 2 & 1 & \\ 3 & 0 & 1 \end{bmatrix} \quad \text{and} \quad \mathbf{L}_2 = \begin{bmatrix} 1 & & \\ 0 & 1 & \\ 0 & 4 & 1 \end{bmatrix}$$

1. Compute $\mathbf{L}_1\mathbf{L}_2$
2. Use Theorem 1.3.7 to write $\mathbf{L}_1^{-1}$ and $\mathbf{L}_2^{-1}$
3. By Theorem 1.3.3
$$(\mathbf{L}_1\mathbf{L}_2)^{-1} = \mathbf{L}_2^{-1}\mathbf{L}_1^{-1}$$
Compute $\mathbf{L}_2^{-1}\mathbf{L}_2^{-1}$ by multiplying the matrices obtained in part 2 in the correct order.
4. Verify that the matrices computed in parts 1 and 3. You should get the identity matrix.

Reflection 1.3.2. What parts of Theorem 1.3.6 are illustrated by Activity 1.3.6?

1.3.5 Solving Matrix Equations

Q 1.3.5. How do we solve the equation $2x = 6$?

Q 1.3.6. What is $\mathbf{A}^{-1}$ if

$$\mathbf{A} = \begin{bmatrix} 3 & 5 \\ 1 & 2 \end{bmatrix}?$$

Suppose we want to solve equations like

$$\begin{aligned} \mathbf{XA} &= \mathbf{B} \quad \text{and} \\ \mathbf{AY} &= \mathbf{C}, \end{aligned}$$

where $\mathbf{A}$, $\mathbf{B}$, and $\mathbf{C}$ are known. If $\mathbf{A}$ is invertible, then we can solve them in the following manner:

$$\mathbf{XA} = \mathbf{B} \implies \mathbf{XAA}^{-1} = \mathbf{BA}^{-1} \implies \mathbf{XI} = \mathbf{BA}^{-1} \implies \mathbf{X} = \mathbf{BA}^{-1}.$$

The idea is to get $\mathbf{A}^{-1}$ to cancel with $\mathbf{A}$. But the order of multiplication matters, so $\mathbf{A}^{-1}$ must be multiplied on the same side as $\mathbf{A}$. Since $\mathbf{XA}$ has $\mathbf{A}$ on the right, then $\mathbf{A}^{-1}$ must be multiplied on the right on both sides of the equation to maintain equality. Similarly, $\mathbf{AY}$ has $\mathbf{A}$ on the left, so we must multiply by $\mathbf{A}^{-1}$ on the left on both sides of the equation:

$$\mathbf{AY} = \mathbf{C} \implies \mathbf{A}^{-1}\mathbf{AY} = \mathbf{A}^{-1}\mathbf{C} \implies \mathbf{IY} = \mathbf{A}^{-1}\mathbf{C} \implies \mathbf{Y} = \mathbf{A}^{-1}\mathbf{C}.$$

Example 1.3.17. Given

$$\mathbf{A} = \begin{bmatrix} 5 & 3 \\ 1 & 2 \end{bmatrix}, \quad \mathbf{B} = \begin{bmatrix} 1 & -1 \end{bmatrix}, \quad \text{and} \quad \mathbf{C} = \begin{bmatrix} 0 & 1 \\ 1 & 1 \end{bmatrix},$$

compute $\mathbf{A}^{-1}$ and use it to solve $\mathbf{XA} = \mathbf{B}$ and $\mathbf{AY} = \mathbf{C}$.

Solution

Using the 2×2 formula $ad - bc = 5 \cdot 2 - 3 \cdot 1 = 10 - 3 = 7$, so

$$\mathbf{A}^{-1} = \frac{1}{7}\begin{bmatrix} 2 & -3 \\ -1 & 5 \end{bmatrix} = \begin{bmatrix} \frac{2}{7} & -\frac{3}{7} \\ -\frac{1}{7} & \frac{5}{7} \end{bmatrix}.$$

To solve for $\mathbf{X}$, we must compute $\mathbf{BA}^{-1}$. Since $\mathbf{B}$ is 1×2 and $\mathbf{A}^{-1}$ is 2×2, there should be no confusion: it is impossible to multiply by $\mathbf{A}^{-1}$ on the left anyway. Since $\mathbf{B} = \begin{bmatrix} 1 & -1 \end{bmatrix}$, we can think of subtracting row 2 from row 1:

$$\begin{array}{cc} & \begin{bmatrix} \frac{2}{7} & -\frac{3}{7} \\ -\frac{1}{7} & \frac{5}{7} \end{bmatrix} \\ \begin{bmatrix} 1 & -1 \end{bmatrix} & \begin{bmatrix} \frac{3}{7} & -\frac{8}{7} \end{bmatrix} = \mathbf{BA}^{-1} = \mathbf{X}. \end{array}$$

To solve for $\mathbf{Y}$, we must compute $\mathbf{A}^{-1}\mathbf{C}$. This time both $\mathbf{C}$ and $\mathbf{A}^{-1}$ are 2×2, so it is possible to compute $\mathbf{CA}^{-1}$, but it will not give the right answer. The only way to know, which side that $\mathbf{A}^{-1}$ must be multiplied on is to look at the equation $\mathbf{AY} = \mathbf{C}$. Since $\mathbf{C}$ is multiplied on the right of $\mathbf{A}^{-1}$, we are taking the dot product of columns of $\mathbf{C}$ with rows of $\mathbf{A}^{-1}$. The first column of $\mathbf{C}$ will have the effect of picking out the second column of $\mathbf{A}^{-1}$, and the second column of $\mathbf{C}$ will have the effect of adding the columns. Hence

$$\begin{array}{cc} & \begin{bmatrix} 0 & 1 \\ 1 & 1 \end{bmatrix} \\ \begin{bmatrix} \frac{2}{7} & -\frac{3}{7} \\ -\frac{1}{7} & \frac{5}{7} \end{bmatrix} & \begin{bmatrix} -\frac{3}{7} & -\frac{1}{7} \\ \frac{5}{7} & \frac{4}{7} \end{bmatrix} = \mathbf{A}^{-1}\mathbf{C} = \mathbf{Y}. \end{array}$$

Activity 1.3.18. Given

$$\mathbf{A} = \begin{bmatrix} 1 & 2 \\ 3 & 4 \end{bmatrix}, \quad \mathbf{B} = \begin{bmatrix} 1 & 2 \end{bmatrix}, \quad \text{and} \quad \mathbf{C} = \begin{bmatrix} 0 & 1 \\ 1 & 1 \end{bmatrix},$$

compute $\mathbf{A}^{-1}$ and use it to solve $\mathbf{XA} = \mathbf{B}$ and $\mathbf{AY} = \mathbf{C}$.

1.3.6 The Invertible Matrix Theorem (Part 1)

Theorem 1.3.8 (The Invertible Matrix Theorem: Part 1)**.** *Let $\mathbf{A}$ be an $n \times n$ matrix with real entries. The following statements are equivalent.*

1. *$\mathbf{A}$ is an invertible matrix.*
2. *$\mathbf{A}^T$ is an invertible matrix.*
3. *The equation $\mathbf{Ax} = \mathbf{0}$ has only the trivial solution.*
4. *There is an $n \times n$ matrix $\mathbf{C}$ such that $\mathbf{CA} = \mathbf{I}_n$.*
5. *There is an $n \times n$ matrix $\mathbf{D}$ such that $\mathbf{AD} = \mathbf{I}_n$.*

Remark 1.3.1. The invertible matrix theorem is spread out over several chapters, so whenever we say "the following statements are equivalent" we mean all statements in all parts. If you compare with Lay's book [15], you should pay attention to the fact that the order of the parts is different. Most of the proof can be found in Lay's book, but this book contains a few parts that Lay does not mention explicitly. The examples and activities in this book also can be regarded as hints of how certain parts are proved.

Remark 1.3.2. The definition of an invertible matrix says that both $\mathbf{C}$ and $\mathbf{D}$ exist and that $\mathbf{C} = \mathbf{D}$. Statements 4 and 5 are saying that the existence of one implies the other, and that both are equal.

Corollary 1.3.9. *Let $\mathbf{A}$ and $\mathbf{B}$ be square matrices. If $\mathbf{AB} = \mathbf{I}$, then $\mathbf{A}$ and $\mathbf{B}$ are both invertible with $\mathbf{B} = \mathbf{A}^{-1}$ and $\mathbf{A} = \mathbf{B}^{-1}$.*

Remark 1.3.3. Actually, it is true, more generally, that if $\mathbf{A_1 A_2} \ldots \mathbf{A_k} = \mathbf{I}$, where all of these are $n \times n$ matrices, then all of them are invertible. But when we have only two matrices in the product, then we get $\mathbf{B} = \mathbf{A}^{-1}$ and $\mathbf{A} = \mathbf{B}^{-1}$ because matrix inverses are unique. The calculation goes like this:

$$\mathbf{AB} = \mathbf{I} \implies \mathbf{A}^{-1}\mathbf{AB} = \mathbf{A}^{-1}\mathbf{I} \implies \mathbf{B} = \mathbf{A}^{-1}$$
$$\mathbf{AB} = \mathbf{I} \implies \mathbf{ABB}^{-1} = \mathbf{IB}^{-1} \implies \mathbf{A} = \mathbf{B}^{-1}.$$

It is for the same reason that we can take $\mathbf{C} = \mathbf{D} = \mathbf{A}^{-1}$ in the Invertible Matrix Theorem.

Example 1.3.19. Show that the statements 1, 6, 3, and 4 in Theorem 1.3.8 are true for

$$\mathbf{A} = \begin{bmatrix} 1 & 1 \\ 0 & 1 \end{bmatrix}.$$

Solution

1. $\mathbf{A}$ *is an invertible matrix.* From Proposition 1.3.2, we obtain

$$\mathbf{A}^{-1} = \begin{bmatrix} 1 & 1 \\ 0 & 1 \end{bmatrix}^{-1} = \frac{1}{1 \cdot 1 - 0 \cdot 1} \cdot \begin{bmatrix} 1 & -1 \\ 0 & 1 \end{bmatrix} = \begin{bmatrix} 1 & -1 \\ 0 & 1 \end{bmatrix}$$

3. *The equation $\mathbf{Ax} = \mathbf{0}$ has only the trivial solution.*

$$\begin{bmatrix} 1 & 1 \\ 0 & 1 \end{bmatrix} \begin{bmatrix} x \\ y \end{bmatrix} = \begin{bmatrix} x + y \\ y \end{bmatrix} = \begin{bmatrix} 0 \\ 0 \end{bmatrix},$$

so $y = 0$ and then $x + y = 0$ gives $x = 0$.

4. *There is an $n \times n$ matrix $\mathbf{C}$ such that $\mathbf{CA} = \mathbf{I}$.*

$$\begin{bmatrix} a & b \\ c & d \end{bmatrix} \begin{bmatrix} 1 & 1 \\ 0 & 1 \end{bmatrix} = \begin{bmatrix} a & a + b \\ c & c + d \end{bmatrix} = \begin{bmatrix} 1 & 0 \\ 0 & 1 \end{bmatrix}$$

gives $a = 1$, $c = 0$, then $d = 1$ and $b = -1$. Check the other side:

$$\begin{bmatrix} 1 & 1 \\ 0 & 1 \end{bmatrix} \begin{bmatrix} 1 & -1 \\ 0 & 1 \end{bmatrix} = \begin{bmatrix} 1 & 0 \\ 0 & 1 \end{bmatrix}.$$

Activity 1.3.20. Show that the statements 2 and 5 in theorem 1.3.8 are true for

$$\mathbf{A} = \begin{bmatrix} 1 & 1 \\ 0 & 1 \end{bmatrix}.$$

1.4 Hermitian Conjugate

1.4.1 Introduction to Hermitian Conjugate

Q 1.4.1. What is the complex conjugate of $3 + 4i$?

Q 1.4.2. What is $\mathbf{A}^T$ if

$$\begin{bmatrix} 1 & 2 \\ 3 & 4 \end{bmatrix}$$

Definition 1.4.1. Given a matrix $\mathbf{A}$, the **Hermitian adjoint** or **Hermitian conjugate** is defined as the complex conjugate of the transpose of $\mathbf{A}$, and is written as $\mathbf{A}^*$ or sometimes $\mathbf{A}^*$.

If $\mathbf{A}^*\mathbf{A} = \mathbf{A}\mathbf{A}^*$ then $\mathbf{A}$ is said to be a **normal matrix**.

If $\mathbf{A} = \mathbf{A}^*$, then $\mathbf{A}$ is said to be **Hermitian** or **self-adjoint**.

Warning 1.4.1. Normal is one of the most overused words in math: normal vectors, normal matrices, normal distribution, normalization, normalizer, etc. Can we stop using this word? Maybe **self-central** would be better for $\mathbf{A}^*\mathbf{A} = \mathbf{A}\mathbf{A}^*$. For now, the best I can say is pay attention to the context.

Example 1.4.1. Given

$$\mathbf{A} = \begin{bmatrix} 0 & 0 \\ 1+2i & 3i \end{bmatrix},$$

then

$$\mathbf{A}^* = \begin{bmatrix} 0 & \overline{1+2i} \\ 0 & \overline{3i} \end{bmatrix} = \begin{bmatrix} 0 & 1-2i \\ 0 & -3i \end{bmatrix}.$$

Clearly $\mathbf{A}^* \neq \mathbf{A}$, so it is not Hermitian.

The concept of a Hermitian conjugate generalizes the transpose, and the concept of a self-adjoint matrix generalizes the concept of a symmetric matrix. In fact, a real Hermitian matrix is symmetric.

Activity 1.4.2. Given the matrices

$$\mathbf{A} = \begin{bmatrix} 0 & i \\ -i & 0 \end{bmatrix} \quad \text{and} \quad \mathbf{B} = \begin{bmatrix} i & 0 \\ 0 & -i \end{bmatrix}$$

compute

$$\mathbf{A}^*, \quad \mathbf{A}\mathbf{A}^*, \quad \mathbf{A}^*\mathbf{A}, \quad \mathbf{B}^*, \quad \mathbf{B}\mathbf{B}^*, \quad \text{and} \quad \mathbf{B}^*\mathbf{B},$$

then use your results to fill in the table below:

	is Hermitian	is normal
A		
B		

The previous activity illustrates a general situation: the fact that a Hermtian matrix is normal. We can also see this algebraically:

Activity 1.4.3. If $\mathbf{A}$ is Hermitian, then $\mathbf{A} = \mathbf{A}^*$. Use this equation to eliminate $\mathbf{A}^*$ in both $\mathbf{A}^*\mathbf{A}$ and $\mathbf{A}\mathbf{A}^*$. Are they equal?

Proposition 1.4.2. *If* $\mathbf{A}$ *is a matrix with complex entries, then* $\mathbf{A}^*\mathbf{A}$ *is Hermitian.*

Activity 1.4.4. Illustrate Proposition 1.4.2 with the matrix $\mathbf{A} = \begin{bmatrix} 0 & 1 & i \\ 0 & 0 & 0 \end{bmatrix}$, by computing both $\mathbf{A}^*\mathbf{A}^*$ and $\mathbf{A}(\mathbf{A}^T)$. The resulting matrices will be different but both should be symmetric.

1.4.2 Properties of Hermitian Conjugate

Q 1.4.3. What of the following is $(\mathbf{AB})^T$ equal to?

(a) $\mathbf{A}^T\mathbf{B}^T$ (b) $\mathbf{B}^T\mathbf{A}^T$ (c) Neither of these

Theorem 1.4.3. *Let* $\mathbf{A}$ *and* $\mathbf{B}$ *denote matrices whose sizes are appropriate for the following sums and products:*

1. $(\mathbf{A}^*)^* = \mathbf{A}$
2. $(\mathbf{A}+\mathbf{B})^* = \mathbf{A}^* + \mathbf{B}^*$
3. *For any scalar* c, $(c\mathbf{A})^* = \overline{c}\mathbf{A}^*$, *where* $\overline{c}$ *is the complex conjugate of* c.
4. $(\mathbf{AB})^* = \mathbf{B}^*\mathbf{A}^*$
5. *If* $\mathbf{A}$ *is invertible, then* $(\mathbf{A}^{-1})^* = (\mathbf{A}^*)^{-1}$

Activity 1.4.5. Verify that all parts of the previous theorem are true for

$$\mathbf{A} = \begin{bmatrix} 1 & 1+i \\ 0 & i \end{bmatrix}, \quad \mathbf{B} = \begin{bmatrix} 0 & -1 \\ 0 & 2 \end{bmatrix}, \quad \text{and} \quad c = 1-i.$$

Note that Proposition 1.3.2 can still be used to compute the inverse of a matrix here. It also helps to remember that $\frac{1}{i} = -i$.

The Hermitian conjugate is fundamental to generalizing the dot product to complex vectors:

Definition 1.4.4. Given two column vectors $\mathbf{u}$ and $\mathbf{v}$ in $\mathbb{C}^n$, their dot product is

$$\mathbf{u}\cdot\mathbf{v} = \mathbf{v}^*\mathbf{u}$$

Theorem 1.4.5. *Let* $\mathbf{u}, \mathbf{v}, \mathbf{w}$ *be real column vectors of the same dimension and let* c *be a scalar. Then the following are true:*

1. $(\mathbf{u}+\mathbf{v})\cdot\mathbf{w} = \mathbf{u}\cdot\mathbf{w} + \mathbf{v}\cdot\mathbf{w}$
2. $(c\mathbf{u})\cdot\mathbf{v} = c(\mathbf{u}\cdot\mathbf{v})$
3. $\mathbf{u}\cdot\mathbf{v} = \overline{\mathbf{v}}\cdot\overline{\mathbf{u}}$ *where* $\mathbf{u}$ *denotes the complex conjugate of* $\mathbf{u}$.

Theorem 1.2.6 follows from the properties of hermitian conjugate, together with the properties of matrix arithmetic. The key thing to realize is that Definition 1.4.4 preserves linearity in the first component, i.e. in the blank $\_\_ \cdot \mathbf{v}$, but not in the second component. That is because the third property in Theorem 1.4.5 says that we must take the complex conjugate when switching components.

Activity 1.4.6. Given

$$\mathbf{u} = \begin{bmatrix} 1 \\ 1 \end{bmatrix}, \quad \mathbf{v} = \begin{bmatrix} 1+i \\ 0 \end{bmatrix}, \quad \mathbf{w} = \begin{bmatrix} 1 \\ -1 \end{bmatrix},$$

and $c = i$, verify each part of Theorem 1.2.6. Additionally, you should check:

1. $\mathbf{u}\cdot(\mathbf{v}+\mathbf{w}) = \mathbf{u}\cdot\mathbf{v} + \mathbf{u}\cdot\mathbf{w}$,
2. $\mathbf{u}\cdot(\overline{c}\mathbf{v}) = c(\mathbf{u}\cdot\mathbf{v})$.

1.5 Block Matrices

Q 1.5.1. What is $\mathbf{AB}$ if

$$\mathbf{A} = \begin{bmatrix} 0 & -1 \\ 1 & 0 \end{bmatrix} \quad \text{and} \quad \mathbf{B} = \begin{bmatrix} 1 & 1 \\ 0 & 1 \end{bmatrix}$$

Q 1.5.2. What is the 2×2 determinant?

Q 1.5.3. What is the 2×2 inverse formula?

Q 1.5.4. How does swapping two rows of a matrix affect the determinant?

There are times when it makes sense to break a matrix up into smaller matrices. Such a matrix is often called a **block matrix** or a **partitioned matrix**. The blocks do not have to be the same size. They do not even have to be square matrices, although it is often the case that the blocks on the diagonal are square matrices. The sizes of the blocks may even have some geometric significance to the problem at hand.

Example 1.5.1. If

$$\mathbf{A} = \begin{bmatrix} 1 & 2 \\ 3 & 4 \end{bmatrix} \quad \text{and} \quad \mathbf{B} = \begin{bmatrix} 5 & 6 \end{bmatrix},$$

then the matrix

$$\mathbf{M}_1 = \begin{bmatrix} \mathbf{A} & \mathbf{0} \\ \mathbf{B} & \mathbf{I}_1 \end{bmatrix} = \left[\begin{array}{cc|c} 1 & 2 & 0 \\ 3 & 4 & 0 \\ \hline 5 & 6 & 1 \end{array}\right]$$

is a block matrix with a 2×2 and a 1×1 block on the diagonal. Note that $\mathbf{0}$ is typically used for a matrix with all zeros, regardless of the shape, while $\mathbf{I}_1$ is the 1×1 identity matrix. When multiplying two block matrices, whole blocks can be multiplied together. For example, if

$$\mathbf{M}_2 = \begin{bmatrix} \mathbf{C} & \mathbf{0} \\ \mathbf{0} & \mathbf{D} \end{bmatrix} = \left[\begin{array}{cc|c} 0 & 1 & 0 \\ 1 & 0 & 0 \\ \hline 0 & 0 & 2 \end{array}\right],$$

then $\mathbf{M}_1\mathbf{M}_2$ can be computed as

$$\begin{array}{cc} & \begin{bmatrix} \mathbf{C} & \mathbf{0} \\ \mathbf{0} & \mathbf{D} \end{bmatrix} \\ \begin{bmatrix} \mathbf{A} & \mathbf{0} \\ \mathbf{B} & \mathbf{I}_1 \end{bmatrix} & \begin{bmatrix} \mathbf{AC} & \mathbf{0} \\ \mathbf{BC} & \mathbf{D} \end{bmatrix} \end{array}.$$

The calculation can be finished by computing with the blocks:

$$\begin{array}{cc} & \begin{bmatrix} 0 & 1 \\ 1 & 0 \end{bmatrix} \\ \begin{bmatrix} 1 & 2 \\ 3 & 4 \end{bmatrix} & \begin{bmatrix} 2 & 1 \\ 4 & 3 \end{bmatrix} = \mathbf{AC} \end{array} \quad \text{and} \quad \begin{array}{cc} & \begin{bmatrix} 0 & 1 \\ 1 & 0 \end{bmatrix} \\ \begin{bmatrix} 5 & 6 \end{bmatrix} & \begin{bmatrix} 6 & 5 \end{bmatrix} = \mathbf{BC}, \end{array}$$

so

$$\mathbf{M}_1\mathbf{M}_2 = \left[\begin{array}{cc|c} 2 & 1 & 0 \\ 4 & 3 & 0 \\ \hline 6 & 5 & 2 \end{array}\right].$$

Warning 1.5.1. Since the blocks are matrices, it is important to remember that matrix multiplication does not commute. In Example 1.5.1, $\mathbf{BC}$ is defined, but $\mathbf{CB}$ is not because $\mathbf{C}$ is 2×2 and $\mathbf{B}$ is 1×2.

Activity 1.5.2. Compute $\mathbf{M}_2\mathbf{M}_1$ with $\mathbf{M}_1$ and $\mathbf{M}_2$ as defined by Example 1.5.1.

Taking the transpose of block matrices is easy.

Example 1.5.3.

$$\begin{bmatrix} \mathbf{A} & \mathbf{B} & \mathbf{C} \\ \mathbf{D} & \mathbf{E} & \mathbf{F} \end{bmatrix}^T = \begin{bmatrix} \mathbf{A}^T & \mathbf{D}^T \\ \mathbf{B}^T & \mathbf{E}^T \\ \mathbf{C}^T & \mathbf{F}^T \end{bmatrix}.$$

Unfortunately, inverses are not. What is the inverse of

$$\mathbf{M}_1 = \begin{bmatrix} \mathbf{A} & \mathbf{B} \\ \mathbf{C} & \mathbf{D} \end{bmatrix}?$$

By analogy with Proposition 1.3.2, you might hope for some multiple of

$$\mathbf{M}_2 = \begin{bmatrix} \mathbf{D} & -\mathbf{B} \\ -\mathbf{C} & \mathbf{A} \end{bmatrix},$$

but just try doing the multiplication! What you get is

$$\mathbf{M}_1\mathbf{M}_2 = \begin{bmatrix} \mathbf{AD} - \mathbf{BC} & \mathbf{BA} - \mathbf{AB} \\ \mathbf{CD} - \mathbf{DC} & \mathbf{DA} - \mathbf{CB} \end{bmatrix},$$

which will only be a multiple of the identity matrix if

$$\mathbf{AD} - \mathbf{BC} = \mathbf{DA} - \mathbf{CB} = k\mathbf{I}, \quad \mathbf{BA} = \mathbf{AB}, \quad \text{and} \quad \mathbf{CD} = \mathbf{DC}$$

for some constant k. The next activity contains two examples: one for which these assumptions are valid and the other for which they are not.

Activity 1.5.4. First, let

$$\mathbf{M} = \begin{bmatrix} \mathbf{A} & \mathbf{B} \\ \mathbf{C} & \mathbf{D} \end{bmatrix} = \left[\begin{array}{cc|cc} 1 & 0 & 5 & 0 \\ 0 & 4 & 0 & 1 \\ \hline 1 & 0 & 6 & 0 \\ 0 & 3 & 0 & 1 \end{array}\right].$$

Show that

1. $\mathbf{AD} - \mathbf{BC} = \mathbf{DA} - \mathbf{CB} = \mathbf{I}_2$,
2. $\mathbf{BA} = \mathbf{AB}$, and
3. $\mathbf{CD} = \mathbf{DC}$.

Now let

$$\mathbf{M} = \begin{bmatrix} \mathbf{A} & \mathbf{B} \\ \mathbf{C} & \mathbf{D} \end{bmatrix} = \left[\begin{array}{cc|cc} 0 & 1 & 1 & 1 \\ 1 & 0 & 0 & 0 \\ \hline 1 & 0 & 0 & 1 \\ 1 & 0 & 1 & 0 \end{array}\right]$$

show that

1. $\mathbf{AD} - \mathbf{BC} \neq \mathbf{I}_2$,
2. $\mathbf{BA} \neq \mathbf{AB}$.

1.6 Algorithm Efficiency Part 1

Throughout this book we will look at the efficiency of various algorithms. There are various important factors to consider, including

1. How much space is needed to represent the calculation at a given step,
2. The extent to which certain parts of the algorithm can be done in parallel,
3. The number of steps needed,
4. The time needed of each step.

In general, while it may be possible to reduce the number of steps needed in an algorithm by doing computations in parallel, this may increase the amount of space needed to represent each step of the calculation. We will generally ignore hardware specifics, but we will look at trade-offs in space efficiency and time efficiency. In my opinion, the parallel algorithms should generally be favored on modern hardware with the understanding that when space runs out, the time must increase instead. One reason that I might not favor a parallel algorithm is that it ends up being slower than a sequential algorithm when space runs out.

Numbers can be represented as ints or floats and may have 32 bits or 64 bits or even 128 bits using AVX instructions on a modern Intel processor or Cuda on a graphics card by Nvidia. Any of these is what I would call a "small number." While the size of a small number depends on hardware, once the hardware is known, the size is known. Any computation done directly to a single small number or pair of small numbers is what I would call a small number computation. This includes things like addition, subtraction, multiplication, or division of small numbers, all of which are generally built in to hardware. I generally include more advanced operations though, such as taking the logarithm of a single small number. Since the size of a small number is fixed for given hardware, the number of possibilities for any small number computation is finite. For this reason, we will regard any small number computation as being bounded by some constant, which enables us to determine algorithm efficiency by:

1. counting the number of ints or floats needed at each step,
2. counting the number of small number operations at each step, and
3. counting the number of steps.

A "big number" computation is one in which the numbers involved require more bits than any of the small numbers available for the given hardware. Hardware is always finite, so there is always some size that will exceed its capacity. For example, say that I want to add two non-negative integers x and y with 1024 bits each. One byte is 8 bits, so each number can be represented by 128 bytes. Essentially, this means we are representing them in base $2^8 = 256$, with 128 places. We then store each place as an int, e.g.

$$x = x_{127} \cdot 256^{127} + \cdots + x_2 \cdot 256^2 + x_1 \cdot 256 + x_0$$

where each $0 \leq x_i < 256$ for all i. For a big number algorithm, the number of places will be denoted with N; in this case $N = 128$. The usual approach of adding is to add in each place and then carry. The places can be added in parallel, but the carrying generally cannot be done in parallel.[1] To compute $x + y$, we must do $N = 128$ small number additions. The

[1] Well, carrying can be done in parallel, but the worst case scenario is no better than the sequential approach. Whether this is better or worse depends on context and is beyond the scope of this book.

maximum number of steps of carrying that is needed is also $N = 128$.[2] So, both the amount of space and time required grow linearly with the number of places N. The situation with subtraction is similar.

It is common to use the notation $O(f(N))$, called big-O notation, which was introduced by Edmund Landau to mean that the growth is bounded by a constant times $f(N)$ as N goes to infinity, where f is a function. Using this notation, we have just justified the first row of the following table:

	Space	Time
Addition/Subtraction	$O(N)$	$O(N)$
Multiplication/Division (sequential)	$O(N)$	$O(N\log(N))$
Multiplication/Division (parallel)	$O(N)$	$O(N)$

FIGURE 1.6: Asymptotic efficiency of basic big number operations

Actually the best multiplication algorithm isn't exactly $O(N\log(N))$, but that gets a little too technical. The best multiplication algorithms use the Fast Fourier Transform (FFT), which is discussed in Section 8.6, and we explain how it leads to the times given in Figure-1.6 for multiplication. Once that algorithm is in hand, then division can be obtained by Newton's algorithm, which is not discussed in this book.[3]

Some of the algorithms that we will encounter require copying. We might hope that large matrices can be copied in parallel, but it depends on hardware. Most modern hardware uses a method called prefetching, which can speed things up, but here we are trying to avoid hardware details. What I will show is that if we assume that addition and multiplication can be done in parallel, then we can make m copies of a vector of length n at least as fast as $O(\log(m+n))$ in parallel. The actual speed of copying might be somewhere in between $O(1)$ and $O(\log(m+n))$, but I will generally use $O(\log(m+n))$ to be safe.

The trick is to use a scan (in the APL sense), which is called accumulate in NumPy. What do we mean by accumulate? Suppose we start with the vector $\begin{bmatrix}0 & 1 & 2 & 3\end{bmatrix}$. If we accumulate with addition, then we should get

$$\begin{bmatrix} 0 & 0+1 & 0+1+2 & 0{+}1{+}2{+}3 \end{bmatrix} = \begin{bmatrix}0 & 1 & 3 & 6\end{bmatrix}.$$

We can accumulate with any binary operation. If the operation is associative, then it can be done in $O(\log(n))$ time using the following algorithm:

Step 0. Given a vector $\mathbf{x}$ of length $n = 2^k$, let $\mathbf{x}_0 = \mathbf{x}$.

Step i+1. For $0 \le i < k$, apply the binary operation between the first and the last $2^k - 2^i$ entries of $\mathbf{x}_i$ to get the last $2^k - 2^i$ entries of $\mathbf{x}_{i+1}$.

Example 1.6.1. Suppose we want to accumulate with addition on the vector

$$\mathbf{x}_0 = \mathbf{x} = \begin{bmatrix} 0 & 1 & 2 & 3 & 4 & 5 & 6 & 7 \end{bmatrix}.$$

Then $n = 8$ and $k = 3$. For step 1, we add the first $8 - 1 = 7$ elements of $\mathbf{x}_0$ to the last 7 elements to get $\mathbf{x}_1$:

$$\begin{array}{l} \begin{bmatrix} 0 & 1 & 2 & 3 & 4 & 5 & 6 & 7 \end{bmatrix} \\ \quad\;\; +\begin{bmatrix} 0 & 1 & 2 & 3 & 4 & 5 & 6 \end{bmatrix} \\ \begin{bmatrix} 0 & 1 & 3 & 5 & 7 & 9 & 11 & 13 \end{bmatrix} = \mathbf{x}_1. \end{array}$$

[2]The maximum for carrying is attained with $x = 256^{128} - 1$ and $y = 1$, which would even force the creation of an additional place.

[3]Newton's algorithm has $\log(N)$ steps and the size doubles at each step. Although big number multiplication is required, we get the same efficiency by using only the size that we need at each step.

For step 2, we add the first $8 - 2 = 6$ elements of $\mathbf{x}_1$ to the last 6 elements to get $\mathbf{x}_2$:

$$\begin{array}{cccccccc} [\ 0 & 1 & 3 & 5 & 7 & 9 & 11 & 13\ ] \\ & +[\ 0 & 1 & 3 & 5 & 7 & 9\ ] \\ [\ 0 & 1 & 3 & 6 & 10 & 14 & 18 & 22\ ] = \mathbf{x}_2. \end{array}$$

For step 3, we add the first $8 - 4 = 4$ elements of $\mathbf{x}_2$ to the last 4 elements to get $\mathbf{x}_3$ (which is the result):

$$\begin{array}{cccccccc} [\ 0 & 1 & 3 & 6 & 10 & 14 & 18 & 22\ ] \\ & & & +[\ 0 & 1 & 3 & 6\ ] \\ [\ 0 & 1 & 3 & 6 & 10 & 15 & 21 & 28\ ] = \mathbf{x}_3. \end{array}$$

Activity 1.6.2. Try applying this algorithm with addition to the vector

$$\mathbf{x}_0 = \mathbf{x} = \begin{bmatrix} 1 & 1 & 1 & 1 & 1 & 1 & 1 & 1 \end{bmatrix}.$$

Suppose we want to make the zero vector. We might start off with a vector of bytes with random values:

$$\begin{bmatrix} 149 & 217 & 157 & 38 & 178 & 36 & 162 & 106 \end{bmatrix}.$$

We set the first byte to zero, then accumulate with multiplication[4] to get the zero vector. Now we will apply these ideas to make several copies of a vector.

Example 1.6.3. Suppose we need 4 copies of the vector $\begin{bmatrix} 1 & 2 & 3 & 4 \end{bmatrix}$. We allocate 3 more rows, which might have random values:

$$\begin{bmatrix} 1 & 2 & 3 & 4 \\ 149 & 217 & 157 & 38 \\ 178 & 36 & 162 & 106 \\ 190 & 94 & 195 & 5 \end{bmatrix}.$$

We then copy in the following steps:

1. Set the 1st byte in the second row to zero.

2. Accumulate with multiplication in the second row to get zero in the entire row.

3. Accumulate with multiplication in the columns to get zeros in all rows below the first.

4. Accumulate with addition[5] in the columns to copy the vector.

If the rows have length n, and the columns have length m, then we have $O(2\log(m)+\log(n))$, which is the same as $O(\log(m+n))$.[6] Of course, this is only helpful if you need several copies.

The bulk of this section concerns the efficiency of various algorithms for matrix multiplication. It must be said that research on matrix multiplication is very active, and we have no intention of illustrating the state of the art. Instead, our goal is to give a basic introduction to some of the ideas and methods, as a foundation to further study. However, before diving

[4] It is faster to accumulate with logical_and, which is bit-wise multiplication.
[5] At the level of bits, it is faster to use logical_or or xor.
[6] They are the same because $2\log(m) + \log(n) \leq 3\log(m+n)$, while $m + n - 1 \leq m^2 n$.

into the main topic, we will address the easier cases of matrix addition and multiplication by a scalar.

Suppose $\mathbf{A}$ and $\mathbf{B}$ are both $m \times n$ matrices, and c is a scalar. Then $\mathbf{A} + \mathbf{B}$ and $c\mathbf{A}$ exist. Each matrix has mn components, and the operations are defined component-wise, thus there are mn additions needed for $\mathbf{A} + \mathbf{B}$ and mn multiplications needed for $c\mathbf{A}$. So, for the sequential algorithm with small nums we get $O(mn)$ in both time and space. For the sequential algorithms with big nums, we suppose that the scalar c and the components of $\mathbf{A}$ and $\mathbf{B}$ all have N places. In that case, we need an extra axis to store the N places for each component, so we have $O(mnN)$ in space. In the big num case we have $O(mnN)$ for addition and $O(mnN\log(N))$ for scalar multiplication because we are applying the big number operations to each component and there are mn components. However, all of these operations can be done in parallel, so in the parallel algorithms we loose the factor of mn from time.

	Space	Time
Sequential with small nums	$O(mn)$	$O(mn)$
Parallel with small nums	$O(mn)$	$O(1)$
Sequential $\mathbf{A} + \mathbf{B}$ with big nums	$O(mnN)$	$O(mnN)$
Sequential $\mathbf{cA}$ with big nums	$O(mnN\log(N))$	$O(mnN\log(N))$
Parallel $\mathbf{A} + \mathbf{B}$ with big nums	$O(mnN)$	$O(N)$
Parallel $\mathbf{cA}$ with big nums	$O(mnN\log(N))$	$O(N)$

FIGURE 1.7: Asymptotic efficiency of $\mathbf{A} + \mathbf{B}$ and $c\mathbf{A}$, for an $m \times n$ matrices $\mathbf{A}$ and $\mathbf{B}$ and a scalar c.

Turning to matrix multiplication, it must be said that most studies focus on square matrices. In some cases it may be fine to take a matrix that is not a square matrix and pad with zeros to get a square matrix. However, it may not be better in all cases. So we begin by establishing a baseline.

If $\mathbf{A}$ is $m \times n$ and $\mathbf{B}$ is $n \times l$, then we must store both matrices so we get $O(n(m + l))$ in space. By the direct approach, we must compute the dot product of ml pairs of vectors. Each dot product has n multiplications and $n - 1$ additions, so the sequential algorithm is $O(mln)$ in time.

In a parallel algorithm, all of the dot products can be computed in parallel, but we need several copies of one of the matrices. For example suppose we want to compute the $\mathbf{AB}$ where

$$\mathbf{A} = \begin{bmatrix} 1 & 2 & 3 & 4 \\ 5 & 6 & 7 & 8 \\ 9 & 10 & 11 & 12 \end{bmatrix} \quad \text{and} \quad \mathbf{B} = \begin{bmatrix} 1 & -2 \\ -3 & 4 \\ 5 & -6 \\ -7 & 8 \end{bmatrix}$$

Each row of $\mathbf{A}$ occurs in two dot products, while each row of $\mathbf{B}$ occurs in three dot products. So, in order to do all of the multiplication in parallel, we actually need to make copies of $\mathbf{A}$ and $\mathbf{B}$. That increases the total space to $O(n(m + l)^2)$. The time needed to make the copies following the method in Example 1.6.3 is $O(\log(l + m + n))$. It is also possible to do some of the addition in parallel by adding in pairs.

Example 1.6.4. Here is the approach for computing a dot product in parallel. Suppose that we want to compute $\mathbf{u} \cdot \mathbf{v}$, where

$$\mathbf{u} = \begin{bmatrix} 1 & 2 & 3 & 4 & 5 \end{bmatrix} \quad \text{and} \quad \mathbf{v} = \begin{bmatrix} 1 & -1 & 1 & -1 & 1 \end{bmatrix}.$$

We begin by multiplying component-wise in parallel:

$$\begin{bmatrix} 1 & -2 & 3 & -4 & 5 \end{bmatrix}.$$

Then we repeat the following steps:

1. pad with 0 to even length,
2. add the first half and last half.

In this case

$$\begin{aligned} \begin{bmatrix}1 & -2 & 3\end{bmatrix} + \begin{bmatrix}-4 & 5 & 0\end{bmatrix} &= \begin{bmatrix}-3 & 3 & 3\end{bmatrix} \\ \begin{bmatrix}-3 & 3\end{bmatrix} + \begin{bmatrix}3 & 0\end{bmatrix} &= \begin{bmatrix}0 & 3\end{bmatrix} \\ \begin{bmatrix}0\end{bmatrix} + \begin{bmatrix}3\end{bmatrix} &= \begin{bmatrix}3\end{bmatrix}, \end{aligned}$$

so $\mathbf{u} \cdot \mathbf{v} = 3$. One could also pad with 0 at the beginning to a length that is a power of 2, which has the advantage that it only has to be done once but the disadvantage of increasing the amount of space. At each step, all of the addition can be done in parallel, and the number of times that the vector must be split in half is equal to the number of bits in n, which is proportional to $\log(n)$. As a consequence we get the following baseline for matrix multiplication.

	Space	Time
Sequential with small nums	$O(n(m+l))$	$O(mln)$
Parallel with small nums	$O(n(m+l)^2)$	$O(\log(m+n+l))$

FIGURE 1.8: Baseline for the asymptotics of **AB**

I find it doubtful that it is possible to improve on the time in the parallel case, but in the sequential case, it depends on the shape. For square matrices $m = l = n$, so the time asymptotics in the sequential case simplify to $O(n^3)$. At the time of this writing, the best known sequential algorithm for square matrices is $O(n^{2.37188})$ in time. But if all you want is the dot product of two vectors, then $m = l = 1$ and so the direct sequential approach to the dot product is $O(n)$ in time. Clearly $O(n)$ is much better than $O(n^{2.37188})$, so padding the vectors to get a square matrix would not only increase the amount of space needed, it would also be slower. Even if you have a square matrix times a vector, the direct approach would be $O(n^2)$ in time.

If we are comparing an algorithm that is $O(n^\alpha)$ with an algorithm that is $O(mln)$, then the $O(mln)$ should be favored when

$$\log(m) + \log(l) \leq (\alpha - 1)\log(n).$$

This works in any base, but in base 2, we can just count the bits of m, l, and n. For the current best algorithm, we have $\alpha = 2.37188$, so $\alpha - 1$ is a little bigger than 4/3. So, if the number of bits of m plus the number of bits of l is less than about 4/3 times the number of bits of n, then I would just use the direct approach.

To get a sense of how some of these algorithms for square matrices work, we will begin by looking at a method of multiplying complex numbers due to Gauss. It is possible to represent a complex number $a + bi$ in matrix form as

$$\begin{bmatrix} a & -b \\ b & a \end{bmatrix} = a \begin{bmatrix} 1 & 0 \\ 0 & 1 \end{bmatrix} + b \begin{bmatrix} 0 & -1 \\ 1 & 0 \end{bmatrix}.$$

The direct approach of multiplying two complex numbers $a + bi$ and $c + di$ is

$$ac - bd + (ad + bc)i,$$

which takes 4 multiplications and 3 additions. Gauss figured out how to do the same thing with 3 multiplications and 5 additions:

1. First compute $x_1 = ac$, $x_2 = bd$ and $x_3 = (a+b)(c+d)$.
2. Then compute $ac - bd = x_1 - x_2$ and $ad + bc = x_3 - (x_1 + x_2)$.

Example 1.6.5. With the direct approach

$$(1+2i)(3+4i) = (3-8) + (6+4)i = -5 + 10i,$$

and with Gauss's approach $(1+2)(3+4) = 3 \cdot 7 = 21$, so

$$(1+2i)(3+4i) = (3-8) + (21 - (3+8))i = -5 + 10i.$$

In terms of matrix multiplication, these approaches give us two different ways of computing

$$\begin{bmatrix} 1 & -2 \\ 2 & 1 \end{bmatrix} \begin{bmatrix} 3 & -4 \\ 4 & 3 \end{bmatrix} = \begin{bmatrix} -5 & 10 \\ 10 & 5 \end{bmatrix}$$

Exercises

Problem 1.1. True or False?

a. A 5×6 matrix has 6 rows.

b. $(\mathbf{A}^T\mathbf{B})^T = \mathbf{B}^T\mathbf{A}$

c. $(\mathbf{A}^{-1}\mathbf{B})^{-1} = \mathbf{A}^{-1}\mathbf{B}$

Problem 1.2. Given the matrices

$$\mathbf{A} = \begin{bmatrix} 2 & 0 & -1 \\ 3 & -4 & 5 \end{bmatrix} \qquad \mathbf{B} = \begin{bmatrix} -3 & 0 \\ 4 & 1 \\ 5 & -2 \end{bmatrix} \qquad \mathbf{C} = \begin{bmatrix} 1 & 2 \\ 0 & 1 \end{bmatrix} \qquad \mathbf{D} = \begin{bmatrix} 0 & -5 \\ 3 & 0 \end{bmatrix}$$

calculate each of the following if it exists:

$$\begin{array}{l} \mathbf{AB} + \mathbf{C}, \quad \mathbf{BA} + \mathbf{C}, \quad \mathbf{A}(\mathbf{B} + \mathbf{C}), \\ \mathbf{B}(\mathbf{C} + \mathbf{D}), \quad \mathbf{BC} + \mathbf{D}, \quad \mathbf{ABC}, \\ \mathbf{CD}, \quad \mathbf{DC}, \quad \mathbf{CA}, \quad -2\mathbf{D}. \end{array}$$

Problem 1.3. Use theorem 1.2.4 to simplify $(\mathbf{u}^T\mathbf{v})^T$. Then by definition 1.2.5 calculation shows that $\mathbf{u} \cdot \mathbf{v} = \mathbf{v} \cdot \mathbf{u}$

Problem 1.4. Suppose that $\mathbf{A}$ and $\mathbf{B}$ are symmetric matrices; that is $\mathbf{A} = \mathbf{A}^T$ and $\mathbf{B} = \mathbf{B}^T$

a. Suppose that $\mathbf{AB} = (\mathbf{AB})^T$. Use theorem 1.2.4 to simplify and show that $\mathbf{AB} = \mathbf{BA}$

b. Suppose that $\mathbf{AB} = \mathbf{BA}$. Replace, $\mathbf{A}$ and $\mathbf{B}$ on one side with $\mathbf{A}^T$ and $\mathbf{B}^T$, then use theorem 1.2.4, to show that $\mathbf{AB} = (\mathbf{AB})^T$.

Problem 1.5. Calculate the product

$$\begin{bmatrix} 0 & 1 \\ 0 & 1 \end{bmatrix} \begin{bmatrix} 1 & 1 \\ 0 & 0 \end{bmatrix}.$$

Use this example to show that given $\mathbf{AB} = \mathbf{AC}$ we cannot conclude $\mathbf{B} = \mathbf{C}$ (this is known as the law of cancellation, and it is true for real numbers but not for matrices).

Problem 1.6. Given the matrix

$$\mathbf{M} = \begin{bmatrix} 1 & 1 \\ 0 & 1 \end{bmatrix},$$

calculate $\mathbf{M}^2$, $\mathbf{M}^3$, and $\mathbf{M}^4$. Is there a pattern?

Problem 1.7. For what value of c is the following system of equations true?

$$\begin{bmatrix} 1 & c \\ 0 & 1 \end{bmatrix} \begin{bmatrix} 1 & 1 \\ 0 & 1 \end{bmatrix} = \begin{bmatrix} 1 & 0 \\ 0 & 1 \end{bmatrix}$$

Problem 1.8. Use row reduction to find the inverse of one of the following 3 matrices.

$$\text{a.} \begin{bmatrix} 1 & -2 & 3 \\ -2 & 5 & -6 \\ 1 & -2 & 4 \end{bmatrix} \qquad \text{b.} \begin{bmatrix} 1 & 2 & -3 \\ -1 & -1 & 3 \\ 2 & 4 & -5 \end{bmatrix} \qquad \text{c.} \begin{bmatrix} 1 & -1 & 3 \\ 1 & 0 & 3 \\ 2 & -2 & 7 \end{bmatrix}$$

Problem 1.9. For each of the 3 matrices to be $\mathbf{A}$

$$\text{a.} \begin{bmatrix} 3 & 2 \\ -1 & 1 \end{bmatrix} \qquad \text{b.} \begin{bmatrix} 5 & -2 \\ -1 & 1 \end{bmatrix} \qquad \text{c.} \begin{bmatrix} 3 & -5 \\ 1 & -1 \end{bmatrix}$$

compute $\mathbf{A}^{-1}$ then use it to solve the equations $\mathbf{XA} = \mathbf{B}$ and $\mathbf{AY} = \mathbf{C}$, where

$$\mathbf{B} = \begin{bmatrix} 1 & -1 \end{bmatrix} \quad \text{and} \quad \mathbf{C} = \begin{bmatrix} 0 & 1 \\ 1 & 1 \end{bmatrix}.$$

Problem 1.10. Given the matrices

$$\mathbf{A} = \begin{bmatrix} 1 & 1 \\ 0 & 1 \end{bmatrix} \quad \text{and} \quad \mathbf{B} = \begin{bmatrix} 0 & -1 \\ 1 & 0 \end{bmatrix},$$

compute the following matrices, indicate which ones are equal, and what propositions or theorems are illustrated by the equalities you find.

a. $\mathbf{AB}$, $(\mathbf{AB})^T$, $\mathbf{B}^T\mathbf{A}^T$, and $\mathbf{A}^T\mathbf{B}^T$

b. $(\mathbf{AB})^{-1}$, $\mathbf{A}^{-1}\mathbf{B}^{-1}$, and $\mathbf{B}^{-1}\mathbf{A}^{-1}$

c. $\mathbf{A}^T$, $\mathbf{A}^T\mathbf{A}$, and $(\mathbf{A}^T\mathbf{A})^T$

d. $\mathbf{A}^{-1}$, $(\mathbf{A}^{-1})^T$, $(\mathbf{A}^T)^{-1}$

Problem 1.11. Show that each part of theorem 1.3.8 is false for the matrix

$$\mathbf{A} = \begin{bmatrix} 1 & 1 \\ 0 & 0 \end{bmatrix}.$$

Problem 1.12. Compute the condition number for each function $f(x)$ below

(a) $f(x) = cx$

(b) $f(x) = x + a$

(c) $f(x) = \frac{1}{x}$

(d) $f(x) = e^{-x}$

(e) $f(x) = e^{-\frac{1}{2}x^2}$

Problem 1.13. Suppose that the condition number for $f(x)$ is $\kappa_1(x)$, the condition number of $g(x)$ is $\kappa_2(x)$ and c is a constant. Use derivative rules to show each of the following:

(a) The condition number of $cf(x)$ is $\kappa_1(x)$

(b) The condition number of $f(x)g(x)$ is $\kappa_1(x) + \kappa_2(x)$

(c) The condition number of $f(g(x))$ is $\kappa_1(g(x))\kappa_2(x)$

2

Systems of Linear Equations

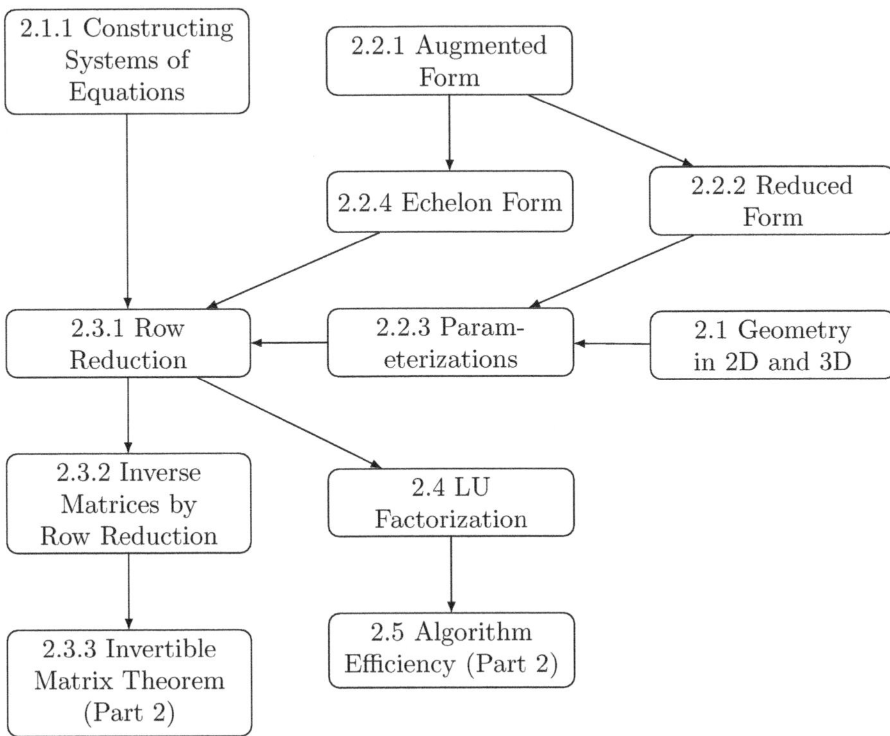

DOI: 10.1201/9781003737490-2

2.1 Geometry in Two or Three Dimensions

Q 2.1.1. Can you give an example of an equation of a line?

Q 2.1.2. What does it mean for two lines to be parallel?

Q 2.1.3. What is the x coordinate of the point $(2, 3)$?

In this section, we will try to build some intuition about what solutions to a system of linear equations can look like. In 2 dimensions, a linear equation represents a line. By a "system" of linear equations, we mean that we have more than one equation. These equations may represent different lines. A "solution" to a system of linear equations in 2 dimensions can be understood in a couple different ways. Geometrically it is a point or collection of points lying on all of the lines defined by the equations. Algebraically, if we look at the coordinates of those points, they must satisfy all of the equations. Only three things can happen:

1. There is exactly 1 solution.
2. There are zero solutions.
3. There are infinitely many solutions.

These remain the only options in higher dimensions as well.

Example 2.1.1. Find the number of solutions to the system of equations below by graphing them.

$$\begin{aligned} x + y &= 2 \\ x - y &= 0. \end{aligned} \tag{2.1}$$

Solution

Two points determine a line. The easiest way to graph a line is to find two points on the line, plot them, then plot the line through those points. When finding points, pick the easiest numbers you can think of. Usually zero is a good start.

For $x+y = 2$ if we take $x = 0$, then $y = 2$, so $(0, 2)$ is a point on the line. By taking $y = 0$, then $x = 2$, so $(2, 0)$ is another point on the line. The dashed diagonal line in Figure 2.1 is the line through them.

For $x - y = 0$ if we take $x = 0$, then $y = 0$, so $(0, 0)$ is a point on a line. But now zero is no longer helpful, so to get another point, we take the second easiest number we can think of: one. If we take $x = 1$, then $y = 1$, so $(1, 1)$ gives us a second point. The solid diagonal line in Figure 2.1 is the line through them.

The two lines intersect at a single point $(1, 1)$. The solution is unique.

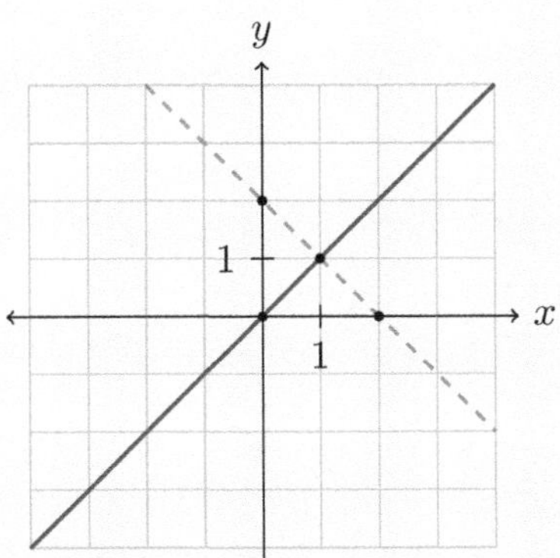

FIGURE 2.1: Graph of the system (2.1)

Sage automatically has the variable x defined, so to solve an equation in one variable, there is no need to define anything. For example:

```
solve(6*x-2==0, x)
```

The first argument specifies the equation $6x - 2 = 0$. The second argument specifies the variable being solved for, x. A $*$ is needed between the 6 and the x, which is standard calculator notation for multiplying. A == is used for equality, while = is used for assignment (defining objects and functions). To solve a system of equations, we first must define the variables.

```
var('y')
solve([x+y==2,x-y==0],[x,y])
```

The first argument of the solve function is a list of equations, and the second argument is a list of variables being solved for.[a] We can therefore define each list, and then insert the lists in the solve function. A list in python is defined with rectangular brackets and separated by commas. So, for example,

```
eqn_list=[x+y==2,x-y==0]
var_list=[x,y]
solve(eqn_list,var_list)
```

If we need other variables as well, they can be separated by commas:

```
var('y,z')
```

[a]For the python literate, a tuple can be used instead of a list.

Activity 2.1.2. Find the number of solutions to the system of equations

$$\begin{aligned} x + y &= 2 \\ x + y &= 0. \end{aligned}$$

Do this by graphing them.

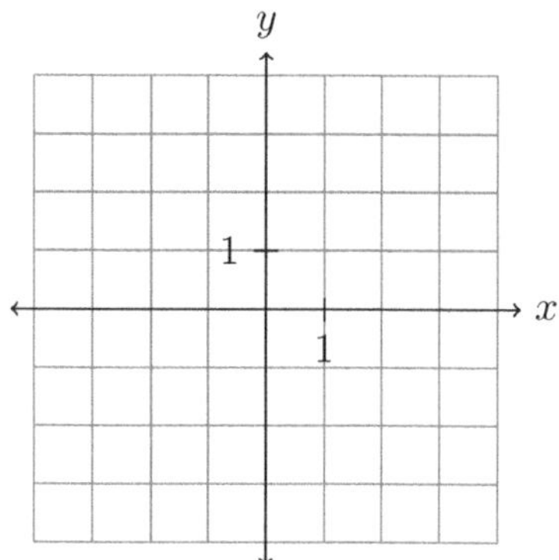

Activity 2.1.3. Find the number of solutions to the system of equations

$$\begin{aligned} x + y &= 2 \\ 2x + 2y &= 4. \end{aligned}$$

Do this by graphing them.

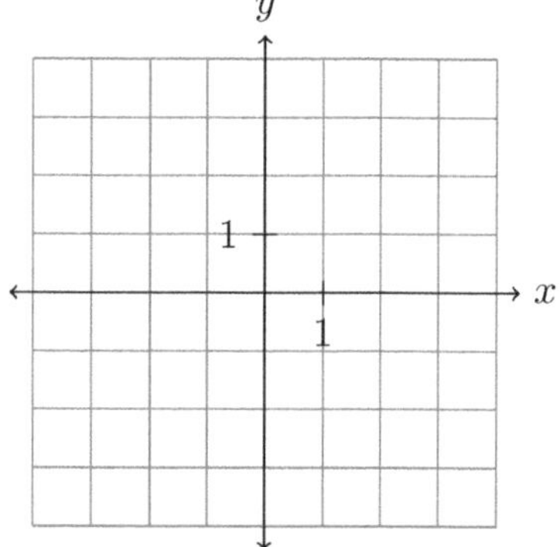

With 3 or more equations the same principles apply, except that the possible types of solutions become a little more interesting. We still have only three possibilities for the number of solutions, but there are more ways they can occur.

1. If there are infinitely many solutions, it is because all of the lines lie on top of each other.

2. If there is exactly one solution, it is because all of the lines go through a common point. Some lines may lie on top of each other, but not all of them will in this case.

3. But, *if there is no solution*, then for more than 2 equations, it is not necessary for the lines to be parallel.

In both Activities 2.1.4 and 2.1.5, there are zero solutions. You should not conclude that there are 3 solutions in Activity 2.1.4 because there is no point common to all three lines. Similarly, in Activity 2.1.5, you should not conclude that there are 2 solutions for the same reason. In terms of the number of solutions, the only options are zero, one, or infinitely many.

Activity 2.1.4. Graph the equations:

$$\begin{aligned} x + y &= 0 \\ x - y &= 0 \\ x &= 1 \end{aligned}$$

For each pair of lines, find the point of intersection (if it exists) and show that it is not on the third line (plug in with the coordinates).

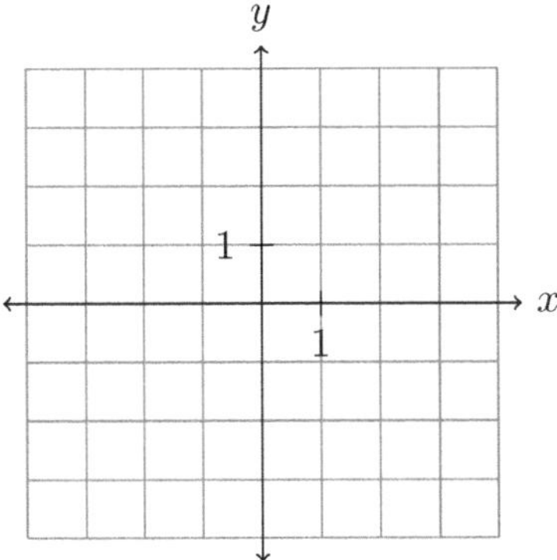

Activity 2.1.5. Graph the following equations.

$$\begin{aligned} y &= 1 \\ y &= -1 \\ x + y &= 0 \end{aligned}$$

For each pair of lines, find the point of intersection (if it exists) and show that it is not on the third line (plug in with the coordinates).

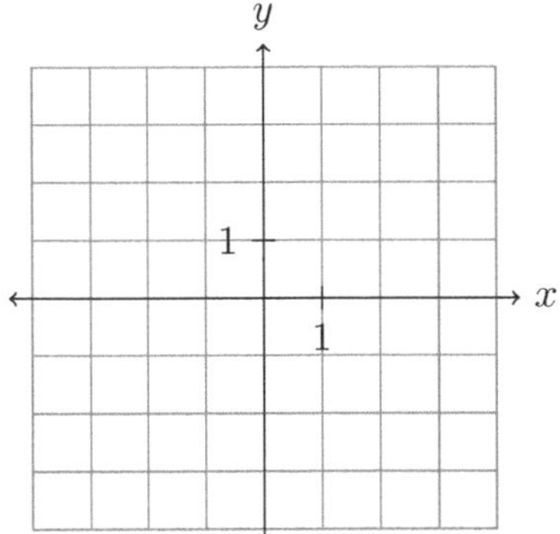

Moving on to 3-dimensions, the first thing that must be understood is that a single equation defines a plane, not a line.[1] For example, using coordinates (x, y, z), the equation

$$z = 0$$

simply sets the z-coordinate equal to zero. Since any point with coordinates $(x, y, 0)$ satisfies this equation, what we have is simply the xy-plane. This can be compared with the fact that in 2-dimensions, the equation $y = 0$ defines the x-axis.

Here are the possibilities for 3 planes (3 equations with 3 unknowns):

1. There is exactly one solution, i.e. a single point where all 3 planes intersect.

2. There are infinitely many solutions:

   (a) The planes intersect along a common line.

[1] Generally in n-dimensions, a single linear equation will define an $n-1$ dimensional linear object, usually called a **hyperplane**.

(a) Sage graph of the system (2.2) in 3D

(b) Sage graph of the system (2.3) in 3D

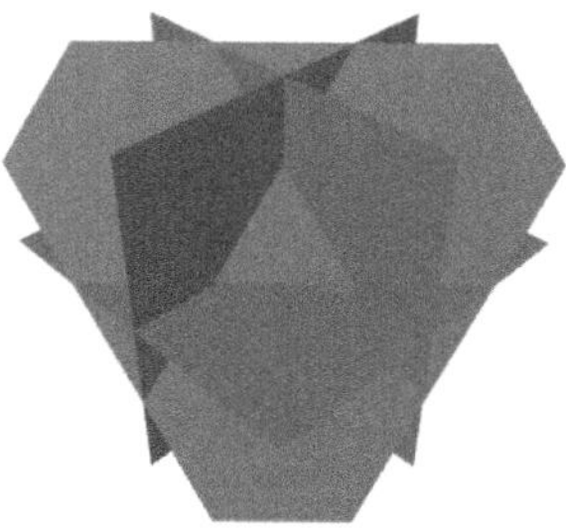

(c) Sage graph of the system (2.4) in 3D

FIGURE 2.2: Examples of systems with no solutions in 3D

   (b) All three planes coincide (are co-planar).

3. There are no solutions:
   (a) The three planes are mutually parallel.
   (b) Two of the planes are parallel, and the third coincides with one of them.
   (c) Two of the planes are parallel, and the third slices through both of those planes. See Figure 2.2b.
   (d) No two planes are parallel, but the lines where pairs of planes intersect are parallel in 3 dimensions. See Figure 2.2a.

Example 2.1.6. Consider the equations in Activities 2.1.4 and 2.1.5 above, except now you should think of them as equations for planes in 3 dimensions. You can just think of this as if the lines were drawn on a wall, and you were to extend them by pulling them out of the wall. How many solutions do these systems have now?

Solution

The equations from Activity 2.1.4 are

$$\begin{aligned} x + y &= 0, \\ x - y &= 0, \\ x &= 1. \end{aligned} \tag{2.2}$$

Considered in three dimensions, they describe 3 planes (one for each equation). Any pair of these planes intersects in a line. The lines formed in this way are parallel and are the edges of an infinitely long triangular tube enclosed by the planes. The tube never comes

to a common point, so there is still no solution. See Figure 2.2a.
The equations from Activity 2.1.5 are

$$\begin{aligned} y &= 1, \\ y &= -1, \\ x + y &= 0. \end{aligned} \tag{2.3}$$

Considered in three dimensions, they describe two parallel planes cut by a plane. The cutting plane intersects each of the parallel planes in a line. The lines formed in this way are parallel. Since they do not intersect, there is no point in common to all three planes. See Figure 2.2b.

Activity 2.1.7. For each system below, describe the solution in terms of the above categorization.

$$\begin{cases} x = 0 \\ y = 0 \\ z = 0 \end{cases} \quad \begin{cases} x = 0 \\ y = 0 \\ x = y \end{cases} \quad \begin{cases} y = 1 \\ 2y = 2 \\ 3y = 3 \end{cases} \quad \begin{cases} y = 1 \\ y = 2 \\ y = 3 \end{cases}$$

If it helps, you may graph them with Sage (see the code at the end of the section).

If we have more than 3 planes, it is not even necessary for the lines where pairs of planes intersect to be parallel. For example, consider the planes

$$\begin{aligned} x &= 0 \\ y &= 0 \\ z &= 0 \\ x + y + z &= 1. \end{aligned} \tag{2.4}$$

These four planes enclose a volume called a "tetrahedron." See Figure 2.2c.

Sage has an enormous number of graphing options. For a system of linear equations, the most natural option is `implicit_plot3d`. For example, to graph $x + y + z = 1$, with x, y, and z between -2 and 2 we can use

```
implicit_plot3d(x+y+z==1,[x,-2,2],[y,-2,2],[z,-2,2])
```

So the first argument specifies that the equation is $x + y + z = 1$, and the next three specify the frame. It is also possible to use a function $f(x, y, z)$ as a first argument. In this case, it is implied that $f(x, y, z) = 0$. So

```
implicit_plot3d(x+y+z-1,[x,-2,2],[y,-2,2],[z,-2,2])
```

is another way of graphing the plane $x + y + z = 1$. A single graph may have many different objects. The first line below defines the graph g (with `=`). Initially, it only has the plane $x = 0$ (again, equality is `==`). The next three lines add the other planes to the graph (with `+=`). The final line shows the graph.

```
g = implicit_plot3d(x==0,[x,-2,2],[y,-2,2],[z,-2,2])
g +=implicit_plot3d(y==0,[x,-2,2],[y,-2,2],[z,-2,2])
```

```
g +=implicit_plot3d(z==0,[x,-2,2],[y,-2,2],[z,-2,2])
g +=implicit_plot3d(x+y+z==1,[x,-2,2],[y,-2,2],[z,-2,2])
g.show(frame=false)
```

In Sage you can zoom and rotate 3D graphs. You can also export graphs.

2.1.1 Setting up a System of Equations

Example 2.1.8. The Train Discount 50 Card costs \$ 120 for students who buy the 2nd class tickets. Using it to buy train tickets gives a 50% savings. What is the least amount that would need to be spent on train tickets in order for it to be cheaper than buying them without the card?

Solution

Step 1: Introduce variables. Let x be the listed cost of the tickets (in dollars), and y be the amount paid (in dollars).

Step 2: Set up equations. In truth, this problem can be done with just one equation, but we set it up with two equations: one with and one without the card. Without the card, we don't get a savings. We must pay the listed price, so $y = x$. With the card, we get a 50% savings. 50% is 0.5 as a decimal, so the cost of the tickets with the savings is

$$x - 0.5x = 0.5x.$$

But we must pay for the card to get the savings, so we must add 120. This reasoning gives us following equations:

$$\begin{aligned} \text{without the card} \quad & y = x, \quad \text{and} \\ \text{with the card} \quad & y = 0.5x + 120. \end{aligned}$$

Step 3: Cancel one of the variables. The variable y can be cancelled by subtracting equation 2 from equation 1:

$$\begin{array}{rl} & y = \quad x \\ - & y = 0.5x + 120 \\ \hline & 0 = 0.5x - 120 \end{array}$$

Of course, we must be careful to subtract on both sides.

Step 4: Solve for the remaining variable by isolating it on one side. This needs to be done in two steps. First, we add 120 to both sides, and then we multiply both sides by 2 (the reciprocal of 0.5)

$$\begin{aligned} 0 + 120 = 0.5x - 120 + 120 &\implies 120 = 0.5x \\ 2 \times 120 = 2 \times 0.5x &\implies 240 = x \end{aligned}$$

Step 5: Plug back into one of the equations with the value of the variable found in order to find the value of the other variable.

$$y = x = 240$$

Activity 2.1.9. With the A-to-Z rewards card, you get a $50 gift card for signing up and an additional $3 for every $100 spent.

1. Set up an equation representing the total amount of rewards y (in dollars) in terms of the amount spent x (in dollars).

2. How much needs to be spent before the total amount of rewards is $110?

Activity 2.1.10. Solve the system of equations:

$$\begin{aligned} x + 2y &= 0, \\ 4x + 9y &= 3 \end{aligned}$$

Proposition 2.1.1. *If we change a system of equations by*

1. *multiplying both sides of an equation by a non-zero constant,*

2. *adding a multiple of one equation to another equation, or*

3. *changing the order of two equations,*

then the new system of equations will have the same solution set.

Remark 2.1.1. We used the second rule of Proposition 2.1.1 in step 3 of Example 2.1.8, and the first rule in step 4. The rule allowing us to change the order of equations may seem boring now, but it is at least true and may seem more significant when phrased in terms of row reduction in Proposition 2.3.1.

While it is true that multiplying both sides of an equation by zero maintains equality, it is unhelpful for solving the system. It is rather like obliterating one of our equations and may lead us to conclude that there are more solutions to the system than we really have.

2.2 Equations in Matrix Form

TL;DR

Figure 2.3 shows how a system of equations can be represented with matrices and vectors.
Pivot columns correspond to variables that have been solved for (see Definition 2.2.4 and Example 2.2.7).
Free variables or parameters can be used to describe all solutions to a system of linear equations (see Definition 2.2.5 and Example 2.2.11). Pivots and free variables will be a fundamental tool used in later sections for quickly determining dimensions.

2.2.1 Augmented Form

Definition 2.2.1. Given a system of equations

$$\begin{aligned} a_{11}x_1 + a_{12}x_2 + \cdots + a_{1n}x_n &= b_1 \\ a_{21}x_1 + a_{22}x_2 + \cdots + a_{2n}x_n &= b_2 \\ \cdots \cdots \cdots \cdots & \cdots \\ a_{m1}x_1 + a_{m2}x_2 + \cdots + a_{mn}x_n &= b_m, \end{aligned} \tag{2.5}$$

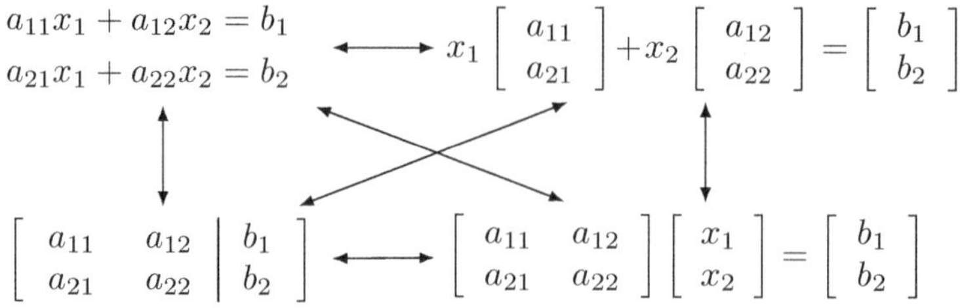

$$\begin{aligned} a_{11}x_1 + a_{12}x_2 &= b_1 \\ a_{21}x_1 + a_{22}x_2 &= b_2 \end{aligned} \longleftrightarrow x_1\begin{bmatrix} a_{11} \\ a_{21} \end{bmatrix} + x_2\begin{bmatrix} a_{12} \\ a_{22} \end{bmatrix} = \begin{bmatrix} b_1 \\ b_2 \end{bmatrix}$$

$$\left[\begin{array}{cc|c} a_{11} & a_{12} & b_1 \\ a_{21} & a_{22} & b_2 \end{array}\right] \longleftrightarrow \begin{bmatrix} a_{11} & a_{12} \\ a_{21} & a_{22} \end{bmatrix}\begin{bmatrix} x_1 \\ x_2 \end{bmatrix} = \begin{bmatrix} b_1 \\ b_2 \end{bmatrix}$$

FIGURE 2.3: Equivalent ways of representing a system of equations

then the **augmented matrix** for this system is

$$[\mathbf{A}|\mathbf{b}] = \left[\begin{array}{cccc|c} a_{11} & a_{12} & \cdots & a_{1n} & b_1 \\ a_{21} & a_{22} & \cdots & a_{2n} & b_2 \\ \vdots & \vdots & & \vdots & \vdots \\ a_{m1} & a_{m2} & \cdots & a_{mn} & b_m \end{array}\right].$$

The vertical line indicates where the equal sign once was. The matrix $\mathbf{A}$, without the augmented column, is called the **coefficient matrix** of the system.

Example 2.2.1. Given the equations

$$\begin{aligned} x + 2y &= 1, \\ 3x + 4y &= -1, \end{aligned} \tag{2.6}$$

the coefficient matrix is

$$\mathbf{A} = \begin{bmatrix} 1 & 2 \\ 3 & 4 \end{bmatrix} \tag{2.7}$$

and the augmented matrix is

$$\left[\begin{array}{cc|c} 1 & 2 & 1 \\ 3 & 4 & -1 \end{array}\right]. \tag{2.8}$$

Activity 2.2.2. Write down the coefficient matrix and the augmented matrix for:

$$\begin{aligned} x + 2y &= 0, \\ 4x + 9y &= 3. \end{aligned}$$

Proposition 2.2.2. *A system of equations* (2.11) *with augmented form* $[\mathbf{A}|\mathbf{b}]$ *can also be represented as:*

1. *a vector equation*

$$x_1\mathbf{a}_1 + x_2\mathbf{a}_2 + \cdots x_n\mathbf{a}_n = \mathbf{b},$$

where $\mathbf{a}_1, \mathbf{a}_2, \ldots \mathbf{a}_n$ *are the columns of* $\mathbf{A}$*; or*

2. *a matrix equation*

$$\mathbf{A}\mathbf{x} = \mathbf{b},$$

where $x_1, x_2 \ldots x_n$ *are the components of* $\mathbf{x}$*.*

In the case of a system of 2 equations in 2 variables, Proposition 2.2.2 is illustrated graphically by Figure 2.3 and numerically by Example 2.2.3. It is often useful to switch between these forms, as they emphasize different things.

When doing a numerical computation by hand, augmented form is often the most concise, and generally will keep work organized while minimizing the number of pencil strokes. When working algebraically by hand, it is more common to work directly with vector equations or matrix equations.

Example 2.2.3. The vector equation

$$x\begin{bmatrix}1\\3\end{bmatrix}+y\begin{bmatrix}2\\4\end{bmatrix}=\begin{bmatrix}1\\-1\end{bmatrix} \tag{2.9}$$

is the same thing as the matrix equation

$$\begin{bmatrix}1&2\\3&4\end{bmatrix}\begin{bmatrix}x\\y\end{bmatrix}=\begin{bmatrix}1\\-1\end{bmatrix}. \tag{2.10}$$

Why? Because, on one hand

$$x\begin{bmatrix}1\\3\end{bmatrix}+y\begin{bmatrix}2\\4\end{bmatrix}=\begin{bmatrix}x\\3x\end{bmatrix}+\begin{bmatrix}2y\\4y\end{bmatrix}=\begin{bmatrix}x+2y\\3x+4y\end{bmatrix},$$

and on the other

$$\begin{array}{cc} & \begin{bmatrix}x\\y\end{bmatrix}\\ \begin{bmatrix}1&2\\3&4\end{bmatrix} & \begin{bmatrix}x+2y\\3x+4y\end{bmatrix}\end{array}.$$

So the left-hand side of equation (2.9) is equal to the left-hand side of (2.10). By matching the components on the left and right-hand side, we get the system of equations (2.6), which has the augmented form (2.8).

Activity 2.2.4. Show that the vector equation

$$x\begin{bmatrix}1\\4\end{bmatrix}+y\begin{bmatrix}2\\9\end{bmatrix}=\begin{bmatrix}0\\3\end{bmatrix}$$

is the same as the matrix equation

$$\begin{bmatrix}1&2\\4&9\end{bmatrix}\begin{bmatrix}x\\y\end{bmatrix}=\begin{bmatrix}0\\3\end{bmatrix}$$

and that both correspond to the system of equations in Activity 2.2.2

Matrices and vectors can be defined in Sage as follows:

```
A=matrix([[1,2],[3,4]])
A
b=vector([1,-1])
b
A.augment(b,subdivide=True)
```

The first line defines **A**. The second line shows what **A** is. If you omit the second line, then **A** will be defined but it won't be shown on the screen. The third and fourth lines are similar in this respect. On the fifth line, nothing is defined because there is no `=` to the left of `A`, so the output is shown. If `subdivide=True` is omitted, then there will be no vertical line in the augmented matrix separating **A** and **b**. Each line of code above is a single "statement." It is possible to split statements across lines like this:

```
A=matrix([[1,2],
          [3,4]])
```

which may make them more readable. Multiple statements can also be combined on one line by separating them with a semicolon:

```
A=matrix([[1,2],[3,4]]); A
```

There is somewhat of a difference between the behavior of Sage when used on the command line versus in a Jupyter notebook:

	To run code	What is displayed?
Command line	Enter key	all variables on the line
Jupyter notebook	Run icon	only the last line of a cell

To display more than one variable in a notebook, you can either

1. put them in different cells, or
2. put them in a list on the last line of a cell.

2.2.2 Reduced Form

If our goal is to solve a system of equations, we need to know what that looks like in augmented form.

Q 2.2.1. The system of equations

$$\begin{aligned} x &= 3 + z \\ y &= 2 + z \end{aligned} \tag{2.11}$$

has been solved for x and y. If z is subtracted from both sides, what is the augmented form for this system?

You should see that the variables that have been solved for are indicated by ones, with zeros above and below them. If they are in lexicographical order (x, y, z), then we obtain a step pattern with the ones leading in each row.

Definition 2.2.3. A row of a matrix has a **leading one** if the first non-zero entry in the row, reading left to right, is one. A matrix is said to be in **reduced echelon form** or **reduced form** if:

1. all rows containing only zeros are at the bottom,
2. every row has a leading one, unless the row has all zeros,
3. the entries in the matrix below and to the left of a leading one are all zero,
4. every other entry in the same column as a leading one is zero, including the entries above it.

Example 2.2.5.

$$\left[\begin{array}{ccccc|c} \boxed{1} & 1 & 0 & 0 & 0 & 3 \\ 0 & 0 & \boxed{1} & 2 & 0 & 0 \\ 0 & 0 & 0 & 0 & 0 & 0 \end{array}\right]$$

- The leading ones are indicated in boxes.
- The zeros are "below and to the left" of the leading one in the second row are indicated with a gray background.
- The columns with leading ones have zeros everywhere else.
- The only row of zeros is at the bottom.
- All rules are satisfied, so this matrix is in reduced form. You should notice the step-like pattern.

Activity 2.2.6. Which matrices below are in reduced form? Which rules are violated for those that are not?

$$A = \left[\begin{array}{ccc|c} 1 & 0 & 0 & 1 \\ 0 & 1 & 0 & 2 \\ 0 & 0 & 1 & 3 \end{array}\right] \qquad B = \left[\begin{array}{ccc|c} 0 & 0 & 1 & 1 \\ 1 & 0 & 0 & 2 \\ 0 & 1 & 0 & 3 \end{array}\right] \qquad C = \left[\begin{array}{ccc|c} 1 & 2 & 0 & 1 \\ 0 & 1 & 0 & 2 \\ 0 & 0 & 1 & 3 \end{array}\right]$$

$$D = \left[\begin{array}{ccc|c} 1 & 0 & 0 & 1 \\ 0 & 1 & 0 & 2 \\ 2 & 0 & 1 & 3 \end{array}\right] \qquad E = \left[\begin{array}{ccc|c} 1 & 0 & 0 & 2 \\ 0 & 1 & 1 & 3 \\ 0 & 0 & 0 & 0 \end{array}\right] \qquad F = \left[\begin{array}{ccc|c} 1 & 0 & 0 & 2 \\ 0 & 0 & 0 & 0 \\ 0 & 0 & 1 & 3 \end{array}\right]$$

It is not necessary for the ones to be leading. For this reason we will often use the following more general terminology:

Definition 2.2.4. The position of a non-zero entry in the matrix that we use to clear the numbers above and below it is sometimes called a **pivot position** or just a **pivot**, and the column that it is in is called a **pivot column**.

Example 2.2.7. For example, the system of equations (2.11) could be solved for x and z instead of x and y:

$$\begin{aligned} x &= 2 + z \\ y &= 3 + z \end{aligned} \quad \longrightarrow \quad \begin{aligned} x &= y - 1 \\ z &= y - 3 \end{aligned}$$

Then after subtracting y from both sides we obtain the following augmented form of the system

$$\left[\begin{array}{ccc|c} \boxed{1} & -1 & 0 & -1 \\ 0 & -1 & \boxed{1} & -3 \end{array}\right]. \tag{2.12}$$

The box in the second row is not leading: -1 is the first non-zero entry in that row. It is not fair to speak of leading ones in this case, so we refer to the boxed numbers as pivots. The pivot columns are the 1st column and the 3rd column, which indicate that x and z have been solved for.

Selecting a pivot amounts to selecting a variable to solve for. The process of selecting is a choice, and the concept of reduce form specifies that the choices be made lexicographically. Reduced form has two advantages that I see:

1. it can be reached systematically, and
2. it is unique (assuming a specific order of the variables).

Reduced form does not seem to have anything to do with a particular strategy of solving a system, and so it will often be used directly to study various concepts. Even the singular value decomposition can be directly connected to reduced form.

2.2.3 Parameterizations

Q 2.2.2. What is the x-coordinate of a point on the y-axis?

Q 2.2.3. In the equation $y = mx + b$, what is the dependent variable and what is the independent variable?

Consider a line in the form

$$y = mx + b \tag{2.13}$$

The value b gives us a point on the line, the y-intercept. In vector form, the y-intercept may be written as

$$\mathbf{b} = \begin{bmatrix} 0 \\ b \end{bmatrix}.$$

The slope m is rise over run, a change in the y direction divided by a change in the x direction. If the change in the x direction is 1, then the change in the y direction is m, giving us the vector

$$\mathbf{m} = \begin{bmatrix} 1 \\ m \end{bmatrix}$$

with the change in x in the x-coordinate and the change in y in the y-coordinate. Then an arbitrary point on the line can be represented in vector form as

$$\begin{bmatrix} x \\ y \end{bmatrix} = \begin{bmatrix} x \\ mx + b \end{bmatrix} = \begin{bmatrix} x \\ mx \end{bmatrix} + \begin{bmatrix} 0 \\ b \end{bmatrix} = x \begin{bmatrix} 1 \\ m \end{bmatrix} + \begin{bmatrix} 0 \\ b \end{bmatrix} = x\mathbf{m} + \mathbf{b} \tag{2.14}$$

Equation (2.14) is an example of what we will call a **parameterization** and the variable x is an example of a **parameter**. The idea is that as the value of the parameter varies, we get all of the points on the line.

Example 2.2.8. The equation

$$y = 2x - 1 \tag{2.15}$$

has been solved for y. So

$$\begin{bmatrix} x \\ y \end{bmatrix} = \begin{bmatrix} x \\ 2x - 1 \end{bmatrix} = x \begin{bmatrix} 1 \\ 2 \end{bmatrix} + \begin{bmatrix} 0 \\ -1 \end{bmatrix} = x\mathbf{m} + \mathbf{b} \tag{2.16}$$

gives us arbitrary points on the line $y = 2x - 1$ as the parameter x varies. In particular, as Figure 2.4 shows, $(-1, -3)$ is a point on the line, and we get the vector for that point using $x = -1$:

$$(-1) \begin{bmatrix} 1 \\ 2 \end{bmatrix} + \begin{bmatrix} 0 \\ -1 \end{bmatrix} = \begin{bmatrix} -1 \\ -2 \end{bmatrix} + \begin{bmatrix} 0 \\ -1 \end{bmatrix} = \begin{bmatrix} -1 \\ -3 \end{bmatrix}.$$

You should also observe in Figure 2.4 that:

1. $\mathbf{b}$ is a vector for a point on the line, and
2. $\mathbf{m}$ is parallel to the line.

Comprehension Check 2.2.1. Using equation (2.16), what is the vector $2\mathbf{m} + \mathbf{b}$? Does the y value match what you would expect by just plugging $x = 2$ into $y = 2x - 1$?

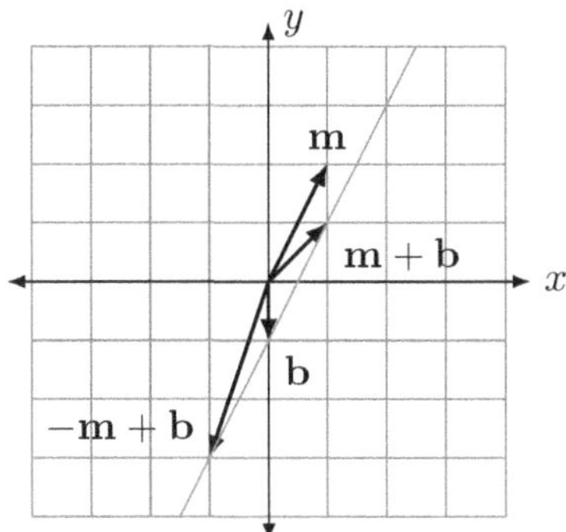

FIGURE 2.4: A parameterization of $y = 2x - 1$ with x as the free-variable

Since equation (2.13) has been solved for y, then after subtracting mx from both sides, we obtain $-mx + y = b$, hence the corresponding augmented form is

$$\left[\begin{array}{cc|c} -m & \boxed{1} & b \end{array}\right]$$

with the pivot boxed. This matrix is not in reduced form unless $m = 0$. Remember, reduced form favors solving for variables in lexicographical order. Since x comes before y alphabetically, reduced form favors solving for x, if possible. Solving for a variable amounts to choosing it to be a dependent variable. The remaining variables, if any, are the independent variables; but we often use other names for them.

Definition 2.2.5. Given an augmented matrix $[\mathbf{A}|\mathbf{b}]$ such that

1. each non-zero row contains a pivot and
2. there is no pivot in the augmented column,

a variable is **free** if the column corresponding to it in the reduced matrix does not contain a pivot. Sometimes a free variable is also called a **parameter**.

What if there is a pivot in the augmented column? The augmented matrix

$$\left[\begin{array}{cc|c} 0 & 0 & \boxed{1} \end{array}\right].$$

corresponds to the equation

$$0x + 0y = 1$$

so $0 = 1$, which is nonsense! As will be seen in Subsection 2.3.1, such an augmented matrix corresponds to situations where there is no solution.

The terminology "free" should make sense in light of the fact that we are free to choose values for them. The number of free variables is equal to the dimension of the solution set:

Object	Dimension (num. of free vars)
Point	0
Line	1
Plane	2

Example 2.2.9. Subtracting $2x$ from both sides of equation (2.15) gives us $-2x + y = -1$, so the augmented form is

$$\left[\begin{array}{cc|c} -2 & \boxed{1} & -1 \end{array}\right]. \tag{2.17}$$

The pivot corresponds to y, which was solved for in $y = 2x - 1$, while the column without pivot corresponds to x, the free variable in this case. The equation $y = 2x - 1$ represents a line, which has dimension 1, and we have only one free variable.

The system of equations

$$\begin{aligned} x &= 2 \\ y &= 3' \end{aligned} \tag{2.18}$$

considered in 2 dimensions, has augmented form

$$\left[\begin{array}{cc|c} \boxed{1} & 0 & 2 \\ 0 & \boxed{1} & 3 \end{array}\right]. \tag{2.19}$$

The pivots indicate that we have solved for both x and y, so they are completely determined: there is no freedom. The equations (2.18) represent a point with coordinates $(2, 3)$, which has dimension zero and there are no free variables.

Activity 2.2.10. For each augmented matrix below, determine:

1. the pivots (leading ones)
2. the free variables (if any)
3. the dimension of the solution set
4. the number of solutions.

$$\left[\begin{array}{ccc|c} 1 & 0 & 0 & 1 \\ 0 & 1 & 0 & 2 \\ 0 & 0 & 1 & 3 \end{array}\right], \quad \left[\begin{array}{ccc|c} 1 & 0 & 0 & 1 \\ 0 & 1 & 0 & 2 \\ 0 & 0 & 0 & 0 \end{array}\right], \quad \left[\begin{array}{ccc|c} 1 & 0 & -1 & 0 \\ 0 & 0 & 0 & 0 \\ 0 & 0 & 0 & 0 \end{array}\right]$$

Equation (2.16) is an example of what we will call a **parameterization**. The parameter was x, and since the top component was left unchanged in the calculation we have $x = x$. Trivial as $x = x$ may seem, it is true and plays a fundamental role in obtaining a parameterization from any augmented form satisfying the conditions of Definition 2.2.5

Example 2.2.11. Compute a parameterization of the solution set of

$$\left[\begin{array}{ccc|c} \boxed{1} & 1 & 0 & 2 \\ 0 & 0 & \boxed{1} & 3 \\ 0 & 0 & 0 & 0 \end{array}\right] \tag{2.20}$$

Solution

First, we identify the pivots, which are boxed. There is a second 1 in the first row, but we can chose only one of them as a pivot. Lexicographical order favors the first.
Using x, y, z as the variable names, the pivots indicate that we have solved for x and z. Thus the augmented form (2.20) gives us:

$$\begin{cases} x + y = 2 \\ \quad\;\; z = 3 \end{cases} \longrightarrow \begin{cases} x = 2 - y \\ z = 3 \end{cases}$$

Since y is free, it is at least true that $y = y$, so as a vector equation

$$\begin{bmatrix} x \\ y \\ z \end{bmatrix} = \begin{bmatrix} 2 - y \\ y \\ 3 \end{bmatrix} = \begin{bmatrix} 2 \\ 0 \\ 3 \end{bmatrix} + \begin{bmatrix} -y \\ y \\ 0 \end{bmatrix} = \begin{bmatrix} 2 \\ 0 \\ 3 \end{bmatrix} + y \begin{bmatrix} -1 \\ 1 \\ 0 \end{bmatrix} = \mathbf{P} + y\mathbf{T}. \tag{2.21}$$

Equation (2.21) gives us a parameterization similar to (2.16):

- Both have one free variable, so they each represent a line.
- Just as **b** in (2.16) gives a point on the line $2x - 1$ (the y-intercept), the vector **P** in equation (2.21) is a point on this line.
- Just as **m** in (2.16) gave us a vector parallel to the line, the vector **T** in equation (2.21) is a parallel to this line. In both cases, the vectors can be moved so that they lie along the line are tangent to it.

Activity 2.2.12. Compute a parameterization of the solution set of

$$\left[\begin{array}{ccc|c} 1 & 0 & 0 & 4 \\ 0 & 1 & 2 & 5 \\ 0 & 0 & 0 & 0 \end{array}\right].$$

Start by putting boxes around the leading ones, to identify them as pivots.

In general if there are d free variables $t_1, t_2, \ldots t_d$, the final parameterization in vector form should look like

$$\mathbf{x} = \mathbf{P} + t_1\mathbf{T}_1 + t_2\mathbf{T}_2 + \cdots t_d\mathbf{T}_d, \tag{2.22}$$

where **P** is a specific solution, and $\mathbf{T}_1, \mathbf{T}_2, \ldots \mathbf{T}_d$ are tangent vectors indicating the directions that you can go from **P** while remaining in the solution set.

Although t is a common choice for the name of a parameter, in the context of augmented matrices the parameters come from columns not containing a pivot. Since those columns already have variable names associated with them, it is natural and meaningful to use those names as we already have been doing. Having said this, I will show you an example where using the variable names is not natural.

Example 2.2.13. Find a parameterization of the plane through the points represented by the vectors

$$\mathbf{P} = \begin{bmatrix} -1 \\ 2 \\ 2 \end{bmatrix}, \quad \mathbf{Q} = \begin{bmatrix} 1 \\ 1 \\ 0 \end{bmatrix}, \quad \mathbf{R} = \begin{bmatrix} 1 \\ 0 \\ 1 \end{bmatrix}.$$

Solution

We can use **P** directly as one point on the line. To get the tangent vectors, we compute the direction from **P** to **Q**

$$\mathbf{T}_1 = \mathbf{Q} - \mathbf{P} = \begin{bmatrix} 1 \\ 1 \\ 0 \end{bmatrix} - \begin{bmatrix} -1 \\ 2 \\ 2 \end{bmatrix} = \begin{bmatrix} 2 \\ -1 \\ 2 \end{bmatrix}$$

and the direction from **P** to **R**

$$\mathbf{T}_2 = \mathbf{R} - \mathbf{P} = \begin{bmatrix} 1 \\ 0 \\ 1 \end{bmatrix} - \begin{bmatrix} -1 \\ 2 \\ 2 \end{bmatrix} = \begin{bmatrix} 2 \\ 2 \\ -1 \end{bmatrix}.$$

This is analogous to getting the slope of a line by subtracting the coordinates and then dividing. Here there is no division because we are using vectors. Then

$$\mathbf{P} + t_1\mathbf{T}_1 + t_2\mathbf{T}_2 = \begin{bmatrix} -1 \\ 2 \\ 2 \end{bmatrix} + t_1 \begin{bmatrix} 2 \\ -1 \\ 2 \end{bmatrix} + t_2 \begin{bmatrix} 2 \\ 2 \\ -1 \end{bmatrix} \tag{2.23}$$

gives us a parameterization of the plane. In particular, for $t_1 = t_2 = 0$ we just get $\mathbf{P}$, and for $t_1 = 1$ and $t_2 = 0$

$$\mathbf{P} + \mathbf{T}_1 = \mathbf{P} + (\mathbf{Q} - \mathbf{P}) = \mathbf{Q}.$$

By keeping $t_2 = 0$ and allowing t_1 to vary, we get the points on the line segment between $\mathbf{P}$ and $\mathbf{Q}$. Similarly, by keeping $t_1 = 0$ and allowing t_2 to vary, we get the points on the line segment between $\mathbf{P}$ and $\mathbf{R}$.

The variables t_1 and t_2 are meaningful to the extent that taking various combinations of 0 and 1 gives us the points $\mathbf{P}$, $\mathbf{Q}$, and $\mathbf{R}$, that we started with, but they are not equal to any of x, y or z. Equation (2.23) gives us the following system of equations

$$\begin{cases} x = -1 + 2t_1 + 2t_2 \\ y = 2 + -t_1 + 2t_2 \\ z = 2 + 2t_1 - t_2 \end{cases}$$

which is a system of three equations in 5 variables. If we put the variables in the order t_1, t_2, x, y, z, then the corresponding augmented form is

$$\left[\begin{array}{ccccc|c} -2 & -2 & \boxed{1} & 0 & 0 & -1 \\ 1 & -2 & 0 & \boxed{1} & 0 & 2 \\ -2 & 1 & 0 & 0 & \boxed{1} & 2 \end{array}\right].$$

But we have a plane in three dimensions, and so we really only need one equation in 3 variables. If this matrix is brought into reduced form by any method (e.g. by Subsection 2.3.1), then the first three columns will contain the pivots, y and z will be free, and the last row will give us an equation in x, y, and z alone.

2.2.4 Echelon Form

Q 2.2.4. What is the first non-zero number in the vector $\begin{bmatrix} 0 & 0 & 2 & 0 & 1 \end{bmatrix}$?

Definition 2.2.6. A **leading entry** is the 1st non-zero entry in the row reading left to right. A matrix is said to be in **echelon form** if:

1. all rows containing only zeros are at the bottom, and
2. the entries in the matrix below and to the left of a leading entry are all zero.

Activity 2.2.14. Which matrices below are in echelon form?

$$A = \left[\begin{array}{ccc|c} 1 & 0 & 0 & 4 \\ 0 & 2 & 0 & 5 \\ 0 & 0 & 3 & 6 \end{array}\right] \quad B = \left[\begin{array}{ccc|c} 0 & 0 & 1 & 4 \\ 1 & 0 & 0 & 5 \\ 0 & 1 & 0 & 6 \end{array}\right] \quad C = \left[\begin{array}{ccc|c} 1 & 2 & 3 & 4 \\ 0 & 1 & 5 & 6 \\ 0 & 0 & 1 & 7 \end{array}\right]$$

$$D = \left[\begin{array}{ccc|c} 1 & 0 & 0 & 1 \\ 0 & 1 & 0 & 2 \\ 2 & 0 & 1 & 3 \end{array}\right] \quad E = \left[\begin{array}{ccc|c} 1 & 0 & 0 & 2 \\ 0 & 1 & 1 & 3 \\ 0 & 0 & 0 & 0 \end{array}\right] \quad F = \left[\begin{array}{ccc|c} 1 & 0 & 0 & 2 \\ 0 & 0 & 0 & 0 \\ 0 & 0 & 1 & 3 \end{array}\right]$$

2.3 Row Reduction

2.3.1 Row Reduction

If our goal is to solve a system of equations, then reduced form gives us a target. How do we get there? One way is by using row reduction.

Proposition 2.3.1. *A matrix in augmented form* $[\mathbf{A}|\mathbf{b}]$ *can be changed to a different augmented form with the same solution set using the following rules:*

1. *A row may be multiplied by an arbitrary non-zero constant.*
2. *A multiple of a row may be added to another row.*
3. *Two rows may be swapped.*

These rules correspond directly with the rules in Proposition 2.1.1. To keep track of what we are doing, we will introduce the following notation.

Definition 2.3.2. We will write R_i to refer to row i.

1. If we want to multiply row i by a constant c, we will write cR_i.
2. If we want to add c times row i to row j, we will write $cR_i + R_j$.
3. If we want to swap row i and row j, we will write $R_i \leftrightarrow R_j$.

Comprehension Check 2.3.1. Given the augmented matrix

$$\left[\begin{array}{ccc|c} 1 & 2 & 3 & 0 \\ 4 & 5 & 6 & 0 \\ 7 & 8 & 9 & 0 \end{array}\right] \tag{2.24}$$

answer each of the following questions:

1. What is R_1?
2. What is $(-2)R_1$?
3. What is $(-2)R_1 + R_2$?
4. Which row is being replaced by $(-2)R_1 + R_2$?

Definition 2.3.3. Two augmented matrices $[\mathbf{A}_1|\mathbf{b}_1]$ and $[\mathbf{A}_2|\mathbf{b}_2]$ are said to be **row equivalent** if it is possible to get from one to the other in some sequence of steps using the rules in Proposition 2.3.1. This process is referred to as row reduction.

Proposition 2.3.4. *Every augmented matrix* $[\mathbf{A}|\mathbf{b}]$ *is row equivalent to a matrix in reduced form.*

If two different matrices in reduced form are row equivalent, they are equal.

The word "equivalence" is being used in this context for two reasons:

1. the solution sets are the same as stated in Proposition 2.1.1, and
2. row equivalence defines an **equivalence relation.**

An equivalence relation must satisfy three properties:

Reflexive. Any augmented matrix $[\mathbf{A}|\mathbf{b}]$ must be equivalent to itself:

The operation $(1)R_1$ does the trick.

Symmetric. Any row operation must be reversible:

1. cR_i is reversed by $\frac{1}{c}R_i$,
2. $cR_i + R_j$ is reversed by $(-c)R_i + R_j$, and
3. $R_i \leftrightarrow R_j$ is reversed by $R_i \leftrightarrow R_j$.

Transitive. If $[\mathbf{A}_1|\mathbf{b}_1]$ is row equivalent to $[\mathbf{A}_2|\mathbf{b}_2]$ and $[\mathbf{A}_2|\mathbf{b}_2]$ is row equivalent to $[\mathbf{A}_3|\mathbf{b}_3]$, then $[\mathbf{A}_1|\mathbf{b}_1]$ must be equivalent to $[\mathbf{A}_3|\mathbf{b}_3]$:

Just join the sequence of steps together.

There are two very practical consequences of this equivalence:

1. You and your friend may arrive at reduced form in a different sequence of steps, but the final reduced form will be the same if you both have the correct answer.
2. It is possible to work against yourself and go backwards. You may do a problem in only 5 steps while your friend uses 20.

Moore [17] describes row reduction as an art. I would say that it is an art of selecting pivots. The considerations for selecting a pivot are very different for a human and for a computer.

For a computer, it is unwise to insist on reduced form or echelon form. The best approach numerically is to select the largest number in absolute value left of the augmented column. This approach is called complete pivoting. In the augmented matrix (2.24) a computer would identify the 9 as the first pivot and start off with $\frac{1}{9}R_3$. No human would do that.

For a human, the main goals are keeping numbers small, avoiding fractions or decimals as long as possible, and reducing the total number of pencil stokes. My own strategy uses the following principles:

1. When life gives you a one, use it: select ones for pivots.
2. When life does not give you a one, try to create a one.
3. Once you have a one, create zeros above and below it.
4. Do not ruin zeros in pivot columns that you already have.
5. Avoid swapping rows except for moving ones into position.

For beginners, it is also strongly advised to work left to right, which leads to reduced form or echelon form. These principles are embodied by the row reduction examples in this section.

One more point should be made before diving into the row reduction examples. Let's say that after some sequence of steps we can get to echelon form in some sequence of steps. Then you could switch back to equations and finish the problem by variable substitution. This technique is what we call **back substitution**.

Example 2.3.1. Given the augmented matrix

$$\left[\begin{array}{ccc|c} \boxed{1} & 4 & 3 & 5 \\ 0 & 0 & \boxed{2} & 6 \end{array}\right]$$

find a parameterization for the solution set by back substitution.

Solution

Using x, y, z as variables, we get the equations

$$\begin{cases} x + 4y + 3z = 5, \\ 2z = 6. \end{cases}$$

Solving the second for z gives $z = 3$. We then substitute back in to the first equation,

$$x + 4y + 3(3) = 5, \quad \longrightarrow \quad x + 4y + 9 = 5$$

and solve for x:

$$x = 5 - 9 - 4y = -4 - 4y.$$

We have solved for x and z, which makes y free. We can also see that y is free from the boxed pivots. Since y is free we get $y = y$ and that gives the parameterization

$$\begin{bmatrix} x \\ y \\ z \end{bmatrix} = \begin{bmatrix} -4 - \\ 4y \\ y \\ 2 \end{bmatrix} = \begin{bmatrix} -4 \\ 0 \\ 2 \end{bmatrix} + y \begin{bmatrix} -4 \\ 1 \\ 0 \end{bmatrix}$$

From the viewpoint of parallel algorithms, back substitution is pointless because, once a pivot has been chosen, the entries both above and below it can be turned into zeros in parallel with a slight modification to LU factorization (see Section 2.4). Sequentially, back substitution is indeed faster than continuing to reduced form.

Humans, on the other hand, do not work in parallel so back substitution is generally faster for a human. But echelon form is not unique. If the goal is to get to a unique form, then you would need to continue to reduced form.

Example 2.3.2. Bring the matrix

$$\left[\begin{array}{cc|c} 3 & 4 & -1 \\ 1 & 2 & 1 \end{array}\right]$$

into reduced form.

Solution

$$\left[\begin{array}{cc|c} 3 & 4 & -1 \\ 1 & 2 & 1 \end{array}\right] \xrightarrow{R_1 \leftrightarrow R_2} \left[\begin{array}{cc|c} 1 & 2 & 1 \\ 3 & 4 & -1 \end{array}\right] \xrightarrow{(-3)R_1 + R_2} \left[\begin{array}{cc|c} 1 & 2 & 1 \\ 0 & -2 & -4 \end{array}\right] \tag{2.25}$$

To get to echelon form systematically, we work left to right, and from the top row down to get the step pattern. There is a 1 in the first column, but it is in the 2nd row. So, we begin by reversing row 1 and row 2.
Next we want to get rid of the 3 below the 1. What adds to 3 to give 0? -3. But we don't have a rule saying that we can add a number. We can add a multiple of a row. Since -3 times 1 is -3, it makes sense to add -3 times row 1 to row 2. Note that we leave row 1 as it is. It can just be copied over to the new matrix. We chose the factor of -3, to get 0 below the 1, so that is an easy spot to fill in. That leaves two spots. Since $(-3)(2) + 4 = -6 + 4 = -2$, then the 4 in row 2 gets replaced by -2. Similarly $(-3)(1) + -1 = -3 - 1 = -4$. The final matrix in (2.25) is in echelon form, so we can stop

there and do back substitution.

$$\left[\begin{array}{cc|c} 1 & 2 & 1 \\ 0 & -2 & -4 \end{array}\right] \xrightarrow{(-\frac{1}{2})R_2} \left[\begin{array}{cc|c} 1 & 2 & 1 \\ 0 & 1 & 2 \end{array}\right] \xrightarrow{(-2)R_2+R_1} \left[\begin{array}{cc|c} 1 & 0 & -3 \\ 0 & 1 & 2 \end{array}\right] \tag{2.26}$$

If we want to get to reduce form, then we need to start working from the bottom up to clear the numbers above the pivots. The first non-zero entry in the second row is -2. We can't use row 1, so to get a 1 there without ruining the zero that we worked so hard to get. We have to multiply row 2 by the reciprocal of -2, which is $-\frac{1}{2}$. Thankfully everything in that row is even, so we still get integers. Zero times anything is zero. We chose $-\frac{1}{2}$ to get $(-\frac{1}{2})(-2) = 1$. That leaves only one number to deal with: $(-\frac{1}{2})(-4) = 2$.
Finally, with the leading one in row 2, we can now get rid of the 2 that is above it. We do this in the same way that we got rid of the 3 in column 1. Since -2 adds to 2 to give zero, we multiply row 2 by -2 and add to row 1. The zero in row 2 is helpful because it does not change the value of anything it is added to, in this case the leading one in row 1. The only remaining calculation to do is $(-2)(2) + 1 = -4 + 1 = -3$. The last matrix in (2.26) is in reduced form.

Activity 2.3.3. Bring the augmented matrix from Activity 2.2.2 into reduced form.

Definition 2.3.5. A system of linear equations with at least one solution is said to be **consistent**. A system of linear equations with no solution is said to be **inconsistent**.

Remark 2.3.1. There is a very good reason for the above definition. With an inconsistent system of equations, we can use the rules for manipulating equations to conclude that $0 = 1$. A contradiction like $0 = 1$ means that there is no solution.

Example 2.3.4. Convert the system of equations

$$\begin{aligned} x + y &= 2 \\ x + y &= 0. \end{aligned}$$

from Activity 2.1.2 into an augmented matrix, and bring this into reduced form.

Solution

$$\left[\begin{array}{cc|c} 1 & 1 & 2 \\ 1 & 1 & 0 \end{array}\right] \xrightarrow{(-1)R_1+R_2} \left[\begin{array}{cc|c} 1 & 1 & 2 \\ 0 & 0 & -2 \end{array}\right]$$

The second row now implies that $0 = -2$, which is false. Therefore, the system is inconsistent. So, there is no solution, which agrees with the fact that the lines plotted in Activity 2.1.2 are parallel.

Example 2.3.5. Convert the following system of equations to matrix form, and solve.

$$\begin{aligned} w + x + y + z &= 1 \\ -3w - 3x - y + 3z &= 1 \\ 2w + 2x + 3y + 5z &= 4 \end{aligned}$$

Solution

$$\left[\begin{array}{cccc|c} 1 & 1 & 1 & 1 & 1 \\ -3 & -3 & -1 & 3 & 1 \\ 2 & 2 & 3 & 5 & 4 \end{array}\right]$$

$$\xrightarrow{3R_1+R_2} \left[\begin{array}{cccc|c} 1 & 1 & 1 & 1 & 1 \\ 0 & 0 & 2 & 6 & 4 \\ 2 & 2 & 3 & 5 & 4 \end{array}\right] \xrightarrow{(-2)R_1+R_3} \left[\begin{array}{cccc|c} 1 & 1 & 1 & 1 & 1 \\ 0 & 0 & 2 & 6 & 4 \\ 0 & 0 & 1 & 3 & 2 \end{array}\right]$$

$$\xrightarrow{R_2 \leftrightarrow R_3} \left[\begin{array}{cccc|c} 1 & 1 & 1 & 1 & 1 \\ 0 & 0 & 1 & 3 & 2 \\ 0 & 0 & 2 & 6 & 4 \end{array}\right] \xrightarrow{(-2)R_2+R_3} \left[\begin{array}{cccc|c} \boxed{1} & 1 & 1 & 1 & 1 \\ 0 & 0 & \boxed{1} & 3 & 2 \\ 0 & 0 & 0 & 0 & 0 \end{array}\right]$$

After converting the equations into matrix form, we must get to echelon form. We have a leading one in the top left corner right where we want it. We use it to get zeros in the rest of the column. To get rid of the -3 in row 2, we multiply row 1 by 3 and add it to row 2. Note that, row 1 and row 3 do not change, only row 2 changes, so row 1 and row 3 can just be copied (carefully).
To get rid of the 2 in row 3, we multiply row 1 by -2 and add it to row 3. Note that row 1 and row 2 do not change, only row 3. We have eliminated everything below the leading one in the top left corner, so we are done with column 1.
Moving on to column 2, the only non-zero value is in row 1. That value is 1, but it is not a leading one because there is a 1 to the left of it. So, we move on to column 3. There is a leading one in column 3, but it is in row 3 not in row 2, so we reverse rows 2 and 3. Note that row 1 does not change.
We now use the leading one in row 2 to get rid of the 2 below it in row 3. To do this, we must multiply row 2 by -2 and add to row 3.
Now that we have echelon form we can see that the system is consistent because there are no pivots in the augmented column. Continuing with back substitution, the non-zero rows give us the equations

$$\begin{aligned} w + x + y + z &= 1 \\ y + 3z &= 2 \end{aligned}$$

Solving the second equation for y gives us

$$y = 2 - 3z$$

Substituting this in the first equation and solving for w gives us

$$w + x + 2 - 3z + z = 1 \quad \longrightarrow \quad w + x + 2 - 2z = 1 \quad \longrightarrow \quad w = -1 - x + 2z.$$

We have solved for w and y, so x and z are free. Since x and z are free we get the redundant equations $x = x$ and $z = z$, which then gives us the parameterization

$$\begin{bmatrix} w \\ x \\ y \\ z \end{bmatrix} = \begin{bmatrix} -1 - x + 2z \\ x \\ 2 - 3z \\ z \end{bmatrix} = \begin{bmatrix} -1 \\ 0 \\ 2 \\ 0 \end{bmatrix} + x \begin{bmatrix} -1 \\ 1 \\ 0 \\ 0 \end{bmatrix} + z \begin{bmatrix} 2 \\ 0 \\ -3 \\ 1 \end{bmatrix}$$

Since there are two free variables, the solution set has dimension 2: it is a plane.

Example 2.3.6. Convert the following system of equations to matrix form, and solve.

$$\begin{aligned} w + \quad\;\; 3y - \;\; z &= 1 \\ x - 2y + 2z &= 1 \\ 2w + \;\; x + 4y + \;\; z &= 4 \\ w + 2x - \;\; y - 2z &= -1 \end{aligned}$$

Solution

$$\left[\begin{array}{cccc|c} 1 & 0 & 3 & -1 & 1 \\ 0 & 1 & -2 & 2 & 1 \\ 2 & 1 & 4 & 1 & 4 \\ 1 & 2 & -1 & -2 & -1 \end{array}\right] \xrightarrow{(-2)R_1+R_3} \left[\begin{array}{cccc|c} 1 & 0 & 3 & -1 & 1 \\ 0 & 1 & -2 & 2 & 1 \\ 0 & 1 & -2 & 3 & 2 \\ 1 & 2 & -1 & -2 & -1 \end{array}\right]$$

$$\xrightarrow{(-1)R_1+R_4} \left[\begin{array}{cccc|c} 1 & 0 & 3 & -1 & 1 \\ 0 & 1 & -2 & 2 & 1 \\ 0 & 1 & -2 & 3 & 2 \\ 0 & 2 & -4 & -1 & -2 \end{array}\right] \xrightarrow{(-1)R_2+R_3} \left[\begin{array}{cccc|c} 1 & 0 & 3 & -1 & 1 \\ 0 & 1 & -2 & 2 & 1 \\ 0 & 0 & 0 & 1 & 1 \\ 0 & 2 & -4 & -1 & -2 \end{array}\right]$$

$$\xrightarrow{(-2)R_2+R_4} \left[\begin{array}{cccc|c} 1 & 0 & 3 & -1 & 1 \\ 0 & 1 & -2 & 2 & 1 \\ 0 & 0 & 0 & 1 & 1 \\ 0 & 0 & 0 & -5 & -4 \end{array}\right] \xrightarrow{5R_3+R_4} \left[\begin{array}{cccc|c} \boxed{1} & 0 & 3 & -1 & 1 \\ 0 & \boxed{1} & -2 & 2 & 1 \\ 0 & 0 & 0 & \boxed{1} & 1 \\ 0 & 0 & 0 & 0 & \boxed{1} \end{array}\right]$$

First we convert to augmented form. There are two ones in column 1, so either of them can be used as a leading one for row reduction. Sometimes there may be an advantage to using one over the other, but if it is not clear, then just use the top one. To get rid of the 2 in row 3 we multiply row 1 by -2 and add to row 3. Note that row 3 is the only row being changed; the others can be copied.

To get rid of the 1 in row 4, we multiply row 1 by -1 and add to row 4. We are now done with column 1.

In column 2, we see two 1's below row 1. Again, we can use either of them, but we pick the top one. To get rid of the 1 in row 3, we multiply row 2 by -1 and add to row 3.

To get rid of the 2 in row 4, we multiply row 2 by -2 and add to row 4. We are now done with column 2.

In column 3, all of the values below the first two rows are zero, so there is nothing we can do. We move on to column 4.

In column 4, there is a leading one in row 3. We can get rid of the -5 below it by multiplying row 3 by 5 and adding to row 4.

In row 4, there are zeros to the left of the vertical line, and a non-zero value to the right. That value happens to be 1, but it doesn't really matter. It could be any non-zero value like 2, and we would still have a problem: $0 = 2$ is just as bad as $0 = 1$. Therefore, we have no solution: the system is inconsistent. One way of looking at this is to say that there is a pivot in the augmented column. Once we see a pivot in the augmented column, we can stop immediately. Of course it is wise to check your work.

Example 2.3.7. Convert the following system of equations to matrix form, and solve.

$$\begin{aligned} x + 2y + 6z &= 7 \\ x + 6y + \;\; z &= -2 \\ 2x + 7y + 4z &= 3 \end{aligned}$$

Solution

$$\left[\begin{array}{ccc|c} 1 & 2 & 6 & 7 \\ 1 & 6 & 1 & -2 \\ 2 & 7 & 4 & 3 \end{array}\right]$$

$$\xrightarrow{(-1)R_1+R_2} \left[\begin{array}{ccc|c} 1 & 2 & 6 & 7 \\ 0 & 4 & -5 & -9 \\ 2 & 7 & 4 & 3 \end{array}\right] \xrightarrow{(-2)R_1+R_3} \left[\begin{array}{ccc|c} 1 & 2 & 6 & 7 \\ 0 & 4 & -5 & -9 \\ 0 & 3 & -8 & -11 \end{array}\right]$$

$$\xrightarrow{(-1)R_3+R_2} \left[\begin{array}{ccc|c} 1 & 2 & 6 & 7 \\ 0 & 1 & 3 & 2 \\ 0 & 3 & -8 & -11 \end{array}\right] \xrightarrow{(-3)R_2+R_3} \left[\begin{array}{ccc|c} \boxed{1} & 2 & 6 & 7 \\ 0 & \boxed{1} & 3 & 2 \\ 0 & 0 & \boxed{-17} & -17 \end{array}\right]$$

Starting with column 1, we have two leading ones to choose between, but there is no sense wasting time switching rows. We use the 1 in row 1 to get rid of the 1 in row 2 by multiplying row 1 by -1 and adding to row 2.
Next we get rid of the 2 by multiplying row 1 by -2 and adding to row 3. We are now done with column 1.
In column 2, we have non-zero values available but not a 1. But fear not, we still can avoid fractions. We can even get a leading one if we multiply row 3 by -1 and add to row 2.
Now that we have a leading one finally, we can use it to eliminate the 3 in row 3. We do this by multiplying row 2 by -3 and adding to row 2.
Yikes! -17? But it will disappear soon enough. We now have echelon form, so we continue with back substitution. The equations we get from the rows are

$$\begin{aligned} x + 2y + 6z &= 7, \\ y + 3z &= 2, \\ -17z &= -17, \end{aligned}$$

Dividing the last equation by -17 gives us $z = 1$.
To find y, we substitute into the second equation with $z = 1$ and solve

$$y + 3(1) = 2 \quad \longrightarrow \quad y + 3 = 2 \quad \longrightarrow \quad y = -1$$

To find x, we substitute into the first equation with $y = -1$ and $z = 1$:

$$x + 2(-1) + 6(1) = 7 \quad \longrightarrow \quad x + 4 = 7 \quad \longrightarrow \quad x = 3$$

So the solution is unique and the dimension is zero. Geometrically this is a point, and the point has coordinates $x = 3$, $y = -1$, and $z = 1$.

Example 2.3.8. Convert the following system of equations to matrix form, and solve.

$$\begin{aligned} 2x + 3y - 2z &= 1 \\ 2x + 2y - z &= 3 \\ -4x - 5y + 4z &= -1 \end{aligned}$$

Solution

$$\left[\begin{array}{ccc|c} 2 & 3 & -2 & 1 \\ 2 & 2 & -1 & 3 \\ -4 & -5 & 4 & -1 \end{array}\right]$$

$$\xrightarrow{2R_1+R_3} \left[\begin{array}{ccc|c} 2 & 3 & -2 & 1 \\ 2 & 2 & -1 & 3 \\ 0 & 1 & 0 & 1 \end{array}\right] \xrightarrow{(-1)R_1+R_2} \left[\begin{array}{ccc|c} 2 & 3 & -2 & 1 \\ 0 & -1 & 1 & 2 \\ 0 & 1 & 0 & 1 \end{array}\right]$$

$$\xrightarrow{R_2 \leftrightarrow R_3} \left[\begin{array}{ccc|c} 2 & 3 & -2 & 1 \\ 0 & 1 & 0 & 1 \\ 0 & -1 & 1 & 2 \end{array}\right] \xrightarrow{R_2+R_3} \left[\begin{array}{ccc|c} \boxed{2} & 3 & -2 & 1 \\ 0 & \boxed{1} & 0 & 1 \\ 0 & 0 & \boxed{1} & 3 \end{array}\right]$$

There are no leading ones in column 1. That is not a problem, though, because there is a 2, and all of the numbers in column 1 are even. We can get rid of the -4 in row 3 by multiplying row 1 by 2 and adding to row 3.
Next we get rid of the 2 in row 2 by multiplying row 1 by -1 and adding to row 2. We are now done with column 1.
In column 2 we see a 1 and a -1 below row 1. The 1 is in row 3, so we move it to row 2 by reversing the rows.
We now have a leading one in row 2. We use it to get rid of the -1 in row 3 by adding row 2 to row 3.
Now that we have echelon form, we use back substitution. The equations we get from the rows are

$$\begin{aligned} 2x + 3y - 2z &= 1, \\ y &= 1, \\ z &= 3. \end{aligned}$$

To find x, we substitute in with the values y and z, and solve:

$$2x + 3(1) - 2(3) = 1 \quad \longrightarrow \quad 2x - 3 = 1 \quad \longrightarrow \quad 2x = 4 \quad \longrightarrow \quad x = 2.$$

So the solution is unique:

$$\begin{bmatrix} 2 \\ 1 \\ 3 \end{bmatrix}$$

Activity 2.3.9. Convert the following system of equations to matrix form, and solve.

$$\begin{aligned} x - y - 2z &= 2 \\ -x + y + 4z &= 2 \\ x - y - z &= 4 \end{aligned}$$

Let's see how to do row reduction with Sage.

```
A=matrix([[ 1, -2,  1],
          [ 0,  3, -1],
          [ 0,  1,  1]])
b=vector([0, 1, -1])
M=A.augment(b,subdivide=True); M
```

```
M.echelon_form()
M.rref()
```

The only new code here are the functions **echelon_form()** and **rref()**. The name **rref** is an abbreviation for "Reduced row echelon form." See Definition 2.2.6.

2.3.2 Inverse Matrices by Row Reduction

To find the inverse of an $n \times n$ square matrix $\mathbf{A}$, we must solve the matrix equation

$$\mathbf{AX} = \mathbf{I} \tag{2.27}$$

for $\mathbf{X}$, where $\mathbf{X}$ is an $n \times n$ equation as well. If an inverse exists, then by multiplying both sides of the equation on the left by $\mathbf{A}^{-1}$ we get

$$\mathbf{A}^{-1}\mathbf{AX} = \mathbf{A}^{-1}\mathbf{I} \quad \Longrightarrow \quad \mathbf{IX} = \mathbf{A}^{-1}. \tag{2.28}$$

We have seen in Subsection 2.2.1 that a matrix equation $\mathbf{Ax} = \mathbf{b}$ can be written in augmented form as $[\mathbf{A}|\mathbf{b}]$. However, it is true more generally that a matrix equation $\mathbf{AX} = \mathbf{B}$ can be written in augmented form as $[\mathbf{A}|\mathbf{B}]$ and solved. It is like solving several systems of equations simultaneously, one for each column of $\mathbf{B}$: the solutions give us each column of $\mathbf{X}$.

The augmented form of equation (2.27) is $[\mathbf{A}|\mathbf{I}]$, while the augmented form of equation (2.28) is $[\mathbf{I}|\mathbf{A}^{-1}]$. But $[\mathbf{I}|\mathbf{A}^{-1}]$ is in reduced form, which leads us to the following strategy for computing $\mathbf{A}^{-1}$:

1. Form the augmented matrix $[\mathbf{A}|\mathbf{I}]$.
2. Row reduce this matrix using the rules in Proposition 2.3.1 until $\mathbf{A}$ is brought into reduced form $\mathbf{R}$, so that the augmented matrix has the form $[\mathbf{R}|\mathbf{B}]$.
3. If $\mathbf{R} = \mathbf{I}$, then $\mathbf{A}$ is invertible and has inverse $\mathbf{B}$, otherwise $\mathbf{A}$ is not invertible.

Example 2.3.10. Using the matrix

$$\mathbf{A} = \begin{bmatrix} 2 & 3 \\ 1 & 2 \end{bmatrix}$$

1. find $\mathbf{A}^{-1}$,
2. verify that $\mathbf{AA}^{-1} = \mathbf{I}$.

Solution

To compute $\mathbf{A}^{-1}$ by row reduction, we must first write down the augmented matrix $[\mathbf{A}|\mathbf{I}]$:

$$\left[\begin{array}{cc|cc} 2 & 3 & 1 & 0 \\ 1 & 2 & 0 & 1 \end{array}\right]$$

$$\xrightarrow{R_1 \leftrightarrow R_2} \left[\begin{array}{cc|cc} 1 & 2 & 0 & 1 \\ 2 & 3 & 1 & 0 \end{array}\right] \xrightarrow{(-2)R_1 + R_2} \left[\begin{array}{cc|cc} 1 & 2 & 0 & 1 \\ 0 & -1 & 1 & -2 \end{array}\right]$$

$$\xrightarrow{(-1)R_2} \left[\begin{array}{cc|cc} 1 & 2 & 0 & 1 \\ 0 & 1 & -1 & 2 \end{array}\right] \xrightarrow{(-2)R_2 + R_1} \left[\begin{array}{cc|cc} 1 & 0 & 2 & -3 \\ 0 & 1 & -1 & 2 \end{array}\right].$$

Since the result of row reduction is $[\mathbf{I}|\mathbf{A}^{-1}]$, then this calculation shows that

$$\mathbf{A}^{-1} = \begin{bmatrix} 2 & -3 \\ -1 & 2 \end{bmatrix}.$$

If we ended up with a row of zeros, the inverse would not exist.
Now for the second part of the question, we must verify that $\mathbf{AA}^{-1} = \mathbf{I}$:

$$\begin{matrix} & \begin{bmatrix} 2 & -3 \\ -1 & 2 \end{bmatrix} \\ \begin{bmatrix} 2 & 3 \\ 1 & 2 \end{bmatrix} & \begin{bmatrix} 1 & 0 \\ 0 & 1 \end{bmatrix} \end{matrix}$$

Activity 2.3.11. In Example 1.3.5, we found that if

$$\mathbf{A} = \begin{bmatrix} 3 & 5 \\ 1 & 2 \end{bmatrix},$$

then

$$\mathbf{A}^{-1} = \begin{bmatrix} 2 & -5 \\ -1 & 3 \end{bmatrix}.$$

Verify this by row reduction.

Warning 2.3.2. If you do not augment with the identity matrix, you will not know what $\mathbf{A}^{-1}$ is, even if you have done all row reduction steps correctly.

The following activity illustrates Theorem 1.3.7 using row reduction.

Activity 2.3.12. Given

$$\mathbf{D} = \begin{bmatrix} 1 & & \\ & 2 & \\ & & 3 \end{bmatrix} \quad \text{and} \quad \mathbf{L}_1 = \begin{bmatrix} 1 & & \\ 2 & 1 & \\ 3 & 0 & 1 \end{bmatrix}$$

compute $\mathbf{D}^{-1}$ and $\mathbf{L}^{-1}$ by row reduction.

2.3.3 Invertible Matrix Theorem (Part 2)

Theorem 2.3.6 (Invertible Matrix Theorem: Part 2)**.** *Let* $\mathbf{A}$ *be an* $n \times n$ *matrix with real entries. The following statements are equivalent:*

6. $\mathbf{A}$ *is row equivalent to the* $n \times n$ *identity matrix.*

See Remark 1.3.1 concerning the Invertible matrix theorem and its proof.

Activity 2.3.13. Use row reduction to show that statement 6 in the theorem is true for the matrix

$$\mathbf{A} = \begin{bmatrix} 1 & 1 \\ 0 & 1 \end{bmatrix}.$$

You do not need to augment.

2.4 LU Factorization

Q 2.4.1. Given the matrix

$$\mathbf{A} = \begin{bmatrix} 1 & 2 \\ 3 & 4 \end{bmatrix}$$

what does each row operation below do to $\mathbf{A}$?

(a) $2R_1$ (b) $2R_2 + R_1$

Q 2.4.2. What is the inverse matrix of each matrix below?

$$\begin{bmatrix} 2 & 0 \\ 0 & 1 \end{bmatrix} \quad \text{and} \quad \begin{bmatrix} 1 & 2 \\ 0 & 1 \end{bmatrix}$$

There is a connection between matrix multiplication and row reduction.

Example 2.4.1. Suppose $\mathbf{A}$ is a $2 \times n$ matrix. Then each row operation can be done by multiplying on the left by a suitable matrix as indicated by the following table:

Row operation	cR_1	cR_2	$cR_1 + R_2$	$cR_2 + R_1$	$R_1 \leftrightarrow R_2$
Matrix	$\begin{bmatrix} c & 0 \\ 0 & 1 \end{bmatrix}$	$\begin{bmatrix} 1 & 0 \\ 0 & c \end{bmatrix}$	$\begin{bmatrix} 1 & 0 \\ c & 1 \end{bmatrix}$	$\begin{bmatrix} 1 & c \\ 0 & 1 \end{bmatrix}$	$\begin{bmatrix} 0 & 1 \\ 1 & 0 \end{bmatrix}$

For instance if

$$\mathbf{A} = \begin{bmatrix} 1 & 2 \\ 3 & 4 \end{bmatrix}$$

and $c = 2$, then

$$\begin{array}{cc} & \begin{bmatrix} 1 & 2 \\ 3 & 4 \end{bmatrix} \\ \begin{bmatrix} 1 & 0 \\ 0 & 2 \end{bmatrix} & \begin{bmatrix} 1 & 2 \\ 6 & 8 \end{bmatrix} \end{array} \quad \text{and} \quad \begin{array}{cc} & \begin{bmatrix} 1 & 2 \\ 3 & 4 \end{bmatrix} \\ \begin{bmatrix} 1 & 2 \\ 0 & 1 \end{bmatrix} & \begin{bmatrix} 7 & 10 \\ 3 & 4 \end{bmatrix} \end{array}.$$

Activity 2.4.2. Check that the matrices listed in the table make sense for cR_1, $cR_1 + R_2$, and $R_1 \leftrightarrow R_2$, using $c = 2$ and

$$\mathbf{A} = \begin{bmatrix} 1 & 2 \\ 3 & 4 \end{bmatrix}.$$

The matrices in the table have two noteworthy features.

1. Since $c \neq 0$ is required by row reduction, they are all invertible.
2. Aside from the matrix for $R_1 \leftrightarrow R_2$, the others are either upper triangular or lower triangular.

It should make sense that the matrices corresponding to row reduction steps are invertible because each row reduction step can be undone by another row reduction step.

Example 2.4.3. If we do cR_1 and then $\frac{1}{c}R_1$, we are back where we started. The corresponding matrices

$$\begin{bmatrix} c & 0 \\ 0 & 1 \end{bmatrix} \quad \text{and} \quad \begin{bmatrix} \frac{1}{c} & 0 \\ 0 & 1 \end{bmatrix}$$

are inverses of each other. Also, if we do $cR_1 + R_2$ and then $(-c)R_1 + R_2$, we are back where we started. The corresponding matrices

$$\begin{bmatrix} 1 & 0 \\ c & 1 \end{bmatrix} \quad \text{and} \quad \begin{bmatrix} 1 & 0 \\ -c & 1 \end{bmatrix}$$

are also inverses of each other. These facts are easy to check by doing the row reduction steps, and using Proposition 1.3.2 to compute the 2×2 inverses. But what is striking is that these are all lower triangular.

If the matrix $\mathbf{B}$ in Activity 1.3.16 is used to do row reduction, then it is doing two row reduction steps in parallel: it does $2R_1 + R_2$ and $3R_1 + R_3$ simultaneously. It is this fact that enables row reduction to be done one whole column at a time in the parallel version of the algorithm. In the parallel implementation given in this book, I used $R_i \leftrightarrow R_j$ as well, but as we have seen from the table, those are not lower triangular. In contrast, LU factorization of a matrix $\mathbf{A}$ is obtained by the following steps:

1. Use row reduction only to echelon form, $\mathbf{U}$, without ever swapping rows if possible.[2]

2. If $\mathbf{L}_i$ is the lower triangular matrix corresponding to doing row reduction on column i in parallel, then the original matrix factors as

$$\mathbf{A} = \mathbf{LU}$$

   where

$$\mathbf{L} = \mathbf{L}_1^{-1}\mathbf{L}_2^{-1} \cdots \mathbf{L}_k^{-1} = (\mathbf{L}_k \cdots \mathbf{L}_2\mathbf{L}_1)^{-1}$$

The best thing is that the inverse matrices $\mathbf{L}_i^{-1}$, and for that matter even $\mathbf{L}$ itself, can just be written down.

Warning 2.4.1. While $\mathbf{L}$ is easy to write down, this cannot usually be done with $\mathbf{L}^{-1}$. Activity 1.3.16 should be enough to convince you of this.

Example 2.4.4. Given

$$\mathbf{A} = \begin{bmatrix} 1 & 1 & 1 & 1 \\ -4 & -3 & -4 & -4 \\ -5 & -6 & -4 & -6 \end{bmatrix},$$

Compute the $\mathbf{LU}$ factorization of $\mathbf{A}$.

Solution

On one hand, in terms of row reduction we have

$$\begin{bmatrix} 1 & 1 & 1 & 1 \\ -4 & -3 & -4 & -4 \\ -5 & -6 & -4 & -6 \end{bmatrix} \xrightarrow[5R_1 + R_3]{4R_1 + R_2} \begin{bmatrix} 1 & 1 & 1 & 1 \\ 0 & 1 & 0 & 0 \\ 0 & -1 & 1 & -1 \end{bmatrix}$$
$$\xrightarrow{R_2+R_3} \begin{bmatrix} 1 & 1 & 1 & 1 \\ 0 & 1 & 0 & 0 \\ 0 & 0 & 1 & -1 \end{bmatrix} = \mathbf{U}.$$

[2] It is not always possible to get echelon form without swapping rows. Even the matrix $\left[\begin{smallmatrix} 0 & 1 \\ 1 & 0 \end{smallmatrix}\right]$ has this problem. Changing the order of the variables amounts to switching columns, which is an alternative.

On the other hand, in terms of matrix multiplication

$$\begin{bmatrix} 1 & 1 & 1 & 1 \\ -4 & -3 & -4 & -4 \\ -5 & -6 & -4 & -6 \end{bmatrix}$$

$$\begin{bmatrix} 1 & 0 & 0 \\ 4 & 1 & 0 \\ 5 & 0 & 1 \end{bmatrix} \begin{bmatrix} 1 & 1 & 1 & 1 \\ 0 & 1 & 0 & 0 \\ 0 & -1 & 1 & -1 \end{bmatrix}$$

$$\begin{bmatrix} 1 & 0 & 0 \\ 0 & 1 & 0 \\ 0 & 1 & 1 \end{bmatrix} \begin{bmatrix} 1 & 1 & 1 & 1 \\ 0 & 1 & 0 & 0 \\ 0 & 0 & 1 & -1 \end{bmatrix} = \mathbf{U}.$$

So

$$\mathbf{L}_1 = \begin{bmatrix} 1 & 0 & 0 \\ 4 & 1 & 0 \\ 5 & 0 & 1 \end{bmatrix} \text{ and } \mathbf{L}_2 = \begin{bmatrix} 1 & 0 & 0 \\ 0 & 1 & 0 \\ 0 & 1 & 1 \end{bmatrix},$$

hence

$$\mathbf{L}_1^{-1} = \begin{bmatrix} 1 & 0 & 0 \\ -4 & 1 & 0 \\ -5 & 0 & 1 \end{bmatrix}, \quad \mathbf{L}_2^{-1} = \begin{bmatrix} 1 & 0 & 0 \\ 0 & 1 & 0 \\ 0 & -1 & 1 \end{bmatrix}, \quad \text{and}$$

$$\mathbf{L}_1^{-1}\mathbf{L}_2^{-1} = \begin{bmatrix} 1 & 0 & 0 \\ -4 & 1 & 0 \\ -5 & -1 & 1 \end{bmatrix}$$

can just be written down. Then

$$\begin{bmatrix} 1 & 1 & 1 & 1 \\ 0 & 1 & 0 & 0 \\ 0 & 0 & 1 & -1 \end{bmatrix}$$

$$\begin{bmatrix} 1 & 0 & 0 \\ -4 & 1 & 0 \\ -5 & -1 & 1 \end{bmatrix} \begin{bmatrix} 1 & 1 & 1 & 1 \\ -4 & -3 & -4 & -4 \\ -5 & -6 & -4 & -6 \end{bmatrix} = \mathbf{LU} = \mathbf{A}$$

Activity 2.4.5. Given

$$\begin{bmatrix} 1 & -1 & 0 & -1 \\ -2 & 2 & 1 & 3 \\ -3 & 3 & -4 & -1 \end{bmatrix}$$

1. show that multiplying on the left by $\mathbf{L}_1$ from Activity 1.3.16 does $2R_1+R_2$ and $3R_1+R_3$ simultaneously, and

2. complete the *LU* factorization.

2.5 Algorithm Efficiency Part 2

In this section we will consider the algorithm efficiency of row reduction via LU factorization. We could have a matrix in which all row reduction steps can be done with small number

arithmetic, or we could have a matrix requiring big number arithmetic. First, suppose we can get away just with small numbers. If the original matrix has integers, it is possible to get to echelon form without ever using floats, so we focus on getting to echelon form. The changes needed to get to reduced form are relatively minor and the asymptotics are the same, however, floats may be inescapable at that point.

Sequential approach. If M is an $m \times n$ matrix, then

1. doing cR_i requires n multiplications,
2. doing $cR_i + R_j$ requires n multiplications and n additions, and
3. $R_i \leftrightarrow R_j$ can be done by making a temporary duplicate of one of those rows, say R_i, then copying R_j into R_i, then copying the duplicate of R_j into R_i (a total of $3n$ copies of a small number).

So far, this gives $O(n)$ for each row operation in time. To get echelon form, we work left to right one column at a time. Suppose that we already have pivots in the first k rows, so that we have already finished with at least k columns. all of which are now to the left. We go down the next column, and have $m - k$ numbers to check below row k. If we find a non-zero value in row i, we choose it, then we do cR_i to get a 1 in that spot, and $cR_j + R_i$ to get zeros in the remaining spots in that column. If a pivot was found, we do $R_i \leftrightarrow R_{k+1}$, unless $i = k + 1$. As we loop through the elements of the column, we will have to do no more than m row operations, each of which is $O(n)$, so we get $O(mn)$ for the entire column. Since there are n columns, then the total is $O(mn^2)$ in time.

How much space does this take? We need to store the matrix, consisting of mn small numbers. We may need a temporary copy of a row, which takes another n. The rest of the information that must be saved does not grow with m or n. So in space, the largest thing is the matrix itself, which gives us $O(mn)$.

Parallel approach. There seems to be two natural parallel algorithms for finding a pivot: a binary search or by using accumulate. As far as I can tell, the second of these is asymptotically faster in parallel, so that is the approach used in the code at the end of this section section. Suppose we have the matrix

$$\mathbf{A} = \begin{bmatrix} 2 & -2 & 0 & 8 & 0 & -4 \\ 0 & 3 & 0 & 0 & 0 & 6 \\ 0 & 0 & 0 & 0 & 5 & 10 \\ 0 & 0 & 0 & 4 & 0 & 8 \\ 0 & 0 & 0 & 4 & 8 & 0 \\ 0 & 0 & 0 & 0 & 0 & 0 \end{bmatrix}.$$

The first two rows already have echelon form. To find a pivot below row 2, we compare the last four rows to zero, which can be done in parallel

$$0 = \begin{bmatrix} 0 & 0 & 0 & 0 & 5 & 10 \\ 0 & 0 & 0 & 4 & 0 & 8 \\ 0 & 0 & 0 & 4 & 8 & 0 \\ 0 & 0 & 0 & 0 & 0 & 0 \end{bmatrix} \longrightarrow \begin{bmatrix} 0 & 0 & 0 & 0 & 1 & 1 \\ 0 & 0 & 0 & 1 & 0 & 1 \\ 0 & 0 & 0 & 1 & 1 & 0 \\ 0 & 0 & 0 & 0 & 0 & 0 \end{bmatrix}.$$

This is a boolean array so now we accumulate with logical_or in the rows:

$$\begin{bmatrix} 0 & 0 & 0 & 0 & 1 & 1 \\ 0 & 0 & 0 & 1 & 1 & 1 \\ 0 & 0 & 0 & 1 & 1 & 1 \\ 0 & 0 & 0 & 0 & 0 & 0 \end{bmatrix}.$$

Then we treat this as a matrix of integers and compute the sum in the rows. If the rows have length n, the sum can be computed in $O(\log(n))$ time using the method explained in Example 1.6.4 in Section 1.6 (adding the first half to the last half repeatedly). In this case, the result is

$$\begin{bmatrix} 2 & 3 & 3 & 0 \end{bmatrix}$$

The last entry is zero because the last row of $\mathbf{A}$ has not pivots. The maximum is 3, which tells us which column to use next. If the columns have length m, then the maximum can also be computed in $O(\log(m))$ using the same method as we did for the sum. By comparing this vector to its maximum, we get a boolean array again:

$$3 = \begin{bmatrix} 2 & 3 & 3 & 0 \end{bmatrix} \longrightarrow \begin{bmatrix} 0 & 1 & 1 & 0 \end{bmatrix}.$$

Finally, accumulating with logical_or, which takes $O(\log(m))$ in time, gives us a bit mask that can be used to rotate the rows.[3] The bit mask is

$$\begin{bmatrix} 0 & 1 & 1 & 1 \end{bmatrix}$$

and the result of applying the rotation to the rows of $\mathbf{A}$ after the first two is

$$\begin{bmatrix} 0 & 0 & 0 & 4 & 0 & 8 \\ 0 & 0 & 0 & 4 & 8 & 0 \\ 0 & 0 & 0 & 0 & 0 & 0 \\ 0 & 0 & 0 & 0 & 5 & 10 \end{bmatrix}. \tag{2.29}$$

The pivot we want is in the top row of the matrix in (2.29), and the column is known by the maximum that was computed. By LU factorization (covered in Section 2.4), we can do the row operations one column at a time. We can do all of the multiplication involved in cR_i and $cR_i + R_j$ in parallel for the pivot column, and all of the addition can be done in parallel after the multiplication is complete. So, those operations do not contribute to the time asymptotics at all. Since accumulate was used in both rows and columns, each step can be done in $O(\log(m + n))$, and the number of steps is equal to the number of pivots, which cannot be bigger than the number of rows or columns. We have justified the time listed in the second row of Figure 2.5. The parallel approach does take more space than the sequential approach, but the total size is still a multiple of mn.

Unfortunately, the analysis above is a little naive because if we are not careful, we may run out of accuracy due to rounding error. We will not consider numerical issues in this section, rather we will deal with the problem by allowing the number of bits used to increase. This is why we are bringing big numbers into the picture.

For big floats, if we start out with N bits, then the growth in the number of bits is proportional to $N + m$. The number of multiplications and additions is the same, so for a matrix with big numbers, it is the multiplications that dominate. This introduces an extra factor coming from the results for multiplication listed in Figure 1.6

For big ints, if the entries in the original matrix take N bits, then after m steps, it is possible that roughly Nm bits are required. Additionally, it should be noted that the number of sequential steps must increase due to the Euclidean algorithm (the idea behind Example 2.3.7), causing an extra factor of $\log(Nm)$.

---

[3]I make no claim that using the bit mask directly is the best approach. Instead, you could compute the sum to get the amount to rotate by and then use that value in some other way. The code at the end of this section is a python adaptation of some code that I first wrote in APL. The bit mask is used to mimic APL's rotate function. The rotation amounts to doing several row operations $R_i \leftrightarrow R_j$ at the same time. Instead of this, we could save all of the operations $R_i \leftrightarrow R_j$ until the very end, and just sort the rows. Sorting is fast, but one advantage to doing rotations is that any pivot that is found is moved to a predictable row.

Hence we obtain the results in the following tables:

Sequential

Type	Space	Time
small nums	$O(mn)$	$O(mn^2)$
big floats	$O(mn(N+m))$	$O(mn^2(N+m)\log(N+m))$
big ints	$O(m^2nN)$	$O(m^2n^2N\log(Nm)^2)$

Parallel

Type	Space	Time
small nums	$O(mn)$	$O(\min(m,n)\log(m+n))$
big floats	$O(mn(N+m))$	$O(\min(m,n)\log(m+n)(N+m))$
big ints	$O(m^2nN)$	$O(\min(m,n)\log(m+n)\log(Nm)Nm)$

FIGURE 2.5: Asymptotic efficiency of row reduction, for an $m \times n$ matrix with N bits required in the original matrix

Remark 2.5.1. There are several details worth pointing out.

1. When people refer to row reduction as "cubic in time," they are referring to the sequential algorithm with small nums applied to square matrices.

2. For large N and m, row reduction with big integers using the euclidean algorithm within row reduction is much slower than with big floats. I would only recommend using the algorithm with integers if an answer with integers is required, or if it is known that the size of the integers remains bounded.

3. While the parallel approach should be preferred, once the total space needed exceeds the hardware limits, then the sequential efficiency takes over again.

We may even face the task of doing row reduction to several different matrices, all of the same size. Then we get a factor coming from the number of matrices. In the sequential algorithm that factor affects both the efficiency in space and time, but in the parallel algorithm it only affects the space.

```
import numpy as np

def div0(A,B):
    C=np.zeros_like(A,dtype=float)
    return np.divide(A,B,out=C,where=B!=0)

def rotate(A,M):
    B=np.empty_like(A)
    B[M[...,::-1]]=A[M]
    M=~M
    B[M[...,::-1]]=A[M]
    return B

def scanor(A,axis=-1):
    return np.logical_or.accumulate(A,axis=axis)

def RotateMask(A):
    B = np.sum(scanor(0!=A).astype(int),axis=-1)
    C = np.max(B,axis=-1)
```

```
    C = np.repeat(C,np.shape(A)[-2]).reshape(np.shape(B))
    return scanor(C == B)

def RotateMask2(A):
    M=np.zeros_like(A,dtype=bool)
    M[...,0,:]=scanor(A[...,0,:]!=0)
    return scanor(M,axis=-2)

def Rotate2(A): return rotate(A,RotateMask2(A))

def rr0(A):
    S = np.shape(A)
    B = np.repeat(A[...,0,0],S[-2]).reshape(S[:-1])
    B = div0(A[...,0],B)
    B[...,0] = 0
    return B

def rr1(A): return rr0(Rotate2(A))

def rr2(A):
    S = np.shape(A)
    B = np.repeat(rr1(A),S[-1]).reshape(S)
    C = np.tile(A[...,0,:],S[-2]).reshape(S)
    return A-B*C

def rr3(A): return rr2(rotate(A,RotateMask(A)))

def rr4(i,A):
    B = np.empty_like(A,dtype=float)
    B[...,:i,:] = A[...,:i,:]
    B[...,i:,:] = rr3(A[...,i:,:])
    return B

def ParallelEchelon(A):
    m,n = np.shape(A)[-2:]
    B = A
    for i in range(min(m,n)):
        B = rr4(i,B)
    return B
```

Exercises

Problem 2.1. True or False?

a. Every elementary row operation is reversible.

b. Two equivalent linear systems can have different solution sets.

c. If a linear system has more equations than variables, then it is inconsistent.

d. A linear system with more variables than equations can be inconsistent.

Problem 2.2. Solve the system of equations graphically.

a $\begin{cases} 2x + y = 4 \\ 2x - y = 0 \end{cases}$ b $\begin{cases} 2x + y = 4 \\ 2x + y = 0 \end{cases}$ c $\begin{cases} 2x + y = 4 \\ -2x - y = -4 \end{cases}$

Problem 2.3. You have a choice of renting a hybrid vehicle at $40 for a day or a regular car at $34 for a day. The regular car gets 30 mi/Gal on the highway, and the hybrid gets 36 mi/Gal on the highway. Assuming the cost of gas is $3.60 per gallon, find

1. equations representing the amount spent on each car in terms of the driving distance, and

2. the distance you would have to drive for the cost to be the same regardless of which car you pick.

Problem 2.4. Usain Bolt and I are competing in the 100m dash, except since I'm no Olympian, Bolt kindly gives me a 30m head start. If Bolt runs at 10m/s and I run at 6m/s, find

1. equations representing the distance x from the finish line after t seconds for both Bolt and myself, and

2. find the distance we will both be from the finish line when Bolt passes me.

Problem 2.5. Determine all values of h such that

$$\begin{aligned} x + hy &= -5 \\ 2x - 8y &= 6 \end{aligned}$$

is a consistent linear system.

Problem 2.6. Choose h and k such that the system has

1. no solution

2. a unique solution

3. infinitely many solutions.

Give a separate answer for each part.

$$\begin{aligned} x - 3y &= 1 \\ 2x + hy &= k \end{aligned}$$

Note: you are not really asked for all solutions h and k, so if it is easier, you may list one pair h and k for each case.

Problem 2.7. Convert each of the following linear systems to an augmented matrix and solve the system by row operations, then answer the following questions.

1. How many solutions are there?

2. What is the dimension of the solution set?

3. What are the free variables (if any)?

4. What are the pivot positions?

$$\text{a}\begin{cases} x + 2y + z = -2 \\ x + 4y - z = 0 \\ 4y - 3z = 3 \end{cases} \qquad \text{b}\begin{cases} x + 2z = -1 \\ 2x + 3y + z = 1 \\ x + 3y = 1 \end{cases} \qquad \text{c}\begin{cases} x + 2y - 6z = 5 \\ -x + y - 2z = 1 \\ y - 3z = 2 \end{cases}$$

$$\text{d}\begin{cases} x - y + 2z = -1 \\ x + z = 1 \\ 2x + y + z = 4 \end{cases} \qquad \text{e}\begin{cases} x + z = 1 \\ x - y + 3z = 0 \\ -x + 2y - 5z = 1 \end{cases} \qquad \text{f}\begin{cases} x + y - 2z = 1 \\ -y + 3z = 1 \\ x - y + 4z = 3 \end{cases}$$

Problem 2.8. Show that each part of theorem 2.3.6 is false for the matrix

$$\mathbf{A} = \begin{bmatrix} 1 & 1 \\ 0 & 0 \end{bmatrix}.$$

3

Linear Independence and Determinants

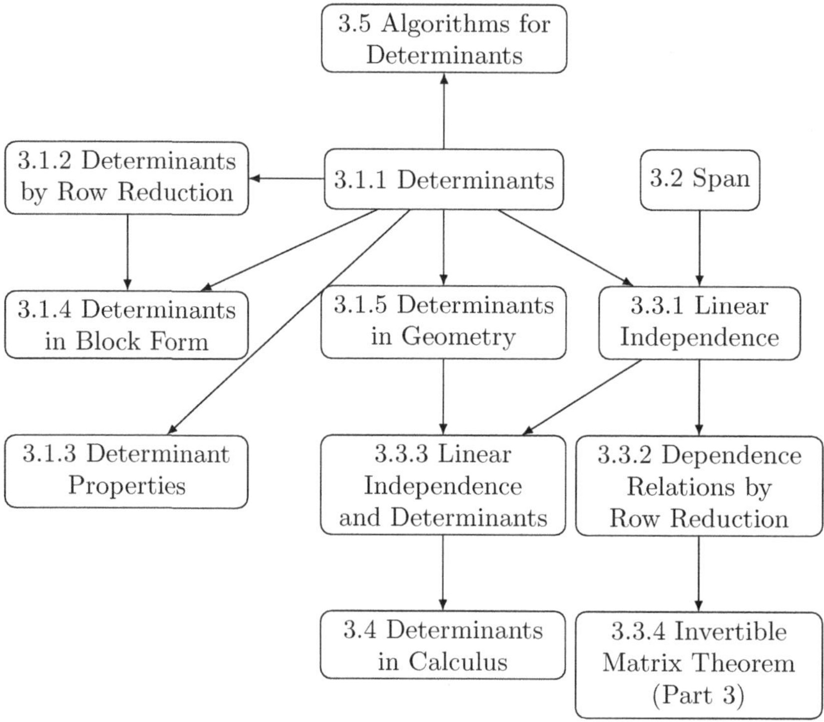

DOI: 10.1201/9781003737490-3

3.1 Determinants

TL;DR

Example 3.1.9 illustrates the connection between determinants and areas. The most practical approaches to compute determinants by hand are given in:

- Definition 3.1.1 (the 1×1 and 2×2 case),
- Proposition 3.1.4 (row reduction),
- Proposition 3.1.3 (triangular form), and
- Theorem 3.1.6 (block triangular form).

The algebraic properties in Theorem 3.1.5 are important for theoretical purposes, but can also sometimes lead to shortcuts in calculations.

3.1.1 Intro to Determinants

Q 3.1.1. For what value of x does x^{-1} not exist?

Q 3.1.2. What is the 2×2 formula for $\mathbf{A}^{-1}$?

Q 3.1.3. If $\mathbf{A} = \left[\begin{smallmatrix}1 & 1\\ 2 & 2\end{smallmatrix}\right]$, why doesn't $\mathbf{A}^{-1}$ exist?

We have already seen a hint of determinants. In the shortcut formula for the inverse of a 2×2 matrix, we are not able to compute a result if the denominator is zero. That denominator is the determinant. A 1×1 matrix $\mathbf{A}$ is essentially just a number, which has an inverse if and only if it is not zero.

Definition 3.1.1 (Determinant of a 2×2 or 1×1 matrix).

$$\text{If} \quad \mathbf{A} = \begin{bmatrix} a & b \\ c & d \end{bmatrix}, \quad \text{then} \quad \det(\mathbf{A}) = ad - bc.$$

$$\text{If} \quad \mathbf{A} = \begin{bmatrix} a \end{bmatrix}, \quad \text{then} \quad \det(\mathbf{A}) = a.$$

For larger square matrices, we take the following recursive approach.[1]

Definition 3.1.2 (Expansion by cofactors (or minors)). Let $\mathbf{A}$ be an $n \times n$ matrix, where $n \geq 2$, and let a_{ij} be the element of the i-th row and j-th column.[2] Let $\mathbf{M}_{ij}$ be the matrix formed by omitting the i-th row and j-th column from $\mathbf{A}$. Then the **determinant** of $\mathbf{A}$ is given by

$$\begin{aligned} \det(\mathbf{A}) &= \sum_{i=1}^{n} (-1)^{i+j} a_{ij} \det(\mathbf{M}_{ij}) \quad \text{(if the sum is along a chosen column } j\text{)} \\ &= \sum_{j=1}^{n} (-1)^{i+j} a_{ij} \det(\mathbf{M}_{ij}) \quad \text{(if the sum is along a chosen row } i\text{)} \end{aligned} \tag{3.1}$$

The determinants $\det(\mathbf{M}_{ij})$ are called **minors** of the matrix $\mathbf{A}$, and the numbers $(-1)^{i+j}\det(\mathbf{M}_{ij})$ are called **cofactors**.

[1] It is also possible to define the determinant using permutations, without a recursion.

[2] This recursive definition can be generalized to $n \geq 1$ if we define the determinant of a 0×0 matrix to be 1.

Remark 3.1.1. $\det(\mathbf{A})$ is only defined for square matrices.

Equation (3.1) allows the determinant of an $n \times n$ matrix to be defined in terms of the determinant of an $(n-1) \times (n-1)$ matrix. By reusing the formula to get to smaller and smaller matrices, eventually we will get to small enough matrices that the result can be computed directly. The values of $(-1)^{i+j}$ follow an alternating pattern:

$$\begin{bmatrix} + & - & + & - & \cdots \\ - & + & - & + & \cdots \\ + & - & + & - & \cdots \\ - & + & - & + & \cdots \\ \vdots & \vdots & \vdots & \vdots & \end{bmatrix}.$$

Example 3.1.1. Compute the determinant of

$$\mathbf{A} = \begin{bmatrix} 0 & 2 & 0 & -1 \\ 0 & 3 & 0 & 2 \\ 1 & 0 & 1 & 1 \\ 1 & 0 & 2 & 5 \end{bmatrix}$$

using expansion by co-factors.

Solution

If you use expansion by co-factors at all, you should pick a row or column with a lot of zeros. Going down the 2nd column, we get:

$$\det(\mathbf{A}) = \begin{vmatrix} 0 & 2 & 0 & -1 \\ 0 & 3 & 0 & 2 \\ 1 & 0 & 1 & 1 \\ 1 & 0 & 2 & 5 \end{vmatrix} = (-2)\begin{vmatrix} 0 & 0 & 2 \\ 1 & 1 & 1 \\ 1 & 2 & 5 \end{vmatrix} + 3\begin{vmatrix} 0 & 0 & -1 \\ 1 & 1 & 1 \\ 1 & 2 & 5 \end{vmatrix}.$$

Now we have two smaller determinants to compute, each of which has two zeros in the top row. So, we apply expansion by co-factors along the top row for each of them, then finish off with the 2×2 formula:

$$\begin{aligned} (-2)\begin{vmatrix} 0 & 0 & 2 \\ 1 & 1 & 1 \\ 1 & 2 & 5 \end{vmatrix} + 3\begin{vmatrix} 0 & 0 & -1 \\ 1 & 1 & 1 \\ 1 & 2 & 5 \end{vmatrix} &= (-2)(2)\begin{vmatrix} 1 & 1 \\ 1 & 2 \end{vmatrix} + (3)(-1)\begin{vmatrix} 1 & 1 \\ 1 & 2 \end{vmatrix} \\ &= -7\begin{vmatrix} 1 & 1 \\ 1 & 2 \end{vmatrix} = -7(2-1) = -7. \end{aligned}$$

Activity 3.1.2. Apply expansion by cofactors to the 3rd row of the matrix

$$\mathbf{A} = \begin{bmatrix} 1 & 0 & 3 \\ 0 & 1 & 1 \\ 2 & -1 & 0 \end{bmatrix}$$

It is shown in the next subsection that expansion by cofactors is computationally inefficient, but it remains useful for theoretical purposes. For example, it may be used to prove the following result.

Proposition 3.1.3. *Let* $\mathbf{A}$ *be an* $n \times n$ *matrix in upper triangular form. Then* $\det A$ *is the product of the diagonal.*

Activity 3.1.3 generalizes to a proof by induction. For a different proof, see [2].

Activity 3.1.3. Given

$$\mathbf{A} = \begin{bmatrix} 1 & 4 & 6 \\ 0 & 2 & 5 \\ 0 & 0 & 3 \end{bmatrix},$$

compute $\det(\mathbf{A})$ by systematically applying expansion by cofactors down the first column repeatedly. Hint: zero times anything is zero.

With Sage, we can get the determinant of a matrix as follows:

```
A=matrix([[ 1, 0, 3],
          [ 0, 1, 1],
          [ 2,-1, 0]])
A.det()
```

AI Prompt 3.1.1. Can you write a function in python to compute the determinant of an upper triangular matrix?

NumPy has a built-in function that called np.prod. If the code you get does not use this function, try asking for it directly:

AI Prompt 3.1.2. Can you write a different implementation using np.prod from NumPy?

3.1.2 Determinants by Row Reduction

Each row operation changes the determinant by a non-zero factor:

1. Adding a multiple of a row to another row leaves the determinant unchanged.
2. Reversing two rows changes the determinant by a factor of -1.
3. If a row is multiplied by a non-constant c, then a factor of c must be removed from the final result.

   For example:
   $$\begin{vmatrix} 3 & 0 \\ 0 & 1 \end{vmatrix} = 3 \quad \text{but after } \frac{1}{3}R_1 \text{ we have} \quad \begin{vmatrix} 1 & 0 \\ 0 & 1 \end{vmatrix} = 1,$$
   so the factor $\frac{1}{3}$ must be removed from the final determinant by dividing by it. Dividing 1 by $\frac{1}{3}$ then gives us the correct answer of 3.

By Proposition 3.1.3, we know that the determinant of a matrix in upper triangular form is the product of the diagonal. In other words, we can stop row reduction when we reach echelon form. It is always possible to obtain echelon form by using just the first two types of row operation. So, if we follow that strategy, then the determinant will only change by a factor of ± 1. Moreover, we can keep track of the sign by counting the number of times we reverse rows.

Proposition 3.1.4. *Let* $\mathbf{A}$ *be a square matrix. If* $\mathbf{B}$ *can be obtained from* $\mathbf{A}$ *using only the first two types of row operations, then:*

$$\det \mathbf{A} = \begin{cases} \det \mathbf{B} & \textit{if two rows have been reversed an even number of times,} \\ -\det \mathbf{B} & \textit{if two rows have been reversed an odd number of times.} \end{cases}$$

Example 3.1.4. Compute the determinant of $\mathbf{A}$ from Example 3.1.1 using row reduction.

Solution

$$\begin{vmatrix} 0 & 2 & 0 & -1 \\ 0 & 3 & 0 & 2 \\ 1 & 0 & 1 & 1 \\ 1 & 0 & 2 & 5 \end{vmatrix} \xrightarrow{(-1)R_3+R_4} \begin{vmatrix} 0 & 2 & 0 & -1 \\ 0 & 3 & 0 & 2 \\ 1 & 0 & 1 & 1 \\ 0 & 0 & 1 & 4 \end{vmatrix} \xrightarrow{(-1)R_1+R_2} \begin{vmatrix} 0 & 2 & 0 & -1 \\ 0 & 1 & 0 & 3 \\ 1 & 0 & 1 & 1 \\ 0 & 0 & 1 & 4 \end{vmatrix}$$

$$\xrightarrow{(-2)R_2+R_1} \begin{vmatrix} 0 & 0 & 0 & -7 \\ 0 & 1 & 0 & 3 \\ 1 & 0 & 1 & 1 \\ 0 & 0 & 1 & 4 \end{vmatrix} \xrightarrow{(-1)R_4+R_3} \begin{vmatrix} 0 & 0 & 0 & -7 \\ 0 & 1 & 0 & 3 \\ 1 & 0 & 0 & -3 \\ 0 & 0 & 1 & 4 \end{vmatrix}$$

$$\xrightarrow{R_1 \leftrightarrow R_3} \begin{vmatrix} 1 & 0 & 0 & -3 \\ 0 & 1 & 0 & 3 \\ 0 & 0 & 0 & -7 \\ 0 & 0 & 1 & 4 \end{vmatrix} \xrightarrow{R_3 \leftrightarrow R_4} \begin{vmatrix} 1 & 0 & 0 & -3 \\ 0 & 1 & 0 & 3 \\ 0 & 0 & 1 & 4 \\ 0 & 0 & 0 & -7 \end{vmatrix} = (1)(1)(1)(-7) = -7$$

Since we reversed rows only in the last two row reduction steps, i.e. an even number of times, then the original determinant is also -7.

Activity 3.1.5. Apply row reduction to compute the determinant of

$$\mathbf{A} = \begin{bmatrix} 1 & 0 & 3 \\ 0 & 1 & 1 \\ 2 & -1 & 0 \end{bmatrix}.$$

AI Prompt 3.1.3. Can you write a function in python to compute the determinant of an upper triangular matrix?

3.1.3 Properties of the Determinant

Theorem 3.1.5. *Let* $\mathbf{A}$ *and* $\mathbf{B}$ *be* $n \times n$ *matrices, and let* c *be a scalar. Then*

1. $\det(\mathbf{A}) \neq 0$ *if and only if* $\mathbf{A}$ *is invertible.*

2. $\det(\mathbf{AB}) = \det(\mathbf{A})\det(\mathbf{B})$.

3. *If* $\mathbf{A}$ *is invertible, then* $\det(\mathbf{A})\det(\mathbf{A}^{-1}) = 1$.

4. $\det(\mathbf{A}^T) = \det(\mathbf{A})$.

5. $\det(c\mathbf{A}) = c^n \det A$.

Most of this theorem is proved in [2].

Note that $c\mathbf{A}$ multiplies all rows by the constant c, unlike in row reduction where just one row gets multiplied by c (i.e. cR_1 means that row 1 gets multiplied by c).

Remark 3.1.2. The transpose switches the rows and columns of $\mathbf{A}$. As a consequence, it is possible to use both row reduction and column reduction to compute determinants. If an entire row (or column) of zeros is obtained, then the determinant is zero. [3]

[3] When I was a student, my linear algebra professor gave us all individualized 7×7 determinants to solve with the rule that we were not to use row reduction. I tried using expansion by cofactors. It took me 6

Activity 3.1.6. Verify the properties above given

$$\mathbf{A} = \begin{bmatrix} 1 & 1 \\ 0 & 1 \end{bmatrix}, \quad \mathbf{B} = \begin{bmatrix} 1 & 1 \\ 0 & 0 \end{bmatrix}, \quad \text{and} \quad c = 3.$$

Check the first property with both $\mathbf{A}$ and $\mathbf{B}$. Use

$$\begin{vmatrix} a & b \\ c & d \end{vmatrix} = ad - bc.$$

3.1.4 Determinants of Block Matrices

In Section 1.5, we were introduced to block form. Unfortunately matrices in block form do not always do what you might expect with determinants. From expansion by cofactors, we can show the following:

Theorem 3.1.6. *Suppose* $\mathbf{A}$ *is a square matrix divided into blocks* $\mathbf{A}_{ij}$ *with* i *and* j *indexing rows and columns in block form. If*

1. $\mathbf{A}_{ii}$ *is a square matrix for all* i, *and*
2. $\mathbf{A}_{ij} = \mathbf{0}$ *for all* $i > j$

then $\mathbf{A}$ *is in* **block upper triangular form** *and* $\det(\mathbf{A})$ *is the product of the determinants of the blocks on the diagonal.*

This theorem is analogous to Proposition 3.1.3 except it applies to matrices in block form.

Activity 3.1.7. Suppose we have

$$\mathbf{A} = \begin{bmatrix} \mathbf{A}_{11} & \mathbf{A}_{12} & \mathbf{A}_{13} \\ \mathbf{A}_{21} & \mathbf{A}_{22} & \mathbf{A}_{23} \\ \mathbf{A}_{31} & \mathbf{A}_{32} & \mathbf{A}_{33} \end{bmatrix} = \left[\begin{array}{cc|c|c} 0 & -1 & 0 & 0 \\ 1 & 0 & 0 & 0 \\ \hline 0 & 0 & 1 & 0 \\ \hline 0 & 0 & 0 & -1 \end{array}\right].$$

1. Compute $\det(\mathbf{A})$ by row reduction, expansion by cofactors, or Sage.
2. Compute $\det(\mathbf{A}_{11})$, $\det(\mathbf{A}_{22})$, and $\det(\mathbf{A}_{33})$ by Definition 3.1.1, then multiply them.

The results should be equal.

On the other hand, given a matrix

$$\mathbf{M} = \begin{bmatrix} \mathbf{A} & \mathbf{B} \\ \mathbf{C} & \mathbf{D} \end{bmatrix},$$

it is not necessarily true that

$$\det(\mathbf{M}) = \det(\mathbf{A})\det(\mathbf{D}) - \det(\mathbf{B})\det(\mathbf{C}).$$

---

pages. I checked with the computer. My answer was wrong. I looked for sign errors and found several. I checked with the computer again. My answer was still wrong. At that point, I decided to use expansion by cofactors together with column reduction, which I knew about from Moore's book. I got the correct answer in only a page, and submitted it with a note pointing out that the use of column reduction had not been ruled out. Thankfully I got full credit.

Activity 3.1.8. Suppose that

$$\mathbf{M} = \begin{bmatrix} \mathbf{A} & \mathbf{B} \\ \mathbf{C} & \mathbf{D} \end{bmatrix} = \left[\begin{array}{cc|cc} 1 & 0 & 0 & 0 \\ 0 & 0 & 1 & 0 \\ \hline 0 & 0 & 0 & -1 \\ 0 & -1 & 0 & 0 \end{array}\right].$$

1. Compute $\det(\mathbf{M})$ by by row reduction, expansion by cofactors, or Sage.
2. Compute $\det(\mathbf{A})$, $\det(\mathbf{B})$, $\det(\mathbf{C})$, and $\det(\mathbf{D})$ from Definition 3.1.1, then compute

$$\det(\mathbf{A})\det(\mathbf{D}) - \det(\mathbf{B})\det(\mathbf{C}).$$

The results should not be the same.

Remark 3.1.3. The 4×4 matrices in Activities 3.1.7 and 3.1.8 differ only by rotating the rows, so only their sign can be different.

3.1.5 Determinants in Geometry

Q 3.1.4. What is the area formula for a rectangle?

Q 3.1.5. What is the determinant of $\mathbf{A} = \left[\begin{smallmatrix} 0 & 2 \\ 3 & 1 \end{smallmatrix}\right]$?

Example 3.1.9. Generally two vectors $\mathbf{u}$ and $\mathbf{v}$ in $\mathbb{R}^2$ describe a parallelogram. On one hand, it is possible to compute the area by the usual formula $A = bh$, where b is the base and h is the height of the parallelogram. Therefore, to make the comparison, we take one of the vectors to be along the x-axis. Consider

$$\mathbf{u} = \begin{bmatrix} 3 \\ 0 \end{bmatrix}, \quad \text{and} \quad \mathbf{v} = \begin{bmatrix} 2 \\ 4 \end{bmatrix}.$$

Since $\mathbf{u}$ is along the x-axis, then the base is just the length of $\mathbf{u}$. Meanwhile, the height is the y-coordinate of $\mathbf{v}$. So $b = 3$ and $h = 4$ with $\mathbf{u}$ and $\mathbf{v}$ as given above. From the base times height formula

$$A = 3 \times 4 = 12.$$

On the other hand, if we build $\mathbf{u}$ and $\mathbf{v}$ into the columns of a square matrix and take the determinant, we find

$$\det([\mathbf{u}|\mathbf{v}]) = \begin{vmatrix} 3 & 2 \\ 0 & 4 \end{vmatrix} = 3 \cdot 4 - 2 \cdot 0 = 12.$$

Since the sign is already positive, there is no need to take the absolute value. But switching rows (or columns) can change the sign (see Proposition 3.1.4).

Activity 3.1.10. The edges of a parallelogram are given by

$$\mathbf{u} = \begin{bmatrix} 2 \\ 0 \end{bmatrix} \quad \text{and} \quad \mathbf{v} = \begin{bmatrix} 1 \\ 3 \end{bmatrix}$$

1. Compute the area of the parallelogram by the usual base times height formula.
2. Form the matrix $\mathbf{A}$ with $\mathbf{u}$ and $\mathbf{v}$ as the columns. Reverse the columns of $\mathbf{A}$ to get $\mathbf{B}$. Then compute

$$|\det(\mathbf{A})| \quad \text{and} \quad |\det(\mathbf{B})|.$$

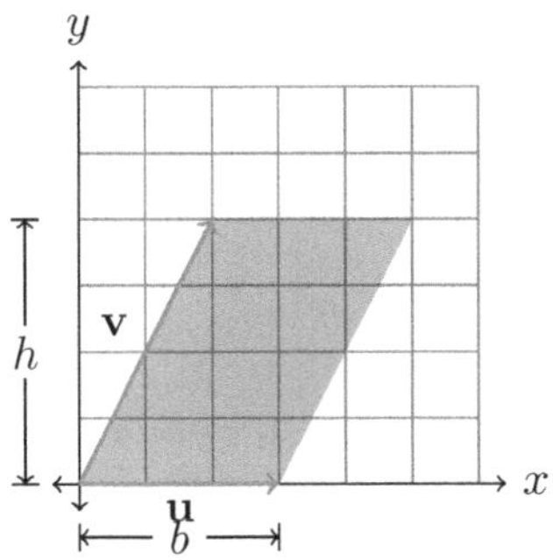

FIGURE 3.1: The parallelogram formed by the vectors u, v

The absolute value here is unavoidable because changing the order of rows or columns changes the sign of the determinant.

Remark 3.1.4. Imagine flattening the parallelogram in Figure 3.1 by moving only the vector $\mathbf{v}$. Then the area becomes zero and the vectors $\mathbf{u}$ and $\mathbf{v}$ will become co-linear: they will both lie on the x-axis. Thus, as a consequence of this application, determinants can be used to detect co-linearity.

Activity 3.1.11. Compute $ad - bc$ for the coefficient matrix of the system of equations

$$\begin{aligned} x + 2y &= 0 \\ 2x + 4y &= 0. \end{aligned}$$

Does this change if the first equation is replaced with $x + 2y = 1$?

3.2 Span

TL;DR

Linear combinations are defined in Definition 3.2.1, and needed for span. Span is defined in Definition 3.2.2 and needed for a basis.

Q 3.2.1. What is $3\begin{bmatrix} 1 \\ 1 \end{bmatrix} + 2\begin{bmatrix} 1 \\ -1 \end{bmatrix}$?

Q 3.2.2. Can you find x and y such that $x\begin{bmatrix} 1 \\ 1 \\ 0 \end{bmatrix} + y\begin{bmatrix} 1 \\ -1 \\ 0 \end{bmatrix} = \begin{bmatrix} 0 \\ 0 \\ 1 \end{bmatrix}$?

Definition 3.2.1. Given a finite number of vectors $\mathbf{v}_1, \mathbf{v}_2, \dots, \mathbf{v}_n$ in $\mathbb{R}^m$ and scalars $c_1, c_2, \dots c_n$, the vector

$$\mathbf{v} = c_1\mathbf{v}_1 + c_2\mathbf{v}_2 + \cdots + c_k\mathbf{v}_n$$

is called a **linear combination** of $\mathbf{v}_1, \mathbf{v}_2, \dots, \mathbf{v}_n$ with weights $c_1, c_2, \dots c_n$.

I am using m and n here because we are used to column vectors: if the vectors $\mathbf{v}_i$ are in $\mathbb{R}^m$, then they have m components, which is the number of rows.

As we have seen, a vector equation does not necessarily yield a solution. The fact that a solution exists in the case of

$$x\begin{bmatrix} 1 \\ 1 \end{bmatrix} + y\begin{bmatrix} 1 \\ -1 \end{bmatrix} = \begin{bmatrix} 5 \\ 1 \end{bmatrix}$$

means that the vector on the right is a linear combination of the vectors on the left-hand side, with the weights being the calculated values of x and y. This motivates the following definition.

Definition 3.2.2. If $\mathbf{v}_1, \ldots, \mathbf{v}_n$ are in $\mathbb{R}^m$, then the set of all linear combinations

$$c_1\mathbf{v}_1 + \cdots + c_n\mathbf{v}_n \tag{3.2}$$

is denoted by $\text{Span}\{\mathbf{v}_1 \ldots \mathbf{v}_n\}$ and is called the **span** of the vectors $\mathbf{v}_1, \ldots, \mathbf{v}_n$.

Here are some important details that you should get out of the definition of span:

1. In the language of this definition, we can now say a vector equation,

$$x_1\mathbf{a}_1 + x_2\mathbf{a}_2 + \cdots + x_n\mathbf{a}_n = \mathbf{b},$$

has a solution if and only if the vector $\mathbf{b}$ lies in the span of the vectors $\mathbf{a}_1, \ldots, \mathbf{a}_n$. As a consequence, the natural way to determine whether a vector lies in the span is to solve the corresponding system of equations. This might be done by row reduction or the SVD.

2. The origin (zero vector) is always in the span because we can get it by taking

$$c_1 = c_2 = \cdots = c_n = 0.$$

3. Equation (3.2) gives us a **parameterization** of the span with **parameters** $c_1, c_2, \ldots c_n$ that is analogous to the parameterization of a solution set as defined by equation (2.22). However, there are two differences:

   (a) If $\mathbf{P} \neq \mathbf{0}$ in equation (2.22), it is possible that $\mathbf{0}$ would not be in the solution set, whereas $\mathbf{0}$ is always in the span.

   (b) The dimension of the span is not necessarily equal to the number of vectors used in equation (3.2). We will study the issue of dimension in more detail in the next section.

   To be clear, when we speak of a parameterization of the span, we will mean writing down equation (3.2) with the given vectors as demonstrated in Example 3.2.1.

Example 3.2.1. Given the vectors

$$\mathbf{v}_1 = \begin{bmatrix} 1 \\ 0 \\ 0 \end{bmatrix}, \quad \mathbf{v}_2 = \begin{bmatrix} 1 \\ 1 \\ 0 \end{bmatrix}, \quad \mathbf{v}_3 = \begin{bmatrix} 5 \\ 2 \\ 0 \end{bmatrix}, \quad \text{and} \quad \mathbf{v}_4 = \begin{bmatrix} 1 \\ 1 \\ 1 \end{bmatrix}$$

do the following:

1. Write down a parameterization for the span of $\mathbf{v}_1$ and $\mathbf{v}_2$.

2. Show that $\mathbf{v}_3$ is in $\text{Span}\{\mathbf{v}_1, \mathbf{v}_2\}$, but $\mathbf{v}_4$ is not.

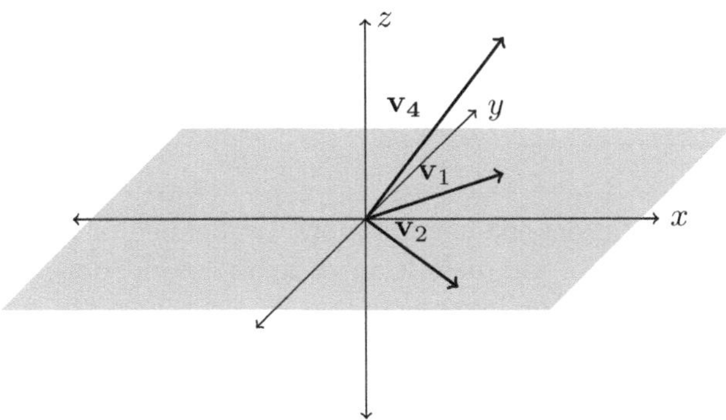

FIGURE 3.2: The span of $\mathbf{v}_1$ and $\mathbf{v}_2$ is the xy-plane

Solution

$$c_1 \begin{bmatrix} 1 \\ 0 \\ 0 \end{bmatrix} + c_2 \begin{bmatrix} 1 \\ 1 \\ 0 \end{bmatrix}$$

is a parameterization for the span of $\mathbf{v}_1$ and $\mathbf{v}_2$. To show that $\mathbf{v}_3$ is in the span is the same as to solve

$$c_1 \begin{bmatrix} 1 \\ 0 \\ 0 \end{bmatrix} + c_2 \begin{bmatrix} 1 \\ 1 \\ 0 \end{bmatrix} = \begin{bmatrix} 5 \\ 2 \\ 0 \end{bmatrix}$$

So, $c_2 = 2$ and $c_1 = 3$. $\mathbf{v}_4$ is in not in span because if

$$c_1 \begin{bmatrix} 1 \\ 0 \\ 0 \end{bmatrix} + c_2 \begin{bmatrix} 1 \\ 1 \\ 0 \end{bmatrix} = \begin{bmatrix} 1 \\ 1 \\ 1 \end{bmatrix}$$

then the last component gives

$$c_1(0) + c_2(0) = 1$$

but $0 \neq 1$. Equivalently, in augmented form

$$\left[\begin{array}{cc|c} 1 & 1 & 1 \\ 0 & 1 & 1 \\ 0 & 0 & \boxed{1} \end{array}\right]$$

there is no solution when a pivot is in the augmented column.

Activity 3.2.2. Given the vectors

$$\mathbf{v}_1 = \begin{bmatrix} 1 \\ 0 \\ 0 \end{bmatrix}, \quad \mathbf{v}_2 = \begin{bmatrix} -1 \\ 1 \\ 0 \end{bmatrix}, \quad \mathbf{v}_3 = \begin{bmatrix} 2 \\ 3 \\ 0 \end{bmatrix}, \quad \text{and} \quad \mathbf{v}_4 = \begin{bmatrix} 1 \\ 1 \\ 1 \end{bmatrix}$$

do the following:

1. Write down a parameterization for the span of $\mathbf{v}_1$ and $\mathbf{v}_2$.
2. Show that $\mathbf{v}_3$ is in $\text{Span}\{\mathbf{v}_1, \mathbf{v}_2\}$, but $\mathbf{v}_4$ is not.

3.3 Linear Independence

TL;DR

Linear independence and dependence relations are defined in Definition 3.3.1. Linear independence is needed for the definition of a basis, and dependence relations are fundamental for computing null spaces.
This section covers three ways of computing dependence relations:

1. stare down (Example 3.3.5),
2. row reduction (Example 3.3.7), and
3. determinants (Example 3.3.9).

3.3.1 Linear Independence

Previously we defined the dimension of a solution set in terms of the number of arbitrary parameters. In the definition of span, we have one parameter for each of the generating vectors, however, there can be redundancy as the following example shows:

Example 3.3.1. Write down a parameterization of the span of

$$\mathbf{v}_1 = \begin{bmatrix} 2 \\ 1 \end{bmatrix}, \mathbf{v}_2 = \begin{bmatrix} 6 \\ 3 \end{bmatrix}, \mathbf{v}_3 = \begin{bmatrix} -4 \\ -2 \end{bmatrix}$$

using as few vectors as possible.

Solution

A vector in the span is just an arbitrary linear combination of the given vectors:

$$\operatorname{Span}\{\mathbf{v}_1, \mathbf{v}_2, \mathbf{v}_3\} = \left\{ c_1 \begin{bmatrix} 2 \\ 1 \end{bmatrix} + c_2 \begin{bmatrix} 6 \\ 3 \end{bmatrix} + c_3 \begin{bmatrix} -4 \\ -2 \end{bmatrix} : c_1, c_2, c_3 \text{ arbitrary} \right\}.$$

This already is a parameterization with parameters c_1, c_2, c_3. However, there is redundancy here. In this case all three vectors are multiples of each other:

$$\begin{bmatrix} 6 \\ 3 \end{bmatrix} = 3 \begin{bmatrix} 2 \\ 1 \end{bmatrix} \quad \text{and} \quad \begin{bmatrix} -4 \\ -2 \end{bmatrix} = (-2) \begin{bmatrix} 2 \\ 1 \end{bmatrix} \tag{3.3}$$

so they all lie along a common line. Since we now have a way of describing the vectors $\mathbf{v}_2$ and $\mathbf{v}_3$ in terms of $\mathbf{v}_1$, we rewrite a linear combination of $\mathbf{v}_1$ and $\mathbf{v}_2$ in terms of $\mathbf{v}_3$ alone. We do this by direct substitution with the equation (3.3):

$$c_1 \begin{bmatrix} 2 \\ 1 \end{bmatrix} + c_2 \begin{bmatrix} 6 \\ 3 \end{bmatrix} + c_3 \begin{bmatrix} -4 \\ -2 \end{bmatrix} = c_1 \begin{bmatrix} 2 \\ 1 \end{bmatrix} + c_2(3) \begin{bmatrix} 2 \\ 1 \end{bmatrix} + c_3(-2) \begin{bmatrix} 2 \\ 1 \end{bmatrix} = (c_1 + 3c_2 - 2c_3) \begin{bmatrix} 2 \\ 1 \end{bmatrix}$$

As a result we get the following simpler description:

$$\operatorname{Span}\{\mathbf{v}_1, \mathbf{v}_2, \mathbf{v}_3\} = \operatorname{Span}\{\mathbf{v}_1\} = \left\{ t \begin{bmatrix} 2 \\ 1 \end{bmatrix} : t \text{ arbitrary} \right\}.$$

The equations in (3.3) above show that $\mathbf{v}_2$ and $\mathbf{v}_3$ are already in the span of $\mathbf{v}_1$ because we can take $t = 3$ and $t = -2$ respectively. There is only one parameter so the dimension is 1, which is less than the dimension of the plane that they lie in ($m = 2$) and the number of vectors ($n = 3$).

In Example 3.3.1 we have several vectors that all lie on the same line, so they are scalar multiples of each other. However, this is not the general picture. The next activity is a very slight generalization.

Activity 3.3.2. Given

$$\mathbf{v}_1 = \begin{bmatrix} 1 \\ 0 \\ 0 \end{bmatrix}, \mathbf{v}_2 = \begin{bmatrix} 0 \\ 1 \\ 0 \end{bmatrix}, \mathbf{v}_3 = \begin{bmatrix} 2 \\ -3 \\ 0 \end{bmatrix},$$

do the following steps:

1. Write down an arbitrary linear combination of $\mathbf{v}_1$, $\mathbf{v}_2$, and $\mathbf{v}_3$.
2. Find constants t_1, t_2 satisfying

$$\mathbf{v}_3 = t_1\mathbf{v}_1 + t_2\mathbf{v}_2, \tag{3.4}$$

   which shows that $\mathbf{v}_3$ is in $\text{Span}\{\mathbf{v}_1, \mathbf{v}_2\}$.
3. Use equation (3.4) to replace $\mathbf{v}_3$ in the linear combination of $\mathbf{v}_1$, $\mathbf{v}_2$, and $\mathbf{v}_3$ written down in step 1, and simplify.
4. Show that no two of the vectors are scalar multiples of each other. For example, it is impossible to find t such that $t\mathbf{v}_1 = \mathbf{v}_2$.

This activity illustrates that if the original vectors belong to a plane in $\mathbb{R}^3$, then the only vectors that can be written as a linear combination of them must also belong to that plane. But this time no two of the original vectors are scalar multiples. Instead, they are linear combinations. We now have sufficient motivation for the following definitions.

Definition 3.3.1. A collection of vectors $\mathbf{v}_1 \ldots \mathbf{v}_k$ is said to be **linearly dependent** if one of the vectors can be written as a linear combination of the others,

$$c_1\mathbf{v}_1 + c_{j-1}\mathbf{v}_{j-1} + c_{j+1}\mathbf{v}_{j+1} + c_k\mathbf{v}_k = \mathbf{v}_j \tag{3.5}$$

or equivalently if the zero vector can be written as a linear combination of all of them,

$$c_1\mathbf{v}_1 + \cdots + c_n\mathbf{v}_n = \mathbf{0}, \tag{3.6}$$

with weights $c_1, \ldots, c_n$ that are not all zero. An equation of this type is called a **dependence relation**.

The collection of vectors $\mathbf{v}_1 \ldots \mathbf{v}_n$ is said to be **linearly independent** if the opposite is true; that is to say, if equation (3.6) is only satisfied by

$$c_1 = \cdots = c_n = 0.$$

Recall that the zero vector can always be written as a linear combination of a set of vectors by taking $c_1, \ldots c_n$ all to be zero, as pointed out immediately after the definition of span. But here we require some of the weights to be non-zero, so that is why linear dependence is possible.

Equations (3.5) and (3.6) provide two equivalent ways of understanding linear dependence. The next example illustrates why they are equivalent.

Example 3.3.3. If

$$\mathbf{v}_1 = \begin{bmatrix} 1 \\ 0 \\ 0 \end{bmatrix}, \quad \mathbf{v}_2 = \begin{bmatrix} 0 \\ 1 \\ 0 \end{bmatrix}, \quad \mathbf{v}_3 = \begin{bmatrix} 0 \\ 0 \\ 1 \end{bmatrix}, \quad \text{and} \quad \mathbf{v}_4 = \begin{bmatrix} 3 \\ 4 \\ 0 \end{bmatrix},$$

then $\mathbf{v}_1, \mathbf{v}_2, \mathbf{v}_4$ are linearly dependent, because of

$$3\mathbf{v}_1 + 4\mathbf{v}_2 = \mathbf{v}_4,$$

which is an example of equation (3.5). I can write zero as a linear combination of $\mathbf{v}_1, \mathbf{v}_2, \mathbf{v}_4$, simply by subtracting $\mathbf{v}_4$ from both sides:

$$3\mathbf{v}_1 + 4\mathbf{v}_2 - \mathbf{v}_4 = \mathbf{0}, \tag{3.7}$$

which is an example of equation (3.6). It is also possible to go in the other direction, from equation (3.6) to equation (3.5), by solving for a vector $\mathbf{v}_j$ for which $c_j \neq 0$. Starting from equation (3.7), I can solve for $\mathbf{v}_2$:

$$4\mathbf{v}_2 = \mathbf{v}_4 - 3\mathbf{v}_1 \quad \longrightarrow \quad \mathbf{v}_2 = \frac{1}{4}\mathbf{v}_4 - \frac{3}{4}\mathbf{v}_1.$$

But I cannot solve for $\mathbf{v}_3$ because $c_3 = 0$ in equation (3.7): $\mathbf{v}_3$ is not involved in that dependence relation.

Activity 3.3.4. In Activity 3.3.2, you found an equation $\mathbf{v}_3 = t_1\mathbf{v}_1 + t_2\mathbf{v}_2$, with specific values of t_1 and t_2.

1. Write $\mathbf{0}$ as a linear combination of $\mathbf{v}_1, \mathbf{v}_2, \mathbf{v}_3$.
2. Write $\mathbf{v}_1$ as a linear combination of $\mathbf{v}_2$ and $\mathbf{v}_3$.

Definition 3.3.2. The **dimension** of the span of a finite set of vectors is equal to the number of vectors in a linearly independent subset that still has the same span.

Remark 3.3.1. The definition of the span of a set of vectors agrees with the dimension of a solution set because the vectors that the free variables are multiplied by are automatically linearly independent. Linear independence of those vectors is caused the redundant equations. For instance, if z is a free variable, then $z = z$ produces a 1 in the coordinate corresponding to z, where the other vectors have a zero. For instance, see the parameterization in Example 2.3.5.

While the dimension of the span can never be larger than n (the number of vectors), it can be smaller. If the vectors lie in $\mathbb{R}^m$, then the dimension also can never be larger than m. If we start out with more than m vectors, then $n > m$ so the dimension is clearly smaller than n. But it can sometimes be smaller than both m and n.

Example 3.3.5 (Dependence relations by stare down)**.** Given the vectors

$$\mathbf{v}_1 = \begin{bmatrix} 1 \\ -1 \\ 0 \end{bmatrix}, \quad \mathbf{v}_2 = \begin{bmatrix} 0 \\ 1 \\ -1 \end{bmatrix}, \quad \mathbf{v}_3 = \begin{bmatrix} -1 \\ 0 \\ 1 \end{bmatrix}, \quad \text{and} \quad \mathbf{v}_4 = \begin{bmatrix} 1 \\ 1 \\ 1 \end{bmatrix}$$

1. find a dependence relation between them, and
2. consider which vectors can be removed without changing the dimension of the span.

Solution

To find a dependence relation, we must find constants c_1, c_2, c_3, c_4 (not all zero) such that

$$c_1 \begin{bmatrix} 1 \\ -1 \\ 0 \end{bmatrix} + c_2 \begin{bmatrix} 0 \\ 1 \\ -1 \end{bmatrix} + c_3 \begin{bmatrix} -1 \\ 0 \\ 1 \end{bmatrix} + c_4 \begin{bmatrix} 1 \\ 1 \\ 1 \end{bmatrix} = \begin{bmatrix} 0 \\ 0 \\ 0 \end{bmatrix}. \tag{3.8}$$

It is hopefully clear that simply adding the first three vectors does the trick because the 1 and -1 cancel in pairs:

$$\begin{bmatrix} 1 \\ -1 \\ 0 \end{bmatrix} + \begin{bmatrix} 0 \\ 1 \\ -1 \end{bmatrix} + \begin{bmatrix} -1 \\ 0 \\ 1 \end{bmatrix} = \begin{bmatrix} 0 \\ 0 \\ 0 \end{bmatrix}.$$

In terms of equation (3.8) we have $c_1 = c_2 = c_3 = 1$ and $c_4 = 0$. This is what we mean by "not all zero": *some* of the constants are allowed to be zero, but at least two of them will not be.

Since $c_4 = 0$, then only $\mathbf{v}_1$, $\mathbf{v}_2$, $\mathbf{v}_3$ are involved in the dependence relation. Any one of them can be removed without changing the dimension of the span. Suppose we eliminate $\mathbf{v}_3$. Does a dependence relation exist between $\mathbf{v}_1$ and $\mathbf{v}_2$ alone? No, because

$$c_1 \begin{bmatrix} 1 \\ -1 \\ 0 \end{bmatrix} + c_2 \begin{bmatrix} 0 \\ 1 \\ -1 \end{bmatrix} = \begin{bmatrix} c_1 \\ c_2 - c_1 \\ -c_2 \end{bmatrix} \tag{3.9}$$

and by setting this equal to the zero vector, we get $c_1 = 0$ from the first component and $c_2 = 0$ from the third component. Hence $\mathbf{v}_1$ and $\mathbf{v}_2$ are linearly independent. Can $\mathbf{v}_4$ be removed? No, because equation (3.9) is an expression for an arbitrary vector in the span of $\mathbf{v}_1$ and $\mathbf{v}_2$, and the components add to zero, regardless of the values of c_1 and c_2. On the other hand the components of $\mathbf{v}_4$ add up to 3, so it cannot possibly be in the span of $\mathbf{v}_1$ and $\mathbf{v}_2$.

We have shown that

$$\text{Span}\{\mathbf{v}_1, \mathbf{v}_2, \mathbf{v}_3, \mathbf{v}_4\} = \text{Span}\{\mathbf{v}_1, \mathbf{v}_2, \mathbf{v}_4\}$$

and that we cannot remove another vector, so the dimension of the span is 3.

Activity 3.3.6. Given the vectors

$$\begin{bmatrix} 1 \\ 0 \\ 0 \\ 0 \end{bmatrix}, \begin{bmatrix} 0 \\ 1 \\ 0 \\ 0 \end{bmatrix}, \begin{bmatrix} 0 \\ 0 \\ 1 \\ 0 \end{bmatrix}, \begin{bmatrix} 2 \\ 0 \\ 3 \\ 0 \end{bmatrix}.$$

1. find a dependence relation between them, and
2. consider which vectors can be removed without changing the dimension of the span.

As Example 3.3.5 shows, just because the dimension of the span is less than the number of the vectors, doesn't mean you can remove any one of them. We were allowed to remove any of $\mathbf{v}_1, \mathbf{v}_2, \mathbf{v}_3$, but not $\mathbf{v}_4$. You might also have to drop more than one vector, of course, which is not illustrated by Example 3.3.5 but is illustrated by Example 3.3.7.

3.3.2 Dependence Relations by Row Reduction

Q 3.3.1. Suppose the variables are x_1, x_2, x_3, x_4. What are the free variables in the matrix

$$\begin{bmatrix} 1 & 0 & 0 & 1 \\ 0 & 0 & 1 & 2 \end{bmatrix}?$$

Since a dependence relation between $\mathbf{a}_1, \dots \mathbf{a}_n$ is a solution to the equation

$$x_1\mathbf{a}_1 + x_2\mathbf{a}_2 + \cdots + x_n\mathbf{a}_n = \mathbf{0},$$

then we can find them by converting to augmented form, $[\mathbf{A}|\mathbf{0}]$, and row reducing.

Example 3.3.7. Given the vectors

$$\mathbf{v}_1 = \begin{bmatrix} 1 \\ 0 \\ -1 \\ 0 \\ 0 \end{bmatrix}, \quad \mathbf{v}_2 = \begin{bmatrix} 0 \\ 1 \\ 0 \\ 1 \\ 0 \end{bmatrix}, \quad \mathbf{v}_3 = \begin{bmatrix} 1 \\ 1 \\ -1 \\ 1 \\ 0 \end{bmatrix}, \quad \text{and} \quad \mathbf{v}_4 = \begin{bmatrix} 1 \\ -1 \\ -1 \\ -1 \\ 0 \end{bmatrix},$$

find a linearly independent subset of them with the same span.

Solution

In order to find a dependence relation, we must find constants c_1, c_2, c_3, c_4 such that

$$c_1 \begin{bmatrix} 1 \\ 0 \\ -1 \\ 0 \\ 0 \end{bmatrix} + c_2 \begin{bmatrix} 0 \\ 1 \\ 0 \\ 1 \\ 0 \end{bmatrix} + c_3 \begin{bmatrix} 1 \\ 1 \\ -1 \\ 1 \\ 0 \end{bmatrix} + c_4 \begin{bmatrix} 1 \\ -1 \\ -1 \\ -1 \\ 0 \end{bmatrix} = \begin{bmatrix} 0 \\ 0 \\ 0 \\ 0 \\ 0 \end{bmatrix},$$

where c_1, c_2, c_3, c_4 are not all zero. But by Proposition 2.2.2, we know that we can convert to augmented form and row reduce.

$$\left[\begin{array}{cccc|c} 1 & 0 & 1 & 1 & 0 \\ 0 & 1 & 1 & -1 & 0 \\ -1 & 0 & -1 & -1 & 0 \\ 0 & 1 & 1 & -1 & 0 \\ 0 & 0 & 0 & 0 & 0 \end{array}\right] \xrightarrow{R_1+R_3} \left[\begin{array}{cccc|c} 1 & 0 & 1 & 1 & 0 \\ 0 & 1 & 1 & -1 & 0 \\ 0 & 0 & 0 & 0 & 0 \\ 0 & 1 & 1 & -1 & 0 \\ 0 & 0 & 0 & 0 & 0 \end{array}\right]$$

$$\xrightarrow{(-1)R_2+R_4} \left[\begin{array}{cccc|c} \boxed{1} & 0 & 1 & 1 & 0 \\ 0 & \boxed{1} & 1 & -1 & 0 \\ 0 & 0 & 0 & 0 & 0 \\ 0 & 0 & 0 & 0 & 0 \\ 0 & 0 & 0 & 0 & 0 \end{array}\right]$$

This is already in reduced form. Reading off the rows, we have

$$c_1 + c_3 + c_4 = 0 \quad \text{and} \quad c_2 + c_3 - c_4 = 0.$$

There are two free variables c_3 and c_4. For $c_3 = -1$, $c_4 = 0$, we get $c_1 = c_2 = 1$, thus

$$(1)\mathbf{v}_1 + (1)\mathbf{v}_2 + (-1)\mathbf{v}_3 = \mathbf{0} \implies \mathbf{v}_1 + \mathbf{v}_2 = \mathbf{v}_3. \tag{3.10}$$

Similarly taking $c_3 = 0$, $c_4 = -1$, we get $c_1 = 1$ and $c_2 = -1$, hence

$$(1)\mathbf{v}_1 + (-1)\mathbf{v}_2 + (-1)\mathbf{v}_4 = \mathbf{0} \implies \mathbf{v}_1 - \mathbf{v}_2 = \mathbf{v}_4. \tag{3.11}$$

Each dependence relation we get in this way allows us to remove one of the vectors. Specifically, we remove the vectors $\mathbf{v}_3$ and $\mathbf{v}_4$ corresponding to the free variables. That leaves us with

$$\mathbf{v}_1 = \begin{bmatrix} 1 \\ 0 \\ -1 \\ 0 \\ 0 \end{bmatrix}, \quad \text{and} \quad \mathbf{v}_2 = \begin{bmatrix} 0 \\ 1 \\ 0 \\ 1 \\ 0 \end{bmatrix}.$$

The dimension of the span is 2 (the number of pivots), and

$$\text{Span}\{\mathbf{v}_1, \mathbf{v}_2, \mathbf{v}_3, \mathbf{v}_4\} = \text{Span}\{\mathbf{v}_1, \mathbf{v}_2\}$$

because the dependence relations (3.10) and (3.11) show how to get $\mathbf{v}_3$ and $\mathbf{v}_4$ in terms of $\mathbf{v}_1$ and $\mathbf{v}_2$.

Activity 3.3.8. Given the vectors

$$\begin{bmatrix} 1 \\ 0 \\ 0 \\ 0 \end{bmatrix}, \begin{bmatrix} 1 \\ 1 \\ 0 \\ 0 \end{bmatrix}, \begin{bmatrix} 0 \\ 0 \\ 1 \\ 1 \end{bmatrix}, \begin{bmatrix} 1 \\ 1 \\ 1 \\ 1 \end{bmatrix}, \begin{bmatrix} 1 \\ 1 \\ -1 \\ -1 \end{bmatrix},$$

do the following parts.

1. Use row reduction to obtain dependence relations by setting one free variable equal to 1, and the remaining free variables equal to zero.

2. What are the vectors corresponding to the pivot columns?

3. What is the dimension of the span?

Reflection 3.3.1. If a matrix has more columns than rows, can every column have a pivot? Can you conclude anything about linear independence of more than n vectors in $\mathbb{R}^n$? What if a matrix has more rows than columns?

3.3.3 Linear Independence and Determinants

Q 3.3.2. Imagine a box being flattened by tipping over the sides while keeping their length the same. What is the area of the flattened box?

In the case where the coefficient matrix – i.e. the part to the left of the augmented column of zeros – is a square matrix, it is possible to verify linear independence by verifying that the determinant is not zero.

Example 3.3.9. The vectors

$$\begin{bmatrix} 1 \\ 1 \end{bmatrix}, \begin{bmatrix} 2 \\ 2 \end{bmatrix}$$

are linearly dependent because

$$\begin{vmatrix} 1 & 2 \\ 1 & 2 \end{vmatrix} = 1 \cdot 2 - 2 \cdot 1 = 0.$$

In terms of the area interpretation, the area between the vectors is zero, meaning that the two vectors are co-linear; they are scalar multiples of each other. In fact, if these vectors are coming from the coefficients of a system of equations such as

$$\begin{aligned} x + 2y &= b_1 \\ x + 2y &= b_2, \end{aligned}$$

then the zero determinant means that the lines represented by the equations are either co-linear or parallel as in Activity 3.1.11. This means that there are either infinitely many solutions or zero solutions; it is not possible in this case to have a unique solution.

Activity 3.3.10. For each collection of vectors below, determine whether or not they are linearly independent by computing the determinant of the square matrix with the given vectors as columns

1. $$\begin{bmatrix} 1 \\ 0 \\ 0 \\ 0 \end{bmatrix}, \begin{bmatrix} 0 \\ 1 \\ 0 \\ 0 \end{bmatrix}, \begin{bmatrix} 0 \\ 0 \\ 1 \\ 0 \end{bmatrix}, \begin{bmatrix} 2 \\ 0 \\ 3 \\ 0 \end{bmatrix}$$

2. $$\begin{bmatrix} 1 \\ 0 \\ 0 \end{bmatrix}, \begin{bmatrix} 1 \\ 1 \\ 0 \end{bmatrix}, \begin{bmatrix} 1 \\ 1 \\ 1 \end{bmatrix}$$

3. $$\begin{bmatrix} 1 \\ 2 \end{bmatrix}, \begin{bmatrix} 3 \\ 6 \end{bmatrix}$$

4. $$\begin{bmatrix} 1 \\ 1 \end{bmatrix}, \begin{bmatrix} -1 \\ 1 \end{bmatrix}$$

3.3.4 The Invertible Matrix Theorem (Part 3)

Theorem 3.3.3 (The Invertible Matrix Theorem: Part 3). *Let* $\mathbf{A}$ *be an* $n \times n$ *matrix with real entries. The following statements are equivalent.*

7. $\det(\mathbf{A}) \neq 0$

8. $\det(\mathbf{A}^T) \neq 0$

9. *The columns of* $\mathbf{A}$ *are linearly independent.*

10. *The rows of* $\mathbf{A}$ *are linearly independent.*

See Remark 1.3.1 concerning the Invertible matrix theorem and its proof.

Example 3.3.11. Show that statements 7 and 9 in the theorem are true for the matrix

$$\mathbf{A} = \begin{bmatrix} 1 & 1 \\ 0 & 1 \end{bmatrix}$$

Solution

7. $\det(A) \neq 0$

$$\det A = 1 \cdot 1 - 1 \cdot 0 = 1 \neq 0$$

9. The columns of **A** are linearly independent. We must show that the only solution to

$$c_1 \begin{bmatrix} 1 \\ 0 \end{bmatrix} + c_2 \begin{bmatrix} 1 \\ 1 \end{bmatrix} = \begin{bmatrix} 0 \\ 0 \end{bmatrix}$$

is $c_1 = c_2 = 0$. We do this by converting to augmented form and row reducing:

$$\left[\begin{array}{cc|c} 1 & 1 & 0 \\ 0 & 1 & 0 \end{array}\right] \xrightarrow{(-1)R_2+R_1} \left[\begin{array}{cc|c} 1 & 0 & 0 \\ 0 & 1 & 0 \end{array}\right].$$

Indeed the solution is unique, and it is $c_1 = c_2 = 0$.

Activity 3.3.12. Show that the statements 8 and 10 in Theorem 3.3.3 are true for

$$\mathbf{A} = \begin{bmatrix} 1 & 1 \\ 0 & 1 \end{bmatrix}.$$

The first of these should be done directly. For the second, you should take the transpose to get column vectors, then augment and do row reduction.[4]

3.4 Determinants in Calculus

Q 3.4.1. What is $\frac{d}{dx} x^3$?

Q 3.4.2. What is $\frac{d}{dx} \sin(x)$?

Q 3.4.3. What is $\frac{\partial}{\partial x}(x^2 y^2)$?

There are many applications of determinants in Calculus, including:

1. linear independence of functions (Wronskians),
2. the curl of a vector field (e.g. in Stoke's theorem),
3. the Jacobian determinant (multi-variable version of u-substitution),
4. the Hessian determinant (multi-variable version of 2nd derivative test),

Definition 3.4.1. The **Wronskian** determinant of a collection of functions $f_1, f_2, \ldots f_n$ of x is given by

$$\begin{vmatrix} f_1(x) & f_2(x) & \cdots & f_n(x) \\ \frac{d}{dx} f_1(x) & \frac{d}{dx} f_2(x) & \cdots & \frac{d}{dx} f_n(x) \\ \vdots & \vdots & & \vdots \\ \frac{d^n}{dx^n} f_1(x) & \frac{d^n}{dx^n} f_2(x) & \cdots & \frac{d^n}{dx^n} f_n(x) \end{vmatrix}$$

Example 3.4.1. Given $f_1(x) = \cos(x)$ and $f_2(x) = \sin(x)$, compute the Wronskian determinant.

[4] Without taking the transpose, you would need an augmented row instead of an augmented column. Then you would need to do column reduction instead of row reduction.

Solution

$$\frac{d}{dx}\cos(x) = -\sin(x) \quad \text{and} \quad \frac{d}{dx}\sin(x) = \cos(x),$$

thus

$$\begin{vmatrix} f_1(x) & f_2(x) \\ f_1'(x) & f_2'(x) \end{vmatrix} = \begin{vmatrix} \cos(x) & \sin(x) \\ -\sin(x) & \cos(x) \end{vmatrix} = \cos^2(x) + \sin^2(x) = 1$$

Activity 3.4.2. Given the functions $f_1(x) = 1, f_2(x) = x, f_3(x) = x^2$, compute the Wronskian determinant.

$$\begin{vmatrix} f_1(x) & f_2(x) & f_3(x) \\ f_1'(x) & f_2'(x) & f_3'(x) \\ f_1''(x) & f_2''(x) & f_3''(x) \end{vmatrix}.$$

Definition 3.4.2. For a function $f : \mathbb{R}^n \to \mathbb{R}^m$, given by

$$\begin{bmatrix} x_1 \\ x_2 \\ \vdots \\ x_n \end{bmatrix} \mapsto \begin{bmatrix} f_1(x_1, x_2, \ldots, x_n) \\ f_2(x_1, x_2, \ldots, x_n) \\ \vdots \\ f_m(x_1, x_2, \ldots, x_n), \end{bmatrix},$$

the **Jacobian matrix** of f is

$$\mathbf{J}_f = \begin{bmatrix} \frac{\partial f_1}{\partial x_1} & \frac{\partial f_1}{\partial x_2} & \cdots & \frac{\partial f_1}{\partial x_n} \\ \frac{\partial f_2}{\partial x_1} & \frac{\partial f_2}{\partial x_2} & \cdots & \frac{\partial f_2}{\partial x_n} \\ \vdots & \vdots & & \vdots \\ \frac{\partial f_m}{\partial x_n} & \frac{\partial f_m}{\partial x_2} & \cdots & \frac{\partial f_m}{\partial x_n} \end{bmatrix}.$$

If $m = n$, then the **Jacobian determinant** for f is $\det(\mathbf{J}_f)$.

Remark 3.4.1. You should avoid simply referring to a Jacobian matrix or Jacobian determinant as the "Jacobian." The problem is that Jacobi did many things, and there is a more advanced concept – the Jacobian of a curve – which is sometimes called just the "Jacobian." If the context is clear, any of these could just be called the "Jacobian," but in my opinion, this is a bad habit. It is not like any of these concepts is rare. With that being said, in Sage and other python implementations, the Jacobian matrix is called just the Jacobian.

The Jacobian matrix provides an approximation of the function f near a point $\mathbf{a}$ in the domain by a linear function. The approximation is given as follows

$$f(\mathbf{x}) \approx f(\mathbf{a}) + \mathbf{J}_f(\mathbf{a})(\mathbf{x} - \mathbf{a}), \tag{3.12}$$

where $\mathbf{J}_f(\mathbf{a})$ is the value of the Jacobian matrix at $\mathbf{a}$ and $\mathbf{J}_f(\mathbf{a})(\mathbf{x} - \mathbf{a})$ is the product with $(\mathbf{x} - \mathbf{a})$ in the sense of matrix multiplication. In the case where $m = n = 1$, the approximation given by (3.12) is just a tangent line approximation. If the determinant is not zero, then we can compute a "local" inverse for the function f given by

$$f^{-1}(\mathbf{y}) \approx \mathbf{a} + (\mathbf{J}_f(\mathbf{a}))^{-1}(\mathbf{y} - f(\mathbf{a}))$$

on a small domain containing $f(\mathbf{a})$.

Example 3.4.3. Consider the function $f : \mathbb{R}^2 \to \mathbb{R}$ given by

$$\begin{bmatrix} x \\ y \end{bmatrix} \mapsto z = \frac{1}{10}(50 - x^2 - y^2),$$

then

$$\frac{\partial z}{\partial x} = -\frac{1}{5}x \quad \text{and} \quad \frac{\partial z}{\partial y} = -\frac{1}{5}y$$

so

$$\mathbf{J}_f = \begin{bmatrix} -\frac{1}{5}x & -\frac{1}{5}y \end{bmatrix}.$$

For $\mathbf{a} = \left[\begin{smallmatrix} 3 \\ 4 \end{smallmatrix}\right]$, that is $x = 3$ and $y = 4$, we have

$$\mathbf{J}_f(\mathbf{a}) = \begin{bmatrix} -\frac{3}{5} & -\frac{4}{5} \end{bmatrix} \quad \text{and} \quad f(\mathbf{a}) = \frac{1}{10}(50 - 3^2 - 4^2) = \frac{25}{10} = 2.5.$$

The approximation of f near $\mathbf{a}$ is the tangent plane:

$$f(\mathbf{a}) + \mathbf{J}_f(\mathbf{a})(\mathbf{x} - \mathbf{a}) = 2.5 + \begin{bmatrix} -\frac{3}{5} & -\frac{4}{5} \end{bmatrix} \left(\begin{bmatrix} x \\ y \end{bmatrix} - \begin{bmatrix} 3 \\ 4 \end{bmatrix} \right) = 2.5 - \frac{3}{5}(x-3) - \frac{4}{5}(y-4)$$

At $\mathbf{x} = \left[\begin{smallmatrix} 3.1 \\ 4.1 \end{smallmatrix}\right]$ the exact value of f is

$$f(\mathbf{x}) = \frac{1}{10}(50 - 3.1^2 - 4.1^2) = 2.358.$$

The approximation from the tangent plane is

$$2.5 - \frac{3}{5}(3.1 - 3) - \frac{4}{5}(4.1 - 4) = 2.5 - 0.6 \times 0.1 - 0.8 \times 0.1 = 2.5 - 0.14 = 2.36.$$

The error of this approximation is 0.002.

Example 3.4.4. Consider the function $f : \mathbb{R}^2 \to \mathbb{R}^2$ given by

$$\begin{bmatrix} x \\ y \end{bmatrix} \mapsto \begin{bmatrix} u \\ v \end{bmatrix} = \begin{bmatrix} x^2 + 3x + 5y + 1 \\ y^2 + xy + x + 2y - 1 \end{bmatrix}.$$

Then

$$\frac{\partial u}{\partial x} = 2x + 3 \qquad \frac{\partial u}{\partial y} = 5$$

$$\frac{\partial v}{\partial x} = y + 2 \qquad \frac{\partial v}{\partial y} = 2y + x + 1$$

so

$$\mathbf{J}_f = \begin{bmatrix} 2x + 3 & 5 \\ y + 1 & 2y + x + 2 \end{bmatrix}.$$

At $\mathbf{a} = \mathbf{0}$, we have

$$f(\mathbf{0}) = \begin{bmatrix} 1 \\ -1 \end{bmatrix} \quad \text{and} \quad \mathbf{J}_f(\mathbf{0}) = \begin{bmatrix} 3 & 5 \\ 1 & 2 \end{bmatrix},$$

because the value of any polynomial evaluated at $\mathbf{0}$ is simply the constant term. The approximation of f near $\mathbf{0}$ is therefore

$$\begin{bmatrix} 1 \\ -1 \end{bmatrix} + \begin{bmatrix} 3 & 5 \\ 1 & 2 \end{bmatrix} \begin{bmatrix} x \\ y \end{bmatrix}.$$

The value of the Jacobian determinant at $\mathbf{0}$ is

$$\det J_f(\mathbf{0}) = 3 \cdot 2 - 5 \cdot 1 = 6 - 5 = 1,$$

so a local inverse exists. The approximation for f^{-1} near $f(\mathbf{0})$ is

$$\begin{bmatrix} 0 \\ 0 \end{bmatrix} + \begin{bmatrix} 2 & -5 \\ -1 & 3 \end{bmatrix} \left(\begin{bmatrix} u \\ v \end{bmatrix} - \begin{bmatrix} 1 \\ -1 \end{bmatrix} \right) = \begin{bmatrix} 2 & -5 \\ -1 & 3 \end{bmatrix} \begin{bmatrix} u - 1 \\ v + 1 \end{bmatrix}.$$

Sage has the ability to compute all of the determinants in this section.
For Wronskians:

```
wronskian(1,x,x^2)
```

The determinant of a square Jacobian matrix can be used to detect when a

```
var('y')
jacobian((x^2+3*x+5*y+1,y^2+x*y+x+2*y-1),(x,y))
f=y^3+x^5-x^2*y^4+7
f.hessian()
```

3.5 Algorithms for Determinants

In this section we will look at the time and space asymptotics of some of the methods of computing determinants. Some of the options are

1. Expansion by cofactors
2. Row reduction
3. Computing a QR-factorization (covered in Section 5.6.1), then computing the determinant by Proposition 3.1.3.
4. Computing the characteristic polynomial by LeVerrier's method (covered in Example 5.2.25).

First, try row reduction:

AI Prompt 3.5.1. Can you write an implementation of the determinant of an arbitrary square matrix using row reduction?

To check that the code works, test it on some examples and check the determinants with Sage. If the code doesn't work, try to get the AI to correct itself. If you get an error. Copy and paste the error into the prompt.

Next, we consider how to implement Proposition 3.1.3. If we can get a vector consisting of the diagonal entries of the matrix, then we can get the product by multiplying in pairs following the idea in Example 1.6.4. That approach would give us $O(\log(n))$ in time in parallel, which is what we are aiming for.

AI Prompt 3.5.2. Can you write a function in python to compute the determinant of an upper triangular matrix?

ChatGPT gave me a for loop that multiplied the entries on the diagonal one at a time. It was correct, but the speed would only be $O(n)$ in parallel. NumPy has a function called `np.diag` that gets the diagonal. Then `np.prod` is the most efficient approach to multiplying the numbers together that NumPy has to offer

AI Prompt 3.5.3. Can you write a different implementation using np.prod from NumPy?

Of course if ChatGPT gives you the NumPy code first, then you should ask it for an implementation with a loop over the indices. Yes, it is slower, but we want both of them so that we can compare the speed. You should also check that they are correct by testing them on a few examples.

AI Prompt 3.5.4. Can you write an implementation of the determinant of an arbitrary square matrix using expansion by cofactors?

Expansion by cofactors, as given in Definition 3.1.2, is a recursion with n steps. ChatGPT gave me a non-tail recursion (see Section 0.2) on its first try, but it is possible to implement expansion by cofactors as a tail recursion. If you are not sure whether ChatGPT gave a tail recursion, put some open and close print statements in the function as I did in Section 0.2. If it is a non-tail recursion, try asking ChatGPT directly for a tail recursion.

My approach to getting a tail recursion for expansion by cofactors is to:

1. keep track of the rows and columns that are omitted from the minors, and

2. keep track of the coefficient of each minor, at each step.

Bear in mind that in going from one step to the next, we must compute minors of minors, so the number of rows and columns accumulates. To keep things clean, we might try working systematically left to right and always doing expansion by cofactors down columns. Then, on step k, it is implied that we omit columns $1, 2, \dots k$, so we do not need to keep track of this explicitly. To keep track of the rows that have been omitted for a given minor, we might use a bit vector with n-bits, with a 1 if the row is omitted and a 0 otherwise. We must omit k-rows at the k-th step. How many ways are there of doing this?

$$\binom{n}{k} = \frac{n!}{k!(n-k)!}$$

This is very bad news! The maximum occurs in the middle with $k = \frac{n}{2}$. To determine the growth, we put $n = 2m$ and look at what happens going from m to $m + 1$

$$\binom{2(m+1)}{m+1} = \frac{(2m+2)!}{(m+1)!^2} = \frac{(2m+1)(2m+2)}{(m+1)^2} \cdot \frac{(2m)!}{m!^2} < 4\binom{2m}{m}.$$

Since we gain a factor of 4 going from n to $n + 2$, then we gain a factor of 2 in going from n to $n + 1$. Hence the number of bit vectors we need to store increases exponentially as a constant times 2^n. Since the bit vectors have n-bits each, then we have $O(n2^n)$ growth so far in space. We also need to store the coefficients of course, but if we suppose they are small numbers, we still have $O(n2^n)$ because the number of bits of precision is bounded and does not grow with n.

What about time efficiency? The number of multiplications performed at each step is proportional to $\binom{n}{k}$. These multiplications can all be done in parallel. Addition and subtraction occur when a minor in the $k + 1$-th step can be obtained from more than one of the minors in the k-th step. In terms of bit vectors, the bit vectors at the k-th step have k ones, and we introduce one more one in each way that is possible to get the bit vectors for the $k + 1$-th step. Going backwards we can imagine removing a one. For example, say $n = 8$ and $k = 2$. Then we can get $(1, 1, 1, 0, 0, 0, 0, 0)$ from three different bit vectors:

$$(1,1,0,0,0,0,0,0), \quad (1,0,1,0,0,0,0,0), \quad \text{and} \quad (0,1,1,0,0,0,0,0),$$

which causes addition/subtraction to be used twice (one less than the total number of ways). Thankfully, this is bounded by the length of the bit vector – i.e. the number of rows – which is n. Some addition can be done in parallel if we add in pairs. For the serial algorithm with small numbers, by summing over k we get

$$n \sum_{k=0}^{n} \binom{n}{k} = n2^n,$$

hence we have $O(n2^n)$ for both space and time. For the parallel algorithm with small numbers we get

$$\log(n) \sum_{k=0}^{n} 1 = n \log n,$$

hence we have $O(n \log(n))$. We summarize these results in the table below:

	Space	Time
Sequential with small nums	$O(n2^n)$	$O(n2^n)$
Parallel with small nums	$O(n2^n)$	$O(n \log(n))$

FIGURE 3.3: Asymptotic efficiency of expansion by cofactors

This is terrible![5] Figure 2.5 shows that row reduction of an $n \times n$ matrix is $O(n^3)$ time sequentially and $O(n \log(n))$ in parallel. But LeVerrier's method is faster sequentially (see Figure 5.3). It may be that the time efficiency is the same for the parallel approach, but that may not be true in practice since it is likely that most hardware will quickly run out of room, causing the sequential asymptotics to take over. As such, there is little reason to use expansion by cofactors on its own to compute determinants unless n is small or in very special cases. I guess it is possible that the space issues might be resolved on a quantum computer, but as of this writing we are not there yet.

Exercises

Problem 3.1. True or False?

1. If 3 vectors lie in the same plane, then there is a dependence relation between them.
2. If a set of vectors is linearly dependent, then a vector in the set is a scalar multiple of one of the others.
3. If a set of vectors is linearly dependent, then each vector in the set can be written as a linear combination of the others.
4. If a set of vectors is linearly dependent, then there exists a vector in the set that can be written as a linear combination of the others.
5. The columns of any 4×5 matrix are linearly dependent.
6. A set of fewer than n vectors in $\mathbb{R}^n$ is linearly independent.

[5] As noted in a previous footnote, it is possible to define determinants using permutations instead. The asymptotics are not any better though. In space it is $O(n!)$. In parallel there are $n - 1$ multiplications and $\log(n!) = n \log n + O(n)$ additions. So with small numbers, the algorithm is $O(n \log(n))$ in time in parallel, and $O(n!)$ in time sequentially.

Problem 3.2. Given the matrices

$$\mathbf{A} = \begin{bmatrix} 1 & 3 \\ 0 & 2 \end{bmatrix} \quad \mathbf{B} = \begin{bmatrix} 0 & 5 \\ 1 & 0 \end{bmatrix} \quad \mathbf{C} = \begin{bmatrix} 1 & 1 \\ 0 & 0 \end{bmatrix}$$

Verify the following facts by calculating.

1. $\det(\mathbf{A}^{-1}) = \dfrac{1}{\det(\mathbf{A})}$
2. $\det(\mathbf{A}^T) = \det(\mathbf{A})$
3. $\det(\mathbf{AB}) = \det(\mathbf{A}) \cdot \det(\mathbf{B})$
4. $\det(\mathbf{C}) = 0$, $\mathbf{CA} = \mathbf{CB}$, and $\det(\mathbf{CA}) = \det(\mathbf{CB}) = 0$, but $\det(\mathbf{A}) \neq \det(\mathbf{B})$ because we would be forced to divide by zero. Note that $\mathbf{B} \neq \mathbf{C}$ but there is nothing to check for this detail. *Thus the law of cancellation fails because* $\det(\mathbf{C}) = 0$.

Problem 3.3. Use row reduction to find the determinant of one of the following 3 matrices.

$$\text{a. } \begin{bmatrix} 1 & 1 & 1 \\ 0 & 2 & 1 \\ 4 & 5 & 3 \end{bmatrix} \qquad \text{b. } \begin{bmatrix} 1 & 1 & 1 \\ 0 & 3 & -1 \\ 2 & 3 & 0 \end{bmatrix} \qquad \text{c. } \begin{bmatrix} 1 & 1 & 1 \\ 0 & 3 & 5 \\ 1 & 2 & 2 \end{bmatrix}$$

Problem 3.4. Use expansion by cofactors to compute the determinant of the matrix

$$\begin{bmatrix} 0 & 0 & 2 & 3 & 0 \\ 0 & 0 & 0 & 2 & 3 \\ 3 & 0 & 0 & 0 & 2 \\ 2 & 3 & 0 & 0 & 0 \\ 0 & 2 & 3 & 0 & 0 \end{bmatrix}.$$

Problem 3.5. Use row reduction to find a dependence relation between the columns of

$$\mathbf{A} = \begin{bmatrix} 2 & 3 & 1 \\ 3 & 4 & 2 \end{bmatrix}.$$

Is any column a multiple of another column?

Problem 3.6. Find a dependence relation between the rows of

$$\mathbf{A} = \begin{bmatrix} 21 & -28 & 35 \\ 3 & -4 & 5 \end{bmatrix}.$$

Problem 3.7. Find a linearly independent subset of the vectors

$$\mathbf{v}_1 = \begin{bmatrix} 1 \\ 1 \\ 0 \\ 0 \end{bmatrix}, \quad \mathbf{v}_2 = \begin{bmatrix} 1 \\ 0 \\ 1 \\ 0 \end{bmatrix}, \quad \mathbf{v}_3 = \begin{bmatrix} 1 \\ -2 \\ 3 \\ 0 \end{bmatrix}, \quad \mathbf{v}_4 = \begin{bmatrix} 2 \\ 1 \\ 1 \\ 1 \end{bmatrix}.$$

Can each of the four vectors be written as a linear combination of the other three?

Problem 3.8. Compute the Wronskian determinant for each collection of functions below.

1. e^{nx}, e^{mx}
2. $\cos(nx), \cos(mx)$
3. $\cos(nx), \sin(nx)$
4. $x, 2x$

Problem 3.9. Show that each part of theorem 3.3.3 is false for the matrix

$$\mathbf{A} = \begin{bmatrix} 1 & 1 \\ 0 & 0 \end{bmatrix}.$$

Problem 3.10. Explore the determinant of $\mathbf{A} + \mathbf{B}$, relative to the determinant of $\mathbf{A}$ and $\mathbf{B}$ as follows. Using 2×2 matrices $\mathbf{A}$ and $\mathbf{B}$, can you find examples where:

1. $\det(\mathbf{A}) + \det(\mathbf{B})$ is not equal to $\det(\mathbf{A} + \mathbf{B})$?
2. $\det(\mathbf{A}) + \det(\mathbf{B}) = \det(\mathbf{A} + \mathbf{B})$?
3. $\det(\mathbf{A})$ and $\det(\mathbf{B})$ are not zero but $\det(\mathbf{A} + \mathbf{B})$ is?
4. $\det(\mathbf{A})$, $\det(\mathbf{B})$, and $\det(\mathbf{A} + \mathbf{B})$ are not zero?

What are your conclusions from this study?

4

Vector Spaces and Subspaces

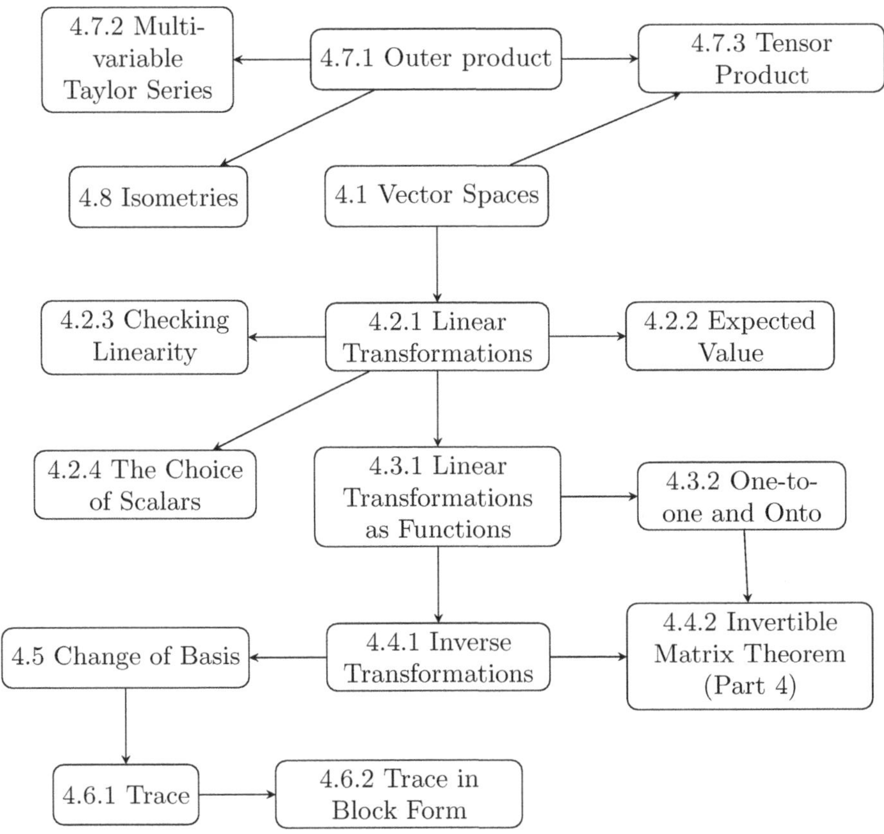

DOI: 10.1201/9781003737490-4

4.1 Vector Spaces

So far we have been working with vectors in $\mathbb{R}^n$ or $\mathbb{C}^n$, and scalars in $\mathbb{R}$ or $\mathbb{C}$, but linear algebra can be used in even more general contexts.

Definition 4.1.1. A vector space consists of a set of vectors V and a field of scalars F, with the following properties:

1. For all $\mathbf{v}$ and $\mathbf{w}$ in V, there exists $\mathbf{v}+\mathbf{w}$ in V (vector addition).
2. For all $\mathbf{v}$ in V and c in F, there exists $c\mathbf{v}$ in V (scalar multiplication).
3. There exists $\mathbf{0}$ in V such that $\mathbf{0}+\mathbf{v}=\mathbf{v}$ for all $\mathbf{v}$ in V.
4. For all $\mathbf{v}$ in V there exists $-\mathbf{v}$ in V such that $\mathbf{v}+(-\mathbf{v})=\mathbf{0}$.
5. $\mathbf{u}+(\mathbf{v}+\mathbf{w})=(\mathbf{u}+\mathbf{v})+\mathbf{w}$ for all $\mathbf{u},\mathbf{v},\mathbf{w}$ in V.
6. $\mathbf{u}+\mathbf{v}=\mathbf{v}+\mathbf{u}$ for all $\mathbf{u},\mathbf{v}$ in V.
7. $a(b\mathbf{v})=(ab)\mathbf{v}$ for all a,b in F and all $\mathbf{v}$ in V.
8. $1\mathbf{v}=\mathbf{v}$, for all $\mathbf{v}$ in V, where 1 is in F.
9. $(a+b)\mathbf{v}=a\mathbf{v}+b\mathbf{v}$ for all a,b in F and $\mathbf{v}$ in V.
10. $a(\mathbf{v}+\mathbf{w})=a\mathbf{v}+b\mathbf{w}$ for all a in F and $\mathbf{v},\mathbf{w}$ in V.

We will not give the formal definition of a field here. Both $\mathbb{R}$ and $\mathbb{C}$ are examples of fields. But there are other fields such as the rational numbers, $\mathbb{Q}$, or finite fields such as $\mathbb{F}_2$ consisting of 0 and 1 with the usual rules for binary arithmetic:

$$\begin{array}{c|cc} + & 0 & 1 \\ \hline 0 & 0 & 1 \\ 1 & 1 & 0 \end{array} \quad \text{and} \quad \begin{array}{c|cc} \times & 0 & 1 \\ \hline 0 & 0 & 0 \\ 1 & 0 & 1 \end{array}$$

Any field will do as a choice of scalars, and any object that satisfies the properties in Definition 4.1.1 will give us a vector space. The first four properties are existential properties, because they all contain the phrase "there exists." The remaining properties might be called structural properties. The structural properties often don't need to be proved directly, but they do when constructing a vector space from scratch.

Example 4.1.1. Here are some of the common examples of vector spaces:

1. $\mathbb{R}^n$ or $\mathbb{C}^n$,
2. polynomials on $[-1,1]$,
3. continuous functions on $[-1,1]$,
4. solutions to differential equations,
5. $m\times n$ matrices over $\mathbb{R}$ or $\mathbb{C}$,
6. a subspace of any of these, and
7. restriction or extension of scalars of any of these.

What do we mean by a subspace?

Definition 4.1.2. Given a vector space $\mathbf{V}$, a **subspace W** is a subset of $\mathbf{V}$ satisfying the 4 existential properties in Definition 4.1.1, with the same addition, scalar multiplication, and the same vector $\mathbf{0}$.

Intuitively, if you think of $\mathbb{R}^3$, then a plane through the origin, or a line through the origin, or even the origin itself is a subspace. But the origin must be included because the vector $\mathbf{0}$ must be shared.

If you are used to thinking of vectors as arrows, it may seem strange at first that functions can be dealt with in the same way. But we can add functions and multiply them by constants. If f and g are functions and c is a constant, then

1. $f + g$ gives us "vector addition,"

$$(f+g)(x) = f(x) + g(x) \quad \text{for all } x$$

   and

2. cf gives us "scalar multiplication,"

$$(cf)(x) = cf(x) \quad \text{for all } x.$$

The next example uses the symbol $\in$. It is supposed to look like a lowercase epsilon, a Greek letter that makes the sound of an E. E is the first letter in element, so $x \in \mathbb{R}$ means that x is an element of $\mathbb{R}$. In other words x is a real number. Even though it is correct to read $x \in \mathbb{R}$ as "x is an element of $\mathbb{R}$," it is easier to read it as "x is in $\mathbb{R}$."

The next example also uses set builder notation. In case you are uncomfortable with set builder notation, think of U as some large box of objects that you are removing things from. You have some conditions that tell you what objects you are allowed to keep. Then you construct the set

$$\{x \in U : \text{conditions on } x\}$$

Example 4.1.2. Consider the set

$$S = \{(x, y) \in \mathbb{R}^2 : x \geq 0 \quad \text{and} \quad y \geq 0\}.$$

We read this as "S is the set of all pairs (x, y) in $\mathbb{R}$ squared such that x is greater than or equal to zero and y is greater than or equal to zero."

Basically, we just removed all points with negative coordinates. What remains is the first quadrant, the positive x and y-axis, and the origin. Now define

$$(x_1, y_1) + (x_2, y_2) = (x_1 + x_2, y_1 + y_2) \quad \text{and} \quad c(x, y) = (cx, cy).$$

Since S borrows $\mathbf{v} + \mathbf{w}$ and $c\mathbf{v}$ from $\mathbb{R}^2$, all of the structural properties are inherited. What about the existential ones?

1. $(0, 0) \in S$ and has the property that if (x, y) is in S, then

$$(x, y) + (0, 0) = (x + 0, y + 0) = (x, y).$$

2. If (x_1, y_1) and (x_2, y_2) are in S, then $x_i \geq 0$ and $y_i \geq 0$. So

$$(x_1, y_1) + (x_2, y_2) = (x_1 + x_2, y_1 + y_2)$$

   is in S because $x_1, x_2 \geq 0$ implies $x_1 + x_2 \geq 0$.

3. $(1,0) \in S$ so suppose $(x,y)+(1,0)=(0,0)$. Using the definition for + we have
$$(x+1, y+0) = (0,0)$$
so $x+1=0$ which means $x=-1$, and $y+0=0$ which means $y=0$. But $(-1,0)$ is not in S because $-1<0$.

4. For $(1,0) \in S$ and $-1 \in \mathbb{R}$, we have
$$(-1)(1,0) = ((-1)1, (-1)0) = (-1,0),$$
which is not in S because $-1<0$.

Activity 4.1.3. Let L be the line $x+y=2$. The line L is a subset of $\mathbb{R}^2$, so take
$$(x_1,y_1)+(x_2,y_2) = (x_1+x_2, y_1+y_2) \quad \text{and} \quad c(x,y) = (cx, cy).$$
Show that none of the four existential properties are valid.

Hint: In my example the set S was defined by $x \geq 0$ and $y \geq 0$, so checking membership in the set involved deciding whether or not x or y was negative. Here L is defined by $x+y=2$, so membership of a specific point (x_1,y_1) means that $x_1+y_1=2$.
See exercise 4.2 for an extension of this activity.

4.2 Linear Transformations

TL;DR

This section contains two critical definitions

- A linear transformation: Definition4.2.1
- A basis: Definition 4.2.2

The standard basis is defined in Proposition 4.2.3.
We can get a matrix for a linear transformation by acting on a basis as illustrated by Examples 4.2.1, 4.2.4, 4.2.7, and 4.2.9.

4.2.1 Intro to Linear Transformations

Q 4.2.1. What does it mean for $\left[\begin{smallmatrix}2\\3\end{smallmatrix}\right]$ to be in the span of $\left[\begin{smallmatrix}1\\0\end{smallmatrix}\right]$ and $\left[\begin{smallmatrix}0\\1\end{smallmatrix}\right]$?

Q 4.2.2. What does it mean for $\left[\begin{smallmatrix}1\\0\end{smallmatrix}\right]$ and $\left[\begin{smallmatrix}0\\1\end{smallmatrix}\right]$ to be linearly independent?

Q 4.2.3. If $f(x)=x$ and $g(x)=x^2$ then what is $(2f+3g)(x)$?

We first encountered the symmetries of the square in Example 1.1.28 where they were used to illustrate the fact that matrix multiplication is not commutative. In Example 1.3.8, we saw that the inverse of the geometric operations of counterclockwise rotation by 90 degrees and mirror reflection correspond to the inverse matrices. But in both those cases, the matrices were just presented without justification. Let's see what is going on.

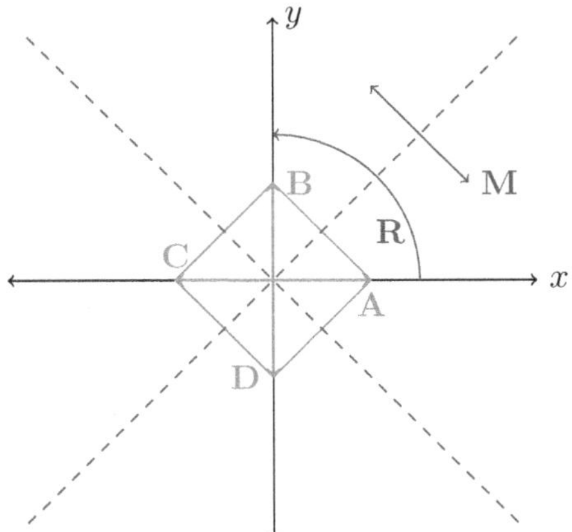

FIGURE 4.1: Symmetries of the square

Example 4.2.1. Figure 4.1 shows a square centered at the origin with vertices at the points $(1,0)$, $(0,1)$, $(-1,0)$ and $(0,-1)$. A symmetry is an operation that sends the square to itself: sending vertices to vertices, and edges to edges. The symmetries of a square can be generated by just two; in this case we use $\mathbf{R}$, rotation counterclockwise by 90 degrees, and $\mathbf{M}$, a mirror reflection across the line $x = y$. If we use column vectors to describe the points in the plane, then the symmetries of the square can be represented by the matrices

$$\begin{aligned} \mathbf{I} &= \begin{bmatrix} 1 & 0 \\ 0 & 1 \end{bmatrix}, & \mathbf{R} &= \begin{bmatrix} 0 & -1 \\ 1 & 0 \end{bmatrix}, & \mathbf{R}^2 &= \begin{bmatrix} -1 & 0 \\ 0 & -1 \end{bmatrix}, & \mathbf{R}^3 &= \begin{bmatrix} 0 & 1 \\ -1 & 0 \end{bmatrix}, \\ \mathbf{M} &= \begin{bmatrix} 0 & 1 \\ 1 & 0 \end{bmatrix}, & \mathbf{MR} &= \begin{bmatrix} 1 & 0 \\ 0 & -1 \end{bmatrix}, & \mathbf{MR}^2 &= \begin{bmatrix} 0 & -1 \\ -1 & 0 \end{bmatrix}, & \mathbf{MR}^3 &= \begin{bmatrix} -1 & 0 \\ 0 & 1 \end{bmatrix}. \end{aligned} \tag{4.1}$$

The top 4 are rotations, and the bottom 4 are reflections.

How do we get the matrix for $\mathbf{R}$? By looking at what it does to the corners of the square. $\mathbf{R}$ is counterclockwise rotation by 90 degrees, so it sends $\mathbf{A}$ to $\mathbf{B}$, $\mathbf{B}$ to $\mathbf{C}$, $\mathbf{C}$ to $\mathbf{D}$, and $\mathbf{D}$ to $\mathbf{A}$. It turns out, however, that the first two already give us enough information.

Step 1: Act on A and B, and write the results as linear combinations of A and B. Since we are using column vectors to represent points in the plane, then

$$\mathbf{A} = \begin{bmatrix} 1 \\ 0 \end{bmatrix}, \quad \mathbf{B} = \begin{bmatrix} 0 \\ 1 \end{bmatrix}, \quad \mathbf{C} = \begin{bmatrix} -1 \\ 0 \end{bmatrix}, \quad \text{and} \quad \mathbf{D} = \begin{bmatrix} 0 \\ -1 \end{bmatrix}. \tag{4.2}$$

It is important to be consistent with the order of the vectors $\mathbf{A}$ and $\mathbf{B}$. We act on $\mathbf{A}$ first and $\mathbf{B}$ second, and in each case we write the result as a linear combination of $\mathbf{A}$ and $\mathbf{B}$ in that order. So,

$$\begin{aligned} \mathbf{R} : \begin{bmatrix} 1 \\ 0 \end{bmatrix} &\mapsto \begin{bmatrix} 0 \\ 1 \end{bmatrix} = \boxed{0} \cdot \begin{bmatrix} 1 \\ 0 \end{bmatrix} + \boxed{1} \cdot \begin{bmatrix} 0 \\ 1 \end{bmatrix}, \quad \text{and} \\ \mathbf{R} : \begin{bmatrix} 0 \\ 1 \end{bmatrix} &\mapsto \begin{bmatrix} -1 \\ 0 \end{bmatrix} = \boxed{-1} \cdot \begin{bmatrix} 1 \\ 0 \end{bmatrix} + \boxed{0} \cdot \begin{bmatrix} 0 \\ 1 \end{bmatrix}. \end{aligned} \tag{4.3}$$

If this is the first time that you have seen the notation $\mapsto$, it can be read as "is sent to," or "maps to." For example, we can also write $f(x) = x^2$ as $x \mapsto x^2$. So here

$$\mathbf{R} : \begin{bmatrix} 1 \\ 0 \end{bmatrix} \mapsto \begin{bmatrix} 0 \\ 1 \end{bmatrix}$$

means that **R** sends **A** to **B**. We then write **B** as a linear combination of **A** and **B**. This can only be done in the way shown by (4.3). Similarly, the 2nd line of (4.3) means **B** is sent to **C**. Then **C** is also written as a linear combination of **A** and **B**. The order matters! Since we acted on **A** in the first line and **B** in the second line, then the linear combination must be in the same order.

Step 2: Form the matrix from the coefficients and take the transpose. The transpose is necessary because we are using column vectors. The coefficients are indicated in boxes in (4.3), so the matrix for **R** is

$$\begin{bmatrix} 0 & 1 \\ -1 & 0 \end{bmatrix}^T = \begin{bmatrix} 0 & -1 \\ 1 & 0 \end{bmatrix},$$

which explains the matrix for **R** in equation (4.1).

What good is this? Well, the point is that now rotation can be described by matrix multiplication, so we should have $\mathbf{RA} = \mathbf{B}$, $\mathbf{RB} = \mathbf{C}$, $\mathbf{RC} = \mathbf{D}$, and $\mathbf{RD} = \mathbf{A}$, which again assumes that we are using column vectors. Let's check this.

$$\begin{array}{cc} & \begin{bmatrix} 1 \\ 0 \end{bmatrix} & & \begin{bmatrix} 0 \\ 1 \end{bmatrix} \\ \begin{bmatrix} 0 & -1 \\ 1 & 0 \end{bmatrix} & \begin{bmatrix} 0 \\ 1 \end{bmatrix} & \begin{bmatrix} 0 & -1 \\ 1 & 0 \end{bmatrix} & \begin{bmatrix} -1 \\ 0 \end{bmatrix} \\ & \begin{bmatrix} -1 \\ 0 \end{bmatrix} & & \begin{bmatrix} 0 \\ -1 \end{bmatrix} \\ \begin{bmatrix} 0 & -1 \\ 1 & 0 \end{bmatrix} & \begin{bmatrix} 0 \\ -1 \end{bmatrix} & \begin{bmatrix} 0 & -1 \\ 1 & 0 \end{bmatrix} & \begin{bmatrix} 1 \\ 0 \end{bmatrix} \end{array}$$

So, as strange as the transpose might seem right now, these calculations show that it is correct to take the transpose.

Activity 4.2.2. If **M** is the mirror reflection across $x = y$, then it has the effect of sending **A** to **B**, **B** to **A**, **C** to **D**, and **D** to **C**.

1. Use the action on **A** and **B** to get a matrix for **M**:

$$\mathbf{M} : \begin{bmatrix} 1 \\ 0 \end{bmatrix} \mapsto \qquad = \underline{\quad} \cdot \begin{bmatrix} 1 \\ 0 \end{bmatrix} + \underline{\quad} \cdot \begin{bmatrix} 0 \\ 1 \end{bmatrix}$$
$$\mathbf{M} : \begin{bmatrix} 0 \\ 1 \end{bmatrix} \mapsto \qquad = \underline{\quad} \cdot \begin{bmatrix} 1 \\ 0 \end{bmatrix} + \underline{\quad} \cdot \begin{bmatrix} 0 \\ 1 \end{bmatrix}$$

2. Verify that $\mathbf{MA} = \mathbf{B}$, $\mathbf{MB} = \mathbf{A}$, $\mathbf{MC} = \mathbf{D}$, and $\mathbf{MD} = \mathbf{C}$.

To summarize, we have just seen that we can think of **R** and **M** as being transformations $\mathbb{R}^2 \to \mathbb{R}^2$ without necessarily having matrices. We were able to get a matrix by acting on **A** and **B** which is an example of a basis: it is a basis for $\mathbb{R}^2$. We do not need to act on **C** and **D**, because $\mathbb{R}^2$ has 2 dimensions, so a basis for $\mathbb{R}^2$ only has 2 vectors. The rest of the vectors can be written as linear combinations of the basis; in this case

$$\mathbf{C} = -\mathbf{A} \quad \text{and} \quad \mathbf{D} = -\mathbf{B}$$

Definition 4.2.1. Let V and W be two vector spaces. A transformation $T : V \to W$ is called a **linear transformation** if both of the following conditions are true

1. $T(\mathbf{v}_1 + \mathbf{v}_2) = T(\mathbf{v}_1) + T(\mathbf{v}_2)$ for all vectors $\mathbf{v}_1, \mathbf{v}_2 \in V$,

2. $T(c\mathbf{v}) = cT(\mathbf{v})$ for all vectors $v \in V$ and all scalars $c \in F$.

The two conditions in Definition 4.2.1 mean that the transformation must send a linear combination to a linear combination with the same coefficients. In particular, if we take $c = 0$, then we see that

$$T(\mathbf{0}) = T(0\mathbf{v}) = 0T(\mathbf{v}) = \mathbf{0},$$

thus T must send the zero vector in V to the zero vector in W. The symmetries of the square are examples of linear transformations from $\mathbb{R}^2$ to $\mathbb{R}^2$. Subsection 4.2.3 discusses how to show that something is a linear transformation.

Definition 4.2.2. Given a vector space V with scalars in F, a collection of vectors in some order is a **basis** for V if both of the following two properties hold:

1. the vectors span V, and
2. the vectors are linearly independent over F.

If the number of basis vectors is finite, then the **dimension** of the vector space V is equal to the number of basis vectors.

Remark 4.2.1. The concept of dimension here matches the one for span, but there are some subtle points. When we say that the basis vectors span V, we mean that we can represent an arbitrary vector $\mathbf{v}$ as a linear combination of the basis vectors. By definition, a linear combination must have a finite number of terms. For more on this point, see Example 4.2.6. Linear independence then gives us uniqueness of those linear combinations. However, what is or is not linearly independent depends on the choice of scalars. For more on this point, see Subsection 4.2.4.

Proposition 4.2.3. *The n vectors*

$$\mathbf{e}_1 = \begin{bmatrix} 1 \\ 0 \\ \vdots \\ 0 \end{bmatrix}, \mathbf{e}_2 = \begin{bmatrix} 0 \\ 1 \\ \vdots \\ 0 \end{bmatrix}, \ldots, \mathbf{e}_n = \begin{bmatrix} 0 \\ 0 \\ \vdots \\ 1 \end{bmatrix},$$

where each vector has n-rows, is a basis for $\mathbb{R}^n$, and is called the **standard basis**.

Example 4.2.3. In Example 4.2.1, we used the basis

$$\mathbf{A} = \mathbf{e}_1 = \begin{bmatrix} 1 \\ 0 \end{bmatrix} \quad \text{and} \quad \mathbf{B} = \mathbf{e}_2 = \begin{bmatrix} 0 \\ 1 \end{bmatrix},$$

which is the standard basis for $\mathbb{R}^2$. But let's check the two properties. An arbitrary linear combination of $\mathbf{A}$ and $\mathbf{B}$ is

$$x\mathbf{A} + y\mathbf{B} = x \begin{bmatrix} 1 \\ 0 \end{bmatrix} + y \begin{bmatrix} 0 \\ 1 \end{bmatrix} = \begin{bmatrix} x \\ y \end{bmatrix}, \tag{4.4}$$

which gives all points in $\mathbb{R}^2$ as x and y vary. So, $\mathbf{A}$ and $\mathbf{B}$ span $\mathbb{R}^2$. They are linearly independent because $x\mathbf{A} + y\mathbf{B} = \mathbf{0}$ means that

$$\begin{bmatrix} x \\ y \end{bmatrix} = \begin{bmatrix} 0 \\ 0 \end{bmatrix},$$

so $x = y = 0$ is the only option.

To get a matrix for a linear transformation from $T : V \to W$, we need a basis for both V and W. We act on the basis vectors $\mathbf{v}_j$ of V, and write them as linear combinations of the basis vectors in W:

$$T(\mathbf{v}_i) = a_{1i}\mathbf{w}_1 + a_{2i}\mathbf{w}_2 + \cdots + a_{mi}\mathbf{w}_m.$$

It is possible to write $T(\mathbf{v}_i)$ as a linear combination of $\mathbf{w}_1, \mathbf{w}_2 \ldots \mathbf{w}_m$ because they span W. There is only one way to do it by linear independence. If there were two ways, we could get a dependence relation by subtracting. In the case of Example 4.2.1, V and W were both $\mathbb{R}^2$. Let's look at an example where V and W are not the same.

Example 4.2.4. Obtain a matrix for $T : \mathbb{R}^2 \to \mathbb{R}^3$ defined by $(x, y) \mapsto (x, y, 0)$, by looking at the action on the standard basis.

Solution

The standard basis on $\mathbb{R}^3$ is

$$\mathbf{e}_1 = \begin{bmatrix} 1 \\ 0 \\ 0 \end{bmatrix}, \quad \mathbf{e}_2 = \begin{bmatrix} 0 \\ 1 \\ 0 \end{bmatrix}, \quad \text{and} \quad \mathbf{e}_3 = \begin{bmatrix} 0 \\ 0 \\ 1 \end{bmatrix}.$$

We also use the standard basis for $\mathbb{R}^2$. The transformation $\mathbf{T}$ is defined by introducing a 3rd component. We act on the basis in $\mathbb{R}^2$, getting a vector in $\mathbb{R}^3$, which we then write in terms of the basis in $\mathbb{R}^3$:

$$T : \begin{bmatrix} 1 \\ 0 \end{bmatrix} \mapsto \begin{bmatrix} 1 \\ 0 \\ 0 \end{bmatrix} = \boxed{1} \cdot \begin{bmatrix} 1 \\ 0 \\ 0 \end{bmatrix} + \boxed{0} \cdot \begin{bmatrix} 0 \\ 1 \\ 0 \end{bmatrix} + \boxed{0} \cdot \begin{bmatrix} 0 \\ 0 \\ 1 \end{bmatrix}, \quad \text{and}$$

$$T : \begin{bmatrix} 0 \\ 1 \end{bmatrix} \mapsto \begin{bmatrix} 0 \\ 1 \\ 0 \end{bmatrix} = \boxed{0} \cdot \begin{bmatrix} 1 \\ 0 \\ 0 \end{bmatrix} + \boxed{1} \cdot \begin{bmatrix} 0 \\ 1 \\ 0 \end{bmatrix} + \boxed{0} \cdot \begin{bmatrix} 0 \\ 0 \\ 1 \end{bmatrix}.$$

Therefore, the matrix that we get for $\mathbf{T}$ is

$$\mathbf{T} = \begin{bmatrix} 1 & 0 & 0 \\ 0 & 1 & 0 \end{bmatrix}^T = \begin{bmatrix} 1 & 0 \\ 0 & 1 \\ 0 & 0 \end{bmatrix}.$$

If it was not clear in Example 4.2.1 why we needed the transpose, it hopefully is clear now: since we are acting on column vectors, matrix multiplication would not be defined unless we take the transpose. Doing the matrix multiplication shows that our calculations are correct:

$$\begin{array}{cc} & \begin{bmatrix} x \\ y \end{bmatrix} \\ \begin{bmatrix} 1 & 0 \\ 0 & 1 \\ 0 & 0 \end{bmatrix} & \begin{bmatrix} x \\ y \\ 0 \end{bmatrix} \end{array}.$$

Activity 4.2.5. Obtain a matrix for $T : \mathbb{R}^3 \to \mathbb{R}^2$ defined by $(x, y, z) \mapsto (x, y)$, by looking at the action on the standard basis.

Check, that the answer makes sense by multiplying $\mathbf{T}$ by an arbitrary vector in $\mathbb{R}^3$.

What happens if you take $x = y = z = 0$?

Example 4.2.6. The set $\mathbb{R}[x]$ is the set of polynomials with real coefficients. It is a vector space, and the powers

$$1, x, x^2, \ldots$$

form a basis. Linear independence can be checked with Wronskians, just as in Activity 3.4.2, and polynomials are finite linear combinations of powers by definition.

On the other hand, $\mathbb{R}[[x]]$ is the set of power series (Taylor series expanded at the origin). It is also a vector space, but the powers do not form a basis. They are still linearly independent, but there are plenty of examples of power series that are not finite linear combinations of powers. For example

$$\sum_{k=0}^{\infty} x^k.$$

But, an infinite sum is a limit anyway. The specific sum above is the limit

$$\sum_{k=0}^{\infty} x^k = \lim_{n\to\infty} \sum_{k=0}^{n} x^k = \lim_{n\to\infty} 1 + x + x^2 + \cdots + x^n.$$

The power series in $\mathbb{R}[[x]]$ are limits of polynomials by definition, so it is often possible to work with a basis for $\mathbb{R}[x]$ without needing a basis for $\mathbb{R}[[x]]$.

While we are generally going to avoid infinite vector spaces, we do want to consider some examples with polynomials. There are many natural ways of getting a linear transformation from a polynomial. For example, we can just plug in.

Example 4.2.7. Let V be the vector space of polynomials of degree less than or equal to 2. Then $T : V \to \mathbb{R}^2$ given by

$$f \mapsto \begin{bmatrix} f(2) \\ f(3) \end{bmatrix}$$

is a linear transformation. Compute a matrix for T with the basis

$$1, x, x^2$$

for V and the standard basis for $\mathbb{R}^2$.

Solution

First, we apply T to 1, x, x^2:

$$T(1) = \begin{bmatrix} 1 \\ 1 \end{bmatrix} = 1 \cdot \begin{bmatrix} 1 \\ 0 \end{bmatrix} + 1 \cdot \begin{bmatrix} 0 \\ 1 \end{bmatrix},$$
$$T(x) = \begin{bmatrix} 2 \\ 3 \end{bmatrix} = 2 \cdot \begin{bmatrix} 1 \\ 0 \end{bmatrix} + 3 \cdot \begin{bmatrix} 0 \\ 1 \end{bmatrix}, \quad \text{and}$$
$$T(x^2) = \begin{bmatrix} 4 \\ 9 \end{bmatrix} = 4 \cdot \begin{bmatrix} 1 \\ 0 \end{bmatrix} + 9 \cdot \begin{bmatrix} 0 \\ 1 \end{bmatrix}.$$

So we get

$$\begin{bmatrix} 1 & 1 \\ 2 & 3 \\ 4 & 9 \end{bmatrix}^T = \begin{bmatrix} 1 & 2 & 4 \\ 1 & 3 & 9 \end{bmatrix}.$$

Note that because of the order of the basis, $ax^2 + bx + c$ is represented by the vector

$$\begin{bmatrix} c \\ b \\ a \end{bmatrix},$$

so the effect on $x^2 - 2$ is

$$\begin{bmatrix} -2 \\ 0 \\ 1 \end{bmatrix}$$

$$\begin{bmatrix} 1 & 2 & 4 \\ 1 & 3 & 9 \end{bmatrix}\begin{bmatrix} 2 \\ 7 \end{bmatrix}.$$

Activity 4.2.8. Let V be the vector space of polynomials of degree less than or equal to 2. Then $T : V \to \mathbb{R}$ given by

$$f \mapsto \begin{bmatrix} f(-2) \end{bmatrix}$$

is a linear transformation. Compute a matrix for T with the basis

$$1, x, x^2$$

for V and the standard basis for $\mathbb{R}$.

Example 4.2.9. Differentiation is also linear because of the rules for differentiation:

$$\frac{d}{dx}(f(x) + g(x)) = \frac{d}{dx}f(x) + \frac{d}{dx}g(x) \quad \text{and} \quad \frac{d}{dx}cf(x) = c\frac{d}{dx}f(x).$$

Let V be the vector space of polynomials of degree less than or equal to 2, with basis

$$1, x, x^2.$$

By applying $\frac{d}{dx}$ to each function in the basis, we get

$$\begin{aligned} \frac{d}{dx}1 &= 0 = \boxed{0} \cdot 1 + \boxed{0} \cdot x + \boxed{0} \cdot x^2 \\ \frac{d}{dx}x &= 1 = \boxed{1} \cdot 1 + \boxed{0} \cdot x + \boxed{0} \cdot x^2 \\ \frac{d}{dx}x^2 &= 2x = \boxed{0} \cdot 1 + \boxed{2} \cdot x + \boxed{0} \cdot x^2. \end{aligned}$$

So, the action of $\frac{d}{dx}$ can be described by the matrix

$$\begin{bmatrix} 0 & 0 & 0 \\ 1 & 0 & 0 \\ 0 & 2 & 0 \end{bmatrix}^T = \begin{bmatrix} 0 & 1 & 0 \\ 0 & 0 & 2 \\ 0 & 0 & 0 \end{bmatrix}.$$

It is worthwhile to check on $ax^2 + bx + c$. Because of the order of the basis, we have

$$\begin{bmatrix} c \\ b \\ a \end{bmatrix}$$

$$\begin{bmatrix} 0 & 1 & 0 \\ 0 & 0 & 2 \\ 0 & 0 & 0 \end{bmatrix}\begin{bmatrix} b \\ 2a \\ 0 \end{bmatrix},$$

which is $2ax + b$ as expected.

4.2.2 Expected Value

In statistics or probability it is often useful to think of a random process with different outcomes. For example, if the process is a coin flip, then the outcomes are "Heads" or "Tails" (H or T). The sample space Ω is the set of all possible outcomes. For one coin flip, it would be

$$\Omega = \{H, T\}.$$

We will use $\mathcal{O}$ to refer to an outcome. Each outcome has a certain probability of occurring. If the coin is fair, H and T both have a probability of $\frac{1}{2}$. If the coin is weighted the probabilities could be different, like $\frac{3}{4}$ for H and $\frac{1}{4}$ for T, but the total probability for a sample space must always be 1. This additional flexibility makes it possible to describe a process like making a free throw in basketball, where the probability of success and the probability of failure are usually not the same.

A **random variable** is a function $X : \Omega \to \mathbb{R}$. Imagine that I enter a free throw competition where I make 1 dollar if I make a basket B, but lose 2 dollars for each free throw that I miss M. This gives me the random variable

$$X(\mathcal{O}) = \begin{cases} 1 & \text{if } \mathcal{O} = B \\ -2 & \text{if } \mathcal{O} = M \end{cases}$$

Definition 4.2.4. The **expected value** of a random variable X is the weighted average

$$\mathrm{E}[X] = X(\mathcal{O}_1)p_1 + X(\mathcal{O}_2)p_2 + \cdots X(\mathcal{O}_n)p_n$$

where p_i is the probability of the outcome $\mathcal{O}_i$, and the sum is over all outcomes.

If I have $\frac{3}{4}$ of making a basket and $\frac{1}{4}$ probability of missing, then

$$\mathrm{E}[X] = (1)\frac{3}{4} + (-2)\frac{1}{4} = \frac{3-2}{4} = \frac{1}{4}$$

so I "expect" to get 1/4 of a dollar for each free throw attempt.

As we have seen, certain sets of functions can be considered as vector spaces, and random variables are functions. If we fix a specific sample space Ω, then the random variables are a vector space.

Proposition 4.2.5. *Expected value is a linear transformation from the vector space of random variables on Ω to $\mathbb{R}$. In particular, if X and Y are two random variables on Ω and a, b are scalars, then*

$$\mathrm{E}[aX + bY] = a\,\mathrm{E}[X] + b\,\mathrm{E}[Y].$$

In real life it is rare for exact probabilities to be known. We collect samples to estimate probabilities. The number N of measurements of a random variable is the sample size. If X is a random variable, the measurements x_i can be collected into a vector in $\mathbb{R}^N$. For example, if have three free throw attempts $N = 3$, and I make the 1st two baskets and miss the third, then that can be represented by the vector

$$\begin{bmatrix} 1 \\ 1 \\ -2 \end{bmatrix}.$$

Then the average of the components gives an estimate for $\mathrm{E}[X]$

$$\frac{1+1-2}{3} = 0.$$

Clearly $0 \neq \frac{1}{4}$. A larger sample would give a better approximation. Averaging in this way also gives a linear transformation from $\mathbb{R}^N$ to $\mathbb{R}$.

Activity 4.2.10. Let $E_3 : \mathbb{R}^3 \to \mathbb{R}$ be defined by averaging the components. Compute a matrix for the transformation by acting on the standard basis in $\mathbb{R}^3$

Activity 4.2.11. Ask an LLM to give you NumPy code that does the following

(a) Sets the sample size N as 100

(b) Creates two random vectors X and Y with different non-zero means

(c) Computes $\mathrm{E}[X]$ and $\mathrm{E}[Y]$ using the built-in NumPy function

(d) Computes $Z = aX + bY$, where $a = 2$ and $b = 3$

(e) Computes $\mathrm{E}[Z]$ using the built-in NumPy function

(f) Computes $a\,\mathrm{E}[X] + b\,\mathrm{E}[X]$

All values computed should be displayed with the print function.

If there are several random variables, the measurements for each one are usually combined into the columns of a data table.

4.2.3 Checking Linearity

We now consider how to show that a function is a linear transformation.

Example 4.2.12. We have claimed that counterclockwise rotation by 90 degrees is a linear transformation. The function

$$R\left(\begin{bmatrix} x \\ y \end{bmatrix}\right) = \begin{bmatrix} -y \\ x \end{bmatrix}$$

for any real x, y. First, consider condition 1, since it is a little easier. In the domain, we have

$$c\begin{bmatrix} x \\ y \end{bmatrix} = \begin{bmatrix} cx \\ cy \end{bmatrix}.$$

We apply the rotation and get

$$R\left(\begin{bmatrix} cx \\ cy \end{bmatrix}\right) = \begin{bmatrix} cy \\ -cx \end{bmatrix}. \tag{4.5}$$

On the other hand

$$cR\left(\begin{bmatrix} x \\ y \end{bmatrix}\right) = c\begin{bmatrix} -y \\ x \end{bmatrix}. \tag{4.6}$$

The vectors in equations (4.5) and (4.6) are now in the range, and they are equal by the definition of multiplication by a scalar.

For condition 1, we must start with two vectors in the domain:

$$\begin{bmatrix} x_1 \\ y_1 \end{bmatrix} + \begin{bmatrix} x_2 \\ y_2 \end{bmatrix} = \begin{bmatrix} x_1 + x_2 \\ y_1 + y_2 \end{bmatrix}$$

Again, we apply the rotation

$$R\left(\begin{bmatrix} x_1 + x_2 \\ y_1 + y_2 \end{bmatrix}\right) = \begin{bmatrix} -y_1 - y_2 \\ x_1 + x_2. \end{bmatrix} \tag{4.7}$$

On the other hand

$$R\left(\begin{bmatrix} x_1 \\ y_1 \end{bmatrix}\right) + R\left(\begin{bmatrix} x_2 \\ y_2 \end{bmatrix}\right) = \begin{bmatrix} -y_1 \\ x_1 \end{bmatrix} + \begin{bmatrix} -y_2 \\ x_2 \end{bmatrix} \tag{4.8}$$

The vectors in equations (4.7) and (4.8) are now in the range and they are equal by the definition vector addition.

Activity 4.2.13. If we define $T : \mathbb{R} \to \mathbb{R}$ by $T(x) = 2x$, show that T is linear.

Activity 4.2.14. Let $T : \mathbb{R}^2 \to \mathbb{R}$ be defined by

$$\begin{bmatrix} x \\ y \end{bmatrix} \mapsto x + y.$$

Show that T is linear.

Proposition 4.2.6. *Expected value is a linear. In particular if X and Y are two random variables, and a and b are constants, then*

$$E[aX + bY] = aE[X] + bE[Y]$$

Activity 4.2.15. Let $N = 2$, and consider

4.2.4 The Choice of Scalars

Changing the choice of scalars can change the dimension.

Example 4.2.16. The complex numbers provide one of the most natural examples that I can think of for the effect of the choice of scalars on the dimension of a vector space:

1. $\mathbb{C}$ is a vector space of dimension 1, with complex scalars, but
2. $\mathbb{C}$ is a vector space of dimension 2, with real scalars.

We draw the complex plane precisely because of our ability to think of the complex numbers as a 2 dimensional real vector space.

Consider the real case first. All complex numbers can be represented in the form

$$a + bi = a1 + bi$$

where a, b are real. We can view $1, i$ as "vectors" and a, b as "scalars," thus

$$\mathbb{C} = \text{Span}\{1, i\}.$$

Does a dependence relation exist between $1, i$?

$$a1 + bi = 0 \implies bi = -a \implies i = -\frac{a}{b},$$

if $b \neq 0$. But a, b are real so $-\frac{a}{b}$ is real, which would make i real, but it is not. Hence $b = 0$, but then $a = 0$. So 1 and i are linearly independent (with real scalars), which means that $\mathbb{C}$ has dimension 2 as a real vector space (a vector space over $\mathbb{R}$).

Now consider what happens when a, b are allowed to be complex. We already get all complex numbers from

$$a + bi = a1 + bi$$

when a, b are real, but if we allow a, b to be complex, then

$$a1 + bi = 0$$

can be satisfied with $a = 1$ and $b = i$:

$$1 \cdot 1 + i \cdot i = 1 - 1 = 0.$$

Since we now have a dependence relation between 1 and i, we must remove one of them to get a linearly independent subset. The most natural thing is to remove i. Every complex number can be represented as a complex number times 1, so 1 gives a basis for the complex numbers as a complex vector space. It follows that $\mathbb{C}$ has dimension 1 as a complex vector space (a vector space over $\mathbb{C}$).

Activity 4.2.17. Consider the set

$$K = \{a + b\sqrt{2} : a, b \in \mathbb{Q}\}.$$

Clearly K is a subset of $\mathbb{R}$. Using the rules from $\mathbb{R}$, we define vector addition by

$$(a_1 + b_1\sqrt{2}) + (a_2 + b_2\sqrt{2}) = (a_1 + a_2) + (b_1 + b_2)\sqrt{2}$$

and scalar multiplication by

$$c(a + b\sqrt{2}) = ca + cb\sqrt{2}.$$

The scalars, however, will be restricted to rational numbers.

1. Show that all four existential properties are valid.
2. Show that $1, \sqrt{2}$ are linearly independent over $\mathbb{Q}$. Do this by supposing that

$$a + b\sqrt{2} = 0$$

can be satisfied with rational a, b, and obtaining a contradiction.

4.3 Linear Transformations as Functions

TL;DR

The column space and rank of a matrix are defined in Definition 4.3.3.
The null space and nullity are defined in Definition 4.3.4. Two natural illustrations of the null space are given

- Example 4.3.3 uses projection onto the x-axis.
- Example 4.3.6 uses differentiation for those who have seen Calculus.

The rank and the nullity are connected by Theorem 4.3.5. Remark 4.3.2 explains the main strategy used in this book for determining the rank and the nullity.
Example 4.3.4 shows how to compute a basis for the null space given a matrix that is already in reduced form. By hand, you would use row reduction to get into reduced form first, but a computer might use the QR algorithm (see Section 5.6).

4.3.1 Column Space and Null Space

Q 4.3.1. If c is a constant, what is $\frac{d}{dx}c$?

Q 4.3.2. What is $\frac{d}{dx}x^n$?

Q 4.3.3. What is the span of $\begin{bmatrix} 1 \\ 0 \\ 0 \end{bmatrix}, \begin{bmatrix} 0 \\ 1 \\ 0 \end{bmatrix}$?

Q 4.3.4. How do we identify free variables in augmented form?

Since a linear transformation $T : \mathbb{R}^n \to \mathbb{R}^m$ can be thought of as a function acting on vectors, the terms domain and range have the usual meaning that they have for functions. The use of n and m here means that if we are acting on column vectors and compute a matrix for T, then the matrix will be $m \times n$.

Definition 4.3.1. For a linear transformation $T : \mathbb{R}^n \to \mathbb{R}^m$, the **domain** is the set of all possible input vectors, namely $\mathbb{R}^n$. The **range** of T is the set of all possible output vectors, specifically

$$\{v \in \mathbb{R}^m : v = T(w) \text{ for some } w \text{ in } \mathbb{R}^n\}.$$

The range may also be written as $T(\mathbb{R}^n)$.

The range is contained in $\mathbb{R}^m$ but may not be all of $\mathbb{R}^m$, so the full space $\mathbb{R}^m$ is given a separate name.

Definition 4.3.2. For a linear transformation $T : \mathbb{R}^n \to \mathbb{R}^m$, the **codomain** is the ambient space containing the range. That is to say $\mathbb{R}^m$ is the codomain.

Since the range may or may not be equal to the codomain, it is the first natural example of a subspace that we have encountered.

Definition 4.3.3. The **column space** of a matrix $\mathbf{A}$ is the vector space spanned by the columns of $\mathbf{A}$. We will write the column space using the notation

$$\mathrm{Col}(\mathbf{A}).$$

In the case where $\mathbf{A}$ is the matrix for a linear transformation acting on column vectors, the column space is the range. The dimension of the column space is the **rank** of $\mathbf{A}$, and we will write it as

$$\dim \mathrm{Col}(\mathbf{A}).$$

Example 4.3.1. In Example 4.2.4 we computed the matrix

$$\mathbf{T} = \begin{bmatrix} 1 & 0 \\ 0 & 1 \\ 0 & 0 \end{bmatrix}$$

for the linear transformation $T : \mathbb{R}^2 \to \mathbb{R}^3$ defined by $(x, y) \mapsto (x, y, 0)$. An arbitrary linear combination of the columns is

$$x \begin{bmatrix} 1 \\ 0 \\ 0 \end{bmatrix} + y \begin{bmatrix} 0 \\ 1 \\ 0 \end{bmatrix} = \begin{bmatrix} x \\ y \\ 0 \end{bmatrix},$$

which means that the column space is the xy-plane inside $\mathbb{R}^3$. We may write the column space as

$$\mathrm{Col}(\mathbf{T}) = \mathrm{span} \left\{ \begin{bmatrix} 1 \\ 0 \\ 0 \end{bmatrix}, \begin{bmatrix} 0 \\ 1 \\ 0 \end{bmatrix} \right\},$$

which connects it with the concept of span, and shows its dimension to be 2. So, the rank is 2.

Remark 4.3.1. The rank can even be less than the dimension of the domain because a linear transformation may or may not preserve linear independence.

Example 4.3.2. The transformation $P_x : \mathbb{R}^3 \to \mathbb{R}$ defined as "projection onto the x-axis" can be seen to have the matrix $\begin{bmatrix} 1 & 0 & 0 \end{bmatrix}$. Given the standard basis for $\mathbb{R}^3$, show that the dimension of the span decreases. That is, the vectors

$$P_x \left(\begin{bmatrix} 1 \\ 0 \\ 0 \end{bmatrix} \right), \quad P_x \left(\begin{bmatrix} 0 \\ 1 \\ 0 \end{bmatrix} \right), \quad \text{and} \quad P_x \left(\begin{bmatrix} 0 \\ 0 \\ 1 \end{bmatrix} \right)$$

are not linearly independent.

Solution

The first is 1, the other two are 0, so we can get a dependence relation between them:

$$2P_x\left(\begin{bmatrix}0\\1\\0\end{bmatrix}\right)+3P_x\left(\begin{bmatrix}0\\0\\1\end{bmatrix}\right)=2\cdot 0+3\cdot 0=0.$$

It is also natural to consider what things get sent to zero.[1]

Definition 4.3.4. If V and W are vector spaces and $T : V \to W$ is a linear transformation, then the **null space** is the subspace of the domain consisting of all vectors $\mathbf{v}$ in V that get sent to $\mathbf{0}$ in W. If $\mathbf{T}$ is a matrix for T acting on column vectors, then we will use the notation

$$\mathrm{Nul}(\mathbf{T}) = \mathrm{Span}\{\mathbf{v} \in \mathbb{R}^n : \mathbf{T}\mathbf{v} = \mathbf{0}\}$$

for the null space. The dimension of the null space is called the **nullity** and will be written as

$$\dim \mathrm{Nul}(\mathbf{T}).$$

Example 4.3.3. Let $P : \mathbb{R}^2 \to \mathbb{R}$ be defined by $(x, y) \mapsto x$. By definition, the null space consists of the vectors in $\mathbb{R}^2$ that get sent to zero. The map P gets the x coordinate, hence the x-coordinate must be zero. So the y-axis is the null space because

$$(0, y) \mapsto 0,$$

not just $(0, 0)$, as illustrated by Figure 4.2. Using the standard basis for both $\mathbb{R}$ and $\mathbb{R}^2$, the matrix for P is

$$\mathbf{P} = \begin{bmatrix}1 & 0\end{bmatrix}.$$

The null space is the set of all vectors $\mathbf{v}$ that are solutions to,

$$\mathbf{P}\mathbf{v} = \mathbf{0} \tag{4.9}$$

which is exactly how we compute dependence relations. So, to compute a basis for the null space, we solve equation (4.9) and compute a parameterization of the solution. In this case the augmented matrix,

$$\left[\begin{array}{cc|c}1 & 0 & 0\end{array}\right],$$

is already in reduced form. There is only one free variable, y, so the dimension of the null space is 1. That is the nullity is 1. To get a basis for the null space, we convert back to equations to get a parameterization:

$$\begin{cases}x = 0\\ y = y\end{cases} \to \begin{bmatrix}x\\y\end{bmatrix} = \begin{bmatrix}0\\y\end{bmatrix} = y\begin{bmatrix}0\\1\end{bmatrix}.$$

Hence we may write

$$\mathrm{Nul}(\mathbf{P}) = \mathrm{Span}\left\{\begin{bmatrix}0\\1\end{bmatrix}\right\}$$

[1]Even when dealing with functions that are not linear, we may be interested in the roots.

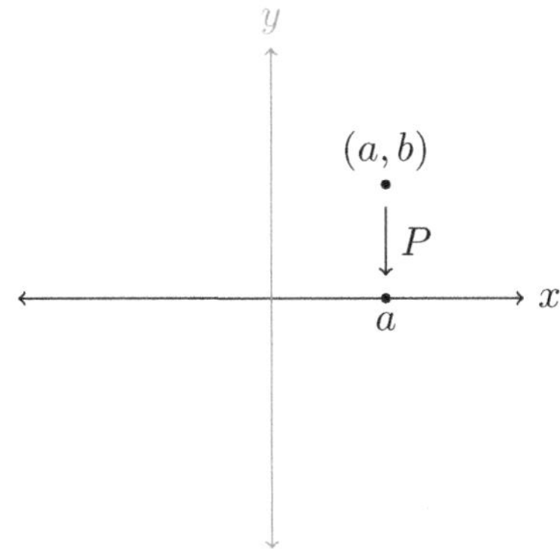

FIGURE 4.2: The null space of projection onto the x-axis is the y-axis

Theorem 4.3.5 (Rank-Nullity Theorem). *Let* $\mathbf{T}$ *be a matrix for a linear transformation from* $\mathbb{R}^n$ *to* $\mathbb{R}^m$ *acting on column vectors. Then*

$$\dim \mathrm{Col}(\mathbf{T}) + \dim \mathrm{Nul}(\mathbf{T}) = n,$$

where n *is the dimension of the domain.*

This theorem is called the Fundamental Theorem of Linear Maps by Axler [2], where a proof can also be found.

Remark 4.3.2. Remember that the column space is a subspace of the codomain while the null space is a subspace of the domain, and yet their dimensions are related by this theorem. We can easily compute all of the values in this theorem using the following observations:

1. The dimension of the column space is the number of pivots in the reduced form of $\mathbf{T}$.
2. The dimension of the null space is the number of free variables in $\mathbf{T}$.
3. The dimension of the domain (i.e. the total number of columns of $\mathbf{T}$).

Since we identify free variables as belonging to columns not having pivots, then the theorem should seem natural.

Example 4.3.4. Suppose a linear transformation $T : \mathbb{R}^5 \to \mathbb{R}^3$ has matrix

$$\mathbf{T} = \begin{bmatrix} \boxed{1} & 1 & 0 & 0 & 3 \\ 0 & 0 & \boxed{1} & 0 & 4 \\ 0 & 0 & 0 & 0 & 0 \end{bmatrix}.$$

1. Compute $\dim \mathrm{Col}(\mathbf{T})$.
2. Compute $\dim \mathrm{Nul}(\mathbf{T})$.
3. Verify the Rank-Nullity Theorem.
4. Find a basis for $\mathrm{Col}(\mathbf{T})$.
5. Find a basis for $\mathrm{Nul}(\mathbf{T})$.

Solution

The first three parts are not bad. The matrix $\mathbf{T}$ is in reduced form, and the pivots are identified in boxes, so

$$\dim \operatorname{Col}(\mathbf{T}) = 2.$$

The remaining columns correspond to free variables, so

$$\dim \operatorname{Nul}(\mathbf{T}) = 3.$$

The dimension of the domain is 5, which can be seen either by the total number of columns or from the power on $\mathbb{R}$ on the left of the $\to$. Since

$$2 + 3 = 5,$$

this verifies the Rank-Nullity Theorem. The columns span the column space, so to get a basis for the column space, we just need to remove vectors from dependence relations as we did in Example 3.3.7. The vectors that are kept correspond to the pivots. In this case those are

$$\begin{bmatrix} 1 \\ 0 \\ 0 \end{bmatrix} \quad \text{and} \quad \begin{bmatrix} 0 \\ 1 \\ 0 \end{bmatrix}.$$

To get a basis for the null space, we need a parameterization of the solutions to $\mathbf{Tx} = \mathbf{0}$. Let the variables in the domain be x_1, x_2, x_3, x_4, and x_5. The equations that we get from reduced form are

$$\begin{cases} x_1 + x_2 + 3x_5 = 0 \\ \qquad\quad x_3 + 4x_5 = 0 \end{cases} \to \begin{cases} x_1 = -x_2 - 3x_5 \\ x_3 = -4x_5 \end{cases}$$

and the free variables x_2, x_4, and x_5 give us the equations $x_2 = x_2$, $x_4 = x_4$, and $x_5 = x_5$. Hence we get the parameterization

$$\begin{bmatrix} x_1 \\ x_2 \\ x_3 \\ x_4 \\ x_5 \end{bmatrix} = \begin{bmatrix} -x_2 - 3x_5 \\ x_2 \\ -4x_5 \\ x_4 \\ x_5 \end{bmatrix} = x_2 \begin{bmatrix} -1 \\ 1 \\ 0 \\ 0 \\ 0 \end{bmatrix} + x_4 \begin{bmatrix} 0 \\ 0 \\ 0 \\ 1 \\ 0 \end{bmatrix} + x_5 \begin{bmatrix} -3 \\ 0 \\ -4 \\ 0 \\ 1 \end{bmatrix}.$$

A basis is obtained by taking one of the free variables to be 1, and the others to be zero, hence

$$\operatorname{Nul}(\mathbf{T}) = \operatorname{Span}\left\{ \begin{bmatrix} -1 \\ 1 \\ 0 \\ 0 \\ 0 \end{bmatrix}, \begin{bmatrix} 0 \\ 0 \\ 0 \\ 1 \\ 0 \end{bmatrix}, \begin{bmatrix} -3 \\ 0 \\ -4 \\ 0 \\ 1 \end{bmatrix} \right\}.$$

Activity 4.3.5. Suppose a linear transformation $T : \mathbb{R}^7 \to \mathbb{R}^5$ has matrix

$$\mathbf{T} = \begin{bmatrix} 1 & 0 & 0 & 0 & 0 & -3 & 0 \\ 0 & 1 & 0 & 0 & 0 & -2 & 0 \\ 0 & 0 & 0 & 1 & 1 & -1 & 0 \\ 0 & 0 & 0 & 0 & 0 & 0 & 1 \\ 0 & 0 & 0 & 0 & 0 & 0 & 0 \end{bmatrix}$$

1. Compute $\dim \operatorname{Col}(\mathbf{T})$.

2. Compute $\dim \mathrm{Nul}(\mathbf{T})$.

3. Verify the Rank-Nullity Theorem.

4. Find a basis for $\mathrm{Col}(\mathbf{T})$.

5. Find a basis for $\mathrm{Nul}(\mathbf{T})$.

Reflection 4.3.1. If the columns of $\mathbf{T}$ are $\mathbf{T}_1, \mathbf{T}_2, \ldots \mathbf{T}_n$, then computing the null space amounts to solving

$$x_1\mathbf{T}_1 + x_2\mathbf{T}_2 + \cdots x_n\mathbf{T}_n = \mathbf{0},$$

which is the same as finding a dependence relation between the columns. A dependence relation exists if the dimension of the null space is positive. What can be said about the dimension of the null space if $\mathbf{T}$ has more columns than rows? What if $\mathbf{T}$ has more rows than columns?

Example 4.3.6. Let V be the vector space of polynomials of degree less than or equal to 2. We will use the basis $x^2, x, 1$.

We have seen in Example 4.2.9 that $\frac{d}{dx} : V \to V$ has the matrix

$$\mathbf{T} = \begin{bmatrix} 0 & 1 & 0 \\ 0 & 0 & 2 \\ 0 & 0 & 0 \end{bmatrix}.$$

1. The null space consists of all constant polynomials c and has dimension 1.

2. The column space consists of all polynomials of degree less than or equal to 1, and has dimension 2.

3. The domain has dimension 3, and $2 + 1 = 3$.

To compute a null space with Sage, you should be aware that by default Sage acts on row vectors not column vectors. Another name for the null space is the kernel, which is what Sage uses. When acting on column vectors, those vectors are on the right-hand side. So, here is the best way to compute the null space with Sage:

```
A=matrix([[ 1, 0, 0],
          [ 0, 1, 0]])
A.right_kernel()
```

The output specifies both a basis for the null space and its "rank" (i.e. its dimension). For clarity compare the following lines:

```
A.rank()
A.right_kernel().basis()
A.right_kernel().basis_matrix()
A.right_kernel().rank()
```

The first line of code shows that the rank of the matrix $\mathbf{A}$ is 2. The second line gives a basis for the null space as a list of vectors. The third line gives a basis for the null space as a matrix. The fourth line gives the rank of the basis matrix for the null space, which is equal to the nullity of $\mathbf{A}$.

4.3.2 One-to-one and Onto

Definition 4.3.6. A mapping $\mathbf{T} : \mathbb{R}^n \to \mathbb{R}^m$ is said to be **onto** $\mathbb{R}^m$ if each $\mathbf{b}$ in $\mathbb{R}^m$ is the image of at least one $\mathbf{x}$ in $\mathbb{R}$.

A mapping $\mathbf{T} : \mathbb{R}^n \to \mathbb{R}^m$ is said to be **one-to-one** if each $\mathbf{b}$ in $\mathbb{R}^m$ is the image of at most one $\mathbf{x}$ in $\mathbb{R}^n$.

Example 4.3.7. We have already seen examples of these from the previous linear transformations discussed

1. $\mathbf{P}_x$ (projection onto the x-axis) from Example 4.3.2 is onto but not one-to-one,
2. $\mathbf{T}$ (defined by $(x, y) \mapsto (x, y, 0)$) from Example 4.2.4 is one-to-one but not onto,
3. $\mathbf{R}$ (counterclockwise rotation by 90 degrees centered at the origin) and $\mathbf{M}$ (reflection across the line $x = y$) from Example 4.2.1 are both one-to-one and onto.

We illustrate how to check these facts.

To see that a function, or linear transformation, is not onto, we must find something that is not in the range. The map $\mathbf{T}$ is not onto because everything in the range has 0 in the last component. For example, it is impossible to get $(0, 0, 1)$. To see that a function is onto, we must take an arbitrary element in the codomain, and find something in the domain that maps to it. The map $\mathbf{P}_x$ is onto because if we take an arbitrary point a on the x-axis, then

$$\mathbf{P}_x \begin{bmatrix} a \\ 0 \\ 0 \end{bmatrix} = a.$$

The method we have just described works even if the dimension of the codomain is infinite, but if it is finite then we can say that a linear transformation is onto if and only if the range and codomain have the same dimension. We know from Remark 4.3.2 that the dimension of the range is equal to the number of pivots. Meanwhile, the dimension of the codomain is equal to the number of rows. So, in terms of matrices, a finite linear transformation is onto if and only if there is a pivot in every row. The matrix for P_x is

$$\begin{bmatrix} 1 & 0 & 0 \end{bmatrix},$$

so it is onto, but the matrix for T is

$$\begin{bmatrix} 1 & 0 \\ 0 & 1 \\ 0 & 0 \end{bmatrix},$$

so it is not onto. Generally if the dimension of the codomain is bigger than the dimension of the domain, then a linear transformation has no hope of being onto. However, it may still not be onto even if this is not the case (for example, send everything to zero).

To see that a function is not one-to-one we must find at least two elements in the domain that map to the same thing in the range. The map P_x is not one-to-one because

$$\mathbf{P}_x \begin{bmatrix} 1 \\ 0 \\ 0 \end{bmatrix} = 1 \quad \text{and} \quad \mathbf{P}_x \begin{bmatrix} 1 \\ 2 \\ 3 \end{bmatrix} = 1.$$

To see that a function is one-to-one, we must take two elements in the domain – not known to be the same at first – and look at their values in the range. Upon setting these values in the range equal to each other, we must be able to conclude that the elements in the domain are actually the same. For example $T : \mathbb{R}^2 \to \mathbb{R}^3$, so we start with (x_1, y_1) and (x_2, y_2). What are the values in the range?

$$\begin{bmatrix} 1 & 0 \\ 0 & 1 \\ 0 & 0 \end{bmatrix} \begin{bmatrix} x_1 \\ y_1 \end{bmatrix} = \begin{bmatrix} x_1 \\ y_1 \\ 0 \end{bmatrix} \quad \text{and} \quad \begin{bmatrix} 1 & 0 \\ 0 & 1 \\ 0 & 0 \end{bmatrix} \begin{bmatrix} x_2 \\ y_2 \end{bmatrix} = \begin{bmatrix} x_2 \\ y_2 \\ 0 \end{bmatrix}.$$

So, then we set the output vectors equal to each other:

$$\begin{bmatrix} x_1 \\ y_1 \\ 0 \end{bmatrix} = \begin{bmatrix} x_2 \\ y_2 \\ 0 \end{bmatrix} \implies x_1 = x_2 \quad \text{and} \quad y_1 = y_2 \implies \begin{bmatrix} x_1 \\ y_1 \end{bmatrix} = \begin{bmatrix} x_2 \\ y_2 \end{bmatrix},$$

so T is one-to-one. Another way to see that a linear transformation is one-to-one is to check if the null space has dimension zero. The matrices for T and P_x are

$$\begin{bmatrix} 1 & 0 \\ 0 & 1 \\ 0 & 0 \end{bmatrix} \quad \text{and} \quad \begin{bmatrix} 1 & 0 & 0 \end{bmatrix}.$$

The first has a pivot in every column, so $\dim \text{Nul}(T) = 0$, hence T is one-to-one. The second has a pivot only in the 1st column, and the other two columns correspond to free variables, hence $\dim \text{Nul}(P_x) = 2$, which means that P_x is not one-to-one.

Activity 4.3.8. Show that

1. $M : (x, y) \mapsto (y, x)$ is one-to-one and onto.
2. $T : (x, y) \mapsto (x, y, 0)$ is not onto.
3. $P_y : (x, y, z) \mapsto y$ is not one-to-one.

4.4 Inverse Transformations

4.4.1 Intro to Invertible Transformations

As with other types of functions, we are often interested in whether or not an inverse of a linear transformation exists and if it exists, how to describe it. So, suppose we have a linear transformation $T : \mathbb{R}^n \to \mathbb{R}^m$. An inverse transformation would need to have $\mathbb{R}^m$ as its domain, so T needs to be onto. That is, we need to get all of $\mathbb{R}^m$ in the range of T. If $n < m$, that is simply not possible. Example 4.2.4 illustrates the problem with $T : \mathbb{R}^2 \to \mathbb{R}^3$: when we start out with 2 linearly independent vectors in the domain, there is no hope of getting more than 2 linearly independent vectors in the range. On the other hand, if $n > m$, the null space must have a positive dimension by the Rank-Nullity Theorem. For P_x in Example 4.3.2, the null space is the entire yz-plane. Any point $(0, y, z)$ is sent to 0 by P_x, and so an inverse of P_x would have to send 0 to the entire yz-plane. Such a thing is not a function. For a linear transformation, the only option is that 0 is sent to 0. So, we must have $m = n$. Even then there still might not be an inverse, but there is no point in looking elsewhere.

Definition 4.4.1. A linear transformation $T : \mathbb{R}^n \to \mathbb{R}^n$ is said to be **invertible** if there exists a function $T^{-1} : \mathbb{R}^n \to \mathbb{R}^n$ such that

$$\begin{aligned} T^{-1}(T(\mathbf{x})) &= \mathbf{x} \quad \text{for all } \mathbf{x} \text{ in } \mathbb{R}^n, \text{ and} \\ T(T^{-1}(\mathbf{x})) &= \mathbf{x} \quad \text{for all } \mathbf{x} \text{ in } \mathbb{R}^n. \end{aligned}$$

Then, T^{-1} is called an **inverse** transformation.

Not surprisingly the concept of inverse transformations is directly connected to inverse matrices. All we have to do is choose a basis.

Theorem 4.4.2. *Let $T : \mathbb{R}^n \to \mathbb{R}^n$ be a linear transformation and let $\mathbf{A}$ be a matrix for T in some basis. Then T^{-1} exists if and only if $\mathbf{A}$ is an invertible matrix. In that case, the matrix for the inverse transformation T^{-1} in the same basis is given by $\mathbf{A}^{-1}$.*

Example 4.4.1. We have already looked at the connection between

$$\mathbf{R} = \begin{bmatrix} 0 & -1 \\ 1 & 0 \end{bmatrix}$$

and its inverse matrix

$$\mathbf{R}^{-1} = \begin{bmatrix} 0 & 1 \\ -1 & 0 \end{bmatrix}$$

back in Example 1.3.8. However, even without matrices, it should make sense that the inverse of counterclockwise rotation by 90 degrees is clockwise rotation by 90 degrees. The standard basis allows them to be expressed as $\mathbf{R}$ and $\mathbf{R}^{-1}$.

Activity 4.4.2. Permutation matrices.

1. Let $\mathbf{C}$ be the cyclic permutation, sending x to y, y to z, and z back to x. Obtain a matrix for $\mathbf{C}$ by acting on the standard basis in $\mathbb{R}^3$ (check that it does what it is supposed to by multiplying).

2. Check (by multiplying) that $\mathbf{C}^T$ sends y to x, z to y, and x to z. Observe that this just reverses the arrows between x, y, and z.

3. Check that $\mathbf{C}^T\mathbf{C} = \mathbf{I}$ so that $\mathbf{C}^{-1} = \mathbf{C}^T$ in this case.

Definition 4.4.3. A **permutation matrix** is an $n \times n$ matrix with exactly one 1 in each row and column, and zeros everywhere else.

Activity 4.4.3. The change between rectangular and polar coordinates is an example of a multi-variable substitution

$$\begin{aligned} &\mathbf{f} : \mathbb{R}^2 \to \mathbb{R}^2 \\ &\begin{bmatrix} r \\ \theta \end{bmatrix} \mapsto \begin{bmatrix} r\cos\theta \\ r\sin\theta \end{bmatrix} = \begin{bmatrix} x \\ y \end{bmatrix} \end{aligned}$$

so $f_1(r,\theta) = r\cos\theta$ and $f_2(r,\theta) = r\sin\theta$ are the components of $\mathbf{f}$.

1. Compute the Jacobian determinant

$$\begin{vmatrix} \frac{\partial f_1}{\partial r} & \frac{\partial f_1}{\partial \theta} \\ \frac{\partial f_2}{\partial r} & \frac{\partial f_2}{\partial \theta} \end{vmatrix}$$

2. Set your result equal to zero. For what values of r and θ is this equation satisfied?

3. At any point (r_0, θ_0) in the domain, the function $\mathbf{f}$ can be approximated by a linear function

$$\begin{bmatrix} x - x_0 \\ y - y_0 \end{bmatrix} = \begin{vmatrix} \frac{\partial f_1}{\partial r} & \frac{\partial f_1}{\partial \theta} \\ \frac{\partial f_2}{\partial r} & \frac{\partial f_2}{\partial \theta} \end{vmatrix} \begin{bmatrix} r - r_0 \\ \theta - \theta_0 \end{bmatrix}$$

where $x_0 = f_1(r_0, \theta_0)$ and $y_0 = f_2(r_0, \theta_0)$. This function has an inverse if and only if the Jacobian determinant is not zero. Given $r_0 = 2$ and $\theta_0 = \frac{\pi}{3}$, compute this linear approximation and its inverse.

4.4.2 The Invertible Matrix Theorem (Part 4)

Theorem 4.4.4 (The Invertible Matrix Theorem: Part 4). *Let* $\mathbf{A}$ *be an* $n \times n$ *matrix with real entries. The following statements are equivalent.*

11. The linear transformation $\mathbf{x} \mapsto \mathbf{Ax}$ *is one-to-one.*

12. The linear transformation $\mathbf{x} \mapsto \mathbf{Ax}$ *is onto.*

13. The columns of $\mathbf{A}$ *span the codomain.*

14. The rows of $\mathbf{A}$ *span the domain.*

15. $\dim \operatorname{Col}(\mathbf{A}) = n$

16. $\dim \operatorname{Nul}(\mathbf{A}) = 0$

See Remark 1.3.1 concerning the Invertible matrix theorem and its proof.

Example 4.4.4. Show that each statement in the theorem is true for the matrix

$$\mathbf{A} = \begin{bmatrix} 1 & 1 \\ 0 & 1 \end{bmatrix}$$

11. The linear transformation $\mathbf{x} \mapsto \mathbf{Ax}$ is one-to-one:

Take two values in the domain, and apply the transformation:

$$\begin{bmatrix} x_1 \\ y_1 \end{bmatrix} \mapsto \begin{bmatrix} 1 & 1 \\ 0 & 1 \end{bmatrix} \begin{bmatrix} x_1 \\ y_1 \end{bmatrix} = \begin{bmatrix} x_1 + y_1 \\ y_1 \end{bmatrix} \quad \text{and}$$
$$\begin{bmatrix} x_2 \\ y_2 \end{bmatrix} \mapsto \begin{bmatrix} 1 & 1 \\ 0 & 1 \end{bmatrix} \begin{bmatrix} x_2 \\ y_2 \end{bmatrix} = \begin{bmatrix} x_2 + y_2 \\ y_2 \end{bmatrix}.$$

Suppose the outputs are equal, then show that the inputs must be equal.

$$\begin{bmatrix} x_1 + y_1 \\ y_1 \end{bmatrix} = \begin{bmatrix} x_2 + y_2 \\ y_2 \end{bmatrix} \implies y_1 = y_2 \quad \text{and} \quad x_1 + y_1 = x_2 + y_2$$
$$\implies x_1 = x_2.$$

12. The linear transformation $\mathbf{x} \mapsto \mathbf{Ax}$ is onto:

Let $\mathbf{b} \in \mathbb{R}^2$ be arbitrary, say

$$\mathbf{b} = \begin{bmatrix} b_1 \\ b_2 \end{bmatrix}.$$

Then the equation $\mathbf{Ax} = \mathbf{b}$ can be solved in augmented form:

$$\left[\begin{array}{cc|c} 1 & 1 & b_1 \\ 0 & 1 & b_2 \end{array}\right] \xrightarrow{(-1)R_2+R_1} \left[\begin{array}{cc|c} 1 & 0 & b_1 - b_2 \\ 0 & 1 & b_2 \end{array}\right].$$

Hence

$$\begin{bmatrix} 1 & 1 \\ 0 & 1 \end{bmatrix} \begin{bmatrix} b_1 - b_2 \\ b_2 \end{bmatrix} = \begin{bmatrix} b_1 \\ b_2 \end{bmatrix},$$

13. The columns of $\mathbf{A}$ span the codomain. Given arbitrary

$$\begin{bmatrix} b_1 \\ b_2 \end{bmatrix} \quad \text{in} \quad \mathbb{R}^2$$

we must find x, y, such that

$$x \begin{bmatrix} 1 \\ 0 \end{bmatrix} + y \begin{bmatrix} 1 \\ 1 \end{bmatrix} = \begin{bmatrix} b_1 \\ b_2 \end{bmatrix}.$$

Again we augment and do row reduction.

$$\left[\begin{array}{cc|c} 1 & 1 & b_1 \\ 0 & 1 & b_2 \end{array}\right] \xrightarrow{(-1)R_2+R_1} \left[\begin{array}{cc|c} 1 & 0 & b_1 - b_2 \\ 0 & 1 & b_2 \end{array}\right]$$

So $x = b_1 - b_2$ and $y = b_2$. In other words

$$(b_1 - b_2) \begin{bmatrix} 1 \\ 0 \end{bmatrix} + b_2 \begin{bmatrix} 1 \\ 1 \end{bmatrix} = \begin{bmatrix} b_1 - b_2 \\ 0 \end{bmatrix} + \begin{bmatrix} b_2 \\ b_2 \end{bmatrix} = \begin{bmatrix} b_1 \\ b_2 \end{bmatrix}.$$

15. $\dim \text{Col}(A) = n$. In this case $n = 2$. In the context of the rank-nullity theorem, we are used to computing $\dim \text{Col}(A)$ by counting the pivots. Here we indicate the pivots in boxes,

$$\left[\begin{array}{cc|c} \boxed{1} & 1 & 0 \\ 0 & \boxed{1} & 0 \end{array}\right] \tag{4.10}$$

and there are two of them.

16. $\dim \text{Nul}(A) = 0$. In the context of the rank-nullity theorem we are also used to counting the number of free variables. In equation (4.10), there is a pivot in both of the columns to the left of the vertical line. So, there are no free variables and $\dim \text{Nul}(A) = 0$.

Remark 4.4.1. $\dim \text{Col}(A) = 2$ means that the columns are a basis for $\mathbb{R}^2$. To be a basis, they must span $\mathbb{R}^2$ and be linearly independent. We have just shown in this example that the columns span $\mathbb{R}^2$, and in Example 3.3.11 we showed that they are linearly independent.

Activity 4.4.5. Using

$$\mathbf{A} = \begin{bmatrix} 1 & 1 \\ 0 & 1 \end{bmatrix}$$

show

14. The rows of $\mathbf{A}$ span the domain.

Hint: Take the transpose to switch rows and columns. Then augment, and work as in part 13 of Example 4.4.4.

4.5 Change of Basis

Example 4.5.1. Suppose that we start in the standard basis

$$\mathbf{e}_1 = \begin{bmatrix} 1 \\ 0 \end{bmatrix} \quad \text{and} \quad \mathbf{e}_2 = \begin{bmatrix} 0 \\ 1 \end{bmatrix}.$$

We will use $\mathcal{B}_1 = (\mathbf{e}_1, \mathbf{e}_2)$ to specify this basis (with order). Suppose we want to work in a new basis with coordinates in the old basis given by

$$\mathbf{v}_1 = \begin{bmatrix} 1 \\ 3 \end{bmatrix}_{\mathcal{B}_1} \quad \text{and} \quad \mathbf{v}_2 = \begin{bmatrix} 2 \\ 1 \end{bmatrix}_{\mathcal{B}_1} \tag{4.11}$$

as shown in Figure 4.3. We will use $\mathcal{B}_2 = (\mathbf{v}_1, \mathbf{v}_2)$ to specify this basis (with order). The thing is that now vectors have two different ways of being represented:

1. as a linear combination of $\mathbf{e}_1$ and $\mathbf{e}_2$, or
2. as a linear combination of $\mathbf{v}_1$ and $\mathbf{v}_2$.

For instance the vector $\mathbf{u}$ in Figure 4.3 has two descriptions:

$$\mathbf{u} = (-3)\mathbf{e}_1 + (1)\mathbf{e}_2 = (1)\mathbf{v}_1 + (-2)\mathbf{v}_2.$$

If we were only using the standard basis, we would probably write $\begin{bmatrix} -3 \\ 1 \end{bmatrix}$. But now that we have two different representations of $\mathbf{u}$, we should specify what basis we are using. So

$$\mathbf{u} = \begin{bmatrix} -3 \\ 1 \end{bmatrix}_{\mathcal{B}_1} = \begin{bmatrix} 1 \\ -2 \end{bmatrix}_{\mathcal{B}_2}. \tag{4.12}$$

In fact, since

$$\mathbf{v}_1 = 1 \cdot \mathbf{v}_1 + 0 \cdot \mathbf{v}_2 \quad \text{and} \quad \mathbf{v}_2 = 0 \cdot \mathbf{v}_1 + 1 \cdot \mathbf{v}_2,$$

then

$$\mathbf{v}_1 = \begin{bmatrix} 1 \\ 0 \end{bmatrix}_{\mathcal{B}_2} \quad \text{and} \quad \mathbf{v}_2 = \begin{bmatrix} 0 \\ 1 \end{bmatrix}_{\mathcal{B}_2}.$$

But that just looks like the standard basis! And it is. From its own viewpoint, $\mathbf{v}_1$ and $\mathbf{v}_2$ is the standard basis, even though from the viewpoint of $\mathbf{e}_1$ and $\mathbf{e}_2$, it is not. It is all a matter of perspective: the standard basis is somehow whichever basis we are using currently.

We can describe the change of basis as a linear transformation $T : \mathbb{R}^2 \to \mathbb{R}^2$, given by

$$\begin{aligned} \mathbf{e}_1 &\mapsto \mathbf{v}_1 = \boxed{1} \cdot \mathbf{e}_1 + \boxed{3} \cdot \mathbf{e}_2 \\ \mathbf{e}_2 &\mapsto \mathbf{v}_2 = \boxed{2} \cdot \mathbf{e}_1 + \boxed{1} \cdot \mathbf{e}_2. \end{aligned}$$

So the matrix for T is

$$\mathbf{T} = \begin{bmatrix} 1 & 3 \\ 2 & 1 \end{bmatrix}^T = \begin{bmatrix} 1 & 2 \\ 3 & 1 \end{bmatrix},$$

which just amounts to taking the vectors in equation (4.11) and building them into the columns of a matrix. The matrix $\mathbf{T}$ acts on column vectors in the new basis $\mathcal{B}_2$, and gives

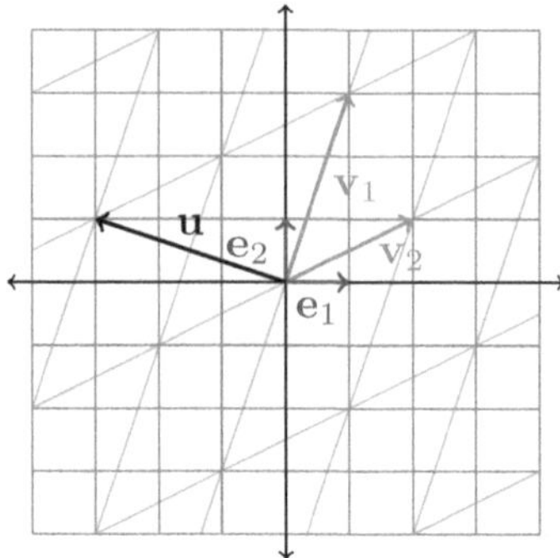

FIGURE 4.3: Change of basis

us a column vector in the original basis $\mathcal{B}_1$. For example, we can verify equation (4.12) as follows

$$\begin{bmatrix} 1 \\ -2 \end{bmatrix}_{\mathcal{B}_2}$$

$$\begin{bmatrix} 1 & 2 \\ 3 & 1 \end{bmatrix} \begin{bmatrix} -3 \\ 1 \end{bmatrix}_{\mathcal{B}_1}.$$

Now suppose we want to rotate the vector $\mathbf{u}$ counterclockwise by 90 degrees. We know that in $\mathcal{B}_1$, we have the matrix

$$\mathbf{R} = \begin{bmatrix} 0 & -1 \\ 1 & 0 \end{bmatrix}.$$

If we start with $\mathbf{u}$ with coordinates in $\mathcal{B}_2$, we have to use T to get them in the basis $\mathcal{B}_1$, then apply the rotation, and finally use T^{-1} to get back to $\mathcal{B}_2$. The combined effect in terms of functions is

$$T^{-1}(R(T(\mathbf{u}))),$$

and that means that matrix for $\mathbf{R}$ in the new basis is

$$\mathbf{T}^{-1}\mathbf{R}\mathbf{T}.$$

Activity 4.5.2. Given

$$\mathbf{u} = \begin{bmatrix} -1 \\ 2 \end{bmatrix}_{\mathcal{B}_2},$$

1. compute $\mathbf{Tu}$ to get coordinates in the standard basis $\mathcal{B}_1$,
2. compute $\mathbf{R}(\mathbf{Tu})$ to get coordinates of the rotated vector in $\mathcal{B}_2$,
3. plot both vectors on a graph,
4. compute $\mathbf{T}^{-1}(\mathbf{R}(\mathbf{Tu}))$,
5. verify that

$$\mathbf{T}^{-1}\mathbf{R}\mathbf{T} = \begin{bmatrix} 1 & 1 \\ -2 & -1 \end{bmatrix}, \quad \text{and}$$

6. compute $(\mathbf{T}^{-1}\mathbf{R}\mathbf{T})\mathbf{u}$.

Definition 4.5.1. Let $\mathbf{A}$ and $\mathbf{Q}$ be $n \times n$ matrices, and suppose $\mathbf{Q}$ is invertible. Then **conjugating A by Q** means computing

$$\mathbf{B} = \mathbf{Q}^{-1}\mathbf{A}\mathbf{Q}. \tag{4.13}$$

Two $n \times n$ matrices $\mathbf{A}$ and $\mathbf{B}$ are said to be **conjugate** or in the same **conjugacy class** if an invertible matrix $\mathbf{Q}$ exists such that equation (4.13) is valid.

Remark 4.5.1. A student once asked me if we use the word conjugating because matrices are somehow like verbs. I would say yes. The matrix $\mathbf{R}$ represents rotation, which is an action, and words used for actions are verbs.

4.6 Trace

4.6.1 Intro to Trace

Definition 4.6.1. The **trace** of a square matrix $\mathbf{A}$ is the sum of the elements on the diagonal and is denoted by $\operatorname{tr}(\mathbf{A})$.

Example 4.6.1. Given the matrix

$$\mathbf{A} = \begin{bmatrix} 1 & 2 \\ 3 & 4 \end{bmatrix}$$

we have $\operatorname{tr}(\mathbf{A}) = 1 + 4 = 5$.

Proposition 4.6.2. *Trace satisfies the following properties:*

1. $\operatorname{tr}(\mathbf{A} + \mathbf{B}) = \operatorname{tr}(\mathbf{A}) + \operatorname{tr}(\mathbf{B})$ *if* $\mathbf{A}$ *and* $\mathbf{B}$ *have the same dimension,*
2. $\operatorname{tr}(c\mathbf{A}) = c\operatorname{tr}(\mathbf{A})$,
3. $\operatorname{tr}(\mathbf{A}^T) = \operatorname{tr}(\mathbf{A})$, *and*
4. $\operatorname{tr}(\mathbf{A}) = \operatorname{tr}(\mathbf{P}^{-1}\mathbf{A}\mathbf{P})$ *for any invertible matrix* $\mathbf{P}$.

Remark 4.6.1. The first property shows that trace is additive. However, while trace is compatible with multiplication by a scalar, it is generally not true that $\operatorname{tr}(\mathbf{AB}) = \operatorname{tr}(\mathbf{A})\operatorname{tr}(\mathbf{B})$; that is to say, trace is not multiplicative. The last property means that trace is well defined on conjugacy classes.

Activity 4.6.2. Let

$$\mathbf{A} = \begin{bmatrix} 3 & 2 \\ 2 & 1 \end{bmatrix}, \quad \mathbf{B} = \begin{bmatrix} 1 & 0 \\ 0 & 1 \end{bmatrix}, \quad \text{and,} \quad \mathbf{P} = \begin{bmatrix} 1 & 1 \\ 0 & 1 \end{bmatrix}$$

and $c = 2$. Do the following steps:

1. Compute $\operatorname{tr}(\mathbf{A})$ and $\operatorname{tr}(\mathbf{B})$.
2. Verify that each of the four properties in Proposition 4.6.2 is true.
3. Show that
$$\operatorname{tr}(\mathbf{AB}) \neq \operatorname{tr}(\mathbf{A})\operatorname{tr}(\mathbf{B}).$$
4. Show that
$$\operatorname{tr}(\mathbf{A}^{-1}) \neq \operatorname{tr}(\mathbf{A}).$$

With Sage, we can get the trace of a matrix as follows:

```
A=matrix([[ 4, 2],
          [ 1, 3]])
A.trace()
```

4.6.2 Trace in Block Form

In Section 1.5 we introduced matrices in block form. If a square matrix is in block form, we have the following result:

Proposition 4.6.3. *Suppose* $\mathbf{A}$ *is a square matrix divided into blocks* $\mathbf{A}_{ij}$ *with* i *and* j *indexing rows and columns in block form. If* $\mathbf{A}_{ii}$ *is a square matrix for all* i, *then*

$$\operatorname{tr}(\mathbf{A}) = \sum_i \operatorname{tr}(\mathbf{A}_{ii}).$$

Activity 4.6.3. Suppose we have

$$\mathbf{A} = \begin{bmatrix} \mathbf{A}_{11} & \mathbf{A}_{12} & \mathbf{A}_{13} \\ \mathbf{A}_{21} & \mathbf{A}_{22} & \mathbf{A}_{23} \\ \mathbf{A}_{31} & \mathbf{A}_{32} & \mathbf{A}_{33} \end{bmatrix} = \left[\begin{array}{cc|c|c} 0 & -1 & 0 & 0 \\ 1 & 0 & 0 & 0 \\ \hline 0 & 0 & 1 & 0 \\ \hline 0 & 0 & 0 & -1 \end{array}\right].$$

1. Compute $\operatorname{tr}(\mathbf{A})$ directly.
2. Compute $\operatorname{tr}(\mathbf{A}_{11})$, $\operatorname{tr}(\mathbf{A}_{22})$, and $\operatorname{tr}(\mathbf{A}_{33})$, then add them up.

The results should be equal.

Remark 4.6.2. If we change the block structure in Activity 4.6.3 to

$$\mathbf{A} = \begin{bmatrix} \mathbf{A}_{11} & \mathbf{A}_{12} & \mathbf{A}_{13} \\ \mathbf{A}_{21} & \mathbf{A}_{22} & \mathbf{A}_{23} \\ \mathbf{A}_{31} & \mathbf{A}_{32} & \mathbf{A}_{33} \end{bmatrix} = \left[\begin{array}{c|cc|c} 0 & -1 & 0 & 0 \\ 1 & 0 & 0 & 0 \\ \hline 0 & 0 & 1 & 0 \\ \hline 0 & 0 & 0 & -1 \end{array}\right],$$

then not all of the blocks on the diagonal are square matrices, so the trace is not defined for them.

4.7 Outer Product

TL;DR

Definition 4.7.1 and Example 4.7.1 are the only things needed for computing reflections, which are used in the QR-algorithm.

4.7.1 Outer Product

In Section 1.1, when we defined matrix multiplication, we began with the special case of a $1 \times n$ row vector times an $n \times 1$ column vector. The result is a 1×1 matrix, essentially just a number, which also goes by the names dot product or inner product. However, the general definition of matrix multiplication allows us to switch the order of the vector: an $n \times 1$ column vector times an $1 \times n$ row vector will result in an $n \times n$ matrix, the outer product of those vectors. Once the order of multiplication is switched, the lengths of the vectors no longer need to match. Therefore, we make the following definition:

Definition 4.7.1. Given an $m \times 1$ column vector $\mathbf{A}$ and a $1 \times n$ row vector $\mathbf{B}$, the **outer product** is the $m \times n$ matrix $\mathbf{AB}$, which in this case is also written as $\mathbf{A} \otimes \mathbf{B}$.

Example 4.7.1. Given

$$\mathbf{A} = \begin{bmatrix} 1 \\ 2 \end{bmatrix} \quad \text{and} \quad \mathbf{B} = \begin{bmatrix} 1 & 2 & 3 \end{bmatrix},$$

compute the outer product $\mathbf{A} \otimes \mathbf{B}$.

Solution

Because of the order of multiplication, no addition occurs at all. Instead we simply multiply the components in pairs

$$\begin{array}{cc} & \begin{bmatrix} 1 & 2 & 3 \end{bmatrix} \\ \begin{bmatrix} 1 \\ 2 \end{bmatrix} & \begin{bmatrix} 1 & 2 & 3 \\ 2 & 4 & 6 \end{bmatrix} = \mathbf{A} \otimes \mathbf{B} \end{array}$$

Activity 4.7.2. Given

$$\mathbf{A} = \begin{bmatrix} 1 \\ -1 \end{bmatrix} \quad \text{and} \quad \mathbf{B} = \begin{bmatrix} 0 & 1 & 2 \end{bmatrix},$$

compute the outer product $\mathbf{A} \otimes \mathbf{B}$.

Remark 4.7.1. If

$$\mathbf{v} = \begin{bmatrix} 0 & 1 & 2 & 3 & 4 & 5 & 6 & 7 & 8 & 9 \end{bmatrix},$$

then the outer product of $\mathbf{v}^T$ and $\mathbf{v}$ is just the base 10 multiplication table.

The best way to compute an outer product in Sage is to use NumPy. The following code shows how to compute the base 10 multiplication table:

```
import numpy as np

v = np.arange(10)
M = np.outer(v,v)
vector(v)
matrix(M)
```

There are a couple subtle details worth noting. First, `v` is neither a row vector nor a column vector. It has shape 10. To interpret `v` as a row vector means to take the shape to be 1×10 and to interpret `v` as a column vector means to take the shape to be 10×1. The function `np.outer` automatically interprets the left argument as a column vector and the right argument as a row vector, so there is no reason to take the transpose of `v`. Second, `v` and `M` are both NumPy arrays, whereas `vector(v)` and `matrix(M)` are vector and matrix objects that are compatible with other Sage code.

Remark 4.7.2. It is standard to use v_i (with a single index) for the element of a vector $\mathbf{v}$, regardless of whether $\mathbf{v}$ is a row vector or a column vector, rather than v_{i1} or v_{1i}. So in terms of indices we can define the outer product of two vectors $\mathbf{v}$ and $\mathbf{w}$ as the matrix with $v_i w_j$ in the i-th row and j-th column. From this viewpoint, the outer product can be defined in a more general way. A vector is an **array** with one index, and a matrix is an array with two indices, but why stop there? Taking the outer product of a vector and a matrix should give us three indices, and taking the outer product of two matrices should give us four. Given a matrix $\mathbf{A}$ with elements a_{ij}, and a matrix $\mathbf{B}$ with elements $b_{k\ell}$, the outer product $\mathbf{A} \otimes \mathbf{B}$ is an array with the elements $a_{ij} b_{k\ell}$. In other words, each element of $\mathbf{A}$ gets multiplied with each element of $\mathbf{B}$. No addition occurs, unlike in $\mathbf{AB}$, for which we would have a matching index: the elements of $\mathbf{B}$ would be written as b_{jk}, and we would sum along the matching index j

$$\sum_j a_{ij} b_{jk}$$

to get the element in the i-th row and k-th column of $\mathbf{AB}$.

The outer product has many important applications including

1. back propagation in a neural network,
2. Householder reflections,
3. the multivariable form of a Taylor series, and
4. tensor products of vector spaces.

We will investigate the last two in this section.

4.7.2 Multi-variable Taylor Series

The multivariable form of Taylor series has a very natural description in terms of the outer product. First, recall that for a single variable function $f(x)$ with infinitely many derivatives, the Taylor series for f at a point $x = a$ is

$$\sum_{k=0}^{\infty} \frac{1}{k!} f^{(k)}(a)(x-a)^k,$$

where $f^{(k)}(x) = \frac{d^k}{dx^k} f(x)$. For a multivariable function, x and a are replaced by vectors $\mathbf{x}$ and $\mathbf{a}$, but when plugged into f we still get just a number (an array with zero indices). However, $\frac{d}{dx}$ must be replaced by

$$\nabla = \left[\begin{array}{cccc} \frac{\partial}{\partial x_1} & \frac{\partial}{\partial x_2} & \cdots & \frac{\partial}{\partial x_n} \end{array}\right].$$

When applied to f and evaluated at $\mathbf{a}$, we obtain a vector (an array with 1 index), which is the gradient of f at $\mathbf{a}$. So far, this gives us approximation by a tangent

$$f(\mathbf{x}) \approx f(\mathbf{a}) + (\nabla f)(\mathbf{a})(\mathbf{x} - \mathbf{a}).$$

For the next term, we need the Hessian matrix, which is obtained by applying $\nabla \otimes \nabla$ to f, then evaluating it at $\mathbf{a}$. We then take two copies of the vector $\mathbf{x} - \mathbf{a}$, and multiply by it once on the left (as a row vector) and once on the right (as a column vector). Each time we multiply by $\mathbf{x} - \mathbf{a}$, the number of indices decreases by one. From there, the pattern continues. For the k-th term, we tensor k copies of ∇, which we write as $\nabla^{\otimes k}$, apply it to f and evaluate it at $\mathbf{a}$, obtaining an array $(\nabla^{\otimes k} f)(\mathbf{a})$ with k indices (see Remark 4.7.2). We then take k copies of $\mathbf{x} - \mathbf{a}$, and multiply by it once along each axis. The resulting form may be written as

$$\sum_{k=0}^{\infty} \frac{1}{k!} (\nabla^{\otimes k} f)(\mathbf{a})(\mathbf{x} - \mathbf{a})^k.$$

It is critical to remember that when we write $(\mathbf{x} - \mathbf{a})^k$, in this case we mean that the vector $\mathbf{x} - \mathbf{a}$ is multiplied once along each axis $(\nabla^{\otimes k} f)(\mathbf{a})$.

Example 4.7.3. Let $\mathbf{x} = \begin{bmatrix} x_1 & x_2 \end{bmatrix}$ and $f(\mathbf{x}) = (2x_1 - x_2 + 1)^2$. Compute the terms of the multivariable Taylor series for f at $\mathbf{a} = \mathbf{0}$.

Solution

For $k = 0$, we just plug into f with $\mathbf{a} = \mathbf{0}$:

$$f(\mathbf{0}) = (2 \cdot 0 - 0 + 1)^2 = 1.$$

For $k = 1$, we first need to compute the partial derivatives of f. From power rule and chain rule, we have

$$\begin{aligned} \frac{\partial}{\partial x_1} f(\mathbf{x}) &= f_{x_1}(\mathbf{x}) = 2(2x_1 - x_2 + 1)(2) \quad \text{and} \\ \frac{\partial}{\partial x_2} f(\mathbf{x}) &= f_{x_2}(\mathbf{x}) = 2(2x_1 - x_2 + 1)(-1). \end{aligned}$$

Evaluating them at $\mathbf{0}$ gives us

$$(\nabla f)(\mathbf{0}) = \begin{bmatrix} f_{x_1}(\mathbf{0}) & f_{x_2}(\mathbf{0}) \end{bmatrix} = \begin{bmatrix} 4 & -2 \end{bmatrix}.$$

Since $\mathbf{a} = \mathbf{0}$, then $\mathbf{x} - \mathbf{a} = \mathbf{x} - \mathbf{0} = \mathbf{x}$, so

$$\begin{array}{cc} & \begin{bmatrix} x_1 \\ x_2 \end{bmatrix} \\ \begin{bmatrix} 4 & -2 \end{bmatrix} & \begin{bmatrix} 4x_1 - 2x_2 \end{bmatrix} = (\nabla f)(\mathbf{0})\mathbf{x} \end{array}$$

For $k = 2$, we now must compute $\nabla \otimes \nabla$:

$$\begin{array}{cc} & \begin{bmatrix} \frac{\partial}{\partial x_1} & \frac{\partial}{\partial x_2} \end{bmatrix} \\ \begin{bmatrix} \frac{\partial}{\partial x_1} \\ \frac{\partial}{\partial x_2} \end{bmatrix} & \begin{bmatrix} \frac{\partial^2}{\partial x_1^2} & \frac{\partial^2}{\partial x_1 \partial x_2} \\ \frac{\partial^2}{\partial x_2 \partial x_1} & \frac{\partial^2}{\partial x_2^2} \end{bmatrix}. \end{array}$$

Next we apply it to f, using power rule and chain rule as before:

$$\begin{bmatrix} \frac{\partial^2}{\partial x_1^2} f(\mathbf{x}) & \frac{\partial^2}{\partial x_1 \partial x_2} f(\mathbf{x}) \\ \frac{\partial^2}{\partial x_2 \partial x_1} f(\mathbf{x}) & \frac{\partial^2}{\partial x_2^2} f(\mathbf{x}) \end{bmatrix} = \begin{bmatrix} 2(2)^2 & 2(-1)(2) \\ 2(2)(-1) & 2(-1)^2 \end{bmatrix} = \begin{bmatrix} 8 & -4 \\ -4 & 2 \end{bmatrix}.$$

In this case the Hessian matrix is constant, so we don't even need to plug in with $\mathbf{0}$. But we still need to multiply by $\mathbf{x}$ twice on the right:

$$\begin{array}{ccc} & \begin{bmatrix} 8 & -4 \\ -4 & 2 \end{bmatrix} & \begin{bmatrix} x_1 \\ x_2 \end{bmatrix} \\ \begin{bmatrix} x_1 & x_2 \end{bmatrix} & \begin{bmatrix} 8x_1 - 4x_2 & -4x_1+2x_2 \end{bmatrix} & \begin{bmatrix} 8x_1^2-8x_1x_2+2x_2^2 \end{bmatrix} \end{array}$$

The higher derivatives are all zero, so the Taylor series for f at $\mathbf{0}$ is

$$1 + (4x_1 - 2x_2) + \frac{1}{2}(8x_1^2 - 8x_1x_2 + 2x_2^2),$$

which can be seen to be equal to $(2x_1 - x_2 + 1)^2$ by multiplying out.

Activity 4.7.4. Let $\mathbf{x} = \begin{bmatrix} x_1 & x_2 \end{bmatrix}$ and $f(\mathbf{x}) = (5x_1 + 2x_2 - 1)^2$. Compute the terms of the multivariable Taylor series for f at $\mathbf{a} = \mathbf{0}$.

To multiply out in Sage, we can use the following code:

```
var('x1,x2')
f = (2*x1-x2+1)^2
f.expand()
```

If a multivariable function has two derivatives in each variable, then the analog of equation (1.6) is

$$f(\widehat{\mathbf{v}}) - f(\mathbf{v}) = (\nabla f)(\mathbf{v}) \cdot \Delta\mathbf{v} + \cdots$$

Now if we take the absolute value of both sides and divide by $|f((v))|$, we have

$$\mathrm{E_{rel}}(f(\mathbf{v})) = \frac{|f(\widehat{\mathbf{v}}) - f(\mathbf{v})|}{|f(\mathbf{v})|} \approx \frac{|(\nabla f)(\mathbf{v}) \cdot \Delta\mathbf{v}|}{|f(\mathbf{v})|}. \tag{4.14}$$

Although we could use this result directly, we do not yet have $\mathrm{E_{rel}}(\mathbf{v})$ on the right-hand side. To get it, we can use Cauchy-Schwarz for the dot product (Proposition 1.2.8):

$$|(\nabla f)(\mathbf{v}) \cdot \Delta\mathbf{v}| \leq \| (\nabla f)(\mathbf{v}) \|_2 \cdot \| \Delta\mathbf{v} \|_2 .$$

By applying this to (4.14) we get

$$\mathrm{E_{rel}}(f(\mathbf{v})) \lesssim \frac{\| (\nabla f)(\mathbf{v}) \|_2 \cdot \| \Delta\mathbf{v} \|_2}{|f(\mathbf{v})|} = \frac{\| (\nabla f)(\mathbf{v}) \|_2 \cdot \| \mathbf{v} \|_2}{|f(\mathbf{v})|} \cdot \frac{\| \Delta\mathbf{v} \|_2}{\| \mathbf{v} \|_2}.$$

Now $\frac{\| \Delta\mathbf{v} \|_2}{\| \mathbf{v} \|_2} = \mathrm{E_{rel}}(\mathbf{v})$, so

$$\kappa(\mathbf{v}) = \frac{\| (\nabla f)(\mathbf{v})) \|_2 \cdot \| \mathbf{v} \|_2}{|f(\mathbf{v})|} \tag{4.15}$$

generalizes the condition number to multi-variable functions. See section 1.3 of [24].

Example 4.7.5. Let $\mathbf{v} = \begin{bmatrix} x & y \end{bmatrix}$ and $f(\mathbf{v}) = x + y$. Then

$$(\nabla f)(\mathbf{v}) = \begin{bmatrix} 1 & 1 \end{bmatrix}.$$

So $\| (\nabla f)(\mathbf{v}) \| = \sqrt{1^2 + 1^2} = \sqrt{2}$, and the condition number is

$$\frac{\| (\nabla f)(\mathbf{v})) \|_2 \cdot \| \mathbf{v} \|_2}{|f(\mathbf{v})|} = \frac{\sqrt{2}\sqrt{x^2 + y^2}}{|x + y|}.$$

We have a problem if $x + y \approx 0$ while $\sqrt{x^2 + y^2}$ remains big.

Activity 4.7.6 (Sage optional)**.** Let $\mathbf{v} = \begin{bmatrix} x & y \end{bmatrix}$ and $f(\mathbf{v}) = xy$.
You can use Sage to compute the gradient if necessary.

(a) Compute the condition number $\kappa(\mathbf{v})$ using equation (4.15).

(b) Determine when $\kappa(\mathbf{v})$ is large.

Activity 4.7.7 (With NumPy)**.** Open a Jupyter notebook, then run the following code in a cell.

```
import numpy as np

def f(x): return np.sum(x)
def kappa(x):
    return np.sqrt(len(x))*np.linalg.norm(x)/np.abs(f(x))
```

This code defines a function which adds the coordinates of a vector, as well as a condition number that works even if the vector $\mathbf{x}$ has more than two components. When $\mathbf{x}$ does have two components, then `len(x)` returns 2.

Next pick:

1. a vector $\mathbf{x}$ for which the sum of the components is very small,
2. a direction $\mathbf{b}$ to move away from $\mathbf{x}$, and
3. a small factor δ to adjust the length of $\mathbf{b}$.

Then $\widehat{\mathbf{x}}$ is computed as $\mathbf{x} + \delta * \mathbf{b}$, and both $f(\mathbf{x})$ and $f(\widehat{\mathbf{x}})$ can be computed as well. All of this can be done by running the code below.

```
x = np.array([1.0, -0.999999999999])
b = np.array([1.0,1.0])
delta = 1e-12

x_hat = x + delta * b
y, y_hat = f(x), f(x_hat)
```

Finally, we define a function for computing the relative error, and print various numbers to compare.

```
def E_rel(X,X_hat):
    return np.linalg.norm(X-X_hat) / np.linalg.norm(X)

print("kappa(x): %s" % kappa(x))
print("Backward error: %s" % E_rel(x,x_hat))
print("Forward error: %s" % E_rel(y,y_hat))
print("backward error * kappa(x)" % kappa(x) * E_rel(x,x_hat))
```

If you kept the original numbers you should see that `kappa(x)` is very big, the backward error is very small, and the forward error is approximately 2. The backward error times `kappa(x)` should be just slightly larger than the forward error.

We can also get the condition number of a dot product in a similar way. Let

$$\mathbf{v} = [\mathbf{x}|\mathbf{y}] \quad \text{where} \quad \mathbf{x} = \begin{bmatrix} x_1 & x_2 \end{bmatrix} \quad \text{and} \quad \mathbf{y} = \begin{bmatrix} y_1 & y_2 \end{bmatrix}.$$

Then

$$f(\mathbf{v}) = x_1 y_1 + x_2 y_2 = \mathbf{x} \cdot \mathbf{y} \tag{4.16}$$

is the dot product of $\mathbf{x}$ and $\mathbf{y}$. If $\widehat{\mathbf{v}} = [\widehat{\mathbf{x}}|\widehat{\mathbf{y}}]$, then

$$\Delta\mathbf{v} = \widehat{\mathbf{v}} - \mathbf{v} = [\widehat{\mathbf{x}}|\widehat{\mathbf{y}}] - [\mathbf{x}|\mathbf{y}] = [\widehat{\mathbf{x}} - \mathbf{x}|\widehat{\mathbf{y}} - \mathbf{y}] = [\Delta\mathbf{x}|\Delta\mathbf{y}].$$

If $\nabla_{\mathbf{x}}$ is the gradient of just the variables in $\mathbf{x}$, then the numerator of (4.14) can be broken up with the help of Cauchy-Schwarz:[2]

$$\begin{aligned} |(\nabla f)(\mathbf{v}) \cdot \Delta\mathbf{v}| &= |(\nabla_{\mathbf{x}} f)(\mathbf{v}) \cdot \Delta\mathbf{x} + (\nabla_{\mathbf{y}} f)(\mathbf{v}) \cdot \Delta\mathbf{y}| \\ &\leq |(\nabla_{\mathbf{x}} f)(\mathbf{v}) \cdot \Delta\mathbf{x}| + |(\nabla_{\mathbf{y}} f)(\mathbf{v}) \cdot \Delta\mathbf{y}| \\ &\leq \| \nabla_{\mathbf{x}} f(\mathbf{v}) \|_2 \cdot \| \Delta\mathbf{x} \|_2 + \| (\nabla_{\mathbf{y}} f)(\mathbf{v}) \|_2 \cdot \| \Delta\mathbf{y} \|_2 . \end{aligned}$$

Next we divide by $|f(\mathbf{v})|$:

$$\mathrm{E}_{\mathrm{rel}}\, f(\mathbf{v}) \lesssim \frac{\| \nabla_{\mathbf{x}} f(\mathbf{v}) \|_2 \cdot \| \mathbf{x} \|_2}{|f(\mathbf{v})|} \cdot \frac{\| \Delta\mathbf{x} \|_2}{\| \mathbf{x} \|_2} + \frac{\| \nabla_{\mathbf{y}} f(\mathbf{v}) \|_2 \cdot \| \mathbf{y} \|_2}{|f(\mathbf{v})|} \cdot \frac{\| \Delta\mathbf{y} \|_2}{\| \mathbf{y} \|_2}$$

Since $\mathrm{E}_{\mathrm{rel}}(\mathbf{x}) = \frac{\| \Delta\mathbf{x} \|_2}{\| \mathbf{x} \|_2}$ and $\mathrm{E}_{\mathrm{rel}}(\mathbf{y}) = \frac{\| \Delta\mathbf{x} \|_2}{\| \mathbf{y} \|_2}$, it makes sense to define the condition number to be

$$\kappa(\mathbf{x}, \mathbf{y}) = \max \left\{ \frac{\| \nabla_{\mathbf{x}} f(\mathbf{v}) \|_2 \cdot \| \mathbf{x} \|_2}{|f(\mathbf{v})|}, \frac{\| \nabla_{\mathbf{y}} f(\mathbf{v}) \|_2 \cdot \| \mathbf{y} \|_2}{|f(\mathbf{v})|} \right\} \tag{4.17}$$

What do we get in the case of the dot product? The gradients are

$$\nabla_{\mathbf{x}} f(\mathbf{v}) = \begin{bmatrix} y_1 & y_2 \end{bmatrix} = \mathbf{y} \quad \text{and} \quad \nabla_{\mathbf{y}} f(\mathbf{v}) = \begin{bmatrix} x_1 & x_2 \end{bmatrix} = \mathbf{x}.$$

By using these results and (4.16), equation (4.17) simplifies to

$$\kappa(\mathbf{x}, \mathbf{y}) = \frac{\| \mathbf{x} \|_2 \cdot \| \mathbf{y} \|_2}{|\mathbf{x} \cdot \mathbf{y}|}. \tag{4.18}$$

Equation (4.18) is the condition number for a dot product of $\mathbf{x}$ and $\mathbf{y}$. By equation (1.14), it is also equal to $\frac{1}{\cos(\theta)}$, where θ is the angle between $\mathbf{x}$ and $\mathbf{y}$. There is a problem if $\mathbf{x} \cdot \mathbf{y} \approx 0$, which happens if $\mathbf{x}$ and $\mathbf{y}$ are almost at right angles to each other.

4.7.3 Tensor Product of Vector Spaces

Given two vector spaces V and W with a common field of scalars F, the **tensor product** $V \otimes W$ is a vector space with the same field of scalars consisting of all linear combinations of outer products $\mathbf{v} \otimes \mathbf{w}$ where $\mathbf{v}$ is in V and $\mathbf{w}$ is in W. Multiplication by a scalar is defined by

$$c(\mathbf{v} \otimes \mathbf{w}) = (c\mathbf{v}) \otimes \mathbf{w} = \mathbf{v} \otimes (c\mathbf{w}),$$

where $\mathbf{v}$ is in V, $\mathbf{w}$ is in V, and c is in F. Addition is easiest to define with a basis

[2]The triangle inequality is also used.

Theorem 4.7.2. *Suppose that V and W are two finite dimensional vector spaces with a common field of scalars F. If $\mathbf{v}_1, \mathbf{v}_2, \dots \mathbf{v}_n$ is a basis for V and $\mathbf{w}_1, \mathbf{w}_2, \dots \mathbf{w}_m$ is a basis for W, then*

$$\mathbf{v}_i \otimes \mathbf{w}_j \quad \text{for} \quad 1 \leq i \leq n \quad \text{and} \quad 1 \leq j \leq m.$$

is a basis for $V \otimes W$. In particular the dimension of $V \otimes W$ is the product of the dimensions of V and W.

Proposition 4.7.3. *Suppose that $\mathbf{v}_1, \mathbf{v}_2$ are in V and $\mathbf{w}_1, \mathbf{w}_2$ are in W, then*

$$(\mathbf{v}_1 + \mathbf{v}_2) \otimes \mathbf{w}_1 = \mathbf{v}_1 \otimes \mathbf{w}_1 + \mathbf{v}_2 \otimes \mathbf{w}_1$$

and

$$\mathbf{v}_1 \otimes (\mathbf{w}_1 + \mathbf{w}_2) = \mathbf{v}_1 \otimes \mathbf{w}_1 + \mathbf{v}_1 \otimes \mathbf{w}_2.$$

Example 4.7.8. Suppose that $V = \mathbb{R}^2$ and $W = \mathbb{R}^3$, where we interpret vectors in V as column vectors and vectors in W as row vectors. The standard basis for V is

$$\mathbf{v}_1 = \begin{bmatrix} 1 \\ 0 \end{bmatrix} \quad \text{and} \quad \mathbf{v}_2 = \begin{bmatrix} 0 \\ 1 \end{bmatrix}$$

and the standard basis for W is

$$\mathbf{w}_1 = \begin{bmatrix} 1 & 0 & 0 \end{bmatrix}, \quad \mathbf{w}_2 = \begin{bmatrix} 0 & 1 & 0 \end{bmatrix}, \quad \mathbf{w}_3 = \begin{bmatrix} 0 & 0 & 1 \end{bmatrix}.$$

The claim made by Theorem 4.7.2 is that the matrices

$$\mathbf{v}_1 \otimes \mathbf{w}_1 = \begin{bmatrix} 1 & 0 & 0 \\ 0 & 0 & 0 \end{bmatrix} \qquad \mathbf{v}_1 \otimes \mathbf{w}_2 = \begin{bmatrix} 0 & 1 & 0 \\ 0 & 0 & 0 \end{bmatrix} \qquad \mathbf{v}_1 \otimes \mathbf{w}_3 = \begin{bmatrix} 0 & 0 & 1 \\ 0 & 0 & 0 \end{bmatrix}$$

$$\mathbf{v}_2 \otimes \mathbf{w}_1 = \begin{bmatrix} 0 & 0 & 0 \\ 1 & 0 & 0 \end{bmatrix} \qquad \mathbf{v}_2 \otimes \mathbf{w}_2 = \begin{bmatrix} 0 & 0 & 0 \\ 0 & 1 & 0 \end{bmatrix} \qquad \mathbf{v}_2 \otimes \mathbf{w}_3 = \begin{bmatrix} 0 & 0 & 0 \\ 0 & 0 & 1 \end{bmatrix}$$

are a basis for $V \otimes W$ (it is not too hard to check that these are actually the outer products). By Definition 4.7.3 the elements of $V \otimes W$ are linear combinations of outer products of vectors in V with vectors in W, which means that they must be 2×3 matrices. In fact, taking an arbitrary linear combination of the matrices $\mathbf{v}_i \otimes \mathbf{w}_j$ gives us

$$\sum_{i=1}^{2} \sum_{j=1}^{3} a_{ij} \mathbf{v}_i \otimes \mathbf{w}_j = \begin{bmatrix} a_{11} & a_{12} & a_{13} \\ a_{21} & a_{22} & a_{23} \end{bmatrix} \tag{4.19}$$

which is an arbitrary 2×3 matrix. So $V \otimes W$ consists of all 2×3 matrices, and the matrices $\mathbf{v}_i \otimes \mathbf{w}_j$ clearly span the space. Linear independence can be shown by setting equation (4.19) equal to the zero vector in $V \otimes W$, which is the zero 2×3 matrix:

$$\sum_{i=1}^{2} \sum_{j=1}^{3} a_{ij} \mathbf{v}_i \otimes \mathbf{w}_j = \begin{bmatrix} a_{11} & a_{12} & a_{13} \\ a_{21} & a_{22} & a_{23} \end{bmatrix} = \begin{bmatrix} 0 & 0 & 0 \\ 0 & 0 & 0 \end{bmatrix}. \tag{4.20}$$

By comparing coefficients, it follows that $a_{ij} = 0$ for all i and j. In other words, the only way to satisfy equation (4.20) is if all coefficients are zero, hence, the matrices $\mathbf{v}_i \otimes \mathbf{w}_j$ are linearly independent.

By the way, it is necessary to specify linear combinations of outer products in Definition 4.7.3 because not every 2×3 matrix can be written as an outer product. For example, suppose

$$\mathbf{A} = \begin{bmatrix} 1 & 2 & 3 \\ 4 & 5 & 6 \end{bmatrix}.$$

A general outer product $\mathbf{v} \otimes \mathbf{w}$ looks like this:

$$\begin{array}{cc} & \begin{bmatrix} a & b & c \end{bmatrix} \\ \begin{bmatrix} x \\ y \end{bmatrix} & \begin{bmatrix} ax & bx & cx \\ ay & by & cy \end{bmatrix} = \mathbf{v} \otimes \mathbf{w}. \end{array}$$

If $x = 0$ then 1st row will be zero, otherwise the 2nd row is $\frac{y}{x}$ times the 1st row. Neither of these is true about $\mathbf{A}$. The number below the 1 in $\mathbf{A}$ is 4 so the 2nd row would have to be 4 times the 1st row, which it is not. So $\mathbf{A}$ is not a pure outer product, but it is possible to write $\mathbf{A}$ as a linear combination of outer products. For example, using the basis above,

$$\mathbf{A} = 1\mathbf{v}_1 \otimes \mathbf{w}_1 + 2\mathbf{v}_1 \otimes \mathbf{w}_2 + 3\mathbf{v}_1 \otimes \mathbf{w}_3 + 4\mathbf{v}_2 \otimes \mathbf{w}_1 + 5\mathbf{v}_2 \otimes \mathbf{w}_2 + 6\mathbf{v}_2 \otimes \mathbf{w}_3.$$

Activity 4.7.9. Suppose that $V = \mathbb{R}^2$ and $W = \mathbb{R}^4$, where we interpret vectors in V as column vectors and vectors in W as row vectors. Do the following steps

1. Use the standard basis for V and W to write down a basis for $V \times W$. Check at least one of the outer products by hand, and convince yourself of the pattern.
2. Show that an arbitrary linear combination of the matrices $\mathbf{v}_i \otimes \mathbf{w}_j$ gives an arbitrary 2×4 matrix.
3. Show that the matrices $\mathbf{v}_i \otimes \mathbf{w}_j$ are linearly independent.
4. Try to find a 2×4 matrix that cannot be written as an outer product. Express the same matrix as a linear combination of the matrices in the basis.

4.8 Isometries

TL;DR

The matrix $\mathbf{P_v}$ defined by equation (4.25) is the reflection that sends $\mathbf{v}$ to $-\mathbf{v}$. Any reflection $\mathbf{P}$ is symmetric

$$\mathbf{P} = \mathbf{P}^T.$$

Two reflections gets us back where we started, so

$$\mathbf{P}^2 = \mathbf{I}.$$

Any linear isometry $\mathbf{Q}$ can be written as a product of reflections:

$$\mathbf{Q} = \mathbf{P}_k \cdots \mathbf{P}_2\mathbf{P}_1.$$

Since $\mathbf{AB}^T = \mathbf{B}^T\mathbf{A}^T$, then

$$\begin{aligned} \mathbf{Q}^T &= \mathbf{P}_1^T\mathbf{P}_2^T \cdots \mathbf{P}_1^T \\ &= \mathbf{P}_1\mathbf{P}_2 \cdots \mathbf{P}_k \end{aligned}$$

is the product of the reflections in reverse order, which means that

$$\mathbf{Q}^T\mathbf{Q} = \mathbf{I} \quad \text{and} \quad \mathbf{Q}^{-1} = \mathbf{Q}^T.$$

Any rotation is the product of just two reflections. The angle of the rotation is equal to 2 times the angle between the reflections.

Q 4.8.1. What is the line perpendicular to $y = x$?

Q 4.8.2. What is the effect of reflecting across the line $y = x$ twice?

Q 4.8.3. What is $\sin^2(\theta) + \cos^2(\theta)$?

Q 4.8.4. What is $\cos(-\theta)$?

Q 4.8.5. What is outer product $\left[\begin{smallmatrix}1\\2\end{smallmatrix}\right] \otimes \left[\begin{smallmatrix}3\\4\end{smallmatrix}\right]$?

The prefix iso means equal and the word metric is another name for a distance. Thus an isometry preserves distance. We are interested primarily in linear isometries, meaning that the zero vector must be fixed. There are two types:

1. Rotation
2. Reflection

First consider rotation. The matrix

$$\mathbf{R}_\theta = \begin{bmatrix} \cos(\theta) & -\sin(\theta) \\ \sin(\theta) & \cos(\theta) \end{bmatrix}, \tag{4.21}$$

when multiplied on the left of a vector, has the effect of rotating the vector by θ in the counter-clockwise direction.

Example 4.8.1. If θ is 90 degrees or $\frac{\pi}{2}$ radians, then

$$\mathbf{R}_\theta = \begin{bmatrix} 0 & -1 \\ 1 & 0 \end{bmatrix}.$$

which may be familiar from the symmetries of the square (see Examples 1.1.28, 1.3.8, and 4.2.1). Given $\mathbf{v} = \left[\begin{smallmatrix}1\\2\end{smallmatrix}\right]$, then

$$\begin{bmatrix} 0 & -1 \\ 1 & 0 \end{bmatrix} \begin{bmatrix} 1 \\ 2 \end{bmatrix} = \begin{bmatrix} -2 \\ 1 \end{bmatrix} = \mathbf{R}_\theta \mathbf{v},$$

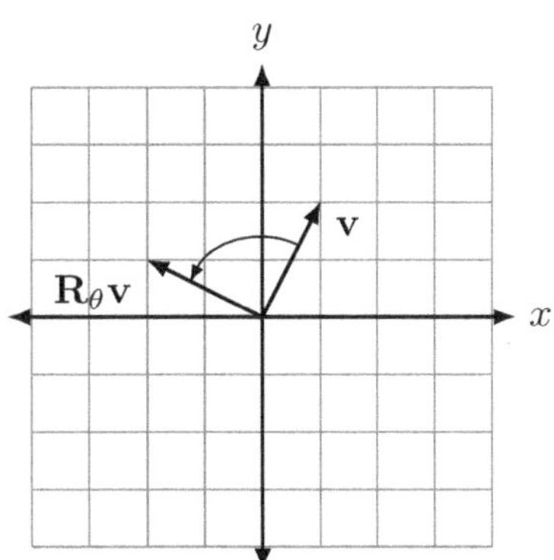

which is rotation counter-clockwise by 90 degrees as shown in the graph. The length does not change:

$$\|\,\mathbf{v}\,\| = \sqrt{1^1 + 2^2} = \sqrt{5} \quad \text{and} \quad \|\,\mathbf{R}_\theta \mathbf{v}\,\| = \sqrt{(-2)^2 + 1^2} = \sqrt{5}$$

In this case, the vectors lie on perpendicular lines since the angle between them is 90 degrees.

Activity 4.8.2. (Sage optional) To use Sage, start with

```
var('theta')
R = matrix([[cos(theta),-sin(theta)],
            [sin(theta),cos(theta)]])
```

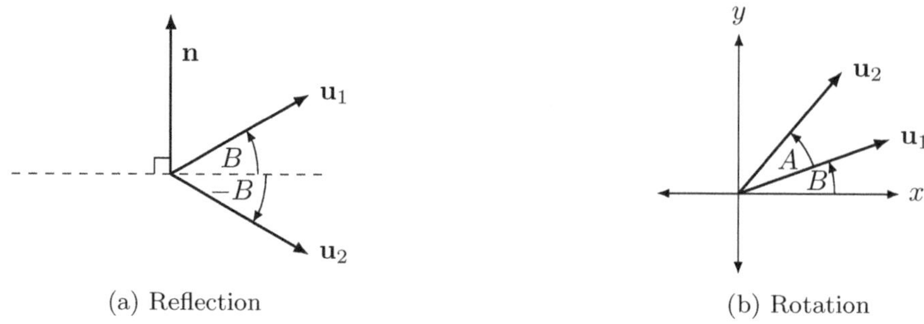

(a) Reflection (b) Rotation

FIGURE 4.4: Isometries

Substitution can be done directly:

```
R(theta=-theta)
```

The following function will get Sage to use trig identities:

```
def simplify_matrix(M):
    return matrix([m.trig_reduce() for m in M])

#Usage:
simplify_matrix(R(theta=-theta))
```

1. Compute $\det(\mathbf{R}_\theta)$.
2. Compute $\mathbf{R}_{-\theta}$ and $\mathbf{R}_\theta^T$. Are they the same?
3. Show, by multiplying, that $\mathbf{R}_\theta^T\mathbf{R}_\theta = \mathbf{I}$.

Reflection 4.8.1. When θ is 90 degrees, do your results from Activity 4.8.2 agree with Example 1.3.8?

An arbitrary unit vector in the plane can be represented as

$$\mathbf{u}_1 = \begin{bmatrix} \cos(B) \\ \sin(B) \end{bmatrix} \tag{4.22}$$

where B is the angle measured from the positive x-axis. If we rotate it counterclockwise by an angle A, then the new angle will be $A+B$ (see Figure 4.4b). Using $\theta = A$, and doing the matrix multiplication gives us:

$$\begin{bmatrix} \cos(A) & -\sin(A) \\ \sin(A) & \cos(A) \end{bmatrix} \begin{bmatrix} \cos(B) \\ \sin(B) \end{bmatrix} \begin{bmatrix} \cos(A)\cos(B)-\sin(A)\sin(B) \\ \sin(A)\cos(B)+\cos(A)\sin(B) \end{bmatrix} = \begin{bmatrix} \cos(A+B) \\ \sin(A+B) \end{bmatrix}$$

for $\mathbf{u}_2$, which are the identities for sine and cosine addition.

Comprehension Check 4.8.2. What do these identities simplify to when $B = A$?

Next, consider reflections. In Figure 4.4a, all vectors are reflected across the dashed line. The vector $\mathbf{n}$ is normal to the line (i.e. it is perpendicular), and all of the vectors $\mathbf{n}$, $\mathbf{u}_1$, and $\mathbf{u}_2$ are unit vectors, so

$$\|\,\mathbf{n}\,\| = \|\,\mathbf{u}_1\,\| = \|\,\mathbf{u}_2\,\| = 1.$$

If we apply equation (1.14) to $\mathbf{n}$ and $\mathbf{u}_1$, then we have

$$\mathbf{n} \cdot \mathbf{u}_1 = \cos(\theta).$$

Then by vector projection (Proposition 1.2.7), we obtain

$$\begin{aligned} \mathbf{u}_2 &= \mathbf{u}_1 - 2(\mathbf{n} \cdot \mathbf{u}_1)\mathbf{n} && (4.23) \\ &= (\mathbf{I} - 2(\mathbf{n} \otimes \mathbf{n}))\mathbf{u}_1. && (4.24) \end{aligned}$$

The 2nd equality is due to associativity of matrix multiplication (Proposition 1.1.16):

$$(\mathbf{n} \otimes \mathbf{n})\mathbf{u}_1 = (\mathbf{n}\mathbf{n}^T)\mathbf{u}_1 = \mathbf{n}(\mathbf{n}^T\mathbf{u}_1) = \mathbf{n}(\mathbf{n} \cdot \mathbf{u}_1),$$

but $\mathbf{n} \cdot \mathbf{u}_1$ is a scalar, and it can be written on either side of $\mathbf{n}$. If $\mathbf{v}$ is any other vector in the same direction as $\mathbf{n}$, then $\frac{\mathbf{v}}{\|\mathbf{v}\|} = \mathbf{n}$ and

$$\mathbf{n} \otimes \mathbf{n} = \frac{\mathbf{v}}{\|\mathbf{v}\|} \otimes \frac{\mathbf{v}}{\|\mathbf{v}\|} = \frac{\mathbf{v} \otimes \mathbf{v}}{\|\mathbf{v}^2\|} = \frac{\mathbf{v} \otimes \mathbf{v}}{\mathbf{v} \cdot \mathbf{v}}.$$

Thus we obtain the following matrix for reflection:

$$\mathbf{P}_\mathbf{v} = \mathbf{I} - 2\frac{\mathbf{v} \otimes \mathbf{v}}{\mathbf{v} \cdot \mathbf{v}} \tag{4.25}$$

Example 4.8.3. Let

$$\mathbf{v} = \begin{bmatrix} 0 \\ 1 \end{bmatrix}.$$

Compute $\mathbf{P}_\mathbf{v}$.

Solution

For the outer product, we have

$$\begin{matrix} & \begin{bmatrix} 0 & 1 \end{bmatrix} \\ \begin{bmatrix} 0 \\ 1 \end{bmatrix} & \begin{bmatrix} 0 & 0 \\ 0 & 1 \end{bmatrix} \end{matrix} = \mathbf{v} \otimes \mathbf{v},$$

and for the dot product we have

$$\mathbf{v} \cdot \mathbf{v} = 0^2 + 1^2 = 1.$$

Thus

$$\mathbf{P}_\mathbf{v} = \begin{bmatrix} 1 & 0 \\ 0 & 1 \end{bmatrix} - \frac{2}{1}\begin{bmatrix} 1 & 0 \\ 0 & 1 \end{bmatrix} = \begin{bmatrix} 1 & 0 \\ 0 & -1 \end{bmatrix}.$$

If we apply this matrix to an arbitrary unit vector in the form (4.22), then

$$\begin{matrix} & \begin{bmatrix} \cos(B) \\ \sin(B) \end{bmatrix} \\ \begin{bmatrix} 1 & 0 \\ 0 & -1 \end{bmatrix} & \begin{bmatrix} \cos(B) \\ -\sin(B) \end{bmatrix} \end{matrix} = \begin{bmatrix} \cos(-B) \\ \sin(-B) \end{bmatrix} = \mathbf{u}_2,$$

See Figure 4.4a.

Activity 4.8.4. Let

$$\mathbf{u} = \begin{bmatrix} 1 \\ -1 \end{bmatrix}.$$

Compute $\mathbf{P_u}$ by the following steps. The result should be familiar.

(a) Compute $\mathbf{u} \cdot \mathbf{u} = \mathbf{u}^T\mathbf{u}$

(b) Compute $\mathbf{u} \otimes \mathbf{u} = \mathbf{u}\mathbf{u}^T$

(c) Compute

$$\mathbf{P_v} = \mathbf{I} - \frac{2}{\mathbf{u} \cdot \mathbf{u}} \mathbf{u} \otimes \mathbf{u}$$

Activity 4.8.5. (Sage optional) To use Sage you will need some of the code from Activity 4.8.2. Let

$$\mathbf{u} = \begin{bmatrix} \cos(\theta) \\ \sin(\theta) \end{bmatrix};$$

in Sage `u = vector((cos(theta),sin(theta))`

1. Compute $\det(\mathbf{P}_{\mathbf{u}(\theta)})$ by equation (4.25) or in Sage, by

```
maxima_calculus('algebraic: true;')

def P(v):
    n = len(v)
    I = matrix.identity(n)
    Pv = I - 2*v.outer_product(v)/v.inner_product(v)
    return Pv.simplify_rational()
simplify_matrix(P(u))
```

2. Show, by multiplying, that $\mathbf{P}_\mathbf{u}^2 = \mathbf{I}$.

3. Consider $\mathbf{u}$ both when $\theta = A$ and $\theta = B$; in Sage you would use `u(theta=A)` and `u(theta=B)`. Let $\mathbf{P}_A$ and $\mathbf{P}_B$ be the corresponding matrices $\mathbf{P_u}$.

   Show, by multiplying, that

   $$\mathbf{P}_B\mathbf{P}_A = \mathbf{R}_{2(B-A)}.$$

   You will need trig identities to simplify. For best results in Sage the `simplify_matrix` function should be applied both before and after multiplying.

Reflection 4.8.3. When happens if A and B are reversed in Activity 4.8.5?

You should observe that $\mathbf{P} = \mathbf{P}_{\mathbf{u}(\theta)}$ is symmetric, thus $\mathbf{P}^2 = \mathbf{I}$ implies

$$\mathbf{P}^T\mathbf{P} = \mathbf{I},$$

which was also true for the rotation matrix $\mathbf{R}_\theta$. Since $(\mathbf{AB})^T = \mathbf{B}^T\mathbf{A}^T$, then

$$\mathbf{R}_{2(B-A)}^T = (\mathbf{P}_B\mathbf{P}_A)^T = \mathbf{P}_A^T\mathbf{P}_B^T = \mathbf{P}_A\mathbf{P}_B.$$

We know from Activity 4.8.2 that $\mathbf{R}_{2(B-A)}^T$ is the inverse of $\mathbf{R}_{2(B-A)}$. This remains true when written as a product of reflections

$$\begin{aligned}
\mathbf{R}_{2(B-A)}^T\mathbf{R}_{2(B-A)} &= (\mathbf{P}_A\mathbf{P}_B)\mathbf{P}_B\mathbf{P}_A \\
&= \mathbf{P}_A\mathbf{P}_B^2\mathbf{P}_A \\
&= \mathbf{P}_A\mathbf{I}\mathbf{P}_A = \mathbf{P}_A^2 = \mathbf{I}
\end{aligned}$$

Exercises

Problem 4.1. True or False?

a. If the codomain and range for a linear transformation T are equal, then T is onto.

b. If the codomain and range for a linear transformation T are equal, then T is one-to-one.

c. It is possible for a linear transformation T to be neither one-to-one nor onto.

Problem 4.2. Let L be the line $x + y = 2$. The point $\mathbf{o} = (1, 1)$ is on L. $\mathbf{R}$ is still a vector space after coordinate translation

$$(x', y') = (x, y) - (1, 1) = (x - 1, y - 1).$$

We use the definition of vector addition and scalar multiplication in the translated coordinates

$$(x_1', y_1') + (x_2', y_2') = (x_1' + x_2', y_1' + y_2') \quad \text{and} \quad c(x', y') = (cx', cy').$$

We then translate back

$$(x, y) = (x', y') + (1, 1) = (x' + 1, y' + 1).$$

The combined result is

$$\begin{aligned} (x_1, y_1) + (x_2, y_2) &= (x_1 + x_2 - 1, y_1 + y_2 - 1) \quad \text{and} \\ c(x, y) &= (c(x - 1) + 1, c(y - 1) + 1). \end{aligned}$$

Show that all four existential properties now hold.

Problem 4.3. Let M_2 denote the 2×2 matrices with real entries. Then M_2 is an $\mathbb{R}$-vector space, i.e. a vector space with $\mathbb{R}$ as the scalars. Given the matrices

$$\mathbf{A} = \begin{bmatrix} 1 & 1 \\ 0 & 1 \end{bmatrix}, \quad \mathbf{B} = \begin{bmatrix} 0 & -1 \\ 1 & 0 \end{bmatrix}, \quad \text{and} \quad \mathbf{C} = \begin{bmatrix} 1 & 0 \\ 1 & 0 \end{bmatrix}$$

and scalars $c = 2$ and $d = -3$, verify the following vector space properties (which is *not* a comprehensive list):

1. $c(\mathbf{A} + \mathbf{B}) = c\mathbf{A} + c\mathbf{B}$.
2. $(c + d)\mathbf{A} = c\mathbf{A} + d\mathbf{A}$.
3. $(cd)\mathbf{A} = c(d\mathbf{A})$.
4. $(\mathbf{A} + \mathbf{B}) + \mathbf{C} = \mathbf{A} + (\mathbf{B} + \mathbf{C})$.

Problem 4.4. Show that the vector space M_2 has a basis

$$\mathbf{e}_{11} = \begin{bmatrix} 1 & 0 \\ 0 & 0 \end{bmatrix}, \quad \mathbf{e}_{12} = \begin{bmatrix} 0 & 1 \\ 0 & 0 \end{bmatrix}, \quad \mathbf{e}_{21} = \begin{bmatrix} 0 & 0 \\ 1 & 0 \end{bmatrix}, \quad \mathbf{e}_{22} = \begin{bmatrix} 0 & 0 \\ 0 & 1 \end{bmatrix}$$

by doing the following steps:

1. Show that the matrices $\mathbf{e}_{11}, \mathbf{e}_{12}, \mathbf{e}_{21}, \mathbf{e}_{22}$ span M_2.
2. Show that the matrices $\mathbf{e}_{11}, \mathbf{e}_{12}, \mathbf{e}_{21}, \mathbf{e}_{22}$ are linearly independent.

3. Let $\mathbf{f}_1, \mathbf{f}_2, \mathbf{f}_3, \mathbf{f}_4$ be the columns of the 4x4 identity matrix. Then M_2 can be identified with $\mathbb{R}^4$ with the identifications

$$\mathbf{e}_{11} \mapsto \mathbf{f}_1 \qquad \mathbf{e}_{12} \mapsto \mathbf{f}_2 \qquad \mathbf{e}_{21} \mapsto \mathbf{f}_3 \qquad \mathbf{e}_{22} \mapsto \mathbf{f}_4.$$

The transpose of a matrix is a linear transformation from $M_2 \to M_2$ (i.e. $\mathbf{A} \mapsto \mathbf{A}^T$). By the identification with $\mathbb{R}^4$ just defined, this induces a linear transformation $T : \mathbb{R}^4 \to \mathbb{R}^4$. Look at the effects of this transformation on the basis vectors $\mathbf{f}_i$, and write down a matrix for the transformation T.

Problem 4.5. The following matrix defines a linear transformation from $\mathbb{R}^5 \to \mathbb{R}^4$:

$$\mathbf{A} = \begin{bmatrix} 1 & 2 & 0 & 0 & 0 \\ 0 & 0 & 1 & 0 & 0 \\ 0 & 0 & 0 & 0 & 1 \\ 0 & 0 & 0 & 0 & 0 \end{bmatrix}.$$

Do the following steps:

1. Give a parameterization of the column space in terms of a basis.
2. Give a parameterization of the null space using a basis.
3. Verify the rank-nullity theorem by adding the dimensions.

Problem 4.6. A linear transformation $T : \mathbb{R}^2 \to \mathbb{R}^2$ is defined by the matrix

$$T = \begin{bmatrix} 2 & 1 \\ 0 & 3 \end{bmatrix}$$

acting on column vectors. Let $\mathbf{e}_1, \mathbf{e}_2$ denote the columns of the 2×2 identity matrix.

1. Calculate $\mathbf{v}_1 = T\mathbf{e}_1$ and $\mathbf{v}_2 = T\mathbf{e}_2$.
2. Graph the parallelogram defined by the two edges $\mathbf{v}_1, \mathbf{v}_2$
3. Calculate the area of the parallelogram in part 2 using the base times height formula.
4. Calculate the area of the parallelogram in part 2 using the determinant of $\mathbf{T}$.

Problem 4.7. *Determinants in geometry.* A linear transformation is "orientation preserving" if the determinant of its matrix is positive, and "orientation reversing" if the determinant of its matrix is negative. Consider the matrices $\mathbf{M}$ and $\mathbf{R}$ from Example 4.2.1

1. Show that $\mathbf{R}$ has positive determinant, and $\mathbf{M}$ has negative determinant.
2. Use the rule that $\det(\mathbf{AB}) = \det(\mathbf{A})\det(\mathbf{B})$ to show that $\mathbf{R}^2$, $\mathbf{R}^3$, and $\mathbf{R}^4 = \mathbf{I}$ also have positive determinant, while multiplying $\mathbf{M}$ by any of them gives negative determinant. Hence any rotation is orientation preserving, and any mirror is orientation reversing.

A linear transformation is "unimodular" if it has a matrix with determinant ± 1. Unimodular transformations preserve area.

3. Use determinant of $[\mathbf{u}|\mathbf{v}]$ to compute the area of the parallelogram with sides

$$\mathbf{u} = \begin{bmatrix} 1 \\ 1 \end{bmatrix} \quad \text{and} \quad \mathbf{v} = \begin{bmatrix} 3 \\ 1 \end{bmatrix}.$$

4. Compute $\mathbf{Ru}$, $\mathbf{Rv}$, $\mathbf{Mu}$, and $\mathbf{Mv}$ where $\mathbf{R}$ and $\mathbf{M}$ are the matrices from example 4.2.1.

5. The 2×2 matrix $[\mathbf{Ru}|\mathbf{Rv}]$ with $\mathbf{Ru}$ and $\mathbf{Rv}$ as columns, satisfies the equation

$$\mathbf{R}[\mathbf{u}|\mathbf{v}] = [\mathbf{Ru}|\mathbf{Rv}].$$

   Check that the area of the parallelogram with sides $\mathbf{Ru}$ and $\mathbf{Rv}$ is the same in two ways:

   i by directly computing the determinant of $[\mathbf{Ru}|\mathbf{Rv}]$, and

   ii by applying $\det(\mathbf{AB}) = \det(\mathbf{A})\det(\mathbf{B})$ to $\mathbf{R}[\mathbf{u}|\mathbf{v}]$

   Repeat with $\mathbf{Mu}$ and $\mathbf{Mv}$.

6. Sketch all three parallelograms in the xy-plane, label the corners with the vector names, e.g. $\mathbf{0}$, $\mathbf{u}$, $\mathbf{v}$, and $\mathbf{u} + \mathbf{v}$ for the original parallelogram. You should observe that the order of the corners is preserved when applying $\mathbf{R}$ and reversed when you apply $\mathbf{M}$.

Problem 4.8. *Polynomial interpolation.* The polynomials x^2, x, and 1 are linearly independent, and thus span a vector space of dimension 3. For real coefficients, the vector space can be identified with $\mathbb{R}^3$, but in more than one way. One such way is by evaluating the functions at different points. We define $T : \text{span}\{x^2, x, 1\} \to \mathbb{R}^3$ by

$$T : f(x) \mapsto (f(-1), f(0), f(1)).$$

1. Construct a matrix $\mathbf{T}$ for T. The order of the basis is important! You should use x^2, x, and 1 in that order.

2. Solve $[\mathbf{T}|\mathbf{y}]$ by row reduction, where

$$\mathbf{y} = \begin{bmatrix} -3 \\ -2 \\ 1 \end{bmatrix}$$

3. Let a, b, c be the components of the solution from the previous step. Check that for $f(x) = ax^2 + bx + c$, you get $f(-1) = -3$, $f(0) = -2$, and $f(1) = 1$. Thus we have computed an interpolation of the points $(-1, -3)$, $(0, -2)$, $(1, 1)$. In general, n points can be interpolated with a polynomial of degree strictly less than n.

Problem 4.9. Let V be the vector space of polynomials of degree less than or equal to 2 with real coefficients. It was shown in example 4.3.6 that this vector space has dimension 3, with basis 1, x, and x^2, and that $\frac{d}{dx}$ is a linear transformation on V. If we were working with Taylor series, multiplication by x would be too, but to get this to work on V, we have to throw out any higher degree terms (so that $x \cdot x^2$ is taken to be zero).

Do the following steps:

1. Calculate a matrix using the basis 1, x, and x^2.

2. Describe the range and null space.

3. If $\mathbf{A}$ denotes the matrix for $\frac{d}{dx}$ computed in example 4.3.6 (see also problem 4.10 above), and $\mathbf{B}$ denotes the matrix for x multiplication by x computed in part 1, show that

$$\mathbf{AB} - \mathbf{BA}$$

   is diagonal and not equal to zero. *In quantum mechanics $\frac{d}{dx}$ corresponds to momentum and multiplication by x corresponds to position. In this way, the calculation in this part is a partial proof of the Heisenberg uncertainty principle for position and momentum.*

Problem 4.10. Let

$$\mathbf{A} = \begin{bmatrix} 0 & 1 & 0 \\ 0 & 0 & 2 \\ 0 & 0 & 0 \end{bmatrix}$$

denote the matrix for differentiation computed in example 4.3.6. Let $\mathbf{C}$ be defined by cyclically permuting the basis $\mathcal{B}_1 = (1, x, x^2)$, that is:

$$\mathbf{C} : 1 \mapsto x, \quad \mathbf{C} : x \mapsto x^2, \quad \mathbf{C} : x^2 \mapsto 1.$$

This defines a *change in basis* to the new basis $\mathcal{B}_2 = (x, x^2, 1)$, i.e. even though we have the same vectors, they occur in a different order.

1. Calculate a matrix for $\mathbf{C}$.
2. Calculate a matrix for $\mathbf{C}^{-1}$ (in this case it can be done by noticing that it must cyclically permute the basis in the reverse order).
3. Write down column vectors $\mathbf{v}_1, \mathbf{v}_2$ corresponding to $3 + 2x + x^2$ in the basis $\mathcal{B}_1$ and $\mathcal{B}_2$ respectively.
4. Calculate the derivative of $3 + 2x + x^2$ and write down the column vectors $\mathbf{w}_1, \mathbf{w}_2$ in the basis $\mathcal{B}_1$ and $\mathcal{B}_2$ respectively.
5. Show that
$$\mathbf{C} : \mathbf{v}_2 \mapsto \mathbf{v}_1 \quad \mathbf{A} : \mathbf{v}_1 \mapsto \mathbf{w}_1 \quad \mathbf{C}^{-1} : \mathbf{w}_1 \mapsto \mathbf{w}_2.$$
6. Calculate $\mathbf{C}^{-1}\mathbf{AC}$ and show that it takes v_2 to w_2 in one step, thus giving a different matrix for the derivative in the new basis $\mathcal{B}_1$.

Problem 4.11. Three different linear transformations are defined as follows. $\mathbf{P} : \mathbb{R}^3 \to \mathbb{R}^2$ is defined as projection onto the xy-plane, $\mathbf{S} : \mathbb{R}^2 \to \mathbb{R}^2$ is defined by tripling the length of a vector, and $\mathbf{T} : \mathbb{R} \to \mathbb{R}^2$ is defined by sending x to $(x, 0)$. For each transformation, do the following steps:

1. Use the standard basis of column vectors to compute a matrix for the transformation.
2. Determine the Domain, Codomain, and Range.
3. State whether the transformation is one-to-one, onto, or both.

Problem 4.12. Show that each part of theorem 4.4.4 is false for the matrix

$$\mathbf{A} = \begin{bmatrix} 1 & 1 \\ 0 & 0 \end{bmatrix}.$$

5

Belonging

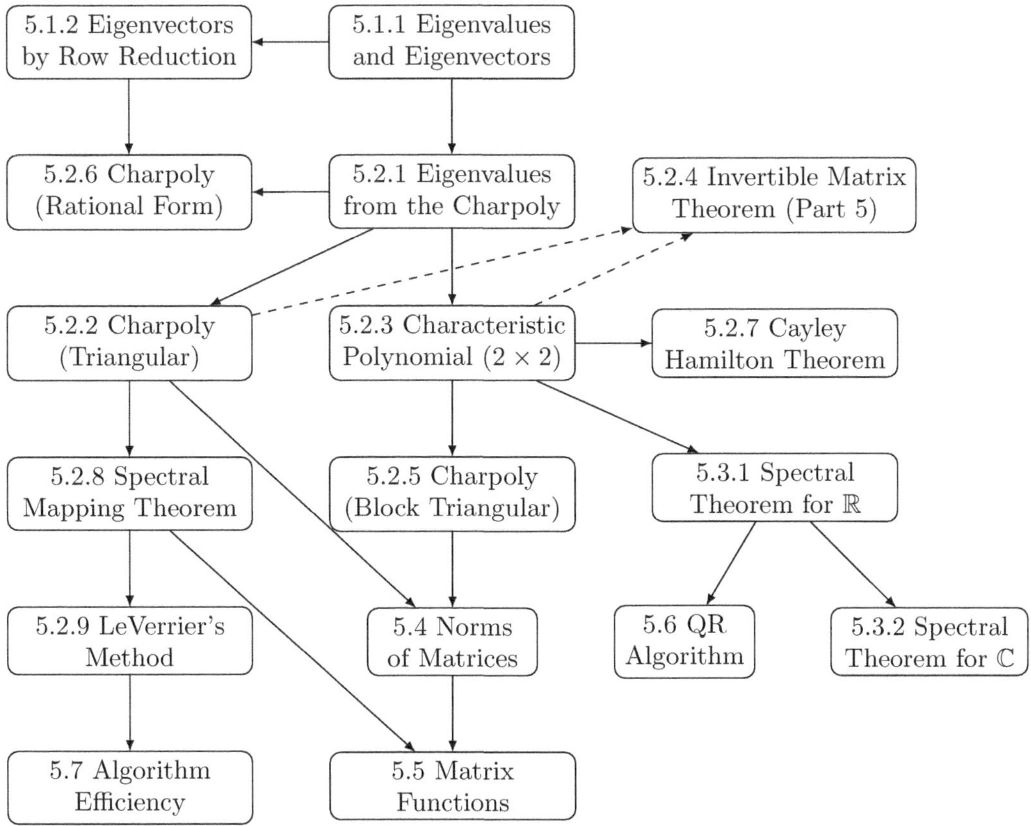

DOI: 10.1201/9781003737490-5

5.1 Eigenvalues and Eigenvectors

5.1.1 The Meaning of Eigenvalues and Eigenvectors

Q 5.1.1. If $\mathbf{I} = \left[\begin{smallmatrix}1 & 0\\ 0 & 1\end{smallmatrix}\right]$ is the 2×2 identity matrix, and $\mathbf{v} = \left[\begin{smallmatrix}x\\ y\end{smallmatrix}\right]$, then what is $\mathbf{Iv}$?

Q 5.1.2. What is $2\mathbf{I}$?

Q 5.1.3. What is $0\left[\begin{smallmatrix}x\\ y\end{smallmatrix}\right]$?

Q 5.1.4. If $A : (x, y) \mapsto (x, 0)$, then what is the null-space of A?

In German the word *eigen* means "own," as in your own or my own. So the name eigenvector or eigenvalue suggests that the vector or value "belongs to" a matrix or a linear transformation in some sense. In fact, I think it would be fair to use the terms "belonging vector" and "belonging value" instead. Whether you call it an eigenvector or a belonging vector, the term suggests that something natural is going on.

Definition 5.1.1. Given an $n \times n$ matrix $\mathbf{A}$, a non-zero vector $\mathbf{v}$ is an **eigenvector** of $\mathbf{A}$ with **eigenvalue** λ if

$$\mathbf{Av} = \lambda\mathbf{v}. \tag{5.1}$$

From the above definition we see that a matrix acts on an eigenvector by *scalar multiplication*, but it is not yet clear why this would be natural. Why would such a vector or value be said to belong to a matrix? Let's look at some examples.

Example 5.1.1. Let $T : \mathbb{R}^2 \to \mathbb{R}^2$ be defined by doubling the length of each vector. Then

$$\mathbf{T} = 2\mathbf{I} = \begin{bmatrix}2 & 0\\ 0 & 2\end{bmatrix}$$

is a matrix for T. All vectors in $\mathbb{R}^2$ are eigenvectors with common eigenvalue $\lambda = 2$:

$$\begin{bmatrix}2 & 0\\ 0 & 2\end{bmatrix}\begin{bmatrix}x\\ y\end{bmatrix} = \begin{bmatrix}2x\\ 2y\end{bmatrix} = 2\begin{bmatrix}x\\ y\end{bmatrix}.$$

Since T doubles the length of vectors, 2 is naturally associated with the operation.

Comprehension Check 5.1.1. Suppose $\mathbf{T}$ is replaced by $5\mathbf{I}$. What is the effect when we multiply by a vector $\left[\begin{smallmatrix}x\\ y\end{smallmatrix}\right]$? Now, what if $\mathbf{T}$ is replaced by $\lambda\mathbf{I}$?

Example 5.1.2. Let $A : \mathbb{R}^2 \to \mathbb{R}^2$ be projection onto the x-axis. Then, in the standard basis, we have

$$\mathbf{A} = \begin{bmatrix}1 & 0\\ 0 & 0\end{bmatrix}.$$

Any vector along the x-axis is an eigenvector with eigenvalue 1 because

$$\begin{bmatrix}1 & 0\\ 0 & 0\end{bmatrix}\begin{bmatrix}x\\ 0\end{bmatrix} = \begin{bmatrix}x\\ 0\end{bmatrix} = (1)\begin{bmatrix}x\\ 0\end{bmatrix}.$$

Any vector along the y-axis is an eigenvector with eigenvalue 0 because

$$\begin{bmatrix}1 & 0\\ 0 & 0\end{bmatrix}\begin{bmatrix}0\\ y\end{bmatrix} = \begin{bmatrix}0\\ 0\end{bmatrix} = (0)\begin{bmatrix}0\\ y\end{bmatrix}.$$

Unlike the first example, not all vectors are eigenvectors. For example

$$\begin{bmatrix}1 & 0\\ 0 & 0\end{bmatrix}\begin{bmatrix}1\\ 1\end{bmatrix} = \begin{bmatrix}1\\ 0\end{bmatrix} \stackrel{?}{=} \lambda\begin{bmatrix}1\\ 1\end{bmatrix}.$$

But in trying to get the last equality to work out, the first component gives $\lambda = 1$, while the second component gives $\lambda = 0$. We cannot have $0 = 1$, so the last equality cannot be satisfied.

There are two distinct eigenvalues. The operation is defined as projection onto the x-axis, so the x-axis is natural to the transformation. Any vector along the x-axis is an eigenvector with eigenvalue 1. It turns out, however, that the y-axis is also natural: y is a free variable for the matrix $\mathbf{A}$, which means that the y-axis is the null space of the transformation $A : \mathbb{R}^2 \to \mathbb{R}^2$.

Activity 5.1.3. Given the matrix

$$\mathbf{A} = \begin{bmatrix}5 & 0 & 0\\ 0 & 5 & 0\\ 0 & 0 & 0\end{bmatrix}$$

1. Check that:

   (a) Any vector in the xy-plane is an eigenvector with eigenvalue 5.

   (b) Any vector along the z-axis is an eigenvector with eigenvalue 0.

2. What is the null space of $\mathbf{A}$?

Definition 5.1.2. Given a fixed eigenvalue λ, the span of the eigenvectors with eigenvalue λ is a vector space called the λ-eigenspace.

- In Example 5.1.1, all vectors in $\mathbb{R}^2$ had eigenvalue 2, so the 2-eigenspace is all of $\mathbb{R}^2$. The dimension of the eigenspace is 2; that is the number of eigenvectors needed to form a basis. Here, we can take the standard basis for $\mathbb{R}^2$

- In Example 5.1.2, any vector along the x-axis has eigenvalue 1, so the 1-eigenspace is the x-axis. It has dimension 1. Any vector along the y-axis has eigenvalue 0, so the 0-eigenspace is the y-axis. It also has dimension 1.

Remark 5.1.1. Note that the dimension does not have to be 1, as 5.1.1 shows, and generally does not match the eigenvalues as 5.1.2 shows because the 0-eigenspace has dimension 1. In fact, eigenvalues do not even have to be integers.

Example 5.1.2 also illustrates that the 0-eigenspace is simply the null space (for a square matrix). Of course, the null space is still defined when the matrix is not square, but these concepts coincide in the case of square matrices. That is because if $\mathbf{v}$ is a vector in the null space of $\mathbf{A}$, then

$$\mathbf{A}\mathbf{v} = \mathbf{0} = 0\mathbf{v}.$$

It turns out that that the concept of a λ-eigenspace is closely connected with the concept of a null space, even when $\lambda \neq 0$.

By Definition 5.1.1, if $\mathbf{v}$ is an eigenvector of a square matrix $\mathbf{A}$ with eigenvalue λ, then it satisfies equation (5.1). If we subtract $\lambda\mathbf{v}$ from both sides, then

$$\mathbf{A}\mathbf{v} - \lambda\mathbf{v} = \mathbf{0}.$$

But $\mathbf{v} = \lambda\mathbf{I}\mathbf{v}$, so by matrix arithmetic properties

$$\mathbf{A}\mathbf{v} - \lambda\mathbf{v} = \mathbf{A}\mathbf{v} - \lambda(\mathbf{I}\mathbf{v}) = \mathbf{A}\mathbf{v} - (\lambda\mathbf{I})\mathbf{v} = (\mathbf{A} - \lambda\mathbf{I})\mathbf{v}.$$

Hence $\mathbf{v}$ must be a solution to

$$(\mathbf{A} - \lambda\mathbf{I})\mathbf{v} = \mathbf{0}. \tag{5.2}$$

In other words, the λ-eigenspace of $\mathbf{A}$ is the null-space of $\mathbf{A} - \lambda\mathbf{I}$.

Remark 5.1.2. To compute $\mathbf{A} - \lambda\mathbf{I}$ all we have to do is subtract λ from the diagonal.

Example 5.1.4. In Example 5.1.1 we saw that all vectors $\left[\begin{smallmatrix} x \\ y \end{smallmatrix}\right]$ were eigenvectors of $\mathbf{T}$ with eigenvalue $\lambda = 2$. We want to see that they are in the null space of $\mathbf{T} - 2\mathbf{I}$.

Thus we begin by subtracting 2 from the diagonal of $\mathbf{T}$:

$$\mathbf{T} - 2\mathbf{I} = \begin{bmatrix} 2-2 & 0 \\ 0 & 2-2 \end{bmatrix} = \begin{bmatrix} 0 & 0 \\ 0 & 0 \end{bmatrix}.$$

There are no pivots, so both x and y are free variables. The free variables give us the equations $x = x$ and $y = y$, hence

$$\begin{bmatrix} x \\ y \end{bmatrix}$$

is in the null space of $\mathbf{T} - 2\mathbf{I}$ (or equivalently the 2-eigenspace of $\mathbf{T}$). To get a basis, we set one of the free variables equal to 1, and the remaining free variable equal to zero. So

$$x = 1, y = 0 \quad \text{gives us} \quad \begin{bmatrix} 1 \\ 0 \end{bmatrix} \quad \text{and} \quad x = 0, y = 1 \quad \text{gives us} \quad \begin{bmatrix} 0 \\ 1 \end{bmatrix},$$

which is just the standard basis for $\mathbf{R}^2$.

Example 5.1.5. In Example 5.1.2, we already observed that the eigenvectors of $\mathbf{A}$ with eigenvalue 0 were in the null space, which we may also call the 0-eigenspace by Definition 5.1.2. But $\lambda = 1$ was also an eigenvalue. Consider

$$\mathbf{A} - (1)\mathbf{I} = \begin{bmatrix} 1-1 & 0 \\ 0 & 0-1 \end{bmatrix} = \begin{bmatrix} 0 & 0 \\ 0 & -1 \end{bmatrix}$$

What is the null space of this matrix? In augmented form, $[\mathbf{A} - (1)\mathbf{I}|\mathbf{0}]$ is

$$\left[\begin{array}{cc|c} 0 & 0 & 0 \\ 0 & -1 & 0 \end{array}\right]$$

which gives us the equation $-y = 0$, while x is free. Since x is free, we must include the equation $x = x$, so

$$\begin{bmatrix} x \\ 0 \end{bmatrix} = x\begin{bmatrix} 1 \\ 0 \end{bmatrix}$$

gives us a parameterization of the null space of $\mathbf{A} - (1)\mathbf{I}$ or equivalently of the 1-eigenspace of $\mathbf{A}$.

Activity 5.1.6. Using the same matrix $\mathbf{A}$ as in Activity 5.1.3, show that any vector in the xy-plane is in the null space of $\mathbf{A} - 5\mathbf{I}$.

5.1.2 Eigenvectors by Row Reduction

If we know a specific eigenvalue λ, then the corresponding eigenvectors can be determined by computing the null space of $\mathbf{A}-\lambda\mathbf{I}$. One way to do that is by row reducing the augmented matrix $[\mathbf{A}-\lambda\mathbf{I}|\mathbf{0}]$. In practice I usually omit the augmented column when calculating eigenvectors since it always remains zero. If you choose to follow the same practice, you must always keep in mind that the augmented column of zeros is still implied, though not written down. The Example 5.1.7 illustrates this strategy.

Example 5.1.7. The matrix

$$\mathbf{A}=\begin{bmatrix}0 & 1\\ 3 & 2\end{bmatrix}$$

has $\lambda=-1$ as an eigenvalue. Use row reduction to compute a parameterization of the -1 eigenspace.

Solution

$$\begin{bmatrix}0-(-1) & 1\\ 3 & 2-(-1)\end{bmatrix}=\begin{bmatrix}1 & 1\\ 3 & 3\end{bmatrix}\xrightarrow{(-3)R_1+R_3}\begin{bmatrix}1 & 1\\ 0 & 0\end{bmatrix}.$$

By remembering the hiding augmented column of zeros, the top row gives us the equation $x+y=0$, so $x=-y$. Since y is free, we have $y=y$, hence

$$\begin{bmatrix}x\\ y\end{bmatrix}=\begin{bmatrix}-y\\ y\end{bmatrix}=y\begin{bmatrix}-1\\ 1\end{bmatrix},$$

which is a parameterization of the -1 eigenspace of $\mathbf{A}$.

For larger matrices, it may help to review Example 4.3.4.

Activity 5.1.8.

$$\mathbf{A}=\begin{bmatrix}3 & 2\\ 1 & 4\end{bmatrix}$$

has $\lambda=2$ as an eigenvalue. Use row reduction to compute a basis for the 2 eigenspace.

In equation (5.1), and all of the above examples, it is assumed that the vector is on the *right*. However, we could just as well consider the case where the vectors are on the *left* instead. That is, we could instead consider

$$\mathbf{vA}=\lambda\mathbf{v}.$$

The eigenvalues remain the same, but the eigenvectors generally do not. For this reason, Sage has two separate functions: `eigenvectors_right()` and `eigenvectors_left()`. For right eigenvectors we use the following code

```
A=matrix([[ 0, 3],
          [ 1,  2]])
A.eigenvectors_right()
```

The output deserves some explanation. We get a list of triples consisting of

0. an eigenvalue
1. a list of basis vectors for the eigenspace

2. the "algebraic multiplicity" of the eigenvalue.

To get a particular triple, we can call it by index:

```
A.eigenvectors_right()[0]
```

Within the triple, the eigenvalue has index 0,

```
A.eigenvectors_right()[0][0]
```

and the list of basis vectors has index 1:

```
A.eigenvectors_right()[0][1]
```

In this case there is only one vector in the basis, and we can get it by its index, which is 0:

```
A.eigenvectors_right()[0][1][0]
```

If we want to use one basis vector at a time, then this is the way to go. If we want to work with several of the basis vectors at once, such as in a parallel algorithm, then sage also provides a function called `eigenmatrix_right()`:

```
A.eigenmatrix_right()
```

The output is a list of two matrices. The 0 index gives a diagonal matrix of eigenvalues, and the 1 index gives a corresponding matrix of eigenvectors.

Activity 5.1.9. The matrix

$$\mathbf{A} = \begin{bmatrix} 2 & 1 & 0 \\ 0 & 2 & 0 \\ 0 & 0 & 3 \end{bmatrix}$$

has 2 as an eigenvalue. Compute a basis for the 2-eigenspace in two steps:

1. Compute $\mathbf{A} - 2\mathbf{I}$.
2. Compute a basis for the null space of $\mathbf{A} - 2\mathbf{I}$.

5.2 The Characteristic Polynomial

TL;DR

The characteristic polynomial of a square matrix $\mathbf{A}$ is defined by equation (5.4). Its roots are the eigenvalues of $\mathbf{A}$.
Several methods are provided to compute the characteristic polynomial by hand:

- For the 2×2 case, see Proposition 5.2.3
- For the upper-triangular case, see Corollary 5.2.2
- For block-upper triangular case, see Corollary 5.2.5
- For rational-canonical form, see Theorem 5.2.6

- For Le Verrier's method, see Subsection 5.2.9 especially Example 5.2.25

The Spectral Mapping Theorem, Theorem 5.2.8, shows how the eigenvalues of $f(\mathbf{A})$ are related to those of $\mathbf{A}$. It can be used to justify both Le Verrier's method and the Cayley-Hamilton Theorem 5.2.7

5.2.1 Computing the Eigenvalues from the Characteristic Polynomial

Q 5.2.1. What is the determinant of $\left[\begin{smallmatrix} a & b \\ c & d \end{smallmatrix}\right]$?

Q 5.2.2. What is $(x-2)(x+3)$ equal to?

In Subsection 5.1.1, we saw that if λ is an eigenvalue of a matrix $\mathbf{A}$, then the λ-eigenspace of $\mathbf{A}$ is the same as the null space of $\mathbf{A} - \lambda\mathbf{I}$. Thus, it should not be surprising that the eigenvalues are the values of λ for which $\mathbf{A} - \lambda\mathbf{I}$ has a null space with positive dimension. If the null space of $\mathbf{A} - \lambda\mathbf{I}$ has positive dimension, then

$$\det(\mathbf{A} - \lambda\mathbf{I}) = 0. \tag{5.3}$$

Example 5.2.1. For Example 5.1.1 we have

$$\det\left(\begin{bmatrix} 2 & 0 \\ 0 & 2 \end{bmatrix} - \lambda\begin{bmatrix} 1 & 0 \\ 0 & 1 \end{bmatrix}\right) = \det\left(\begin{bmatrix} 2-\lambda & 0 \\ 0 & 2-\lambda \end{bmatrix}\right) = (2-\lambda)^2 = 0.$$

The only root is $\lambda = 2$, hence it is the only eigenvalue.

Example 5.2.2. For Example 5.1.2 we have

$$\det\left(\begin{bmatrix} 1 & 0 \\ 0 & 0 \end{bmatrix} - \lambda\begin{bmatrix} 1 & 0 \\ 0 & 1 \end{bmatrix}\right) = \det\left(\begin{bmatrix} 1-\lambda & 0 \\ 0 & -\lambda \end{bmatrix}\right) = -\lambda(1-\lambda) = 0.$$

The roots $\lambda = 0$ and $\lambda = 1$ are the eigenvalues.

For an $n \times n$ matrix $\mathbf{A}$, the determinant of $\mathbf{A} - \lambda\mathbf{I}$ is a polynomial

$$\det(\mathbf{A} - \lambda\mathbf{I}) = (-\lambda)^n + c_{n-1}(-\lambda)^{n-1} + \cdots c_1(-\lambda) + c_0,$$

which is a degree n polynomial with coefficients in the field of scalars. If n is odd, then $(-\lambda)^n = -\lambda^n$, but we would like the coefficient of λ^n to be 1, which we can can accomplish if we multiply through by $(-1)^n$:

$$(-1)^n \det(\mathbf{A} - \lambda\mathbf{I}) = f(\lambda) = 0. \tag{5.4}$$

Definition 5.2.1. Equation (5.4) is called the **characteristic equation**, and f is called the **characteristic polynomial**.

Example 5.2.3. Suppose that a matrix $\mathbf{A}$ has characteristic polynomial

$$f(x) = x^2 - 2x - 3$$

Compute the eigenvalues of $\mathbf{A}$.

Solution

While you can use the quadratic formula, in this case we can apply the rational root theorem to find the roots: 3 factors only as $1 \cdot 3$, so if there is a rational root, then $x^2 - 2x - 3$ must factor either as

$$(x-3)(x+1) \quad \text{or} \quad (x+3)(x-1).$$

This is necessary to match the constant term. But what about the middle term? The middle term is -2, and it must be the sum of -3 and 1 or of 3 and -1. It is the first of these. So we have

$$\begin{aligned} x^2 - 2\lambda - 3 = (x-3)(x+1) &= 0 \\ \implies \quad x - 3 = 0 \quad &\text{or} \quad x + 1 = 0 \\ \implies \quad x = 3 \quad &\text{or} \quad x = -1. \end{aligned}$$

Thus the eigenvalues of $\mathbf{A}$ are 3 and -1.

Activity 5.2.4. Suppose that a matrix $\mathbf{A}$ has characteristic polynomial

$$f(x) = x^2 - 4x + 3$$

Compute the eigenvalues of $\mathbf{A}$.

It is generally not easy to obtain the characteristic polynomial by using equation (5.4) directly. You can use Sage, but this section covers several practical ways of computing the characteristic polynomial by hand.

With Sage, we can get the characteristic polynomial and factor it as follows:

```
A=matrix([[ 1, 0],
          [ 0, 0]])
f=A.charpoly()
f
f.factor()
```

Sage can also compute the eigenvalues directly:

```
A.eigenvalues()
```

Note that the result is returned as a list.

5.2.2 Characteristic Polynomial in Triangular Form

Q 5.2.3. What is the determinant of $\begin{bmatrix} 1 & 2 & 3 \\ 0 & 5 & 6 \\ 0 & 0 & 9 \end{bmatrix}$?

Proposition 3.1.3 says that for an upper triangular matrix, the determinant is the product of the diagonal. If we apply that to $\det(\mathbf{A} - \lambda\mathbf{I})$, we get the following corollary:

Corollary 5.2.2. *Let* $\mathbf{A}$ *be an* $n \times n$ *upper triangular matrix. Then the the characteristic polynomial of* $\mathbf{A}$ *is the product*

$$\prod_{i=1}^{n}(x - a_{ii}),$$

where the a_{ii} *are the diagonal entries. In particular the eigenvalues of* $\mathbf{A}$ *are the values on the diagonal.*

The proof using Proposition 3.1.3 is straight forward, but for another proof, see [2].

Example 5.2.5. Compute the characteristic polynomial of

$$\mathbf{A} = \begin{bmatrix} 2 & 1 & 0 \\ 0 & 2 & 0 \\ 0 & 0 & 3 \end{bmatrix}.$$

Solution

By Corollary 5.2.2, we have

$$(x-2)^2(x-3).$$

Activity 5.2.6. Compute the characteristic polynomial of

$$\mathbf{A} = \begin{bmatrix} 5 & 1 & 0 & 0 \\ 0 & 5 & 0 & 0 \\ 0 & 0 & -5 & 0 \\ 0 & 0 & 0 & 0 \end{bmatrix}.$$

5.2.3 The Characteristic Polynomial in the 2×2 case.

In the 2×2 case, the trace and determinant are sufficient for computing the characteristic polynomial.

Proposition 5.2.3. *For a* 2×2 *matrix* $\mathbf{A}$*, the characteristic polynomial is*

$$f(x) = x^2 - \operatorname{tr}(\mathbf{A})x + \det(\mathbf{A}).$$

Example 5.2.7. Given

$$\mathbf{A} = \begin{bmatrix} 0 & 3 \\ 1 & 2 \end{bmatrix},$$

we have

$$\operatorname{tr}(\mathbf{A}) = 2 \quad \text{and} \quad \det(\mathbf{A}) = -3,$$

so the characteristic polynomial is

$$f(x) = x^2 - 2x - 3.$$

As we have seen in Example 5.2.3, this factors as

$$f(x) = (x-3)(x+1).$$

Activity 5.2.8. Given the matrix

$$\mathbf{A} = \begin{bmatrix} 2 & 4 \\ 1 & 2 \end{bmatrix},$$

use Proposition 5.2.3 to compute the characteristic polynomial, then factor it.

5.2.4 Invertible Matrix Theorem (Part 5)

Theorem 5.2.4 (Invertible Matrix Theorem: Part 5). *Let* $\mathbf{A}$ *be an* $n \times n$ *matrix with real entries. The following statements are equivalent:*

17. $\mathbf{A}$ *does not have* 0 *as an eigenvalue.*

See Remark 1.3.1 concerning the Invertible matrix theorem and its proof.

Activity 5.2.9. Use either Corollary 5.2.2 or Proposition 5.2.3 to show that statement 17 in the theorem is true for the matrix

$$\mathbf{A} = \begin{bmatrix} 1 & 1 \\ 0 & 1 \end{bmatrix}.$$

5.2.5 Block Triangular Form

Theorem 3.1.6 also can be applied to give us the following corollary

Corollary 5.2.5. *If* $\mathbf{A}$ *is an* $n \times n$ *matrix in block upper triangular form, then the characteristic polynomial of* $\mathbf{A}$ *is the product of the characteristic polynomials of the blocks on the diagonal.*

Example 5.2.10. Compute the characteristic polynomial for each matrix below:

$$\mathbf{A} = \begin{bmatrix} 0 & 2 & 1 \\ 1 & 0 & 1 \\ 0 & 0 & 5 \end{bmatrix} \quad \text{and} \quad \mathbf{B} = \begin{bmatrix} 0 & 2 & 0 & 1 \\ 1 & 0 & 1 & 0 \\ 0 & 0 & 5 & 1 \\ 0 & 0 & 0 & 5 \end{bmatrix}.$$

Solution

The blocks on the diagonal have been highlighted for clarity. For $\mathbf{A}$, the 2×2 block has trace 0 and determinant -2. The characteristic polynomial of that block is x^2-2. The 1×1 block has the characteristic polynomial $x-5$. So, by Corollary 5.2.5, the characteristic polynomial of $\mathbf{A}$ is

$$(x^2-2)(x-5).$$

For $\mathbf{B}$, the 2×2 block in the upper left corner is the same as in $\mathbf{A}$. The 2×2 block in the lower left corner is upper triangular, so, by Corollary 5.2.2, it has characteristic polynomial $(x-5)^2$. Then, by Corollary 5.2.5, the characteristic polynomial of $\mathbf{B}$ is

$$(x^2-2)(x-5)^2.$$

Activity 5.2.11. Compute the characteristic polynomial for each matrix below:

$$\mathbf{A} = \begin{bmatrix} 0 & -3 & 1 \\ 1 & 0 & 1 \\ 0 & 0 & 0 \end{bmatrix} \quad \text{and} \quad \mathbf{B} = \begin{bmatrix} 0 & -3 & 0 & 1 \\ 1 & 0 & 1 & 0 \\ 0 & 0 & 0 & 1 \\ 0 & 0 & 0 & 0 \end{bmatrix}.$$

5.2.6 Rational Canonical Form

As a final example of when it is not to hard to compute the characteristic polynomial directly from the definition, consider the following theorem:

Theorem 5.2.6. *Given a matrix in the form*

$$\mathbf{A} = \begin{bmatrix} 0 & \cdots & & & -c_0 \\ 1 & 0 & \cdots & & -c_1 \\ 0 & 1 & 0 & \cdots & -c_2 \\ \vdots & \ddots & \ddots & & \vdots \\ 0 & \cdots & 0 & 1 & -c_{n-1} \end{bmatrix}$$

the characteristic polynomial of **A** *is*

$$x^n + c_{n-1}x^{n-1} + \cdots c_1 x + c_0$$

The following example generalizes to a proof of this theorem:

Example 5.2.12. Verify Theorem 5.2.6 for the matrix

$$\mathbf{A} = \begin{bmatrix} 0 & 0 & -1 \\ 1 & 0 & -2 \\ 0 & 1 & -3 \end{bmatrix}$$

by row reducing $\det(\lambda\mathbf{I} - \mathbf{A})$.

Solution

In this case, the most efficient approach is to work from the bottom to the top:

$$\det(\lambda\mathbf{I} - \mathbf{A}) = \begin{vmatrix} \lambda & 0 & 1 \\ -1 & \lambda & 2 \\ 0 & -1 & \lambda+3 \end{vmatrix} \xrightarrow{\lambda R_3 + R_2} \begin{vmatrix} \lambda & 0 & 1 \\ -1 & 0 & \lambda^2+3\lambda+2 \\ 0 & -1 & \lambda+3 \end{vmatrix}$$

$$\xrightarrow{\lambda R_2 + R_1} \begin{vmatrix} 0 & 0 & \lambda^3+3\lambda^2+2\lambda+1 \\ -1 & 0 & \lambda^2+3\lambda+2 \\ 0 & -1 & \lambda+3 \end{vmatrix}.$$

So far, the row reduction steps we have used do not change the determinant, but to get to upper-triangular form, we need to move the top row to the bottom. This requires us to use $R_i \leftrightarrow R_j$ several times. First, we do $R_1 \leftrightarrow R_2$, then we do $R_2 \leftrightarrow R_3$. The result is

$$\begin{vmatrix} -1 & 0 & \lambda^2+3\lambda+2 \\ 0 & -1 & \lambda+3 \\ 0 & 0 & \lambda^3+3\lambda^2+2\lambda+1 \end{vmatrix} = (-1)^2(\lambda^3+3\lambda^2+2\lambda+1).$$

Since we have swapped rows twice, we get two more factors of -1, hence

$$\det(\lambda\mathbf{I} - \mathbf{A}) = \lambda^3 + 3\lambda^2 + 2\lambda + 1.$$

Activity 5.2.13. Verify Theorem 5.2.6 for the matrix

$$\mathbf{A} = \begin{bmatrix} 0 & 0 & 0 & -1 \\ 1 & 0 & 0 & -2 \\ 0 & 1 & 0 & -3 \\ 0 & 0 & 1 & -4 \end{bmatrix}$$

by row reducing $\det(\lambda\mathbf{I} - \mathbf{A})$.

5.2.7 The Cayley-Hamilton Theorem

The next theorem provides an indirect way of determining the characteristic polynomial, and it is also an important result in its own right.

Theorem 5.2.7 (Cayley-Hamilton Theorem). *If* $\mathbf{A}$ *is an* $n \times n$ *matrix, and* f *is the characteristic polynomial of* $\mathbf{A}$*, then*

$$f(\mathbf{A}) = 0.$$

A proof can be found in [2], however, the Cayley-Hamilton Theorem can also be seen as a corollary of the Spectral Mapping Theorem (Theorem 5.2.8).

Example 5.2.14. We now return to our story about the symmetries of the square. This time however, the setting will be the complex plane. $\mathbb{C}$ is a 2 dimensional vector space over $\mathbb{R}$, with basis $\mathbf{1}, \mathbf{i}$ (written in bold to indicate that we are thinking of them as vectors). Multiplication by i acts as rotation by 90 degrees ($\frac{\pi}{2}$ radians), which may be easier to see in polar form because $i = e^{\frac{\pi}{2}i}$. We get a matrix in the usual manner:

$$\begin{aligned} i \cdot \mathbf{1} &= \mathbf{i} = 0 \cdot \mathbf{1} + 1 \cdot \mathbf{i}, \\ i \cdot \mathbf{i} &= -\mathbf{1} = (-1) \cdot \mathbf{1} + 0 \cdot \mathbf{i} \end{aligned}$$

hence

$$\mathbf{R} = \begin{bmatrix} 0 & -1 \\ 1 & 0 \end{bmatrix},$$

which is exactly the same as what we computed before. The characteristic equation can be computed as follows:

$$\operatorname{tr}(\mathbf{R}) = 0 + 0 = 0 \quad \text{and} \quad \det(\mathbf{R}) = 0 \cdot 0 - (-1)(1) = 1\,,$$

hence

$$f(x) = x^2 - \operatorname{tr}(\mathbf{R})x + \det(\mathbf{R}) = x^2 + 1\,.$$

The trace, again, is just the sum of the elements on the main diagonal, which are both zero in this case. The eigenvalues are the roots of the characteristic polynomial,

$$x^2 + 1 = (x + i)(x - i) = 0\,,$$

so the eigenvalues are i and $-i$. We obtained the matrix $\mathbf{R}$ from multiplication by i in the complex plane, and that matrix has i as an eigenvalue. An eigenvalue satisfies the characteristic equation, but according to Theorem 5.2.7 so does the matrix. The key is to interpret the constant term as a constant times the identity matrix. Since $f(x) = x^2 + 1$ for the matrix $\mathbf{R}$, then

$$f(\mathbf{R}) = \mathbf{R}^2 + \mathbf{I}. \tag{5.5}$$

To verify that this is zero we must compute $\mathbf{R}^2$:

$$\mathbf{R}^2 = \begin{bmatrix} 0 & -1 \\ 1 & 0 \end{bmatrix} \begin{bmatrix} 0 & -1 \\ 1 & 0 \end{bmatrix} = \begin{bmatrix} -1 & 0 \\ 0 & -1 \end{bmatrix} = -\mathbf{I}.$$

Hence

$$f(\mathbf{R}) = \mathbf{R}^2 + \mathbf{I} = -\mathbf{I} + \mathbf{I} = \mathbf{0}.$$

The story goes deeper. The eigenvalues of $\mathbf{R}$ are imaginary, so there are no real eigenvectors. This makes sense geometrically. Eigenspaces are "fixed" by the action of a matrix. That is, if we start with a vector in the eigenspace and multiply by $\mathbf{R}$, the result stays in the eigenspace (its components just get multiplied by a scalar). The matrix $\mathbf{R}$ is rotation counterclockwise by 90 degrees in the complex plane (viewed as a real vector space). What subspace is fixed by counterclockwise rotation by 90 degrees? Well the center of rotation is fixed, namely $\mathbf{0}$, a.k.a. the origin. But any line through the origin gets rotated. So, there are no eigenvectors in the plane corresponding to the eigenvalues.

The problem is that thinking of $\mathbb{C}$ as the complex plane requires us to use only real numbers as scalars. From this viewpoint, we are using $\mathbf{i}$ as a vector not as a scalar. However, the eigenvalues of $\mathbf{R}$ are not real. Thus to get an eigenvector, we must extend scalars to include imaginary numbers. To avoid confusion with our use of $\mathbf{i}$ as a vector, it is wise to give the new imaginary unit a different name. Exercise 5.9 takes precisely this approach, and finds eigenvectors in the Hamilton quaternions viewed as a 2 dimensional complex vector space.

Trying to mentally picture the spaces that are fixed in the extended vector space is not easy. To picture things, we have the most success when a vector space can be viewed as a real vector space with no more than 3 dimensions. We are able to picture complex numbers because $\mathbb{C}$ can be viewed as a two dimensional real vector space. But viewing the Hamilton quaternions as a real vector space requires us to go up to dimension 4. Let's ratchet down to 3 real dimensions and consider counterclockwise rotation by 90 degrees in the xy-plane. Then the z-axis is fixed. Going back up to 4 real dimensions gives a plane that is fixed. Four real dimensions is not the same thing, as 2 complex dimensions – what we need to imagine in 2 complex dimensions is that there are two fixed lines, not one fixed plane. But this is the best we can do to picture things with our limited human faculties. The rest must be done by computation.

Activity 5.2.15. Let $V = \{a+b\sqrt{2} : a, b \in \mathbb{Q}\}$. Let $T : V \to V$ be defined as multiplication $\sqrt{2}$. Using $1, \sqrt{2}$ as a basis

1. compute a matrix for T,
2. compute the characteristic equation, and
3. show that the matrix for T satisfies its own characteristic equation.

5.2.8 Spectral Mapping Theorem

Theorem 5.2.8 (Spectral Mapping Theorem)**.** *Let $\mathbf{A}$ be an $n \times n$ matrix with real or complex entries, and let $f(x)$ be a polynomial with complex coefficients. Then the eigenvalues of $f(\mathbf{A})$ are the values $f(\lambda)$ such that λ is an eigenvalue of $\mathbf{A}$.*

We will analyze this theorem in stages, starting with the simplest case and building up in steps. The simplest case is where $f(x)$ is a power, and $\mathbf{A}$ is a diagonal matrix.

Example 5.2.16. Let $f(x) = x^2$ and

$$\mathbf{A} = \begin{bmatrix} 2 & 0 \\ 0 & 3 \end{bmatrix},$$

which has eigenvalues 2 and 3. Then

$$\begin{bmatrix} 2 & 0 \\ 0 & 3 \end{bmatrix}$$

$$f(\mathbf{A}) = \mathbf{A}^2 = \begin{bmatrix} 2 & 0 \\ 0 & 3 \end{bmatrix} \begin{bmatrix} 4 & 0 \\ 0 & 9 \end{bmatrix},$$

which has eigenvalues 4 and 9. Since $4 = f(2)$ and $9 = f(3)$, then that verifies the Spectral Mapping Theorem for this example.

Activity 5.2.17. Let $f(x) = x^3$ and

$$\mathbf{A} = \begin{bmatrix} 2 & 0 \\ 0 & 3 \end{bmatrix}.$$

Verify that $f(\mathbf{A})$ is diagonal with $f(2)$ and $f(3)$ on the diagonal. You may use the fact that $\mathbf{A}^3 = \mathbf{A}^2\mathbf{A}$ and the result for $\mathbf{A}^2$ from Example 5.2.16.

Hopefully you are now convinced that this generalizes to arbitrary diagonal matrices $\mathbf{A}$ with arbitrary powers $f(x) = x^k$. Next we look at what happens when we multiply by a constant.

Example 5.2.18. Let $f(x) = 5x^2$ and $\mathbf{A}$ as in Example 5.2.16. In that example, we computed $\mathbf{A}^2$, so with this new $f(x)$

$$f(\mathbf{A}) = 5\mathbf{A}^2 = 5\begin{bmatrix} 4 & 0 \\ 0 & 9 \end{bmatrix} = \begin{bmatrix} 20 & 0 \\ 0 & 45 \end{bmatrix}.$$

Since $f(2) = 5(2)^2 = 5 \times 4 = 20$ and $f(3) = 5(3)^2 = 5 \times 9 = 45$, the Spectral Mapping Theorem is verified for this case as well.

Activity 5.2.19. Let $f(x) = 2x^3$ and $\mathbf{A}$ as in Example 5.2.17. Compute $f(\mathbf{A})$ with this new $f(x)$, then check that its eigenvalues are equal to $f(2)$ and $f(3)$.

So, we can see that a coefficient times a power works. We now need to add.

Example 5.2.20. Let $f(x) = 4 + 5x^2$ and let $\mathbf{A}$ be as in Examples 5.2.16 and 5.2.18. Then

$$f(\mathbf{A}) = 4\mathbf{I} + 5\mathbf{A}^2 = \begin{bmatrix} 4 & 0 \\ 0 & 4 \end{bmatrix} + \begin{bmatrix} 20 & 0 \\ 0 & 45 \end{bmatrix} = \begin{bmatrix} 24 & 0 \\ 0 & 49 \end{bmatrix},$$

So, the eigenvalues of $f(\mathbf{A})$ are 24 and 49. Since $f(2) = 4+5(2)^2 = 24$ and $f(3) = 4+5(3)^2 = 49$, the Spectral Mapping Theorem checks out once again.

Activity 5.2.21. Let $f(x) = 2x^3 - 4$ and $\mathbf{A}$ as in Examples 5.2.17 and 5.2.19. Compute $f(\mathbf{A})$ with this new $f(x)$, then check that its eigenvalues are equal to $f(2)$ and $f(3)$.

By now, you should be convinced that the Spectral Mapping Theorem works for arbitrary polynomials, so long as $\mathbf{A}$ is diagonal. But, what if $\mathbf{A}$ isn't diagonal? As we will see in Section 5.3, in many cases it is possible to find $\mathbf{Q}$ such that

$$\mathbf{Q}^{-1}\mathbf{A}\mathbf{Q}$$

is diagonal. When it is not possible to get diagonal form, it is always possible to get block diagonal form, where blocks looking like

$$\mathbf{J} = \begin{bmatrix} \lambda & 1 & 0 & \cdots & 0 \\ 0 & \lambda & 1 & \cdots & 0 \\ 0 & 0 & \lambda & \cdots & 0 \\ \vdots & \vdots & \vdots & \ddots & \vdots \\ 0 & 0 & 0 & \cdots & \lambda \end{bmatrix}$$

are allowed, which are called Jordan blocks. That leaves us with two things to convince ourselves of:

1. that the Spectral Mapping Theorem is true for $f(\mathbf{J})$, and

2. that the Spectral Mapping Theorem is compatible with a change of basis

$$\mathbf{Q}^{-1}f(\mathbf{A})\mathbf{Q} = f(\mathbf{Q}^{-1}\mathbf{A}\mathbf{Q}). \tag{5.6}$$

Example 5.2.22. Let $f(x) = 4 + 5x^2$ and

$$\mathbf{J} = \begin{bmatrix} -1 & 1 \\ 0 & -1 \end{bmatrix}.$$

Compute $f(\mathbf{J})$ and show that the Spectral Mapping Theorem is valid in this case.

Solution

First, we compute $\mathbf{J}^2$:

$$\begin{array}{cc} & \begin{bmatrix} -1 & 1 \\ 0 & -1 \end{bmatrix} \\ \begin{bmatrix} -1 & 1 \\ 0 & -1 \end{bmatrix} & \begin{bmatrix} 1 & -2 \\ 0 & 1 \end{bmatrix}. \end{array}$$

Then

$$f(\mathbf{J}) = 4\mathbf{I} + 5\mathbf{J}^2 = \begin{bmatrix} 4 & 0 \\ 0 & 4 \end{bmatrix} + \begin{bmatrix} 5 & -10 \\ 0 & 5 \end{bmatrix} = \begin{bmatrix} 9 & -10 \\ 0 & 9 \end{bmatrix}.$$

Both $\mathbf{J}$ and $f(\mathbf{J})$ are in upper triangular form, so by Corollary 5.2.2 their eigenvalues are on the diagonal. So $\lambda = -1$ is the only eigenvalue of $\mathbf{J}$, and $\lambda = 9$ is the only eigenvalue of $f(\mathbf{J})$. Since

$$f(-1) = 4 + 5(-1)^2 = 4 + 5 = 9,$$

the Spectral Mapping Theorem checks out.

Activity 5.2.23. Let $f(x) = 2x^3 - 4$ and

$$\mathbf{J} = \begin{bmatrix} 2 & 1 \\ 0 & 2 \end{bmatrix}.$$

Compute $f(\mathbf{J})$ and show that the Spectral Mapping Theorem is valid in this case.

Now we will try to make sense of equation (5.6). The reason this happens is that $\mathbf{Q}\mathbf{Q}^{-1} = \mathbf{I}$, and multiplying a matrix changes nothing. For example,

$$\mathbf{Q}^{-1}\mathbf{A}^2\mathbf{Q} = \mathbf{Q}^{-1}\mathbf{A}\mathbf{Q}\mathbf{Q}^{-1}\mathbf{A}\mathbf{Q} = (\mathbf{Q}^{-1}\mathbf{A}\mathbf{Q})^2.$$

While the order of matrix multiplication cannot be changed, a scalar factor can be moved around in a product, so

$$\mathbf{Q}^{-1}5\mathbf{A}^2\mathbf{Q} = 5\mathbf{Q}^{-1}\mathbf{A}^2\mathbf{Q} = 5(\mathbf{Q}^{-1}\mathbf{A}\mathbf{Q})^2.$$

And because of the distributive properties (see Proposition 1.1.16),

$$\mathbf{Q}^{-1}(\mathbf{B} + \mathbf{C})\mathbf{Q} = \mathbf{Q}^{-1}\mathbf{B}\mathbf{Q} + \mathbf{Q}^{-1}\mathbf{C}\mathbf{Q}.$$

So if $\mathbf{B} = 4\mathbf{I}$ and $\mathbf{C} = 5\mathbf{A}^2$, then that completes the task of verifying (5.6) with $f(x) = 4 + 5x^2$. For a more numerical illustration, try the next activity.

Activity 5.2.24. Let

$$\mathbf{A} = \begin{bmatrix} 2 & 1 \\ 0 & 3 \end{bmatrix} \quad \text{and} \quad \mathbf{Q} = \begin{bmatrix} 1 & 1 \\ 0 & 1 \end{bmatrix}, \quad \text{so} \quad \mathbf{Q}^{-1} = \begin{bmatrix} 1 & -1 \\ 0 & 1 \end{bmatrix}$$

1. Show that

$$\mathbf{Q}^{-1}\mathbf{A}\mathbf{Q} = \begin{bmatrix} 2 & 0 \\ 0 & 3 \end{bmatrix}.$$

2. Compute $\mathbf{A}^2$, then show that

$$\mathbf{Q}^{-1}\mathbf{A}^2\mathbf{Q} = \begin{bmatrix} 4 & 0 \\ 0 & 9 \end{bmatrix}.$$

3. Compute $5\mathbf{A}^2$, then show that

$$\mathbf{Q}^{-1}\mathbf{A}^2\mathbf{Q} = \begin{bmatrix} 20 & 0 \\ 0 & 45 \end{bmatrix}.$$

5.2.9 Le Verrier's Method

One very useful application of the Spectral Mapping Theorem is that it gives us a fast algorithm for computing the characteristic polynomial of a square matrix $\mathbf{A}$. Here is the idea. Suppose that $\mathbf{A}$ is an $n \times n$ matrix with eigenvalues

$$\lambda_1, \lambda_2 \ldots, \lambda_n,$$

which are the roots of its characteristic polynomial. Then the eigenvalues of $\mathbf{A}^k$ are

$$\lambda_1^k, \lambda_2^k \ldots, \lambda_n^k,$$

and so the trace of $\mathbf{A}^k$ is

$$t_k = \lambda_1^k + \lambda_2^k + \cdots + \lambda_n^k. \tag{5.7}$$

On the other hand, if $f(x)$ is a degree n polynomial with roots $\lambda_1, \lambda_2 \cdots \lambda_n$, then

$$\begin{aligned} f(x) &= \prod_{k=1}^{n}(x - \lambda_k) = (x - \lambda_1)(x - \lambda_2)\cdots(x - \lambda_n) \\ &= x^n - s_1 x^{n-1} + s_2 x^{n-1} - \cdots + (-1)^n s_n = \sum_{k=0}^{n}(-1)^k s_k x^{n-k}, \end{aligned}$$

where s_k is the k-th symmetric polynomial of the roots. Note that $s_0 = 1$ since it is the coefficient of x^n. Also

$$s_1 = \lambda_1 + \lambda_2 + \cdots + \lambda_n = \operatorname{tr}(\mathbf{A}) \quad \text{and} \quad s_n = \lambda_1\lambda_2\cdots\lambda_n = \det(\mathbf{A}). \tag{5.8}$$

In general s_k is the sum of all products of k roots. As an example for $n = 3$, the symmetric polynomials are

$$\begin{aligned} s_0 &= 1, \\ s_1 &= \lambda_1 + \lambda_2 + \lambda_3, \\ s_2 &= \lambda_1\lambda_2 + \lambda_2\lambda_3 + \lambda_3\lambda_1, \text{ and} \\ s_3 &= \lambda_1\lambda_2\lambda_3. \end{aligned}$$

There are identities relating s_k with t_k attributed to Newton:

$$s_k = \frac{1}{k}\sum_{i=1}^{k}(-1)^{i-1}s_{k-i}t_i \tag{5.9}$$

for $n \geq k \geq 1$. Equation (5.9) together with the Spectral Mapping Theorem enables us to take the following method, first published by Urbain Le Verrier, to compute the characteristic polynomial of $\mathbf{A}$:

Step 1: Compute $\mathbf{A}, \mathbf{A}^2, \ldots, \mathbf{A}^n$.

Step 2: Compute $t_k = \operatorname{tr}(\mathbf{A}^k)$ for $k = 1, 2, \ldots n$.

Step 3: Use equation (5.9) to compute the values of s_k recursively starting with $k = 1$.

Example 5.2.25. Compute the characteristic polynomial of

$$\mathbf{A} = \begin{bmatrix} 1 & 0 & -1 \\ -1 & 1 & 0 \\ 0 & -1 & 1 \end{bmatrix}.$$

Solution

First, we compute $\mathbf{A}^2$ and $\mathbf{A}^3$:

$$\begin{bmatrix} 1 & 0 & -1 \\ -1 & 1 & 0 \\ 0 & -1 & 1 \end{bmatrix} \begin{bmatrix} 1 & 0 & -1 \\ -1 & 1 & 0 \\ 0 & -1 & 1 \end{bmatrix}$$

$$\begin{bmatrix} 1 & 0 & -1 \\ -1 & 1 & 0 \\ 0 & -1 & 1 \end{bmatrix} \begin{bmatrix} 1 & 1 & -2 \\ -2 & 1 & 1 \\ 1 & -2 & 1 \end{bmatrix} \begin{bmatrix} 0 & 3 & -3 \\ -3 & 0 & 3 \\ 3 & -3 & 0 \end{bmatrix}.$$

Next, we compute the traces:

$$t_1 = \operatorname{tr}(\mathbf{A}) = 1 + 1 + 1 = 3, \quad t_2 = \operatorname{tr}(\mathbf{A}^2) = 1 + 1 + 1 = 3,$$
$$t_3 = \operatorname{tr}(\mathbf{A}^3) = 0 + 0 + 0 = 0.$$

For $k = 1$, equation (5.9) says that $s_1 = s_0 t_1$, but $s_0 = 1$ so $s_1 = 3$, which should not be surprising since equation (5.8) also shows that $s_1 = \operatorname{tr}(A)$.
For $k = 2$, equation (5.9) gives us

$$s_2 = \frac{1}{2}(s_1 t_1 - s_0 t_2) \frac{3^2 - 3}{2} = \frac{6}{2} = 3.$$

For $k = 3$, equation (5.9) gives us

$$s_3 = \frac{1}{3}(s_2 t_1 - s_1 t_2 + s_0 t_3) = \frac{3^2 - 3^2 + 1 \cdot 0}{3} = \frac{0}{3} = 0.$$

It follows that the characteristic polynomial of $\mathbf{A}$ is

$$f(x) = x^3 - s_1 x^2 + s_2 x - s_3 = x^3 - 3x^2 + 3x.$$

Activity 5.2.26. Compute the characteristic polynomial of

$$\mathbf{A} = \begin{bmatrix} 0 & 1 & 1 \\ 1 & 0 & 1 \\ 1 & 1 & 0 \end{bmatrix}$$

by doing the following steps:

1. Compute $\mathbf{A}^2$ and $\mathbf{A}^3$.

2. Compute $t_k = \mathrm{tr}(\mathbf{A}^k)$ for $k = 1, 2, 3$.
3. Compute s_k recursively by equation (5.9).

5.3 Diagonalization

5.3.1 Spectral Theorem (Real Case)

Q 5.3.1. What is the condition for **A** to be symmetric?

The first two examples of eigenvalues and eigenvectors used matrices that were already in diagonal form, specifically

$$\begin{bmatrix} 2 & 0 \\ 0 & 2 \end{bmatrix} \quad \text{and} \quad \begin{bmatrix} 1 & 0 \\ 0 & 0 \end{bmatrix}.$$

This is no accident. Part of what makes eigenvectors natural is that they often enable us to obtain a basis for which the matrix becomes diagonal. I say often because the eigenvectors must span the space. This does not always happen, as will be seen in Example 5.3.3, but it does happen for symmetric matrices.

Theorem 5.3.1 (Spectral theorem (real case))**.** *If* $\mathbf{A} = \mathbf{A}^T$ *is a symmetric matrix in which all values in* **A** *are real, then it can be diagonalized using an eigenbasis (a basis of eigenvectors).*

For proof, see [2].

Example 5.3.1. The matrix

$$\mathbf{A} = \begin{bmatrix} 0 & 3 \\ 3 & 8 \end{bmatrix}$$

is symmetric, and has eigenvalues -1 and 9. The -1 eigenspace is generated by the eigenvector

$$\mathbf{v}_1 = \begin{bmatrix} -3 \\ 1 \end{bmatrix}.$$

The 9 eigenspace is generated by the eigenvector

$$\mathbf{v}_2 = \begin{bmatrix} 1 \\ 3 \end{bmatrix}.$$

It can be seen in several ways that the eigenvectors $\mathbf{v}_1$ and $\mathbf{v}_2$ form a basis of $\mathbb{R}^2$. If we build them into a matrix

$$\mathbf{Q} = [\mathbf{v}_1 | \mathbf{v}_2] = \begin{bmatrix} -3 & 1 \\ 1 & 3 \end{bmatrix},$$

then $\det \mathbf{Q} = 9 - (-1) = 10$, so

$$\mathbf{Q}^{-1} = \frac{1}{-10} \begin{bmatrix} 3 & -1 \\ -1 & -3 \end{bmatrix}.$$

Conjugating by **Q** give us a change of basis that diagonalizes the matrix:

$$\mathbf{Q}^{-1}\mathbf{A}\mathbf{Q} = \frac{1}{-10} \begin{bmatrix} 3 & -1 \\ -1 & -3 \end{bmatrix} \begin{bmatrix} \begin{bmatrix} 0 & 3 \\ 3 & 8 \end{bmatrix} \\ \begin{bmatrix} -3 & 1 \\ -9 & -27 \end{bmatrix} \end{bmatrix} \begin{bmatrix} \begin{bmatrix} -3 & 1 \\ 1 & 3 \end{bmatrix} \\ \begin{bmatrix} 10 & 0 \\ 0 & -90 \end{bmatrix} \end{bmatrix} = \begin{bmatrix} -1 & 0 \\ 0 & 9 \end{bmatrix}.$$

Note that the order of the eigen-values on the diagonal matches the order of the eigenvectors in the columns of $\mathbf{Q}$: the first column of $\mathbf{Q}$ is $\mathbf{v}_1$ and its eigenvalue is -1, which is in the first spot of the diagonal of $\mathbf{Q}^{-1}\mathbf{AQ}$.

Activity 5.3.2. The matrix

$$\mathbf{A} = \begin{bmatrix} 1 & 2 \\ 2 & 4 \end{bmatrix}$$

has eigenvalues 0 and 5. The 5-eigenspace is generated by the eigenvector

$$\begin{bmatrix} 1 \\ 2 \end{bmatrix}$$

and the 0-eigenspace is generated by the eigenvector

$$\begin{bmatrix} -2 \\ 1 \end{bmatrix}.$$

1. Write down a matrix $\mathbf{Q}$ with the eigenvectors as columns.
2. Compute $\mathbf{Q}^{-1}$.
3. Compute $\mathbf{QAQ}^{-1}$.

Your final answer should be either

$$\begin{bmatrix} 5 & 0 \\ 0 & 0 \end{bmatrix} \quad \text{or} \quad \begin{bmatrix} 0 & 0 \\ 0 & 5 \end{bmatrix}$$

depending on the order you chose for the eigenvectors. The order of the eigenvectors in $\mathbf{Q}$ must match the order of the eigenvalues on the diagonal of $\mathbf{QAQ}^{-1}$.

It is not always possible to diagonalize a matrix.

Example 5.3.3. The matrix

$$\mathbf{J} = \left[\begin{array}{cc} 1 & 1 \\ 0 & 1 \end{array}\right]$$

is a 2×2 Jordan block. By Corollary 5.2.2 its characteristic polynomial is

$$(x-1)^2.$$

This means that 1 is the only eigenvalue. The power of two here is what is called the **algebraic multiplicity** of the eigenvalue. What are the eigenvectors?

$$\left[\begin{array}{cc|c} 1-1 & 1 & 0 \\ 0 & 1-1 & 0 \end{array}\right] = \left[\begin{array}{cc|c} 0 & 1 & 0 \\ 0 & 0 & 0 \end{array}\right]$$

So, $y = 0$ and x is a free variable, hence

$$\begin{bmatrix} x \\ y \end{bmatrix} = \begin{bmatrix} x \\ 0 \end{bmatrix} = x \begin{bmatrix} 1 \\ 0 \end{bmatrix}$$

is a parameterization of the eigenspace. But that means that the 1-eigenspace is the x-axis, which is 1 dimensional. There is no eigenbasis because we need two linearly independent vectors to form a basis of $\mathbb{R}^2$.

5.3.2 Spectral Theorem (Complex Case)

Q 5.3.2. What is the condition for **A** to be Hermitian?

What is the condition for **A** to be normal?

Theorem 5.3.2 (Spectral Theorem (Complex Case)). *Let* **A** *be an* $n \times n$ *matrix with complex entries. If* **A** *is a normal matrix, i.e.* $\mathbf{A}^*\mathbf{A} = \mathbf{A}\mathbf{A}^*$ *then* **A** *can be diagonalized with respect to an eigenbasis (a basis of eigenvectors).*

If in addition **A** *is Hermitian, then all eigenvalues are real.*

For proof, see [2].

Remark 5.3.1. If **A** is Hermitian, the diagonalized matrix is also Hermitian. Once diagonalized, the eigenvalues are on the diagonal and stay on the diagonal after the transpose. So the eigenvalues must be equal to their own complex conjugates, which can only happen for real numbers. For instance, the matrix

$$\mathbf{A} = \begin{bmatrix} i & 0 \\ 0 & -i \end{bmatrix} \tag{5.10}$$

is diagonal but not Hermitian because

$$\begin{bmatrix} i & 0 \\ 0 & -i \end{bmatrix}^* = \begin{bmatrix} -i & 0 \\ 0 & i \end{bmatrix}^T = \begin{bmatrix} -i & 0 \\ 0 & i \end{bmatrix} \neq \begin{bmatrix} i & 0 \\ 0 & -i \end{bmatrix}.$$

There are plenty of matrices though that do have complex eigenvalues, including the matrix **A** in equation (5.10), which being diagonal has its eigenvalues on the diagonal by Corollary 5.2.2. The matrix **R** from the symmetries of the square also has complex eigenvalues, as seen in Example 5.2.14, and it can be brought into diagonal form. Both of these are examples of normal matrices.

Example 5.3.4. Consider the matrix

$$\mathbf{A} = \begin{bmatrix} 0 & 1+i \\ 1-i & 1 \end{bmatrix}.$$

The eigenvalues are $\lambda = 2$ and -1. The 2-eigenspace is generated by the eigenvector

$$\mathbf{v}_1 = \begin{bmatrix} 1 \\ 1-i \end{bmatrix}$$

The -1-eigenspace is generated by the eigenvector

$$\mathbf{v}_2 = \begin{bmatrix} -1-i \\ 1 \end{bmatrix}.$$

The vectors $\mathbf{v}_1$ and $\mathbf{v}_2$ are a basis of $\mathbb{C}$. If we build the eigenvectors into a matrix

$$\mathbf{Q} = [\mathbf{v}_1|\mathbf{v}_2] = \begin{bmatrix} 1 & -1-i \\ 1-i & 1 \end{bmatrix},$$

then $\det(\mathbf{Q}) = 1 - (-2) = 3$, so

$$\mathbf{Q}^{-1} = \frac{1}{3}\begin{bmatrix} 1 & 1+i \\ -1+i & 1 \end{bmatrix}.$$

Computing $\mathbf{Q}^{-1}\mathbf{AQ}$ give us a change of basis that diagonalizes the matrix:

$$\frac{1}{3}\begin{bmatrix} 1 & 1+i \\ -1+i & 1 \end{bmatrix} \begin{bmatrix} 0 & 1+i \\ 1-i & 1 \end{bmatrix} \begin{bmatrix} 1 & -1-i \\ 1-i & 1 \end{bmatrix} = \frac{1}{3}\begin{bmatrix} 1 & 1+i \\ -1+i & 1 \end{bmatrix} \begin{bmatrix} 2 & 2+2i \\ 1-i & -1 \end{bmatrix} = \frac{1}{3}\begin{bmatrix} 6 & 0 \\ 0 & -3 \end{bmatrix} = \begin{bmatrix} 2 & 0 \\ 0 & -1 \end{bmatrix}.$$

Note that the order of the eigen-values on the diagonal matches the order of the eigenvectors in the columns of $\mathbf{Q}$: the first column of $\mathbf{Q}$ is $\mathbf{v}_1$ and its eigenvalue is 2, which is in the first spot of the diagonal of $\mathbf{Q}^{-1}\mathbf{AQ}$.

Activity 5.3.5. Given the matrix

$$\mathbf{A} = \begin{bmatrix} 0 & i \\ -i & 0 \end{bmatrix}$$

has eigenvalues 1 and -1. The -1 eigenspace is generated by the eigenvector

$$\begin{bmatrix} 1 \\ -i \end{bmatrix}$$

The -1 eigenspace is generated by the eigenvector

$$\begin{bmatrix} 1 \\ i \end{bmatrix}$$

1. Check that $\mathbf{A}^* = \mathbf{A}$
2. Write down a matrix $\mathbf{Q}$ with the eigenvectors as columns.
3. Compute $\mathbf{Q}^{-1}$.
4. Compute $\mathbf{Q}^{-1}\mathbf{AQ}$.

Your final answer should be either

$$\begin{bmatrix} 1 & 0 \\ 0 & -1 \end{bmatrix} \quad \text{or} \quad \begin{bmatrix} -1 & 0 \\ 0 & 1 \end{bmatrix}$$

depending on the order you chose for the eigenvectors. The order of the eigenvectors in $\mathbf{Q}$ must match the order of the eigenvalues on the diagonal of $\mathbf{Q}^{-1}\mathbf{AQ}$.

5.4 Norms of Matrices

Just as there are many different norms for vectors, there are many matrix norms and relationships between them.

First, $m \times n$ matrices over $\mathbb{R}$ or $\mathbb{C}$ are a vector space of dimension mn. Proposition 1.1.15 covers some of the structural properties, and exercise 4.4 studies the 2×2 case. So any norm for $\mathbb{R}^{mn}$ or $\mathbb{C}^{mn}$ can be applied directly to matrices. In particular, the ∞-norm leads to the max norm

$$\| \mathbf{A} \|_{max} = \max_{i,j} |a_{ij}|$$

and the 2-norm leads to the Frobenius norm

$$\| \mathbf{A} \|_F = \sqrt{\sum_{ij} |a_{ij}|^2}.$$

We do not call them the ∞-norm or 2-norm for matrices, because these names are reserved for a completely different concept.

If $\mathbf{x}$ is a vector in $\mathbb{R}^n$ and $\mathbf{A}$ is an $m \times n$ matrix of real numbers, then $\mathbf{Ax}$ is a vector in $\mathbb{R}^m$. The same applies to $\mathbb{C}$ instead of $\mathbb{R}$. In both cases, we can compute

$$\| \mathbf{x} \|_p \quad \text{and} \quad \| \mathbf{Ax} \|_p .$$

Then we can define the p-norm of $\mathbf{A}$ as

$$\| \mathbf{A} \|_p = \max_{\mathbf{x} \neq 0} \frac{\| \mathbf{Ax} \|_p}{\| \mathbf{x} \|_p}, \tag{5.11}$$

or equivalently

$$\| \mathbf{A} \|_p = \max_{\| \mathbf{x} \|_p = 1} \| \mathbf{Ax} \|_p . \tag{5.12}$$

Evaluating these norms may not be easy, but at least for $p = 1$ or $p = \infty$ we have the following result:

Theorem 5.4.1. *Let* $\mathbf{A}$ *be an* $m \times n$ *matrix. Then*

$$\| \mathbf{A} \|_\infty = \max_{1 \leq i \leq m} \sum_{j=1}^{n} |a_{ij}| \tag{5.13}$$

and

$$\| \mathbf{A} \|_1 = \max_{1 \leq j \leq n} \sum_{i=1}^{m} |a_{ij}|. \tag{5.14}$$

The 2-norm of a matrix is probably best understood in terms of the singular value decomposition (SVD). However, we explain it here with the help of the following concept.

Definition 5.4.2. For an $n \times n$ matrix $\mathbf{A}$, the **Spectral radius** $\rho(\mathbf{A})$ is

$$\rho(\mathbf{A}) = \max_i |\lambda_i|,$$

where $\lambda_1, \lambda_2, \ldots \lambda_n$ are the eigenvalues of $\mathbf{A}$.

The word "radius" makes the most sense in the complex plane, because the eigenvalues of $\mathbf{A}$ are contained in a circle with radius $\rho(\mathbf{A})$ centered at the origin. However, as the next theorem shows, it is also possible to cover the eigenvalues with several circles.

Theorem 5.4.3 (Gershgorin Circle Theorem)**.** *Let* $\mathbf{A}$ *be an* $n \times n$ *complex matrix. Let*

$$r_i = \sum_{i \neq j} |a_{ij}|$$

and let D_i *be the set of points* x *in the complex plane with* $|a_{ii} - x| \leq r_i$*, which is a closed disc with center* a_{ii} *and radius* r_i*. Then the discs* D_i *cover the spectrum of* $\mathbf{A}$*.*

For proof, see [2] or [25]

Corollary 5.4.4. *Let* $\mathbf{A}$ *be an* $n \times n$ *complex matrix. Then the spectral radius is less than or equal to the maximum of* $r_i + |a_{ii}|$*:*

$$\rho(\mathbf{A}) \leq \max_i r_i + |a_{ii}| = \max_i \sum_j |a_{ij}|$$

Remark 5.4.1. Since $\mathbf{A}$ and $\mathbf{A}^T$ have the same eigenvalues, then both Gershgorin's theorem and its corollary remain true with i and j reversed. So

$$\rho(\mathbf{A}) \leq \min\{\| \mathbf{A} \|_1, \| \mathbf{A} \|_\infty\}. \tag{5.15}$$

Example 5.4.1. Suppose

$$\mathbf{A} = \begin{bmatrix} -2 & -1 & 0 \\ 0 & 0 & -2 \\ 0 & 1 & 2 \end{bmatrix}.$$

First, we compute the spectrum (the eigenvalues). Since $\mathbf{A}$ is in block upper triangular form, we can use Corollary 5.2.5. The 1×1 block gives us the factor $x + 2$. For the 2×2 block we have trace 2, and determinant 2, which gives us the factor $x^2 - 2x + 2$. It follows that the characteristic polynomial of $\mathbf{A}$ is

$$(x + 2)(x^2 - 2x + 2).$$

From the $x + 2$ factor, we can see that $\lambda = -2$ is an eigenvalue. However, $x^2 - 2x + 2$ does not have real roots. In this case, we must to use the quadratic formula

$$\lambda = \frac{2 \pm \sqrt{(-2)^2 - 4(1)(2)}}{2} = \frac{2 \pm \sqrt{4 - 8}}{2} = \frac{2 \pm \sqrt{-4}}{2} = 1 \pm i.$$

So, the spectrum consists of -2, $1 + i$, and $1 - i$. Next we compute r_1. We must add the absolute values in the first row that are not on the diagonal:

$$\begin{bmatrix} -2 & -1 & 0 \\ 0 & 0 & -2 \\ 0 & 1 & 2 \end{bmatrix}.$$

The -2 is on the diagonal, so it is not included in the sum:

$$r_1 = |-1| + |0| = 1 + 0 = 1.$$

Then we repeat with the 2nd and 3rd rows, omitting the number on the diagonal in each case:

$$r_2 = |0| + |-2| = 2, \quad \text{and}$$
$$r_3 = |0| + |1| = 1.$$

Since the matrix $\mathbf{A}$ has all real values, then the centers of the circle are all on the real axis in the complex plane, as shown in Figure 5.1. Each of the discs D_1, D_2, and D_3 is drawn as a shaded circle. They overlap in this case, but do not need to. The disc D_1 is centered at $a_{11} = -2$ and has radius $r_1 = 1$. It is the leftmost shaded circle in Figure 5.1. The disc D_2 is centered $a_{22} = 0$, the origin of the complex plane, and has radius $r_2 = 2$. The disc D_3 is centered at $a_{33} = 2$ and has radius $r_3 = 1$. Each eigenvalue lies in at least one disc:

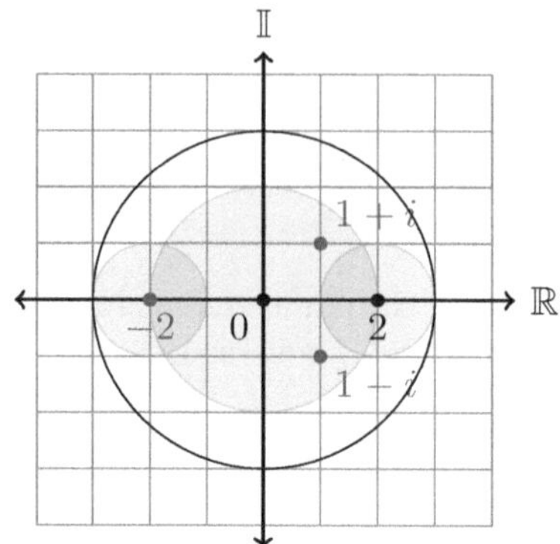

FIGURE 5.1: Gershgorin circle theorem

1. -2 is in both D_1 and D_2 (it is at the edge of D_2, but the edge is included), and
2. $1+i$ and $1-i$ are only in D_2.

So, as it turns out, D_3 does not contain any of the eigenvalues. To compute the spectral radius, we must compute the absolute value of each eigenvalue. For a complex number in rectangular coordinates

$$|a+bi| = \sqrt{(a-bi)(a+bi)} = \sqrt{a^2+b^2},$$

so

$$|1+i| = \sqrt{1^2+1^2} = \sqrt{2}.$$

But $|-2| = 2$, and $2 > \sqrt{2}$ so

$$\rho(\mathbf{A}) = \max\{|-2|, |1+i|, |1-i|\} = 2.$$

On the other hand, the estimate that we get from the corollary to Gershgorin's theorem is 3 because

$$\begin{aligned} |a_{11}| + r_1 &= |-2| + 1 = 3, \\ |a_{22}| + r_2 &= |0| + 2 = 2, \quad \text{and} \\ |a_{33}| + r_3 &= |2| + 1 = 3. \end{aligned}$$

This is illustrated in Figure 5.1 by the large circle with radius 3 centered at the origin, which contains all of the discs and the eigenvalues. None of the eigenvalues lie on the large circle, so its radius is larger than the spectral radius.

Activity 5.4.2. Given the matrix

$$\mathbf{A} = \begin{bmatrix} 2 & -1 & -1 \\ 0 & 0 & -1 \\ 0 & 1 & 0 \end{bmatrix},$$

do the following:

1. Compute the eigenvalues of $\mathbf{A}$.
2. Compute r_1, r_2, r_3.

3. Plot and label the eigenvalues, and sketch the discs D_1, D_2, D_3. Label the centers of the discs as well.

4. Compute the spectral radius.

5. Compute the estimate for the spectral radius from the corollary to Gershgorin's Theorem.

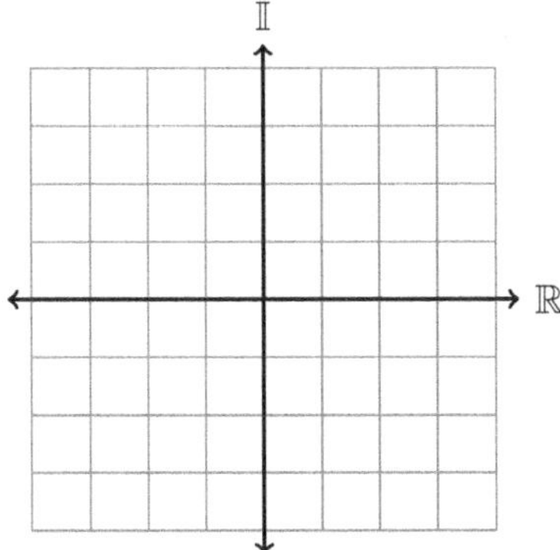

Now we can explain how to compute the 2-norm of a matrix in terms of the spectral radius.

Theorem 5.4.5. *If* $\mathbf{A}$ *is an* $m \times n$ *matrix over* $\mathbb{R}$*, then*

$$\| \mathbf{A} \|_2 = \sqrt{\rho(\mathbf{A}^T\mathbf{A})}.$$

If $\mathbf{A}$ *is an* $m \times n$ *matrix over* $\mathbb{C}$*, then*

$$\| \mathbf{A} \|_2 = \sqrt{\rho(\mathbf{A}^*\mathbf{A})}.$$

Corollary 5.4.6. *If* $\mathbf{A}$ *is a real symmetric matrix or a complex hermitian matrix then*

$$\| \mathbf{A} \|_2 = \rho(\mathbf{A})$$

This corollary follows from the fact that if $\mathbf{A}$ is symmetric, then

$$\mathbf{A}^T\mathbf{A} = \mathbf{A}^2.$$

Then by the spectral mapping theorem (see Subsection 5.2.8) $\rho(\mathbf{A}^2) = \rho(\mathbf{A})^2$, which allows the square root and square to cancel.

However, $\| \mathbf{A} \|_2$ may not be equal to $\rho(\mathbf{A})$. If $\mathbf{A}$ is not a square matrix, then $\rho(\mathbf{A})$ is not even defined, but $\| \mathbf{A} \|_2$ is. The next example shows what happens if $\mathbf{A}$ is upper triangular.

Example 5.4.3. Consider the matrix

$$\mathbf{A} = \begin{bmatrix} a & b \\ 0 & d \end{bmatrix}$$

where a, b, d are real. Since $\mathbf{A}$ is upper triangular, the eigenvalues are on the diagonal: a and d. So

$$\rho(\mathbf{A}) = \max\{|a|, |d|\}.$$

However, $\mathbf{A}^T\mathbf{A}$ is

$$\begin{array}{cc} & \begin{bmatrix} a & b \\ 0 & d \end{bmatrix} \\ \begin{bmatrix} a & 0 \\ b & d \end{bmatrix} & \begin{bmatrix} a^2 & ab \\ ab & b^2+d^2 \end{bmatrix} \end{array}.$$

Let

$$T = \text{tr}(\mathbf{A}^T\mathbf{A}) = a^2 + b^2 + d^2$$
$$D = \det(\mathbf{A}^T\mathbf{A}) = a^2(b^2 + d^2) - a^2b^2 = a^2d^2.$$

Then the characteristic polynomial of $\mathbf{A}^T\mathbf{A}$ is

$$x^2 - Tx + D$$

so the eigenvalues are

$$\frac{T \pm \sqrt{T^2 - 4D}}{2}.$$

Since T is positive, we get the larger eigenvalue by adding the square root. So by Theorem 5.4.5

$$\| \mathbf{A} \|_2 = \sqrt{\rho(\mathbf{A}^T\mathbf{A})} = \sqrt{\frac{T + \sqrt{T^2 - 4D}}{2}}$$

If $b = 0$, this simplifies to $\max\{|a|, |d|\}$, which is also equal to $\rho(\mathbf{A})$. However,This usually they are not equal as can be seen from the next activity.

Activity 5.4.4 (with Desmos). Let

$$\mathbf{A} = \begin{bmatrix} a & b \\ 0 & d \end{bmatrix} \quad \text{and} \quad \mathbf{x} = \begin{bmatrix} \cos(x) \\ \sin(x) \end{bmatrix}$$

Then

$$\mathbf{Ax} = \begin{bmatrix} a\cos(x)+b\sin(x) \\ d\sin(x) \end{bmatrix}.$$

Since $\| \mathbf{x} \|_2 = 1$, then $\| \mathbf{A} \|_2$ is equal to the maximum of

$$f(x) = \| \mathbf{Ax} \|_2 = \sqrt{(a\cos(x) + b\sin(x))^2 + d^2\sin^2(x)}$$

by equation (5.12). Go to desmos.com.

1. In the 1st cell enter $f(x) = \sqrt{(a\cos(x) + b\sin(x))^2 + d^2\sin^2(x)}$. Create sliders for all of a, b, and d. The sliders will appear in cells 2, 3, and 4.

2. In the 5th cell enter $T = a^2 + b^2 + d^2$

3. In the 6th cell enter $D = a^2d^2$

4. In the 7th cell enter $y = \sqrt{\frac{T+\sqrt{T^2-4D}}{2}}$. You will get a horizontal line with the y-value equal to $\sqrt{\rho(\mathbf{A}^T\mathbf{A})}$. No matter how you move the sliders for a, b, and d, the line will always match the maximum of $f(x)$. This illustrates Theorem 5.4.5.

5. In the 8th cell enter $y = \max(|a|, |d|)$. You will get a second horizontal line, which will generally be lower. If you set the slider for b equal to zero, the two lines will coincide. This illustrates Corollary 5.4.6.

We are now in a position to state the condition number of an invertible matrix. Suppose that we are trying to solve $\mathbf{Ax} = \mathbf{b}$. Then we are looking for $\widehat{\mathbf{x}}$ as an output. So

$$\mathrm{E}_{\mathrm{rel}}(\mathbf{x}) = \frac{\| \widehat{\mathbf{x}} - \mathbf{x} \|_p}{\| \mathbf{x} \|_p} = \frac{\| \Delta \mathbf{x} \|_p}{\| \mathbf{x} \|_p}$$

is the forward error. The inputs are the matrix $\mathbf{A}$ and the vector $\mathbf{b}$, so the backward error should involve both of them. We will use the calculus-based approach in Section 7.8 of [11] to show that

$$\frac{\| \Delta \mathbf{x} \|_p}{\| \mathbf{x} \|_p} \lesssim \kappa(\mathbf{A}) \left(\frac{\| \Delta \mathbf{A} \|_p}{\| \mathbf{A} \|_p} + \frac{\| \Delta \mathbf{b} \|_p}{\| \mathbf{b} \|_p} \right) \tag{5.16}$$

where the condition number is

$$\kappa_p(\mathbf{A}) = \| \mathbf{A} \|_p \cdot \| \mathbf{A}^{-1} \|_p . \tag{5.17}$$

For an arbitrary matrix $\mathbf{A}$, even one that isn't square, we can replace $\mathbf{A}^{-1}$ by the generalized inverse (see Subsection 6.5.3).

Here is the basic idea to the calculus-based approach. We connect the vectors $\mathbf{x}$ and $\widehat{\mathbf{x}}$ by a curve in $\mathbb{R}^n$. In this way we get a parametric function $\mathbf{x}(t)$ in which

$$\mathbf{x}(0) = \mathbf{x} \quad \text{and} \quad \mathbf{x}(1) = \widehat{\mathbf{x}}.$$

Then by the Taylor series viewpoint

$$\mathbf{x}(t) = \mathbf{x}(0) + \mathbf{x}'(0)(t - 0) + \cdots .$$

If we subtract $\mathbf{x}(0)$ and take $t = 0$, we get

$$\Delta \mathbf{x} \approx \mathbf{x}'(0).$$

We do the same with the vector $\mathbf{b}$ and with the matrix $\mathbf{A}$, by considering $\mathbf{A}$ and $\widehat{\mathbf{A}}$ as a points connected by a curve in $\mathbb{R}^{n^2}$. Then we use product rule[1] to differentiate $\mathbf{A}(t)\mathbf{x}(t) = \mathbf{b}(t)$:

$$\frac{d}{dt} (\mathbf{A}(t)\mathbf{x}(t)) = \mathbf{A}'(t)\mathbf{x}(t) + \mathbf{A}(t)\mathbf{x}'(t) = \mathbf{b}'(t).$$

Taking $t = 0$ gives

$$\Delta \mathbf{A} \mathbf{x} + \mathbf{A} \Delta \mathbf{x} \approx \Delta \mathbf{b}.$$

Then we subtract $\Delta \mathbf{A} \mathbf{x}$ from both sides and multiply by $\mathbf{A}^{-1}$ to get

$$\Delta \mathbf{x} \approx -\mathbf{A}^{-1} \Delta \mathbf{A} \mathbf{x} + \mathbf{A}^{-1} \Delta \mathbf{b}.$$

Here is where we take norms:[2]

$$\begin{aligned} \frac{\| \Delta \mathbf{x} \|_p}{\| \mathbf{x} \|_p} &\lesssim \frac{\| \mathbf{A}^{-1} \Delta \mathbf{A} \mathbf{x} \|_p}{\| \mathbf{x} \|_p} + \frac{\| \mathbf{A}^{-1} \Delta \mathbf{b} \|_p}{\| \mathbf{x} \|_p} \\ &\lesssim \frac{\| \mathbf{A}^{-1} \|_p \| \Delta \mathbf{A} \|_p \| \mathbf{x} \|_p}{\| \mathbf{x} \|_p} + \frac{\| \mathbf{A}^{-1} \|_p \| \Delta \mathbf{b} \|_p}{\| \mathbf{x} \|_p} \\ &\lesssim \| \mathbf{A} \|_p \| \mathbf{A}^{-1} \|_p \left(\frac{\| \Delta \mathbf{A} \|_p}{\| \mathbf{A} \|_p} + \frac{\| \Delta \mathbf{b} \|_p}{\| \mathbf{A} \|_p \| \mathbf{x} \|_p} \right). \end{aligned}$$

[1] If it bothers you that matrix multiplication is defined by a sum, try writing the sum down. Product rule applies to each term of the sum.

[2] This calculation uses the triangle inequality and $\| \mathbf{AB} \|_p \leq \| \mathbf{A} \|_p \| \mathbf{B} \|_p$.

Now you can almost see where equations (5.16) and (5.17) come from, except for one more important fact. Since $\mathbf{b} = \mathbf{A}\mathbf{x}$, then

$$\| \mathbf{b} \|_p = \| \mathbf{A}\mathbf{x} \|_p \leq \| \mathbf{A} \|_p \| \mathbf{x} \|_p$$

by the definition of a p-norm as a maximum. The next activity is a numerical demonstration of equations (5.16) and (5.17).

Activity 5.4.5 (with Sage and NumPy)**.** Start by defining $\mathbf{A}$ and $\mathbf{b}$ as NumPy arrays

```
import numpy as np
A = np.array([[5.0, 0.0],
              [0.0, 4.0]])
b = np.array([1.0, 1.0])
```

Then get the exact solution and the condition number using the built-in NumPy functions

```
    x = np.linalg.solve(A, b);      print(x)
kappa = np.linalg.cond(A,2);        print(kappa)
```

You should see that the condition number is 1.25. Example 5.4.7 shows how to calculate the condition number by hand. Next we construct approximate values

```
delta = 1e-6
B = np.array([[0.0, 0.0],
              [0.0, 1.0]])
A_hat = A + delta*B;                print(A_hat)
x_hat = np.linalg.solve(A_hat, b);  print(x_hat)
```

You should see that $\widehat{\mathbf{A}}$ and $\widehat{\mathbf{x}}$ are not exactly the same as $\mathbf{A}$ and $\mathbf{x}$. Now we want to get a graph. To generate the graph, we will use the idea of connecting $\mathbf{x}$ and $\widehat{\mathbf{x}}$ by a parametric curve. But not just any curve, we will use a line segment.

```
def E_rel(X,X_hat):
    return np.linalg.norm(X-X_hat)/np.linalg.norm(X)
ts = np.linspace(0, 1, 100)

pairs = []
for t in ts:
    pairs+=[[E_rel(A,t*A_hat+(1-t)*A),E_rel(x,t*x_hat+(1-t)*x)]]

plt = scatter_plot(pairs)
plt += line([(0,0),(E_rel(A,A_hat),kappa*E_rel(A,A_hat))])
plt.show()
```

First we define a function for computing the relative error. Then we divide the interval $[0, 1]$ into 100 pieces. The for loop constructs a list of pairs

$$(\text{backward error}, \text{forward error}).$$

The remaining lines graph the points together with a line of slope κ. The idea of the condition number is that the points should never be above the line. With the current plot you should see that the points hide the line completely: the slope matches. But try changing the matrix **B** and you will see that sometimes the points are below the line. Some good choices for **B** include

$$\begin{bmatrix} 1 & 0 \\ 0 & 0 \end{bmatrix}, \quad \begin{bmatrix} 1 & -1 \\ 0 & 1 \end{bmatrix}, \quad \text{and} \quad \begin{bmatrix} 1 & -1 \\ -1 & 1 \end{bmatrix}.$$

In each of these cases, you should see that the individual dots lie below the solid line. But since the slope of the solid line is the condition number it is not possible to choose **B** in such a way that the dots are above the line. The dots appear to lie on a line. The condition number is supposed to be the maximum slope. The original choice of **B** gets the slope to match? As for why that choice of **B** works, it may make more sense if you look at Example 5.4.7 and Activity 5.4.8.

Any student who has learned power series from me has heard me say that the radius of convergence of a power series is equal to the distance to the nearest singularity. The condition number of an invertible matrix can be understood from a similar viewpoint using the following concept of distance.

$$\operatorname{dist}_p(\mathbf{A}) = \min \left\{ \frac{\| \Delta \mathbf{A} \|_p}{\| \mathbf{A} \|_p} : \mathbf{A} + \Delta \mathbf{A} \text{ singular} \right\}. \tag{5.18}$$

In this context "singular" means not invertible. Then we can say that the condition number of an invertible matrix is equal to the reciprocal of the distance to the nearest singularity:

Theorem 5.4.7 (Gastinel, Kahan). *If* **A** *is an invertible* $n \times n$ *matrix, then*

$$\operatorname{dist}_p(\mathbf{A}) = \frac{1}{\| \mathbf{A} \|_p \| \mathbf{A}^{-1} \|_p} = \frac{1}{\kappa_p(\mathbf{A})}.$$

See [11] for proof.

Example 5.4.6. A 1×1 matrix is just a number, and the 2-norm is just the absolute value. Consider 3. Then the inverse is $\frac{1}{3}$, and

$$\kappa_2(3) = |3| \cdot \left| \frac{1}{3} \right| = 1.$$

Zero is the only number without an inverse. You may be thinking "the distance from 3 to 0 is 3 not 1." But look carefully at equation (5.18) defining distance here. It is relative, not absolute. We must consider what ΔA is possible. Only -3 is possible because

$$3 + (-3) = 0.$$

so the distance is

$$\frac{|-3|}{|3|} = \frac{3}{3} = 1$$

which indeed agrees with $\frac{1}{\kappa_2(3)}$. Of course, it also agrees with $\kappa_2(3)$, but see the next example.

Example 5.4.7. Let

$$\mathbf{A} = \begin{bmatrix} 5 & 0 \\ 0 & 4 \end{bmatrix}.$$

Then

$$\mathbf{A}^{-1} = \begin{bmatrix} \frac{1}{5} & 0 \\ 0 & \frac{1}{4} \end{bmatrix} \quad \| \mathbf{A} \|_2 = 5 \quad \text{and} \quad \| \mathbf{A}^{-1} \|_2 = \frac{1}{4} \tag{5.19}$$

so

$$\kappa_2(\mathbf{A}) = \frac{5}{4}. \tag{5.20}$$

Let

$$\Delta\mathbf{A} = \begin{bmatrix} a & b \\ c & d \end{bmatrix} \quad \text{so} \quad \mathbf{A} + \Delta\mathbf{A} = \begin{bmatrix} 5+a & b \\ c & 4+d \end{bmatrix}.$$

For $\mathbf{A} + \Delta\mathbf{A}$ to be singular, we need $\det(\mathbf{A} + \Delta\mathbf{A}) = 0$. So $(5+a)(4+d) - bc = 0$, which means that

$$(5+a)(4+d) = bc \tag{5.21}$$

If $c = 0$, then $(5+a)(4+d) = 0$, so

$$a = -5 \quad \text{or} \quad d = -4$$

This leads to the options

$$\Delta\mathbf{A} = \begin{bmatrix} a & b \\ 0 & -4 \end{bmatrix} \quad \text{or} \quad \Delta\mathbf{A} = \begin{bmatrix} -5 & b \\ 0 & d \end{bmatrix}. \tag{5.22}$$

The norms are complicated, but can be studied with the help of Sage in the next activity. For the first, the minimum value is $|-4| = 4$ and occurs when $b = 0$ and $|a| \leq 4$. For the second, the minimum value is $|-5| = 5$ and occurs when $b = 0$ and $|d| \leq 5$. So using these values for $\| \Delta\mathbf{A} \|_2$ and the value of $\| \mathbf{A} \|_2$ from (5.19) the distance is

$$\min\left\{\frac{\| \Delta\mathbf{A} \|_2}{\| \mathbf{A} \|_2}\right\} = \min\left\{\frac{4}{5}, \frac{5}{5}\right\} = \frac{4}{5},$$

which agrees with the reciprocal of the condition number in (5.20).

Activity 5.4.8 (with Sage)**.** Open a Jupyter Notebook with the Sage kernel running. Next define the variables a, b, c, and d, the matrix $\Delta\mathbf{A}$, and the trace T and determinant D of $(\Delta\mathbf{A})^T\Delta\mathbf{A}$, by running the following code:

```
var('a,b,c,d')

deltaA = matrix([[a,b],
                 [c,d]])

T = (deltaA.T*deltaA).trace()
D = (deltaA.T*deltaA).det().factor()
```

You should see that T and D both have a very simple form if you run

```
T,D
```

We need to consider the 2-norm of $\Delta\mathbf{A}$ as a function of a, b, c, and d. For this purpose, we define the norm in Sage as

```
f = sqrt((T+sqrt(T^2-4*D))/2)
```

The function f is a function of 4 variables, but the special cases given by (5.22) can be plotted in 3 dimensions. For the first matrix in (5.22), you run the code

```
plot3d(f(c=0,d=-4),(a,-8,8),(b,-8,8))
```

If you spin the graph around and look at the labels on wire frame, you should see that the minimum value is 4 and that it occurs along a line segment where $b = 0$ and $|a| \leq 4$. For the second matrix in (5.22), you run the code

```
plot3d(f(c=0,a=-5), (b,-8,8),(d,-8,8))
```

The minimum value is 5 and that it occurs when $b = 0$ and $|d| \leq 5$. The smaller minimum is 4. But we have assumed that $c = 0$. What if c is not zero? Can we do better? The short answer is no because Theorem 5.4.5 is a Theorem. In general, a minimum must occur at a critical point. A critical point for a multi-variable function occurs if the gradient is zero or does not exist (DNE), so all cases would need to be considered systematically. The DNE case matters because of examples like $|x|y^2$ and $|x| + y^2$.

5.5 Matrix Functions

If $f(x)$ is a polynomial and $\mathbf{A}$ is a square matrix, it is easy to evaluate $f(\mathbf{A})$ because we only need the rules for matrix arithmetic. But what are we to make of $\sqrt{\mathbf{A}}$, $e^{\mathbf{A}}$, or $\ln(\mathbf{A})$? One general idea is to try using a Taylor series for f. The Taylor series for f centered at a point c is given by

$$\sum_{k=0}^{\infty} \frac{f^{(k)}(c)}{k!}(x-c)^k = \lim_{n\to\infty} \sum_{k=0}^{n} \frac{f^{(k)}(c)}{k!}(x-c)^k, \tag{5.23}$$

where $f^{(k)}(x) = \frac{d^k}{dx^k} f(x)$ is the derivative of order n. The limit in equation (5.23) converges if $|x - c| < R$ where R is the radius of convergence. For example

$$e^x = \sum_{k=0}^{\infty} \frac{1}{k!} x^k \qquad \text{for all } x \tag{5.24}$$

$$\sqrt{x} = c^{\frac{1}{2}} \sum_{k=0}^{\infty} \binom{\frac{1}{2}}{k} c^{-k}(x-c)^k \qquad \text{for } |x-c| < c, \tag{5.25}$$

where

$$\binom{m}{k} = \frac{m(m-1)(m-2)\cdots(m-k+1)}{k(k-1)(k-2)\cdots 1}.$$

What happens if we plug in with a matrix? First, consider what happens with a diagonal matrix. If

$$\mathbf{A} = \begin{bmatrix} \lambda_1 & 0 & \cdots & 0 \\ 0 & \lambda_2 & \cdots & 0 \\ \vdots & \vdots & & \vdots \\ 0 & 0 & \cdots & \lambda_n \end{bmatrix},$$

then

$$(\mathbf{A} - c\mathbf{I})^n = \begin{bmatrix} (\lambda_1 - c)^n & 0 & \cdots & 0 \\ 0 & (\lambda_2 - c)^n & \cdots & 0 \\ \vdots & \vdots & & \vdots \\ 0 & 0 & \cdots & (\lambda_n - c)^n \end{bmatrix}.$$

If we then insert these in the Taylor series, the result is a diagonal matrix in which the eigenvalues λ_i are plugged into the Taylor series on the diagonal. As a consequence, the convergence of the Taylor series with matrices is equivalent to the convergence of the Taylor series for all of the eigenvalues λ_i. This means that the limit in equation (5.23) converges for an $n \times n$ matrix $\mathbf{A}$, if $\rho(\mathbf{A} - c\mathbf{I}) < R$.

The next example illustrates that sometimes the series can be reduced to a finite sum.

Example 5.5.1. Given

$$\mathbf{A} = \begin{bmatrix} 2 & 1 \\ 0 & 2 \end{bmatrix}$$

1. compute $\sqrt{\mathbf{A}}$, and
2. verify that Theorem 5.2.8 remains true with $f(x) = \sqrt{x}$.

Solution

In this case we will use the series (5.25) with $c = 2$. Then

$$\mathbf{A} - 2\mathbf{I} = \begin{bmatrix} 2 & 1 \\ 0 & 2 \end{bmatrix} - \begin{bmatrix} 2 & 0 \\ 0 & 2 \end{bmatrix} = \begin{bmatrix} 0 & 1 \\ 0 & 0 \end{bmatrix}.$$

Zero is the only eigenvalue, so $\rho(\mathbf{A} - 2\mathbf{I}) = 0$. When the spectral radius of an $n \times n$ matrix is zero, then the n-th power is the zero matrix and any power greater than n is also zero. In this case

$$\begin{matrix} & \begin{bmatrix} 0 & 1 \\ 0 & 0 \end{bmatrix} \\ \begin{bmatrix} 0 & 1 \\ 0 & 0 \end{bmatrix} & \begin{bmatrix} 0 & 0 \\ 0 & 0 \end{bmatrix}, \end{matrix}$$

so all of the terms for $k \geq 2$ are zero. Hence, we need only the terms $k < 2$:

$$\begin{aligned} \sqrt{\mathbf{A}} &= \sqrt{2}\left(\binom{\frac{1}{2}}{0}\mathbf{I} + \binom{\frac{1}{2}}{1}(\mathbf{A} - 2\mathbf{I})\right) \\ &= \sqrt{2}\left(\begin{bmatrix} 1 & 0 \\ 0 & 1 \end{bmatrix} + \frac{1}{4}\begin{bmatrix} 0 & 1 \\ 0 & 0 \end{bmatrix}\right) = \begin{bmatrix} \sqrt{2} & \frac{\sqrt{2}}{4} \\ 0 & \sqrt{2} \end{bmatrix}. \end{aligned}$$

It can be checked that multiplying $\sqrt{\mathbf{A}}$ by itself gives us back $\mathbf{A}$.
As for Theorem 5.2.8, $\mathbf{A}$ is in upper triangular form, so the eigenvalues are on the diagonal, which means that 2 is the only eigenvalue of $\mathbf{A}$. But $\sqrt{\mathbf{A}}$ is also in upper triangular form, so $\sqrt{2}$ is the only eigenvalue of $\mathbf{A}$.

Activity 5.5.2. Given

$$\mathbf{A} = \begin{bmatrix} 0 & 1 & 0 \\ 0 & 0 & 1 \\ 0 & 0 & 0 \end{bmatrix},$$

1. compute $e^{\mathbf{A}}$ using (5.24), and
2. verify that Theorem 5.2.8 remains true with $f(x) = e^x$.

Example 5.5.3. Given

$$\mathbf{A} = \begin{bmatrix} 5 & 3 \\ 3 & 5 \end{bmatrix},$$

1. compute $\sqrt{\mathbf{A}}$ using (5.23), and

2. verify that Theorem 5.2.8 remains true with $f(x) = \sqrt{x}$.

Solution

If we take $c = 5$, then

$$\mathbf{A} - 5\mathbf{I} = \begin{bmatrix} 5 & 3 \\ 3 & 5 \end{bmatrix} - 5\begin{bmatrix} 1 & 0 \\ 0 & 1 \end{bmatrix} = \begin{bmatrix} 0 & 3 \\ 3 & 0 \end{bmatrix}.$$

By Gershgorin's Theorem $\rho(\mathbf{A} - 5\mathbf{I}) \leq 3 < 5$. To handle powers, there are two cases:

$$(\mathbf{A} - 5\mathbf{I})^k = \begin{cases} 3^k \begin{bmatrix} 1 & 0 \\ 0 & 1 \end{bmatrix} & \text{if } k \text{ is even,} \\ \\ 3^k \begin{bmatrix} 0 & 1 \\ 1 & 0 \end{bmatrix} & \text{if } k \text{ is odd.} \end{cases} \tag{5.26}$$

So far, this gives us

$$\sqrt{A} = \sqrt{5} \sum_{k \text{ even}} \binom{\frac{1}{2}}{k} 5^{-k} 3^k \begin{bmatrix} 1 & 0 \\ 0 & 1 \end{bmatrix} + \sqrt{5} \sum_{k \text{ odd}} \binom{\frac{1}{2}}{k} 5^{-k} 3^k \begin{bmatrix} 0 & 1 \\ 1 & 0 \end{bmatrix}$$

But, how do we simplify the even and odd parts? Imagine that instead of 3^k in equation (5.26), we had $(-3)^k$. Then $3^k + (-3)^k$ is equal to $2(3^k)$ whenever k is even, and is zero whenever k is odd. The situation reverses for $3^k - (-3)^k$. So, how do we get $(-3)^k$? Without matrices, the sum over all even and odd parts is

$$\sqrt{8} = \sqrt{5+3} = \sum_{k=0}^{\infty} \binom{\frac{1}{2}}{k} 5^{-k} 3^k.$$

Basically, since $c = 5$, if $3 = x - 5$, then x is 8. If $-3 = x - 5$ then $x = 2$, so

$$\sqrt{8} + \sqrt{2} = \sqrt{5}\sum_{k=0}^{\infty} \binom{\frac{1}{2}}{k} 5^{-k} 3^k + \sqrt{5}\sum_{k=0}^{\infty} \binom{\frac{1}{2}}{k} 5^{-k} (-3)^k = 2\sqrt{5} \sum_{k \text{ even}} \binom{\frac{1}{2}}{k} 5^{-k} 3^k$$

$$\sqrt{8} - \sqrt{2} = \sqrt{5}\sum_{k=0}^{\infty} \binom{\frac{1}{2}}{k} 5^{-k} 3^k - \sqrt{5}\sum_{k=0}^{\infty} \binom{\frac{1}{2}}{k} 5^{-k} (-3)^k = 2\sqrt{5} \sum_{k \text{ odd}} \binom{\frac{1}{2}}{k} 5^{-k} 3^k.$$

Dividing by 2 gives us exactly what we need:

$$\sqrt{\mathbf{A}} = \frac{\sqrt{8}+\sqrt{2}}{2} \begin{bmatrix} 1 & 0 \\ 0 & 1 \end{bmatrix} + \frac{\sqrt{8}-\sqrt{2}}{2} \begin{bmatrix} 0 & 1 \\ 1 & 0 \end{bmatrix} = \frac{\sqrt{2}}{2} \begin{bmatrix} 3 & 1 \\ 1 & 3 \end{bmatrix}, \tag{5.27}$$

where we have used $\sqrt{8} = 2\sqrt{2}$ to simplify.
The appearance of $\sqrt{8}$ and $\sqrt{2}$ should not be surprising, because 8 and 2 are the eigenvalues of $\mathbf{A}$. We have

$$\text{tr}(\mathbf{A}) = 5 + 5 = 10 \quad \text{and} \quad \det(\mathbf{A}) = 5^2 - 3^2 = 4^2 = 16,$$

so the characteristic polynomial is

$$x^2 - 10x + 16 = (x-8)(x-2).$$

For Theorem 5.2.8 to be true in this case, the matrix computed in equation (5.27) should have eigenvalues $\sqrt{2}$ and $\sqrt{8}$, so the characteristic polynomial for $\sqrt{\mathbf{A}}$ should be

$$(x - \sqrt{2})(x - \sqrt{8}) = x^2 - 3\sqrt{2}x + 4.$$

This checks out with the trace and determinant:

$$\operatorname{tr}(\sqrt{\mathbf{A}}) = \frac{\sqrt{2}}{2}(3+3) = 3\sqrt{2} \text{ and } \det(\sqrt{\mathbf{A}}) = \left(\frac{\sqrt{2}}{2}\right)^2 (3^2 - 1^2) = \frac{1}{2}(9-1) = 4.$$

Activity 5.5.4. Given the matrix

$$\mathbf{A} = \begin{bmatrix} 0 & -t \\ t & 0 \end{bmatrix},$$

1. compute $e^{\mathbf{A}}$ using (5.24), and
2. verify that Theorem 5.2.8 remains true with $f(x) = e^x$.

Hint: You will need to separate even and odd parts. The Taylor series for $\cos(x)$ and $\sin(x)$ can be used to simplify.

Sage has the built in capability of computing Taylor series. Unfortunately, it is not currently possible to plug in directly with a matrix. The following code provides a workaround:

```
def TaylorEval(f,c,n,X0):
    R.<z> = QQ[]
    g = f.taylor(x,c,n)
    return R(g(x=z))(z=X0)
```

It might be worthwhile to use this code to see what happens when the matrix norm is not less than the radius of convergence. For example, if $f(x) = \frac{1}{1-x}$, then the Taylor series at $c = 0$ is

$$\frac{1}{1-x} = \sum_{k=0}^{\infty} x^k \quad \text{for all} \quad |x| < 1,$$

which is to say that the radius of convergence is 1. Take any matrix with spectral radius greater than 1, such as

$$\mathbf{A} = \begin{bmatrix} 0.4 & 0.8 \\ 0.8 & 0.4 \end{bmatrix} = 0.4 \begin{bmatrix} 1 & 2 \\ 2 & 1 \end{bmatrix}.$$

The eigenvalues of $\mathbf{A}$ are 0.4 and 1.2, so the $\rho(\mathbf{A}) = 1.2 > 1$. But 1 is not an eigenvalue, so $\mathbf{I} - \mathbf{A}$ is invertible and

$$(\mathbf{I} - \mathbf{A})^{-1} = \begin{bmatrix} -\frac{1}{12} & \frac{1}{6} \\ \frac{1}{6} & -\frac{1}{12} \end{bmatrix}.$$

But the following code illustrates that the approximations from the Taylor series generally get worse as n increases:

```
A = matrix([[0.4,0.8],
            [0.8,0.4]])
f = 1/(1-x)
for n in range(10):
    print(n)
    print(TaylorEval(f,0,n,A))
```

5.6 QR Factorization

5.6.1 Householder Reflections

Q 5.6.1. If $\mathbf{v} = \left[\begin{smallmatrix}1\\2\end{smallmatrix}\right]$, what is $\mathbf{v} \otimes \mathbf{v}$?

Q 5.6.2. What is the formula to compute the reflection matrix $\mathbf{P_v}$?

Q 5.6.3. When does it mean to say that $\mathbf{P_v}$ is an isometry?

Given an $n \times n$ matrix $\mathbf{A}$, our goal is to compute a factorization

$$\mathbf{A} = \mathbf{QR} \tag{5.28}$$

where $\mathbf{Q}$ is a linear isometry and $\mathbf{R}$ is upper triangular. Such a factorization is called a QR-factorization . The idea of Householder is to construct $\mathbf{Q}$ in $n-1$ steps using reflections.

Suppose

$$\mathbf{A} = \begin{bmatrix} a & b \\ c & d \end{bmatrix}.$$

We must find a reflection $\mathbf{P}$ such that $\mathbf{PA}$ has a 0 in the bottom left corner. Linear isometries do not change the lengths of vectors; that is $\| \mathbf{Pv} \| = \| \mathbf{v} \|$. The length of the first column of $\mathbf{A}$ is

$$L = \sqrt{a^2 + c^2}$$

First we define

$$\mathrm{Sgn}(x) = \begin{cases} 1 & \text{if } x \geq 0 \\ -1 & \text{if } x < 0, \end{cases}$$

which is not the usual sign function. Usually zero would be zero, but that causes the QR factorization algorithm to get stuck. Next, if we define

$$\mathbf{v} = \begin{bmatrix} a \\ c \end{bmatrix} - \mathrm{Sgn}(a) \begin{bmatrix} L \\ 0 \end{bmatrix}$$

then $\mathbf{v}$ is perpendicular to the dashed line and $\mathbf{P_v}$, as defined by equation (4.25), gives us the reflection across the dashed line. For $n = 2$ we are finished after only one step, so $\mathbf{Q} = \mathbf{P_v}$ in this case.

Activity 5.6.1. (Sage optional) If you are using Sage, you can borrow the function `P` defined in Activity 4.8.5.

Given

$$\mathbf{A} = \begin{bmatrix} -3 & 4 \\ 4 & 8 \end{bmatrix}$$

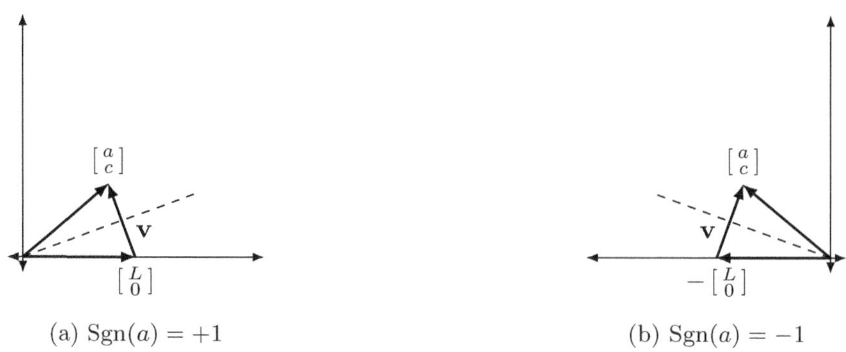

(a) Sgn(a) = +1 (b) Sgn(a) = −1

FIGURE 5.2: Householder construction

1. Compute the length L of the first column,
2. Compute $\mathbf{v}$ by subtracting L from the top component.
3. Compute
$$\mathbf{P} = \mathbf{I} - 2\frac{\mathbf{v} \otimes \mathbf{v}}{\mathbf{v} \cdot \mathbf{v}}$$
4. Compute $\mathbf{PA}$.

For $n > 2$ we must work left to right. We will use zero indexing be consistent with Python, so the left most column has $j = 0$. We initialize $\mathbf{Q}$ and $\mathbf{R}$ with

$$\mathbf{Q}_0 = \mathbf{I} \quad \text{and} \quad \mathbf{R}_0 = \mathbf{A}. \tag{5.29}$$

For each j, starting from 0, up to $n - 2$, we compute a reflection $\mathbf{P}_{\mathbf{v}_j}$, and then update the values of $\mathbf{Q}$ and $\mathbf{R}$ as follows

$$\mathbf{Q}_{j+1} = \mathbf{Q}_j \mathbf{P}_{\mathbf{v}_j} \quad \text{and} \quad \mathbf{R}_{j+1} = \mathbf{P}_{\mathbf{v}_j} \mathbf{R}_j. \tag{5.30}$$

The reflection $\mathbf{P}_{\mathbf{v}_j}$ is computed following a generalization of the strategy in Activity 5.6.1. Specifically, let $\mathbf{w}_j$ be the j-th column of $\mathbf{R}_j$ with the first j components set to zero, let $L_j = \| \mathbf{w}_j \|$, and let

$$\mathbf{v}_j = \mathbf{w}_j - L_j \mathbf{e}_j, \tag{5.31}$$

where $\mathbf{e}_j$ is the vector with 1 in the j-th component, and zeros everywhere else. Then $\mathbf{P}_{\mathbf{v}_j}$ us computed by equation (4.25).

Example 5.6.2. Given

$$\mathbf{A} = \begin{bmatrix} 1 & 5 & 3 \\ 2 & 0 & 1 \\ 2 & -1 & -1 \end{bmatrix},$$

compute $\mathbf{Q}$ and $\mathbf{R}$ using Householder reflections.

Solution

For $j = 0$, $\mathbf{R}_0 = \mathbf{A}$. With 0 indexing, $\mathbf{w}_0$ is the left most column of $\mathbf{A}$, so

$$L_0 = \| \mathbf{w}_0 \| = \sqrt{1^2 + 2^2 + 2^2} = \sqrt{9} = 3$$

and, by equation (5.31),

$$\mathbf{v}_0 = \mathbf{w}_0 - L_0\mathbf{e}_0 = \begin{bmatrix}1\\2\\2\end{bmatrix} - 3\begin{bmatrix}1\\0\\0\end{bmatrix} = \begin{bmatrix}-2\\2\\2\end{bmatrix}.$$

For $\mathbf{v}_0 \otimes \mathbf{v}_0$, we have

$$\begin{array}{cc} & \begin{bmatrix} -2 & 2 & 2 \end{bmatrix} \\ \begin{bmatrix}-2\\2\\2\end{bmatrix} & \begin{bmatrix} 4 & -4 & -4 \\ -4 & 4 & 4 \\ -4 & 4 & 4 \end{bmatrix} \end{array}$$

and

$$\mathbf{v}_0 \cdot \mathbf{v}_0 = (-2)^2 + 2^2 + 2^2 = 12.$$

Therefore, by equation (4.25), we have

$$\mathbf{P}_{\mathbf{v}_0} = \begin{bmatrix} 1 & 0 & 0 \\ 0 & 1 & 0 \\ 0 & 0 & 1 \end{bmatrix} - \frac{2}{12}\begin{bmatrix} 4 & -4 & -4 \\ -4 & 4 & 4 \\ -4 & 4 & 4 \end{bmatrix} = \begin{bmatrix} \frac{1}{3} & \frac{2}{3} & \frac{1}{3} \\ \frac{2}{3} & \frac{1}{3} & -\frac{2}{3} \\ \frac{2}{3} & -\frac{2}{3} & \frac{1}{3} \end{bmatrix}.$$

To compute $\mathbf{Q}_1$ and $\mathbf{R}_1$, we use equation (5.30). Since $\mathbf{Q}_0 = \mathbf{I}$, then $\mathbf{Q}_1 = \mathbf{P}_{\mathbf{v}_0}$, so no extra computation is needed for it. But $\mathbf{R}_0 = \mathbf{A}$, so $\mathbf{R}_1 = \mathbf{P}_{\mathbf{v}_0}\mathbf{A}$:

$$\begin{array}{cc} & \begin{bmatrix} 1 & 5 & 3 \\ 2 & 0 & 1 \\ 2 & -1 & -1 \end{bmatrix} \\ \begin{bmatrix} \frac{1}{3} & \frac{2}{3} & \frac{1}{3} \\ \frac{2}{3} & \frac{1}{3} & -\frac{2}{3} \\ \frac{2}{3} & -\frac{2}{3} & \frac{1}{3} \end{bmatrix} & \begin{bmatrix} 3 & 1 & 1 \\ 0 & 4 & 3 \\ 0 & 3 & 1 \end{bmatrix} = \mathbf{R}_1. \end{array}$$

For $j = 1$, we now move to the next column, the middle, but we use the new matrix $\mathbf{R}_1$. Since $j = 1$, we set the top component to zero when creating $\mathbf{w}_1$. Thus

$$\mathbf{w}_1 = \begin{bmatrix}0\\4\\3\end{bmatrix} \quad \text{and} \quad L_1 = \|\, \mathbf{w}_1 \,\| = \sqrt{4^2 + 3^2} = 5.$$

Then

$$\mathbf{v}_1 = \begin{bmatrix}0\\4\\3\end{bmatrix} - 5\begin{bmatrix}0\\1\\0\end{bmatrix} = \begin{bmatrix}0\\-1\\3\end{bmatrix},$$

so $\mathbf{v}_1 \otimes \mathbf{v}_1$ is

$$\begin{array}{cc} & \begin{bmatrix} 0 & -1 & 3 \end{bmatrix} \\ \begin{bmatrix}0\\-1\\3\end{bmatrix} & \begin{bmatrix} 0 & 0 & 0 \\ 0 & 1 & -3 \\ 0 & -3 & 9 \end{bmatrix} \end{array}$$

and

$$\mathbf{v}_1 \cdot \mathbf{v}_1 = 0^2 + (-1)^2 + 3^2 = 10$$

Therefore

$$\mathbf{P}_{\mathbf{v}_1} = \begin{bmatrix} 1 & 0 & 0 \\ 0 & 1 & 0 \\ 0 & 0 & 1 \end{bmatrix} - \frac{2}{10}\begin{bmatrix} 0 & 0 & 0 \\ 0 & 1 & -3 \\ 0 & -3 & 9 \end{bmatrix} = \begin{bmatrix} 1 & 0 & 0 \\ 0 & \frac{4}{5} & \frac{3}{5} \\ 0 & \frac{3}{5} & -\frac{4}{5} \end{bmatrix}$$

Finally, $\mathbf{R}_2 = \mathbf{P}_{\mathbf{v}_1}\mathbf{R}_1$ is

$$\begin{bmatrix} 3 & 1 & 1 \\ 0 & 4 & 3 \\ 0 & 3 & 1 \end{bmatrix}$$

$$\begin{bmatrix} 1 & 0 & 0 \\ 0 & \frac{4}{5} & \frac{3}{5} \\ 0 & \frac{3}{5} & -\frac{4}{5} \end{bmatrix}\begin{bmatrix} 3 & 1 & 1 \\ 0 & 5 & 3 \\ 0 & 0 & 1 \end{bmatrix} = \mathbf{R}_2.$$

which is indeed upper triangular and therefore is $\mathbf{R}$. Since $\mathbf{Q}_1 = \mathbf{P}_{\mathbf{v}_0}$, then $\mathbf{Q}_2 = \mathbf{P}_{\mathbf{v}_0}\mathbf{P}_{\mathbf{v}_1}$ by equation (5.30). It is definitely not easy to compute $\mathbf{P}_{\mathbf{v}_0}\mathbf{P}_1$ by hand, but if you ask Sage to do it, you will find

$$\mathbf{Q} = \mathbf{Q}_2 = \mathbf{P}_{\mathbf{v}_0}\mathbf{P}_{\mathbf{v}_1} = \begin{bmatrix} \frac{1}{3} & \frac{14}{15} & -\frac{2}{15} \\ \frac{2}{3} & -\frac{2}{15} & \frac{11}{15} \\ \frac{2}{3} & -\frac{1}{3} & -\frac{2}{3} \end{bmatrix}.$$

The following code can be used to do the same steps.

```
def Sgn(x): return 1-2*(x<0)

def QRfactor(A):
    n = A.ncols()
    I = matrix.identity(n)
    Q,R = I,A
    for j in range(n-1):
        print("j = "+ str(j))
        print("Q"+ str(j)) ; print(Q)
        print("R"+ str(j)) ; print(R)
        w = vector(j*(0,)+tuple(R.T[j][j:]))
        L = w.norm()
        v = w - Sgn(w[j])*L*I[j]
        Pv = P(v)
        R = Pv*R
        Q = Q*Pv
        print("w"+ str(j)) ; print(w)
        print("L"+ str(j)) ; print(L)
        print("sgn(w" + str(j) + ")") ; print(sgn(w[j]))
        print("v"+ str(j)) ; print(v)
        print("Pv"+ str(j)) ; print(Pv)
    return Q,R
```

Remark 5.6.1. If you did Section 1.5, then you may have observed that in block form

$$\mathbf{P}_j = \begin{bmatrix} \mathbf{I}_j & \\ & \mathbf{P}_{v_j}, \end{bmatrix}$$

except that $\mathbf{v}_j$ is computed in a slightly different way. To get $\mathbf{w}_j$ we drop off the first j-components of the column instead of replacing them by zeros. This gives us a smaller vector, but L_j remains the same. We then get $\mathbf{v}_j$ by subtracting L_j from the top component. This observation leads to a different but equivalent approach to coding the algorithm.

Activity 5.6.3. (Sage recommended) Given

$$\mathbf{A} = \begin{bmatrix} 3 & 6 & 5 \\ 4 & 3 & 0 \\ 0 & 4 & 12 \end{bmatrix},$$

compute $\mathbf{Q}$ and $\mathbf{R}$ using Householder reflections.

Proposition 5.6.1. *Let* $\mathbf{A}$ *be an* $n \times n$ *matrix, and let* $\mathbf{A} = \mathbf{QR}$*, where* $\mathbf{Q}$ *is computed by Householder reflections. Then*

$$\det(\mathbf{A}) = (-1)^{n-1} \det(\mathbf{R}),$$

where $\det(\mathbf{R})$ *is the product of the diagonal elements of* $\mathbf{R}$.

This proposition follows from the following facts:

1. $\det(\mathbf{A}) = \det(\mathbf{Q}) \det(\mathbf{R})$ by Theorem 3.1.5.
2. $\mathbf{Q}$ is the product of $n - 1$ reflection matrices, each of which has determinant -1. So $\det(\mathbf{Q}) = (-1)^{n-1}$ by Theorem 3.1.5.
3. $\mathbf{R}$ is upper triangular, so $\det(\mathbf{R})$ is the product of the diagonal elements by Proposition 3.1.3.

Activity 5.6.4. Consider the $n = 2$ case. Let

$$\mathbf{A} = \begin{bmatrix} -3 & 7 \\ 4 & -1 \end{bmatrix}, \quad \mathbf{Q} = \begin{bmatrix} -\frac{3}{5} & \frac{4}{5} \\ \frac{4}{5} & \frac{3}{5} \end{bmatrix}, \quad \text{and} \quad \mathbf{R} = \begin{bmatrix} 5 & -5 \\ 0 & 5 \end{bmatrix}.$$

1. Show that $\mathbf{A} = \mathbf{QR}$ by multiplying.
2. Compute $\det(\mathbf{A})$ by the 2×2 formula.
3. Compute $\det(\mathbf{R})$ by Proposition 3.1.3.
4. Check that $\det(\mathbf{A}) = (-1)^{n-1} \det(\mathbf{R})$.

Activity 5.6.5. (with Sage) In Sage the following code can be used to compute the determinant by QR factorization

To use Sage, you will need the code

```
def QRdet(A):
    Q,R = QRfactor(A)
    R = np.array(R)
    n = len(R)-1
    return (-1)^n*np.prod(np.diag(R)).simplify_rational()
```

To use it you can create a 3×3 matrix with random digits by

```
import numpy as np
A = matrix(np.random.randint(0,10,(3,3)))
```

Then you can compare the determinant as computed by Sage's built-in function with the QRdet function:

```
A.det()   #built-in Sage function
QRdet(A)
```

You can insert print functions if you want to see individual parts. If you want a larger matrix, just change 3 to a larger number. If you want a different range of integers, then change the values 0 and 10. If you want a 3×3 matrix with random decimals in $[0, 1]$ you can use

```
import numpy as np
A = matrix(np.random.rand(3,3))
```

but you will also need to remove `.simplify_rational()` from both `QRdet(A)` and from $P(v)$.

5.6.2 The QR-Algorithm

If we have a QR-factorization of $\mathbf{A}$ where $\mathbf{Q}^{-1} = \mathbf{Q}.T$, then

$$\mathbf{A} = \mathbf{QR} \implies \mathbf{Q}^T\mathbf{A} = \mathbf{R} \implies \mathbf{Q}^T\mathbf{AQ} = \mathbf{RQ}.$$

But now $\mathbf{A}$ and $\mathbf{Q}^T\mathbf{AQ}$ are conjugate, so they have the same eigenvalues and eigenvectors. Thus we obtain a sequence

$$\mathbf{A}_0 = \mathbf{A} \qquad \mathbf{U}_0 = \mathbf{I} \tag{5.32}$$

$$\mathbf{A}_k = \mathbf{Q}_k\mathbf{R}_k \quad \text{for } k \geq 0 \tag{5.33}$$

$$\mathbf{A}_{k+1} = \mathbf{Q}_k^T\mathbf{A}_i\mathbf{Q}_k = \mathbf{R}_k\mathbf{Q}_k \qquad \mathbf{U}_{k+1} = \mathbf{U}_k\mathbf{Q}_k. \tag{5.34}$$

This algorithm is called the **QR-algorithm.**. These equations imply that:

$$\mathbf{A}_k = \mathbf{U}_k\mathbf{A}\mathbf{U}_k^T$$

for all k. If

$$\lim_{k\to\infty} \mathbf{A}_k \tag{5.35}$$

converges to a diagonal matrix $\mathbf{A}$, then the eigenvalues will be on the diagonal. Even if the limit (5.35) converges, the limit

$$\lim_{k\to\infty} \mathbf{U}_k$$

may not exist as can be seen from Example 5.6.6. But if we stop with some large value of k for which $\mathbf{A}_k$ is close to diagonal, then the columns of $\mathbf{U}_k$ will approximate eigenvectors of $\mathbf{A}$. If $\mathbf{A}$ is a real matrix, then so are all of the matrices. So the limit is real, if it exists, which means the eigenvalues would have to be real. If $\mathbf{A}$ is a real symmetric matrix, then the eigenvalues are real by Theorem 5.3.1.

Example 5.6.6. Suppose t is a positive integer, and let

$$\mathbf{A} = \begin{bmatrix} t & 1 \\ 1 & 0 \end{bmatrix}. \tag{5.36}$$

Since $\text{tr}(\mathbf{A}) = t$ and $\det(\mathbf{A}) = -1$, the characteristic polynomial is

$$f(x) = x^2 - tx - 1.$$

By the quadratic formula, the roots are

$$\lambda_{\pm} = \frac{1 \pm \sqrt{t^2+4}}{2}.$$

It can be shown,[3] that for $k \geq 0$

$$\mathbf{A}^k = \begin{bmatrix} q_{k+1} & q_k \\ q_k & q_{k-1} \end{bmatrix} \tag{5.37}$$

where

$$q_{-1} = 1, \quad q_0 = 0, \quad \text{and} \tag{5.38}$$

$$q_k = tq_{k-1} + q_{k-2} \quad \text{for} \quad k \geq 1. \tag{5.39}$$

In the case where $t = 1$, then $q_k = F_k$ are the Fibonacci numbers. It can be shown that if the QR-algorithm is applied to $\mathbf{A}$ with Householder reflections, then

$$\mathbf{A}_k = \frac{1}{q_{2k+1}} \begin{bmatrix} q_{2k+2} & 1 \\ 1 & -q_{2k} \end{bmatrix}, \tag{5.40}$$

$$\mathbf{Q}_k = \frac{1}{\sqrt{q_{2k+1}q_{2k+3}}} \begin{bmatrix} q_{2k+2} & 1 \\ 1 & -q_{2k+2} \end{bmatrix}, \tag{5.41}$$

$$\mathbf{R}_k = \frac{1}{\sqrt{q_{2k+1}q_{2k+3}}} \begin{bmatrix} q_{2k+3} & t \\ 0 & q_{2k+1} \end{bmatrix}, \tag{5.42}$$

$$\mathbf{U}_k = \frac{1}{\sqrt{q_{2k+1}}} \begin{bmatrix} q_{k+1} & (-1)^{k+1}q_k \\ q_k & (-1)^k q_{k+1} \end{bmatrix}. \tag{5.43}$$

Then

$$\lim_{k\to\infty} \mathbf{A}_k = \mathbf{D}$$

where

$$\mathbf{D} = \begin{bmatrix} \lambda_+ & 0 \\ 0 & \lambda_- \end{bmatrix}.$$

The alternating sign in the second column of $\mathbf{U}_k$, causes $\lim_{k\to\infty} \mathbf{U}_k$ not to exist, however, if we divide that column by $(-1)^k$, then

$$\lim_{k\to\infty} \frac{1}{\sqrt{q_{2k+1}}} \begin{bmatrix} q_{k+1} & -q_k \\ q_k & q_{k+1} \end{bmatrix}$$

does converge a matrix $\mathbf{U}$ whose columns are normalized eigenvectors of $\mathbf{A}$.

The hand calculations are not easy, so the next activity will study this example with Sage. Example and Activity does explore some of the hand calculations if you are curious.

Activity 5.6.7 (With Sage)**.** You will need to borrow the code for `QRfactor` after Example 5.6.2. The following code, can be used to define the QR algorithm

```
def QR(A,steps):
    n = A.ncols()
    Ak,Uk = A,matrix.identity(n)
    for k in range(steps):
        print("k = "+str(k))
```

[3]By induction.

```
        print("U"+str(k));print(N(Uk))
        print("A"+str(k));print(Ak)
        Qk,Rk = QRfactor(Ak)
        print("Q"+str(k));print(N(Qk))
        print("R"+str(k));print(N(Rk))
        Ak = (Rk*Qk).simplify_rational()
        Uk = (Uk*Qk).simplify_rational()
    return Ak,Uk
```

Now define

```
A = matrix([[1,1],
            [1,0]])
```

which is the $t = 1$ case in Example 5.6.6. Now run the code

```
[A^k for k in range(7) ]
```

You should observe that $\mathbf{A}^k$ follows the pattern in equation (5.37), and that the components are Fibonacci numbers; in fact, this is the fastest way to compute them. Now run the code

```
Ak,Uk = QR(A,4)
```

You should observe that $\mathbf{A}_k$ follows the pattern in (5.40). The function `N()` is used to get numerical estimates for the others, making them easier to read. You should observe the alternating pattern in the second column of $\mathbf{U}$. In particular, `Uk` is symmetric when k is odd, but not when k is even. If you run the code

```
N(Uk.T*Uk)
Ak
(Uk.T*A*Uk).simplify_rational()
(Uk*A*Uk.T).simplify_rational()
```

each in a separate cell, you should see that:

1. Numerically `Uk.T*Uk` is approximately the identity matrix $\mathbf{A}_k = \mathbf{U}_k^T$ (there is some round-off error).

2. `Ak` is equal to `Uk.T*A*Uk` not to `Uk*A*Uk.T`.

To see convergence, it is easier to look at the numerical approximations. To do this, you will need to do several things. First, redefine $\mathbf{A}$:

```
A = N(A)
```

Next comment out `.simplify_rational()` everywhere by putting `#` immediately in front of the dot. Finally, comment out the print statements. Then run the code:

```
Ak,Uk = QR(A,10)
```

If this proves to be troublesome, an alternative is to construct `Ak` directly from the functions you have for Fibonacci numbers. Either way, you will also need to define the matrix $\mathbf{D}$:

```
phi = (1+sqrt(5))/2
D = N(matrix([[phi,0],
              [0,-1/phi]]))
```

Then all components of

```
Ak-D
```

should be small.

Activity 5.6.8. Let

$$\mathbf{A} = \begin{bmatrix} 1 & 1 \\ 1 & 0 \end{bmatrix}$$

as in Activity 5.6.6. The characteristic polynomial is $f(x) = x^2 - x - 1$.

1. If λ is an eigenvalue of $\mathbf{A}$, then $f(\lambda) = 0$. Solve $f(\lambda) = 0$ for λ^2.
2. Show that

$$\mathbf{v} = \begin{bmatrix} \lambda \\ 1 \end{bmatrix}$$

is an eigenvector of $\mathbf{A}$ with eigenvalue λ, by computing $\mathbf{Av}$ and simplifying.

3. Use the quadratic formula to show that the roots of $f(x)$ are

$$\lambda_+ = \frac{1+\sqrt{5}}{2} \quad \text{and} \quad \lambda_- \frac{1-\sqrt{5}}{2}.$$

4. Check that $\lambda_+ + \lambda_- = 1$ and $\lambda_+ \lambda_- = -1$.

Example 5.6.9. In this example and subsequent activity, we look at some of the algebra behind Example 5.6.6.

Suppose m and n are both non-negative. Then $\mathbf{A}^m \mathbf{A}^n = \mathbf{A}^{m+n}$. By equation (5.37), if we do the matrix multiplication on the left and equate coefficients, we obtain

$$q_{m+1}q_{n+1} + q_m q_n = q_{m+n+1} \tag{5.44}$$

$$q_{m+1}q_n + q_m q_{n-1} = q_{m+n} \tag{5.45}$$

By Theorem 3.1.5, $\det(\mathbf{A}^k) = \det(\mathbf{A})^k$ so by applying the 2×2 formula to both $\det(\mathbf{A}^k)$ and $\det(\mathbf{A})$, we obtain

$$q_{k+1}q_{k-1} - q_k^2 = (-1)^k. \tag{5.46}$$

Then by the 2×2 inverse formula

$$\mathbf{A}^{-k} = (-1)^k \begin{bmatrix} q_{k-1} & q_k \\ q_k & q_{k+1} \end{bmatrix}$$

when $k \geq 0$. Now suppose $m \geq n \geq 0$. Then $\mathbf{A}^m \mathbf{A}^{-n} = \mathbf{A}^{m-n}$.

$$q_m q_{n+1} - q_{m+1} q_n = (-1)^n q_{m-n} \tag{5.47}$$

$$q_m q_{n-1} - q_{m-1} q_m = (-1)^n q_{m-n} \tag{5.48}$$

We now use this to show that $\mathbf{R}_k = \mathbf{Q}_k \mathbf{A}_k$, as given by the formulas (5.40), (5.41), and (5.42). Temporarily ignoring the fractions in front of $\mathbf{Q}_k$ and $\mathbf{A}_k$, we have

$$\begin{matrix} & \begin{bmatrix} q_{2k+2} & 1 \\ 1 & -q_{2k} \end{bmatrix} & \\ \begin{bmatrix} q_{2k+2} & 1 \\ 1 & -q_{2k+2} \end{bmatrix} & \begin{bmatrix} q_{2k+2}^2 + 1 & q_{2k+2} - q_{2k} \\ 0 & 1+q_{2k}q_{2k+2} \end{bmatrix} & = \begin{bmatrix} q_{2k+1}q_{2k+3} & tq_{2k+1} \\ 0 & q_{2k+1}^2 \end{bmatrix} \end{matrix} \tag{5.49}$$

The component in the top right corner was simplified using (5.39) with k replaced by $2k+2$. The components on the diagonal were simplified using equation (5.46). If we replace k in equation (5.46) with $2k+2$, then

$$q_{2k+3}q_{2k+1} - q_{2k+2}^2 = 1$$

hence

$$q_{2k+2}^2 + 1 = q_{2k+3}q_{2k+1} \tag{5.50}$$

If instead we replace k in equation (5.46) with $2k+1$, then

$$q_{2k+2}q_{2k} - q_{2k+1}^2 = -1$$

hence

$$1 + q_{2k+2}q_{2k} = q_{2k+1}^2. \tag{5.51}$$

We now complete the calculation of $\mathbf{R}_k$. Multiplying the fractions in (5.40) and (5.41) by the result in (5.49) gives us

$$\frac{1}{q_{2k+1}\sqrt{q_{2k+1}q_{2k+3}}}\begin{bmatrix} q_{2k+1}q_{2k+3} & tq_{2k+1} \\ 0 & q_{2k+1}^2 \end{bmatrix} = \frac{1}{\sqrt{q_{2k+1}q_{2k+3}}}\begin{bmatrix} q_{2k+3} & t \\ 0 & q_{2k+1} \end{bmatrix}$$

which justifies equation (5.42).

Activity 5.6.10. Using equation (5.40) for $\mathbf{A}^k$:

1. compute the product $\mathbf{A}^m\mathbf{A}^n$.
2. write down $\mathbf{A}^{m+n}$.

Equating coefficients should give equations (5.44) and (5.45).

Compute

$$\mathbf{R}_k\mathbf{Q}_k.$$

Simplify following the strategy of Example 5.6.9. The final result should match $\mathbf{A}_{k+1}$.

Compute $\mathbf{Q}_k^2$ and simplify. The result should be $\mathbf{I}$.

Example 5.6.11. It is also worth looking at the convergence of $\lim_{k\to\infty} \mathbf{A}_k$ from an algebraic viewpoint. There is a lot of continued fraction theory hiding here (see [19] for instance). The simple continued fraction for λ_+ is

$$t + \cfrac{1}{t + \frac{1}{t+\cdots}}$$

or $[t, t, t, \ldots]$ in shorthand. The fractions $\frac{q_{n+1}}{q_n}$ called convergents because

$$\lim_{n\to\infty} \frac{q_{n+1}}{q_n} = \lambda_+.$$

If you plot the convergents, they alternate above and below the limit. For $n = 2k$, they approach from above and are decreasing

$$t = \frac{q_2}{q_1} < \frac{q_4}{q_3} < \frac{q_6}{q_5} < \cdots < \lambda_+ < \cdots < \frac{q_7}{q_6} < \frac{q_5}{q_4} < \frac{q_3}{q_2} = t + \frac{1}{t}.$$

The theory of continued fractions also tells us that

$$\left|\lambda_+ - \frac{q_{2k+2}}{q_{2k+1}}\right| < \frac{1}{q_{2k+1}^2}, \tag{5.52}$$

for all k. Since $\lambda_- = -\frac{1}{\lambda_+}$, then

$$\left|\lambda_- - -\frac{q_{2k}}{q_{2k+1}}\right| = \left|-\frac{1}{\lambda_+} + \frac{1}{\frac{q_{2k+1}}{q_{2k}}}\right| = \frac{1}{\frac{q_{2k+1}}{q_{2k}}\lambda_+}\left|\frac{q_{2k+2}}{q_{2k+1}} - \lambda_+\right| < \frac{1}{q_{2k+1}^2}. \tag{5.53}$$

The inequalities (5.52) and (5.53) give us bounds for the diagonal components of

$$\mathbf{D} - \mathbf{A}_k = \begin{bmatrix} \lambda_+ - \frac{q_{2k+2}}{q_{2k+1}} & -\frac{1}{q_{2k+1}} \\ -\frac{1}{q_{2k+1}} & \lambda_- + \frac{q_{2k}}{q_{2k+1}} \end{bmatrix},$$

so it follows that

$$\| \mathbf{D} - \mathbf{A}_k \|_{\max} < \max\left\{\frac{1}{q_{2k+1}^2}, \frac{1}{q_{2k+1}}\right\} = \frac{1}{q_{2k+1}}.$$

and the denominator q_{2k+1} grows exponentially.

5.7 Algorithm Efficiency

Although there the characteristic polynomial of a matrix $\mathbf{A}$ can be computed directly from the definition, it turns out that the method using the spectral mapping theorem is faster. We begin by analyzing the efficiency in space and time.

First, we must compute the powers $\mathbf{A}^k$. These matrices are all square matrices, so the faster algorithms of matrix multiplication apply (see Section 1.6). In the sequential case we have $O(n^\alpha)$ where $2 \le \alpha < 3$. We must multiply $n - 1$ times to get all of the matrices that we need, which leads to $O(n^{\alpha+1})$. In the parallel case we can use accumulate algorithm described in Section 2.5. Commutativity is not usually true for matrix multiplication (see Remark 1.1.6), but it is true for powers of a square matrix. Even so, commutativity is not assumed in the accumulate algorithm given in Section 2.5, only associativity, which is true for matrix multiplication (see Proposition 1.1.16). As indicated by Figure 1.8 we have $O(\log(n))$ for matrix multiplication in parallel, and the accumulate algorithm takes $\log_2(n)$ steps, so we get $O(\log(n)^2)$ in time.

The second step requires us to compute the trace of each of the matrices $\mathbf{A}^k$. In both cases, the time of that computation is less than the time to get the matrices in $\mathbf{A}^k$ in the first place. Sequentially each $\text{tr}(\mathbf{A}^k)$ takes $O(n)$, and there are n matrices, so we get $O(n^2)$ which is less than $O(n^{\alpha+1})$. In parallel, the traces can be computed simultaneously in $O(\log(n))$ time.

The third step is a recursive computation with n steps. The computation of s_k takes k multiplications and $k - 1$ additions or subtractions. The total number of operations grows quadratically, so we get $O(n^2)$ sequentially, which once again is less than $O(n^{\alpha+1})$. If we do each step of the recursive calculation using the parallel approach to the dot product, then we must estimate

$$\sum_{k=1}^{n} \log(k).$$

Those familiar with the integral test should recognize that

$$\sum_{k=1}^{n} \log(k) \le \int_1^{n+1} \ln(x)\, dx.$$

By integration by parts, with

$$u = \ln(x) \qquad dv = dx$$

$$du = \frac{1}{x}\, dx \qquad v = x,$$

then

$$\int_1^{n+1} \ln(x)\, dx = x\ln(x)\Big|_1^{n+1} - \int_1^{n+1} dx = n\ln(n+1) - n.$$

Hence we get $O(n\log(n))$ in parallel.

How much space is required? Using the values in Figure 1.8 with $m = l = n$ and remembering that we have n matrices total, gives us $O(n^3)$ in space sequentially and $O(n^4)$ in parallel. The results are summarized in the table below:

	Space	Time
Sequential with small nums	$O(n^3)$	$O(n^{\alpha+1})$
Parallel with small nums	$O(n^4)$	$O(n\log(n))$

FIGURE 5.3: Computing the characteristic polynomial of **A**

Exercises

Problem 5.1. True or False.

1. The eigenvectors of an $n \times n$ matrix over $\mathbb{R}$ always span $\mathbb{R}^n$.
2. The eigenvectors of an $n \times n$ matrix over $\mathbb{C}$ always span $\mathbb{C}^n$.
3. A matrix **A** is invertible if and only if zero is not an eigenvalue.
4. If λ is an eigenvalue for a matrix, then the dimension of the λ eigenspace is always 1.

Problem 5.2. What are the eigenvalues of the identity matrix? What are the eigenvectors?

Problem 5.3. Construct a 2×2 matrix such that

$$\mathbf{v}_1 = \begin{bmatrix} 1 \\ -1 \end{bmatrix}$$

is an eigenvector with eigenvalue 0, and

$$\mathbf{v}_2 = \begin{bmatrix} 1 \\ 1 \end{bmatrix}$$

is an eigenvector with eigenvalue 4.

Problem 5.4. Do the following for each matrix below:

1. Compute the characteristic polynomial.
2. Compute the eigenvalues.

a. $\begin{bmatrix} 0 & 0 & 1 \\ 1 & 0 & 0 \\ 0 & 1 & 0 \end{bmatrix}$ b. $\begin{bmatrix} -2 & 1 & 1 \\ 1 & -2 & 1 \\ 1 & 1 & -2 \end{bmatrix}$ c. $\begin{bmatrix} 0 & 1 & -1 \\ -1 & 0 & 1 \\ 1 & -1 & 0 \end{bmatrix}$

d. $\begin{bmatrix} 0 & 0 & 0 & 1 \\ 1 & 0 & 0 & 0 \\ 0 & 1 & 0 & 0 \\ 0 & 0 & 1 & 0 \end{bmatrix}$ e. $\begin{bmatrix} 0 & 1 & 1 & 0 \\ 0 & 0 & 1 & 1 \\ 1 & 0 & 0 & 1 \\ 1 & 1 & 0 & 0 \end{bmatrix}$ f. $\begin{bmatrix} 0 & 1 & 1 & 1 \\ 1 & 0 & 1 & 1 \\ 1 & 1 & 0 & 1 \\ 1 & 1 & 1 & 0 \end{bmatrix}$

g. $\begin{bmatrix} 1 & 1 & 0 & 0 \\ 1 & -1 & 0 & 0 \\ 0 & 0 & 1 & 1 \\ 0 & 0 & 1 & -1 \end{bmatrix}$ h. $\begin{bmatrix} 2 & 1 & 0 & 0 \\ 1 & -2 & 0 & 0 \\ 0 & 0 & 2 & 1 \\ 0 & 0 & 1 & -2 \end{bmatrix}$ i. $\begin{bmatrix} 2 & 1 & 0 & 0 \\ -1 & -2 & 0 & 0 \\ 0 & 0 & 2 & 1 \\ 0 & 0 & -1 & -2 \end{bmatrix}$

Problem 5.5. Given the matrix:

$$\mathbf{A} = \begin{bmatrix} 0 & 1 \\ 2 & -1 \end{bmatrix}$$

1. calculate the characteristic polynomial,
2. calculate the eigenvalues λ,
3. for each eigenvalue, calculate an eigenvector.
4. write down a parameterization of each eigenspace.

Problem 5.6. For each matrix below, compute an eigenbasis and diagonalize the matrix

a. $\begin{bmatrix} 2 & 1 \\ 1 & 2 \end{bmatrix}$ b. $\begin{bmatrix} 3 & 2 \\ 2 & 3 \end{bmatrix}$ c. $\begin{bmatrix} 4 & 3 \\ 3 & 4 \end{bmatrix}$

d. $\begin{bmatrix} 2 & i \\ -i & 2 \end{bmatrix}$ e. $\begin{bmatrix} 0 & 1+2i \\ 1-2i & 4 \end{bmatrix}$ f. $\begin{bmatrix} 5 & 3+4i \\ 3-4i & 5 \end{bmatrix}$

Problem 5.7. The transformation $\mathbf{P} : \mathbb{R}^2 \to \mathbb{R}^2$ is defined as projection onto the line $x = y$. In the standard basis, it has the matrix

$$\mathbf{P} = \begin{bmatrix} \frac{\sqrt{2}}{2} & \frac{\sqrt{2}}{2} \\ \frac{\sqrt{2}}{2} & \frac{\sqrt{2}}{2} \end{bmatrix}.$$

1. calculate the eigenvalues of $\mathbf{P}$
2. for each eigenvalue calculate an eigenvector.
3. Setup a change of basis from the standard basis in $\mathbb{R}^2$ to the eigenbasis, and show that in the eigenbasis the matrix for $\mathbf{P}$ becomes diagonal with the eigenvalues on the diagonal.

Problem 5.8. Given the matrix

$$\mathbf{C} = \begin{bmatrix} 0 & 0 & 1 \\ 1 & 0 & 0 \\ 0 & 1 & 0 \end{bmatrix}$$

Calculate the characteristic polynomial, and factor it by finding a rational root. *Hint: if it is not obvious what the rational root is, look up the rational root theorem.*

1. How many real eigenvalues are there?

2. How many complex eigenvalue are there (which includes the real ones)?
3. Calculate the real eigenvectors.

Problem 5.9. We have seen that multiplication by i in the complex plane amounts to rotation by 90 degrees counterclockwise. We saw that by thinking of the complex plane as a real vector space, 1 and i are linearly independent, span $\mathbb{C}$, and are thus a *real* basis for $\mathbf{C}$. The matrix for multiplication by i with respect to this basis is

$$\mathbf{R} = \begin{bmatrix} 0 & -1 \\ 1 & 0 \end{bmatrix}$$

We have also seen in example 5.2.14 (for rotation by 90 clockwise) that the characteristic polynomial is λ^2+1, thus there are no real eigenvalues. Since we are thinking of the complex plane as a real vector space, i.e. only scalars in $\mathbb{R}$ are allowed, then any eigenvector *in the complex plane* corresponds to a real eigenvalue so there are none: it is rotation by 90 degrees after all. Therefore, to find an eigenvector we need to introduce a new imaginary unit, say j, satisfying $j^2 = -1$. Enter the Hamilton quaternions. Hamilton was a 19th century mathematician and physicist who introduced "quaternions"

$$a + bi + cj + dk$$

where a, b, c, d are real, and i, j, k satisfy the following rules

$$i^2 = -1 \quad j^2 = -1 \quad k^2 = -1$$
$$ij = -ji = k \quad jk = -kj = i \quad ki = -ik = j.$$

Also i, j, k are allowed to commute with the real numbers. Using $ij = k$ we have

$$a + bi + cj + dk = a + bi + cj + dij = (a + cj) + i(b + dj),$$

so if we allow for scalars in the form $a + cj$ and $b + dj$ (being careful to multiply the scalars *only on the right*), then the Hamilton quaternions provide an extension of the complex plane within which we can find eigenvectors. Since j is the imaginary unit being used for the scalars (and eigenvalues are scalars), then the characteristic polynomial factors as

$$\lambda^2 + 1 = (\lambda - j)(\lambda + j)$$

so the eigenvalues λ are $\pm j$. Your job is to calculate vectors $(a + cj) + i(b + dj)$, which are eigenvectors for $\lambda = \pm j$, that is you must solve

$$i((a + cj) + i(b + dj)) = ((a + cj) + i(b + dj))j$$

and

$$i((a + cj) + i(b + dj)) = ((a + cj) + i(b + dj))(-j)$$

with real numbers a, b, c, d. Note that these equation are supposed to be solved separately, so you will get a different set of values a, b, c, d for each equation. *Hint: multiply out on both sides.*

Problem 5.10. Show that each part of theorem 5.2.4 is false for the matrix

$$\mathbf{A} = \begin{bmatrix} 1 & 1 \\ 0 & 0 \end{bmatrix}$$

Problem 5.11. Compute the condition number for $\mathbf{M}$ and $\mathbf{R}$ from the symmetries of the square (Example 4.2.1).

The symmetries of the square are isometries, so b

6

Orthogonality

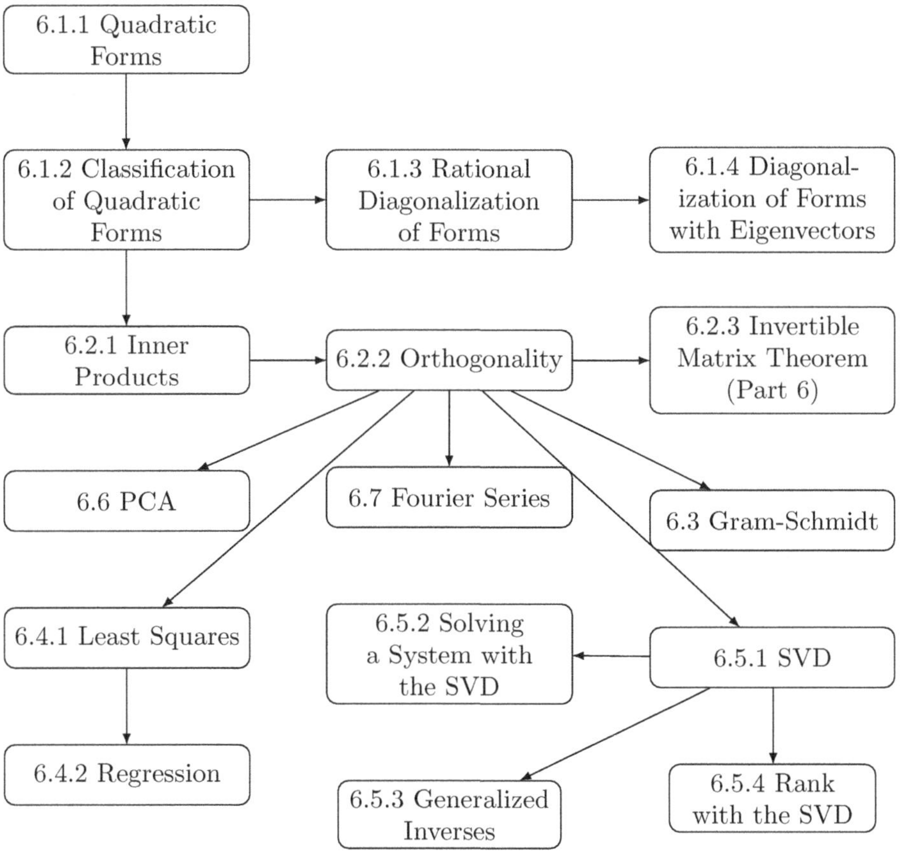

DOI: 10.1201/9781003737490-6

6.1 Quadratic Forms

6.1.1 Intro to Quadratic Forms

We often define the (Euclidean) length of a vector $\mathbf{v}$ in $\mathbb{R}^n$ to be

$$\| v \| = \sqrt{\mathbf{v} \cdot \mathbf{v}}.$$

Instead of looking at the length, we might look at the squared length and not have to worry about the square root. The following idea generalizes this concept.

Definition 6.1.1. A **quadratic form** is a function

$$Q(\mathbf{x}) = \mathbf{x}^T \mathbf{A} \mathbf{x}.$$

In the 2 dimensional case it is called a binary quadratic form, and often is written as

$$Q(x, y) = ax^2 + 2bxy + cy^2.$$

A quadratic form can always be represented by a symmetric matrix. For example, in the 2 dimensional case

$$ax^2 + 2bxy + cy^2 = \begin{bmatrix} x & y \end{bmatrix} \begin{bmatrix} a & b \\ b & c \end{bmatrix} \begin{bmatrix} x \\ y \end{bmatrix}.$$

Example 6.1.1. The squared euclidean length of a vector can now be seen as a quadratic form with $\mathbf{A}$ as the identity matrix:

$$\| v \|^2 = \mathbf{v}^T \mathbf{I} \mathbf{v} = \mathbf{v}^T \mathbf{v}.$$

For the 2 dimensional case, $a = c = 1$ and $b = 0$ is just $Q(x, y) = x^2 + y^2$.

Activity 6.1.2. Suppose we define a quadratic form $Q(x, y)$ by

$$Q(x, y) = \begin{bmatrix} x & y \end{bmatrix} \begin{bmatrix} 3 & 6 \\ 0 & 5 \end{bmatrix} \begin{bmatrix} x \\ y \end{bmatrix}.$$

Compute the product to get $Q(x, y)$ written in the form $ax^2 + 2bxy + cy^2$. Then check that

$$Q(x, y) = \begin{bmatrix} x & y \end{bmatrix} \begin{bmatrix} a & b \\ b & c \end{bmatrix} \begin{bmatrix} x \\ y \end{bmatrix}.$$

Finally, show that the discriminant of the quadratic form is -4 times the determinant of the matrix.

6.1.2 Classification of Quadratic Forms

Q 6.1.1. What is the quadratic formula?

Q 6.1.2. Given $\mathbf{A} = \left[\begin{smallmatrix} 1 & 2 \\ 3 & 4 \end{smallmatrix}\right]$

1. What is the trace of $\mathbf{A}$?
2. What is the determinant of $\mathbf{A}$?
3. What is the characteristic polynomial of $\mathbf{A}$?

Q 6.1.3. What are the eigenvalues of the identity matrix?

Definition 6.1.2. A quadratic form $Q(\mathbf{v})$ is called

1. **positive definite** if $Q(\mathbf{v}) > 0$ for any $\mathbf{v} \neq \mathbf{0}$,
2. **negative definite** if $Q(\mathbf{v}) < 0$ for any $\mathbf{v} \neq \mathbf{0}$,
3. **positive semi-definite** if $Q(\mathbf{v}) \geq 0$ for any $\mathbf{v} \neq \mathbf{0}$,
4. **negative semi-definite** if $Q(\mathbf{v}) \leq 0$ for any $\mathbf{v} \neq \mathbf{0}$, and
5. **indefinite** if both $Q(\mathbf{v}) > 0$ and $Q(\mathbf{v}) < 0$ are possible.

We mostly care about the positive definite case because they are the most similar to the Euclidean length. Since $\mathbf{A}$ can always be taken as a symmetric matrix, then the eigenvalues are all real. There is an equivalent characterization in terms of eigenvalues.

Theorem 6.1.3. *A quadratic form $Q(\mathbf{v}) = \mathbf{v}^T\mathbf{A}\mathbf{v}$ is*

1. *positive definite if and only if all eigenvalues of* $\mathbf{A}$ *are positive,*
2. *negative definite if and only if all eigenvalues of* $\mathbf{A}$ *are negative,*
3. *positive semi-definite if and only if all eigenvalues of* $\mathbf{A}$ *are non-negative,*
4. *negative semi-definite if and only if all eigenvalues of* $\mathbf{A}$ *are non-positive, and*
5. *indefinite if and only if both positive and negative eigenvalues of* $\mathbf{A}$ *occur.*

In the case of binary quadratic forms, there is yet another way.

Theorem 6.1.4. *A binary quadratic form $Q(x, y) = ax^2 + 2bxy + cy^2$ has discriminant*

$$(2b)^2 - 4ac = 4(b^2 - ac).$$

Then $Q(x, y)$ is

1. *positive definite if and only if the discriminant is negative and $a > 0$,*
2. *negative definite if and only if the discriminant is negative and $a < 0$,*
3. *positive semi-definite if and only if the discriminant is zero and $a + c > 0$,*
4. *negative semi-definite if and only if the discriminant is zero and $a + c < 0$,*
5. *indefinite if and only if the discriminant is positive.*

The connection between the discriminant and the determinant is commonly found in books on number theory. My preferred sources for the theory of binary quadratic forms are [6] and [12]. A more modern presentation can be found in [26]. Once the connection between the discriminant and the determinant is known, then this theorem is can be obtained by applying Theorem 6.1.3 to the characteristic polynomial in the 2×2 case (Proposition 5.2.3).

Example 6.1.3. Classify each form below, as positive definite, negative definite, or indefinite. Use both criteria.

$$Q(x, y) = x^2 + y^2 = \begin{bmatrix} x & y \end{bmatrix} \begin{bmatrix} 1 & 0 \\ 0 & 1 \end{bmatrix} \begin{bmatrix} x \\ y \end{bmatrix}$$

$$Q(x, y) = -x^2 - y^2 = \begin{bmatrix} x & y \end{bmatrix} \begin{bmatrix} -1 & 0 \\ 0 & -1 \end{bmatrix} \begin{bmatrix} x \\ y \end{bmatrix}$$

$$Q(x, y) = x^2 - y^2 = \begin{bmatrix} x & y \end{bmatrix} \begin{bmatrix} 1 & 0 \\ 0 & -1 \end{bmatrix} \begin{bmatrix} x \\ y \end{bmatrix}$$

Solution

The form $Q(x, y) = x^2 + y^2$ is positive definite because:

1. $\lambda = 1$ is the only eigenvalue, and it is positive.
2. The discriminant is $0^2 - 4 \cdot 1 \cdot 1 = -4$, which is negative, and the coefficient of x^2 is 1, which positive.

The form $Q(x, y) = -x^2 - y^2$ is negative definite because:

1. $\lambda = -1$ is the only eigenvalue, and it is negative.
2. The discriminant is $0^2 - 4 \cdot (-1) \cdot (-1) = -4$ and the coefficient of x^2 is -1.

The form $Q(x, y) = x^2 - y^2$ is indefinite because:

1. The eigenvalues are ± 1.
2. The discriminant is $0^2 - 4 \cdot (-1) \cdot (-1) = -4$ and the coefficient of x^2 is -1.

It is also possible to show that $Q(x, y) = x^2 - y^2$ is indefinite directly by using the definition. Both positive and negative values occur because $Q(1, 0) = 1$ and $Q(0, 1) = -1$. This approach is only practical for indefinite forms.

Activity 6.1.4. Classify each form below, as positive definite, negative definite, or indefinite. Use both criteria. The eigenvalues can be computed by factoring the characteristic polynomial.

$$Q(x, y) = 2xy = \begin{bmatrix} x & y \end{bmatrix} \begin{bmatrix} 0 & 1 \\ 1 & 0 \end{bmatrix} \begin{bmatrix} x \\ y \end{bmatrix}$$

$$Q(x, y) = 2x^2 + 2xy + 2y^2 = \begin{bmatrix} x & y \end{bmatrix} \begin{bmatrix} 2 & 1 \\ 1 & 2 \end{bmatrix} \begin{bmatrix} x \\ y \end{bmatrix}$$

$$Q(x, y) = -3x^2 + 2xy - 3y^2 = \begin{bmatrix} x & y \end{bmatrix} \begin{bmatrix} -3 & 1 \\ 1 & -3 \end{bmatrix} \begin{bmatrix} x \\ y \end{bmatrix}$$

Activity 6.1.5. Let $\mathbf{v} = \begin{bmatrix} x \\ y \end{bmatrix}$ and $Q(\mathbf{v}) = \mathbf{v}^T \mathbf{I} \mathbf{v} = \mathbf{v}^T \mathbf{v} = x^2 + y^2$. Show:

1. $Q\left(2\begin{bmatrix} 1 \\ 0 \end{bmatrix}\right) \neq 2Q\left(\begin{bmatrix} 1 \\ 0 \end{bmatrix}\right)$
2. $Q\left(\begin{bmatrix} 1 \\ 0 \end{bmatrix}\right) + Q\left(\begin{bmatrix} 0 \\ 1 \end{bmatrix}\right) = Q\left(\begin{bmatrix} 1 \\ 0 \end{bmatrix} + \begin{bmatrix} 0 \\ 1 \end{bmatrix}\right)$
3. $Q\left(\begin{bmatrix} 1 \\ 0 \end{bmatrix}\right) + Q\left(\begin{bmatrix} 1 \\ 1 \end{bmatrix}\right) < Q\left(\begin{bmatrix} 1 \\ 0 \end{bmatrix} + \begin{bmatrix} 1 \\ 1 \end{bmatrix}\right)$
4. $Q\left(\begin{bmatrix} -1 \\ 0 \end{bmatrix}\right) + Q\left(\begin{bmatrix} 1 \\ 1 \end{bmatrix}\right) > Q\left(\begin{bmatrix} -1 \\ 0 \end{bmatrix} + \begin{bmatrix} 1 \\ 1 \end{bmatrix}\right)$

Moral: quadratic forms are not linear.

6.1.3 Rational Diagonalization of Forms

Q 6.1.4. What is $(x+a)^2$ equal to?

There are two approaches to diagonalizing a quadratic form. Certainly we should expect it to be possible by the spectral theorem, but the eigenvalues and eigenvectors involved are not always rational. It is possible to diagonalize using rational numbers alone. There are times when we want the eigenvalues and eigenvectors, and there are times when the rational approach is fine or even preferred. First we look at the rational approach.

Example 6.1.6. Completing the square relies on the identity

$$(x+ny)^2 = x^2 + 2nxy + n^2y^2.$$

So, given

$$Q(x,y) = x^2 + 6xy - 7y^2,$$

we observe that $2n = 6$ so $n = 3$. Then

$$x^2 + 6xy - 7y^2 = x^2 + 6xy + 9y^2 - 9y^2 - 7y^2 = (x+3y)^2 - 16y^2.$$

The change of variables $u = x + 3y$ and $v = y$ diagonalizes the form:

$$Q(u,v) = u^2 - 16v^2 = \begin{bmatrix} u & v \end{bmatrix} \begin{bmatrix} 1 & 0 \\ 0 & -16 \end{bmatrix} \begin{bmatrix} u \\ v \end{bmatrix}.$$

Example 6.1.7.

$$2x^2 + 4xy + 5y^2 = 2\left(x^2 + 2xy\right) + 5y^2.$$

Looking only at the xy term we observe that $2 = 2n$ so $n = 1$. Then we continue:

$$\begin{aligned} 2x^2 + 2xy + 5y^2 &= 2\left((x+y)^2 - y^2\right) + 5y^2 \\ &= 2\,(x+y)^2 - 2y^2 + 5y^2 \\ &= 2\,(x+y)^2 + 3y^2. \end{aligned}$$

The change of variables $u = x + y$ and $v = y$ diagonalizes the form:

$$Q(u,v) = 2u^2 + 3v^2 = \begin{bmatrix} u & v \end{bmatrix} \begin{bmatrix} 2 & 0 \\ 0 & 3 \end{bmatrix} \begin{bmatrix} u \\ v \end{bmatrix}.$$

Activity 6.1.8. Complete the square for each of the following quadratic forms:

(a) $Q(x,y) = x^2 - 8xy + 25y^2$

(b) $Q(x,y) = 3x^2 + 6xy + 5y^2$

Then do a change of variables to complete the diagonalization.

In Example 6.1.6, the original quadratic form was

$$Q(x,y) = x^2 + 6xy - 7y^2 = \mathbf{x}^T\mathbf{A}\mathbf{x} \quad \text{where} \quad \mathbf{x} = \begin{bmatrix} x \\ y \end{bmatrix} \quad \text{and} \quad \mathbf{A} = \begin{bmatrix} 1 & 3 \\ 3 & -7 \end{bmatrix},$$

and we made the change of variables $u = x + 3y$ and $v = y$. In matrix form this is

$$\mathbf{u} = \begin{bmatrix} u \\ v \end{bmatrix} = \begin{bmatrix} 1 & 3 \\ 0 & 1 \end{bmatrix} \begin{bmatrix} x \\ y \end{bmatrix} \quad \text{or} \quad \mathbf{S}\mathbf{u} = \mathbf{x} \quad \text{where} \quad \mathbf{u} = \begin{bmatrix} u \\ v \end{bmatrix} \quad \text{and} \quad \mathbf{S} = \begin{bmatrix} 1 & -3 \\ 0 & 1 \end{bmatrix},$$

which is possible since $\mathbf{S}$ is invertible. So, by the properties of transpose,

$$Q(x,y) = \mathbf{x}^T\mathbf{A}\mathbf{x} = (\mathbf{S}\mathbf{u})^T\mathbf{A}\mathbf{S}\mathbf{u} = \mathbf{u}^T\mathbf{S}^T\mathbf{A}\mathbf{S}\mathbf{u}.$$

Note that the matrix $\mathbf{S}$ is upper triangular. This remains true if several steps of completing the square is done for larger matrices. For this reason, Examples 6.1.6 and 6.1.7 give a concrete illustration of how to compute a diagonalization with upper triangular matrices (see [9]). For matrices larger than 2×2, inverse matrices are still easy to compute by Subsection 1.3.4.

If $\mathbf{A}$ is positive definite, we can compute the **Cholesky Decomposition** with just one extra step. The matrix

$$\mathbf{D} = \mathbf{S}^T\mathbf{A}\mathbf{S}$$

is diagonal, and even though the elements on the diagonal may not be the eigenvalues they still must be positive. Thus we can take the square root. Then

$$\begin{aligned}\mathbf{A} &= (\mathbf{S}^T)^{-1}\mathbf{D}\mathbf{S}^{-1}\\ &= (\mathbf{S}^{-1})^T\sqrt{\mathbf{D}}\sqrt{\mathbf{D}}\mathbf{S}^{-1} = \mathbf{C}\mathbf{C}^T\end{aligned}$$

where

$$\mathbf{C} = (\mathbf{S}^{-1})^T\sqrt{\mathbf{D}}.$$

Example 6.1.9. In Example 6.1.7, the original quadratic form was given by

$$Q(x,y) = 2x^2 + 4xy + 5y^2 = \mathbf{x}^T\mathbf{A}\mathbf{x} \quad \text{where} \quad \mathbf{x} = \begin{bmatrix} x \\ y \end{bmatrix} \quad \text{and} \quad \mathbf{A} = \begin{bmatrix} 2 & 2 \\ 2 & 5 \end{bmatrix}.$$

We found the diagonal form

$$\mathbf{D} = \begin{bmatrix} 2 & 0 \\ 0 & 3 \end{bmatrix},$$

by the change of variable $u = x + y$ and $v = y$. In matrix form this is

$$\mathbf{u} = \begin{bmatrix} u \\ v \end{bmatrix} = \begin{bmatrix} 1 & 1 \\ 0 & 1 \end{bmatrix}\begin{bmatrix} x \\ y \end{bmatrix} \quad \text{so} \quad \mathbf{S}^{-1} = \begin{bmatrix} 1 & 1 \\ 0 & 1 \end{bmatrix} \quad \text{and} \quad (\mathbf{S}^{-1})^T = \begin{bmatrix} 1 & 0 \\ 1 & 1 \end{bmatrix}.$$

For a diagonal matrix $\mathbf{D}$ we get $\sqrt{\mathbf{D}}$ by taking the square root of the elements on the diagonal:

$$\sqrt{\mathbf{D}} = \begin{bmatrix} \sqrt{2} & 0 \\ 0 & \sqrt{3} \end{bmatrix}.$$

Then

$$\begin{array}{cc} & \begin{bmatrix} \sqrt{2} & 0 \\ 0 & \sqrt{3} \end{bmatrix} \\ \begin{bmatrix} 1 & 0 \\ 1 & 1 \end{bmatrix} & \begin{bmatrix} \sqrt{2} & 0 \\ \sqrt{2} & \sqrt{3} \end{bmatrix} = (\mathbf{S}^{-1})^T\sqrt{\mathbf{D}} = \mathbf{C}, \end{array}$$

and we can verify that this satisfies the Cholesky decomposition $\mathbf{A} = \mathbf{C}\mathbf{C}^T$:

$$\begin{array}{cc} & \begin{bmatrix} \sqrt{2} & \sqrt{2} \\ 0 & \sqrt{3} \end{bmatrix} \\ \begin{bmatrix} \sqrt{2} & 0 \\ \sqrt{2} & \sqrt{3} \end{bmatrix} & \begin{bmatrix} 2 & 2 \\ 2 & 5 \end{bmatrix} \end{array}$$

Activity 6.1.10. Use your results from Activity 6.1.8 part b to compute the Cholesky decomposition of

$$\mathbf{A} = \begin{bmatrix} 3 & 3 \\ 3 & 5 \end{bmatrix}$$

6.1.4 Diagonalization of Forms with Eigenvectors

Q 6.1.5. What is the length of $\left[\begin{smallmatrix}1\\3\end{smallmatrix}\right]$?

In Section 5.3 we diagonalized matrices by $\mathbf{S}^{-1}\mathbf{A}\mathbf{S}$, while we have just seen that a quadratic form can be diagonalized with $\mathbf{S}^T\mathbf{A}\mathbf{S}$. We can get both at the same time if $\mathbf{S}^T = \mathbf{S}^{-1}$, which is to say $\mathbf{S}^T\mathbf{S} = \mathbf{I}$. If the eigenvectors are computed by the QR-algorithm, then this condition is automatic. Between this subsection and Section 6.3, we consider how to obtain $\mathbf{S}$ when it isn't.

Example 6.1.11. For $Q(x, y) = x^2 + 6xy - 7y^2$

$$\det(\mathbf{A} - \lambda\mathbf{I}) = \begin{vmatrix} 1-\lambda & 3 \\ 3 & -7-\lambda \end{vmatrix} = \lambda^2 + 6\lambda - 16 = (\lambda + 8)(\lambda - 2) = 0$$

so the eigenvalues are 2 and -8.

The 2-eigenspace is generated by

$$\mathbf{v}_1 = \begin{bmatrix} 3 \\ 1 \end{bmatrix}$$

and the -8-eigenspace is generated by

$$\mathbf{v}_2 = \begin{bmatrix} -1 \\ 3 \end{bmatrix}.$$

The lengths of $\mathbf{v}_1$ and $\mathbf{v}_2$ are the same,

$$\|\,\mathbf{v}_1\,\|_2 = \sqrt{3^2 + 1^2} = \sqrt{10} \quad \text{and} \quad \|\,\mathbf{v}_2\,\|_2 = \sqrt{(-1)^2 + 3^2} = \sqrt{10},$$

so to normalize them we divide each of them by $\sqrt{10}$:

$$\frac{1}{\sqrt{10}}\begin{bmatrix} 3 \\ 1 \end{bmatrix} \quad \text{and} \quad \frac{1}{\sqrt{10}}\begin{bmatrix} -1 \\ 3 \end{bmatrix}.$$

If we now define

$$\mathbf{S} = \frac{1}{\sqrt{10}}\begin{bmatrix} 3 & -1 \\ 1 & 3 \end{bmatrix} = \begin{bmatrix} \frac{3}{\sqrt{10}} & -\frac{1}{\sqrt{10}} \\ \frac{1}{\sqrt{10}} & \frac{3}{\sqrt{10}} \end{bmatrix},$$

then $\mathbf{S}^{-1} = \mathbf{S}^T$, which can be check as follows:

$$ad - bc = \frac{3}{\sqrt{10}} \cdot \frac{3}{\sqrt{10}} - \frac{1}{\sqrt{10}} \cdot \frac{-1}{\sqrt{10}} = \frac{9+1}{10} = 1,$$

so

$$\mathbf{S}^{-1} = \frac{1}{1}\begin{bmatrix} \frac{3}{\sqrt{10}} & \frac{1}{\sqrt{10}} \\ -\frac{1}{\sqrt{10}} & \frac{3}{\sqrt{10}} \end{bmatrix} = \begin{bmatrix} \frac{3}{\sqrt{10}} & \frac{1}{\sqrt{10}} \\ -\frac{1}{\sqrt{10}} & \frac{3}{\sqrt{10}} \end{bmatrix} = \mathbf{S}^T.$$

Reality Check 6.1.1. If you set up $\mathbf{S}$ correctly, then you should get $\det \mathbf{S} = \pm 1$.

Finally, we can check that the substitution $\mathbf{S}\mathbf{u} = \mathbf{x}$ diagonalizes the form. We have

$$\mathbf{A} = \begin{bmatrix} 1 & 3 \\ 3 & -7 \end{bmatrix} \quad \text{and} \quad \mathbf{S} = \frac{1}{\sqrt{10}}\begin{bmatrix} 3 & -1 \\ 1 & 3 \end{bmatrix}.$$

We must check that $\mathbf{S}^T\mathbf{AS}$ diagonalizes the matrix:

$$\mathbf{AS} = \begin{bmatrix} 1 & 3 \\ 3 & -7 \end{bmatrix} \frac{1}{\sqrt{10}} \begin{bmatrix} 3 & -1 \\ 1 & 3 \end{bmatrix} = \frac{1}{\sqrt{10}} \begin{bmatrix} 6 & 8 \\ 2 & -24 \end{bmatrix}$$

$$\mathbf{S}^T\mathbf{AS} = \frac{1}{\sqrt{10}} \begin{bmatrix} 3 & 1 \\ -1 & 3 \end{bmatrix} \cdot \frac{1}{\sqrt{10}} \begin{bmatrix} 6 & 8 \\ 2 & -24 \end{bmatrix} = \frac{1}{10} \begin{bmatrix} 20 & 0 \\ 0 & -80 \end{bmatrix} = \begin{bmatrix} 2 & 0 \\ 0 & -8 \end{bmatrix}.$$

Hence in the new coordinates we have $Q(u, v) = 2u^2 - 8v^2$.

Activity 6.1.12. The quadratic form

$$Q(x, y) = 2x^2 + 2xy + 2y^2 = \begin{bmatrix} x & y \end{bmatrix} \begin{bmatrix} 2 & 1 \\ 1 & 2 \end{bmatrix} \begin{bmatrix} x \\ y \end{bmatrix}$$

the eigenvalues are 3 and 1. The 3-eigenspace is generated by

$$\mathbf{v}_1 = \begin{bmatrix} 1 \\ 1 \end{bmatrix},$$

and the 1-eigenspace is generated by

$$\mathbf{v}_2 = \begin{bmatrix} 1 \\ -1 \end{bmatrix}.$$

Do all of the following:

1. Normalize the eigenvectors.
2. Build the normalized eigenvectors into the columns of a matrix $\mathbf{S}$
3. Check that $\mathbf{S}^{-1} = \mathbf{S}^T$.
4. Check that $\mathbf{S}^T\mathbf{AS}$ diagonalizes the matrix. Write down the corresponding quadratic form.

6.2 Inner Products and Orthogonality

6.2.1 Intro to Inner Products

Q 6.2.1. What is the dot product of $\mathbf{u} = \left[\begin{smallmatrix} 1 \\ 2 \end{smallmatrix}\right]$ and $\mathbf{v} = \left[\begin{smallmatrix} -2 \\ 1 \end{smallmatrix}\right]$?

Q 6.2.2. What is the length of $\mathbf{u} = \left[\begin{smallmatrix} 1 \\ 2 \end{smallmatrix}\right]$?

The dot product for a real vector space should already be a familiar example of an inner product:

$$\mathbf{u} \cdot \mathbf{v} = \sum_{i=1}^{n} u_i v_i.$$

Here, the sum is finite, but we can generalize to an infinite sum, or even an integral:

$$\langle f, g \rangle = \int_{-1}^{1} f(x)g(x)\,dx.$$

The index i in the sum is over a finite set of values from 1 to n. The variable x plays an analogous role and takes on values in the infinite set $[-1, 1]$. Also $f(x)$ and $g(x)$ are multiplied with the same x-coordinate, just as u_i and v_i are multiplied with the same index. In fact u_i and v_i can be regarded as real valued functions on the set $\{1, 2, \ldots n\}$ just as $f(x)$ and $g(x)$ are real valued functions on $[-1, 1]$.

Definition 6.2.1. If V is real vector space, then an **inner product** V is a function

$$\begin{aligned} V \times V &\to \mathbb{R} \\ (\mathbf{u}, \mathbf{v}) &\mapsto \langle \mathbf{u}, \mathbf{v} \rangle \end{aligned}$$

satisfying the following properties:

1. Linearity in the first component:

$$\begin{aligned} \langle c\mathbf{u}, \mathbf{v} \rangle &= c\langle \mathbf{u}, \mathbf{v} \rangle \\ \langle \mathbf{u} + \mathbf{v}, \mathbf{w} \rangle &= \langle \mathbf{u}, \mathbf{w} \rangle + \langle \mathbf{v}, \mathbf{w} \rangle \end{aligned}$$

   for all $\mathbf{u}, \mathbf{v}, \mathbf{w} \in V$ and for all $c \in \mathbb{R}$.

2. The symmetric property:

$$\langle \mathbf{u}, \mathbf{v} \rangle = \langle \mathbf{v}, \mathbf{u} \rangle$$

   for all $\mathbf{u}, \mathbf{v} \in V$.

3. Positive definiteness:

$$\langle \mathbf{u}, \mathbf{u} \rangle > 0$$

   for all $\mathbf{u} \in V$ such that $\mathbf{u} \neq 0$.

Activity 6.2.1. Given

$$\langle \mathbf{u}, \mathbf{v} \rangle = \mathbf{u}^T \mathbf{A} \mathbf{v} \quad \text{where} \quad \mathbf{A} = \begin{bmatrix} 2 & 0 \\ 0 & 3 \end{bmatrix},$$

let

$$\mathbf{u} = \begin{bmatrix} -1 \\ 1 \end{bmatrix}, \quad \mathbf{v} = \begin{bmatrix} 1 \\ 0 \end{bmatrix}, \quad \text{and} \quad \mathbf{w} = \begin{bmatrix} 1 \\ 1 \end{bmatrix}$$

and $c = 2$. Verify the following properties:

1. $\langle \mathbf{u}, \mathbf{0} \rangle = 0$
2. $\langle \mathbf{u}, \mathbf{u} \rangle > 0$
3. $\langle \mathbf{u}, \mathbf{v} \rangle = \langle \mathbf{v}, \mathbf{u} \rangle$
4. $\langle c\mathbf{u}, \mathbf{v} \rangle = c\langle \mathbf{u}, \mathbf{v} \rangle$
5. $\langle \mathbf{u}, \mathbf{w} \rangle + \langle \mathbf{v}, \mathbf{w} \rangle = \langle \mathbf{u} + \mathbf{v}, \mathbf{w} \rangle$

Notice that given a real inner product $\langle \mathbf{u}, \mathbf{v} \rangle$ on $\mathbb{R}$, we can define a positive definite quadratic form by taking $\mathbf{u} = \mathbf{v}$:

$$Q(\mathbf{u}) = \| \mathbf{u} \|^2 = \langle \mathbf{u}, \mathbf{u} \rangle.$$

It is possible to do the reverse by

$$\| \mathbf{u} - \mathbf{v} \|^2 = \| \mathbf{u} \|^2 + \| \mathbf{v} \|^2 - 2\langle \mathbf{u}, \mathbf{v} \rangle. \tag{6.1}$$

Reflection 6.2.1. In the case where $\langle u, v\rangle$ is the dot product, then equation (6.1) is just the law of cosines. However, for the inner product in Activity 6.2.1, it is not true that

$$\langle \mathbf{u}, \mathbf{v}\rangle = \| \mathbf{u} \| \| \mathbf{v} \| \cos(\theta),$$

regardless of whether the norm is understood to be $\langle \mathbf{u}, \mathbf{u}\rangle$ or to be the 2-norm. For example try computing both sides with

$$\mathbf{u} = \begin{bmatrix} 1 \\ 1 \end{bmatrix} \quad \text{and} \quad \mathbf{v} = \begin{bmatrix} 2 \\ 0 \end{bmatrix},$$

the same vectors as in Activity 1.2.8. You might also want to consider the fact that

$$\langle \mathbf{u}, \mathbf{v}\rangle = \mathbf{u} \cdot (\mathbf{A}\mathbf{v}).$$

With equation (1.15) in hand, we can define an inner product by solving for $\langle u, v\rangle$:[1]

$$\langle \mathbf{u}, \mathbf{v}\rangle = \frac{1}{2}\left(\| \mathbf{u} - \mathbf{v} \|^2 - \| \mathbf{u} \|^2 - \| \mathbf{v} \|^2\right).$$

Proposition 6.2.2 (Cauchy-Schwarz). *Given a vector space V over $\mathbb{R}$ or $\mathbb{C}$, with an inner product $\langle \cdot, \cdot \rangle$, we have*

$$|\langle \mathbf{u}, \mathbf{v}\rangle|^2 \leq \langle \mathbf{u}, \mathbf{u}\rangle \cdot \langle \mathbf{v}, \mathbf{v}\rangle$$

For proof, see [2].

Activity 6.2.2. Given

$$\mathbf{u} = \begin{bmatrix} 1 \\ -1 \end{bmatrix} \quad \text{and} \quad \mathbf{v} = \begin{bmatrix} 1 \\ 2 \end{bmatrix},$$

1. use the inner product defined in Activity 6.2.1 to compute $\langle \mathbf{u}, \mathbf{v}\rangle$, $\langle \mathbf{u}, \mathbf{u}\rangle$, and $\langle \mathbf{v}, \mathbf{v}\rangle$
2. verify the inequality

$$|\langle \mathbf{u}, \mathbf{v}\rangle|^2 \leq \langle \mathbf{u}, \mathbf{u}\rangle\langle \mathbf{v}, \mathbf{v}\rangle.$$

This is the same **u** and **v** as in Activity 1.2.9. Only the inner product has changed, but the Cauchy-Schwarz inequality remains valid.

In statistics, covariance is almost an inner product except that it is only semi-definite.

Definition 6.2.3. Let X and Y be two random variables. The **covariance** of X and Y is

$$\mathrm{Cov}(X, Y) = \mathrm{E}[(X - \mathrm{E}[X])(Y - \mathrm{E}[Y])]$$

If $X = Y$, then this formula reduces to the **variance**

$$\mathrm{Var}(X) = \mathrm{E}[(X - \mathrm{E}[X])^2]$$

the **covariance matrix** is

$$\mathbf{C} = \begin{bmatrix} \mathrm{Var}(X) & \mathrm{Cov}(X, Y) \\ \mathrm{Cov}(X, Y) & \mathrm{Var}(Y) \end{bmatrix}$$

[1]A different identity must be used in the case of complex inner products.

Suppose that we have two samples of X and Y each with size n, and tabulate them in the columns of a data table. Then the data table can be regarded as an $n \times 2$ matrix. We can compute the covariance matrix by the following steps.

Step 1 Compute the vector of means by averaging down the columns.

Step 2 Subtract the vector of means from each row. The result is an $n \times 2$ matrix $\mathbf{A}$ where the means are moved to 0.

Step 3 Compute

$$\mathbf{C} = \mathbf{A}^T\mathbf{A}$$

Example 6.2.3. Suppose that the sample size is $n = 3$ and the data is

$$\begin{bmatrix} 1 & 1 \\ 2 & 1 \\ 3 & 1 \end{bmatrix}$$

The average of the first column is 2, and the average of the second column is 1. If we subtract

$$\begin{bmatrix} 2 & 1 \end{bmatrix}$$

from each row, then we get the matrix

$$\begin{bmatrix} 1 & 1 \\ 2 & 1 \\ 3 & 1 \end{bmatrix} - \begin{bmatrix} 2 & 1 \\ 2 & 1 \\ 2 & 1 \end{bmatrix} = \begin{bmatrix} -1 & 0 \\ 0 & 0 \\ 1 & 0 \end{bmatrix} = \mathbf{A}.$$

Then

$$\begin{array}{cc} & \begin{bmatrix} -1 & 0 \\ 0 & 0 \\ 1 & 0 \end{bmatrix} \\ \begin{bmatrix} -1 & 0 & 1 \\ 0 & 0 & 0 \end{bmatrix} & \begin{bmatrix} 2 & 0 \\ 0 & 0 \end{bmatrix} = \mathbf{A}^T\mathbf{A} = \mathbf{C} \end{array} \tag{6.2}$$

Example 6.2.3 illustrates not only how to compute the covariance matrix, it also suggests why covariance is almost as good as an inner product in most practical situations. The zero on the diagonal of $\mathbf{C}$ in (6.2) means that $\mathrm{Var}(Y)$ is zero. It is computed from the non-zero vector

$$\begin{bmatrix} 1 \\ 1 \\ 1 \end{bmatrix}.$$

So Property 3 in the definition of an inner product fails. But the components of the vector are all the same: the random variable Y is always producing the value of 1 (at least for this sample). While $\mathrm{Var}(Y) = 0$ does not imply that Y is 0, it does imply that Y is constant. So long as the data isn't measuring a constant, or something that is nearly constant, the covariance matrix will be positive-definite.

6.2.2 Orthogonality

Definition 6.2.4. Given an inner product $\langle \mathbf{u}, \mathbf{v} \rangle$, two vectors are called **orthogonal** if

$$\langle \mathbf{u}, \mathbf{v} \rangle = 0.$$

For the dot product, this has the usual meaning (there is a right angle between the vectors).

Definition 6.2.5. Let V be a subspace of $\mathbb{R}^n$. Then the **perpendicular subspace** $V^\perp$ is defined as

$$\{\mathbf{u} \in \mathbb{R}^n : \langle \mathbf{u}, \mathbf{v} \rangle = 0 \quad \text{for all} \quad \mathbf{v} \in V\}.$$

We can compute a basis for $V^\perp$ by the following steps:

1. Take a basis for V and build the vectors into the columns of a matrix.
2. Take the transpose, augment with zero, and bring it into reduced form.
3. The rest is just like computing a basis for the null space.

Example 6.2.4. Let $V = \text{Span}\left\{\begin{bmatrix}1\\2\end{bmatrix}\right\}$. Then $V^\perp$ consists of all $\begin{bmatrix}x\\y\end{bmatrix}$ such that

$$\begin{bmatrix}1 & 2\end{bmatrix}\begin{bmatrix}x\\y\end{bmatrix} = 0.$$

This can be written in augmented form as

$$\left[\begin{array}{cc|c}1 & 2 & 0\end{array}\right].$$

It is already reduced, giving us $x + 2y = 0$. Hence

$$\begin{bmatrix}-2y\\y\end{bmatrix}$$

gives a parameterization of $V^\perp$. This answer should make sense in terms of what we know about perpendicular lines. V is a line with slope 2 through the origin. The slope of any line perpendicular to V is the negative reciprocal, so $-\frac{1}{2}$ in this case. Our computation shows that $V^\perp$ is the line with slope $-\frac{1}{2}$ through the origin.

Example 6.2.5. Let $V = \text{Span}\left\{\begin{bmatrix}1\\0\\0\\1\end{bmatrix}, \begin{bmatrix}1\\1\\0\\0\end{bmatrix}\right\}$. Then $V^\perp$ consists of all solutions to

$$\left[\begin{array}{cccc|c}1 & 0 & 0 & 1 & 0\\1 & 1 & 0 & 0 & 0\end{array}\right] \xrightarrow{(-1)R_1+R_2} \left[\begin{array}{cccc|c}1 & 0 & 0 & 1 & 0\\0 & 1 & 0 & -1 & 0\end{array}\right].$$

So $w + z = 0$, $x - z = 0$, and y and z are free ($y = y$ and $z = z$), yielding the parameterization

$$w = -z, x = z, y = y, z = z$$

of $V^\perp$. We obtain a basis for $V^\perp$ as follows

$$\begin{bmatrix}w\\x\\y\\z\end{bmatrix} = \begin{bmatrix}-z\\z\\y\\z\end{bmatrix} = y\begin{bmatrix}0\\0\\1\\0\end{bmatrix} + z\begin{bmatrix}-1\\1\\0\\1\end{bmatrix}.$$

Activity 6.2.6. Compute $V^\perp$ for each V given below, using the standard dot product.

1.

$$V = \text{Span}\left\{\begin{bmatrix}1\\0\\0\end{bmatrix}\right\}$$

2.

$$V = \text{Span}\left\{ \begin{bmatrix} 1 \\ 0 \\ -1 \end{bmatrix}, \begin{bmatrix} -1 \\ 1 \\ 0 \end{bmatrix} \right\}$$

Theorem 6.2.6. *Let* **A** *be an* $m \times n$ *matrix. Then*

$$(\text{Row}(\mathbf{A}))^{\perp} = \text{Nul}(\mathbf{A})$$

and

$$(\text{Col}(\mathbf{A}))^{\perp} = \text{Nul}(\mathbf{A}^T).$$

It is easiest to check on basis vectors obtained from reduced form.

Example 6.2.7. Verify that Theorem 6.2.6 is valid for

$$\begin{bmatrix} 1 & 0 & 0 & 0 & -2 \\ 0 & 0 & 1 & 0 & 0 \\ 0 & 0 & 0 & 1 & 0 \\ 0 & 0 & 0 & 0 & 0 \end{bmatrix}.$$

Solution

First, we show that $(\text{Row}(\mathbf{A}))^{\perp} = \text{Nul}(\mathbf{A})$. In order to do this, we need to know what the null space of **A** is. We use the same method as Example 4.3.4. Let x_1 through x_5 be variables for the domain. Since **A** is already in reduced form, we get equations immediately after augmenting with 0:

$$\left[\begin{array}{ccccc|c} 1 & 0 & 0 & 0 & -2 & 0 \\ 0 & 0 & 1 & 0 & 0 & 0 \\ 0 & 0 & 0 & 1 & 0 & 0 \\ 0 & 0 & 0 & 0 & 0 & 0 \end{array}\right] \longrightarrow \begin{array}{r} x_1 - 2x_5 = 0 \\ x_3 = 0 \\ x_4 = 0 \end{array} \quad \begin{array}{l} \longrightarrow \quad x_1 = 2x_5 \\ \\ \\ \end{array}$$

and x_2 and x_5 are free, so we need $x_2 = x_2$ and $x_5 = x_5$. Hence

$$\begin{bmatrix} 2x_5 \\ x_2 \\ 0 \\ 0 \\ x_5 \end{bmatrix} = x_2 \begin{bmatrix} 0 \\ 1 \\ 0 \\ 0 \\ 0 \end{bmatrix} + x_5 \begin{bmatrix} 2 \\ 0 \\ 0 \\ 0 \\ 1 \end{bmatrix} = \begin{bmatrix} 0 & 2 \\ 1 & 0 \\ 0 & 0 \\ 0 & 0 \\ 0 & 1 \end{bmatrix} \begin{bmatrix} x_2 \\ x_5 \end{bmatrix}$$

gives a parameterization of $\text{Nul}(\mathbf{A})$. The row space of **A** is the span of the rows. Thus we can check orthogonality as follows:

$$\begin{array}{cc} & \begin{bmatrix} 0 & 2 \\ 1 & 0 \\ 0 & 0 \\ 0 & 0 \\ 0 & 1 \end{bmatrix} \\ \begin{bmatrix} 1 & 0 & 0 & 0 & -2 \\ 0 & 0 & 1 & 0 & 0 \\ 0 & 0 & 0 & 1 & 0 \\ 0 & 0 & 0 & 0 & 0 \end{bmatrix} & \begin{bmatrix} 0 & 0 \\ 0 & 0 \\ 0 & 0 \\ 0 & 0 \end{bmatrix}. \end{array}$$

Checking $(\text{Col}(\mathbf{A}))^\perp = \text{Nul}(\mathbf{A}^T)$ is similar, except that to get a basis for $\text{Nul}(\mathbf{A}^T)$, we first must reduce $[\mathbf{A}^T|\mathbf{0}]$:

$$\mathbf{A}^T = \left[\begin{array}{cccc|c} 1 & 0 & 0 & 0 & 0 \\ 0 & 0 & 0 & 0 & 0 \\ 0 & 1 & 0 & 0 & 0 \\ 0 & 0 & 1 & 0 & 0 \\ -2 & 0 & 0 & 0 & 0 \end{array}\right] \xrightarrow{2R_1+R_5} \left[\begin{array}{cccc|c} 1 & 0 & 0 & 0 & 0 \\ 0 & 0 & 0 & 0 & 0 \\ 0 & 1 & 0 & 0 & 0 \\ 0 & 0 & 1 & 0 & 0 \\ 0 & 0 & 0 & 0 & 0 \end{array}\right]$$

Let y_1 to y_4 be variables for the domain of $\mathbf{A}^T$, which is the same as the co-domain of $\mathbf{A}$. Then y_4 is free, so

$$\begin{aligned} y_1 &= 0 \\ y_2 &= 0 \\ y_3 &= 0 \\ y_4 &= y_4 \end{aligned} \longrightarrow \begin{bmatrix} 0 \\ 0 \\ 0 \\ y_4 \end{bmatrix} = y_4 \begin{bmatrix} 0 \\ 0 \\ 0 \\ 1 \end{bmatrix}$$

gives a parameterization of $\text{Nul}(\mathbf{A}^T)$. Orthogonality can be checked the same way as before:

$$\begin{array}{c} \\ \begin{bmatrix} 1 & 0 & 0 & 0 \\ 0 & 0 & 0 & 0 \\ 0 & 1 & 0 & 0 \\ 0 & 0 & 1 & 0 \\ -2 & 0 & 0 & 0 \end{bmatrix} \end{array} \begin{array}{c} \begin{bmatrix} 0 \\ 0 \\ 0 \\ 1 \end{bmatrix} \\ \begin{bmatrix} 0 \\ 0 \\ 0 \\ 0 \\ 0 \end{bmatrix} \end{array}.$$

Activity 6.2.8. Verify the previous theorem for:

$$\mathbf{A} = \begin{bmatrix} 1 & 0 & 0 & 0 & 2 & 0 \\ 0 & 1 & 1 & 0 & 0 & 0 \\ 0 & 0 & 0 & 1 & 0 & -3 \\ 0 & 0 & 0 & 0 & 0 & 0 \\ 0 & 0 & 0 & 0 & 0 & 0 \end{bmatrix}.$$

6.2.3 Invertible Matrix Theorem (Part 6)

Theorem 6.2.7 (Invertible Matrix Theorem: Part 6)**.** *Let* $\mathbf{A}$ *be an* $n \times n$ *matrix with real entries. The following statements are equivalent.*

18. $\dim(\text{Col}(\mathbf{A}))^\perp = 0$

19. $\dim(\text{Row}(\mathbf{A}))^\perp = 0$

20. $\dim(\text{Nul}(\mathbf{A}))^\perp = n$

See Remark 1.3.1 concerning the Invertible matrix theorem and its proof.

Example 6.2.9. Show that statements 18 and 20 in the theorem are true for the matrix

$$\mathbf{A} = \begin{bmatrix} 1 & 1 \\ 0 & 1 \end{bmatrix}.$$

18. $\dim(\mathrm{Col}(\mathbf{A}))^{\perp} = 0$. To compute the perpendicular subspace to the column space of $\mathbf{A}$, we must compute the null space of the transpose:

$$\left[\begin{array}{cc|c} 1 & 0 & 0 \\ 1 & 1 & 0 \end{array}\right] \xrightarrow{(-1)R_1+R_2} \left[\begin{array}{cc|c} \boxed{1} & 0 & 0 \\ 0 & \boxed{1} & 0 \end{array}\right].$$

There are no free variables, so $\dim(\mathrm{Col}(\mathbf{A}))^{\perp} = 0$.

20. $\dim(\mathrm{Nul}(\mathbf{A}))^{\perp} = n$. Here $n = 2$. We showed already that $\dim \mathrm{Nul}(\mathbf{A}) = 0$. This means that the zero vector is the only element in the null space. What is orthogonal to the zero vector? Take the transpose, and augment with zero:

$$\left[\begin{array}{cc|c} 0 & 0 & 0 \end{array}\right].$$

There are no pivots, so both columns to the left of the vertical line correspond to free variables: the dimension is 2.

Activity 6.2.10. Using

$$\mathbf{A} = \begin{bmatrix} 1 & 1 \\ 0 & 1 \end{bmatrix},$$

show:

19. $\dim(\mathrm{Row}(\mathbf{A}))^{\perp} = 0$

Hint: taking the transpose switches rows and columns.

6.3 Gram-Schmidt

The vectors

$$\begin{bmatrix} 1 \\ 1 \\ 1 \end{bmatrix}, \begin{bmatrix} 1 \\ -1 \\ 1 \end{bmatrix}, \begin{bmatrix} 1 \\ 1 \\ -1 \end{bmatrix}$$

give a basis for $\mathbb{R}^3$, but they are not orthogonal. For instance

$$\begin{bmatrix} 1 & 1 & 1 \end{bmatrix} \begin{bmatrix} 1 \\ -1 \\ 1 \end{bmatrix} = 1 \cdot 1 + (-1)(1) + 1 \cdot 1 = 1.$$

We would like to obtain a set of orthogonal vectors from these that is still a basis for $\mathbb{R}^3$. The Gram-Schmidt process accomplishes this goal with the help of vector projection (Proposition 1.2.7).

Example 6.3.1. Step 1: Start with one vector

$$\mathbf{v}_1 = \begin{bmatrix} 1 \\ 1 \\ 1 \end{bmatrix}.$$

Step 2: Take another vector, project onto $\mathbf{v}_1$ and subtract:

$$\begin{bmatrix} 1 & -1 & 1 \end{bmatrix} \begin{bmatrix} 1 \\ 1 \\ 1 \end{bmatrix} = 1 - 1 + 1 = 1 \quad \text{and} \quad \| \mathbf{v}_1 \|^2 = \begin{bmatrix} 1 & 1 & 1 \end{bmatrix} \begin{bmatrix} 1 \\ 1 \\ 1 \end{bmatrix} = 1 + 1 + 1 = 3$$

so then

$$\mathbf{v}_2 = \begin{bmatrix} 1 \\ -1 \\ 1 \end{bmatrix} - \frac{1}{3} \begin{bmatrix} 1 \\ 1 \\ 1 \end{bmatrix} = \begin{bmatrix} \frac{2}{3} \\ -\frac{4}{3} \\ \frac{2}{3} \end{bmatrix}.$$

The vector $\mathbf{v}_2$ is orthogonal to $\mathbf{v}_2$.

Step 3: Take the third vector, project onto both $\mathbf{v}_1$ and $\mathbf{v}_2$, and subtract:

$$\begin{bmatrix} 1 & 1 & -1 \end{bmatrix} \begin{bmatrix} 1 \\ 1 \\ 1 \end{bmatrix} = 1 + 1 - 1 = 1 \quad \text{and} \quad \| \mathbf{v}_1 \|^2 = 3 \quad \text{as before}$$

$$\begin{bmatrix} 1 & 1 & -1 \end{bmatrix} \begin{bmatrix} \frac{2}{3} \\ -\frac{4}{3} \\ \frac{2}{3} \end{bmatrix} = \frac{2}{3} - \frac{4}{3} - \frac{2}{3} = -\frac{4}{3} \quad \text{and}$$

$$\| \mathbf{v}_2 \|^2 = \begin{bmatrix} \frac{2}{3} & -\frac{4}{3} & \frac{2}{3} \end{bmatrix} \begin{bmatrix} \frac{2}{3} \\ -\frac{4}{3} \\ \frac{2}{3} \end{bmatrix} = \frac{4}{9} + \frac{16}{9} + \frac{4}{9} = \frac{8}{3}$$

so then

$$\mathbf{v}_3 = \begin{bmatrix} 1 \\ 1 \\ -1 \end{bmatrix} - \frac{1}{3} \begin{bmatrix} 1 \\ 1 \\ 1 \end{bmatrix} - \frac{-\frac{4}{3}}{\frac{8}{3}} \begin{bmatrix} \frac{2}{3} \\ -\frac{4}{3} \\ \frac{2}{3} \end{bmatrix} = \begin{bmatrix} \frac{2}{3} \\ \frac{2}{3} \\ -\frac{4}{3} \end{bmatrix} + \begin{bmatrix} \frac{1}{3} \\ -\frac{2}{3} \\ \frac{1}{3} \end{bmatrix} = \begin{bmatrix} 1 \\ 0 \\ -1 \end{bmatrix}.$$

It follows that

$$\begin{bmatrix} 1 \\ 1 \\ 1 \end{bmatrix}, \begin{bmatrix} \frac{2}{3} \\ -\frac{4}{3} \\ \frac{2}{3} \end{bmatrix}, \begin{bmatrix} 1 \\ 0 \\ -1 \end{bmatrix}$$

gives an orthogonal basis for $\mathbb{R}^3$.

Remark 6.3.1. The original set of vectors does not need to be linearly independent. If you apply Gram-Schmidt to a set of vectors that is not linearly independent, then some vectors will be removed in the course of the algorithm by getting the zero vector. To see this, try applying Gram-Schmidt to the vectors

$$\begin{bmatrix} 1 \\ -1 \\ 0 \end{bmatrix}, \begin{bmatrix} -1 \\ 0 \\ 1 \end{bmatrix}, \begin{bmatrix} 0 \\ 1 \\ -1 \end{bmatrix}.$$

Activity 6.3.2. Given the vectors

$$\begin{bmatrix} -1 \\ 2 \\ 2 \end{bmatrix}, \begin{bmatrix} 1 \\ 1 \\ 4 \end{bmatrix}, \begin{bmatrix} 5 \\ -1 \\ -1 \end{bmatrix}$$

1. Apply Gram-Schmidt to obtain an orthogonal basis (*going in order is recommended*).
2. Check orthogonality.

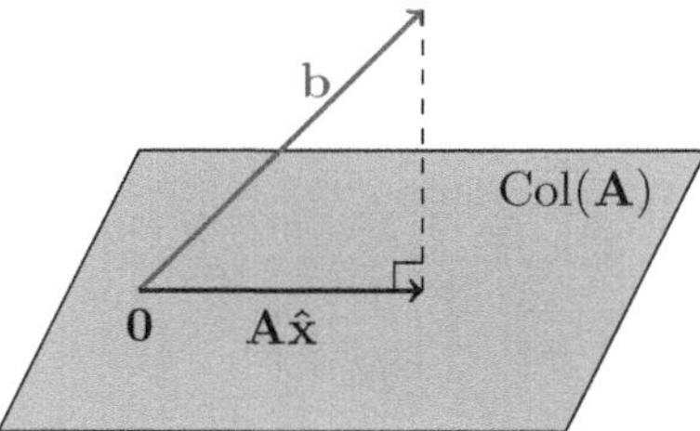

FIGURE 6.1: Projection onto Col(**A**) (the range)

6.4 Least-Squares

6.4.1 Introduction to Least Squares

Given a matrix equation

$$\mathbf{Ax} = \mathbf{b}$$

an exact solution may not exist. This means that **b** does not lie in the range of the linear transformation defined by **A**. The goal is to construct an approximate solution $\hat{\mathbf{x}}$, such that

$$\| \mathbf{A}\hat{\mathbf{x}} - \mathbf{b} \|_2$$

is minimized.

As Figure 6.1 shows, $\mathbf{b} - \mathbf{A}\hat{\mathbf{x}}$ is orthogonal to Col(**A**), which means that:

$$\begin{aligned}
\mathbf{A}^T(\mathbf{b} - \mathbf{A}\hat{\mathbf{x}}) &= \mathbf{0} \\
\mathbf{A}^T\mathbf{b} - \mathbf{A}^T\mathbf{A}\hat{\mathbf{x}} &= \mathbf{0} \\
\mathbf{A}^T\mathbf{A}\hat{\mathbf{x}} &= \mathbf{A}^T\mathbf{b}.
\end{aligned} \tag{6.3}$$

We can now solve for $\hat{\mathbf{x}}$ in the last equation by row reducing

$$\left[\mathbf{A^T A} \,\middle|\, \mathbf{A}^T\mathbf{b}\right].$$

There are some important facts to keep in mind:

1. If $\mathbf{Ax} = \mathbf{b}$ has an exact solution, then the least squares approach yields the same solution.
2. A least squares solution always exists.
3. A least squares solution is not always unique.
4. $\mathbf{A}^T\mathbf{A}$ is a symmetric matrix.
5. A least squares solution is unique if and only if $\mathbf{A}^T\mathbf{A}$ is invertible.
6. $\mathbf{A}^T\mathbf{A}$ is not invertible if $\dim \mathrm{Nul}(\mathbf{A}) > 0$ (e.g. if **A** has more columns than rows).
7. If **A** has more rows than columns, then $\mathbf{A}^T\mathbf{A}$ is *invertible with probability 1.*

The phrase "invertible with probability 1" is subtle. What does it mean and why is it true? We know $\mathbf{A}^T\mathbf{A}$ is symmetric, so consider the 2×2 case:

$$\mathbf{A}^T\mathbf{A} = \begin{bmatrix} x & y \\ y & z \end{bmatrix}.$$

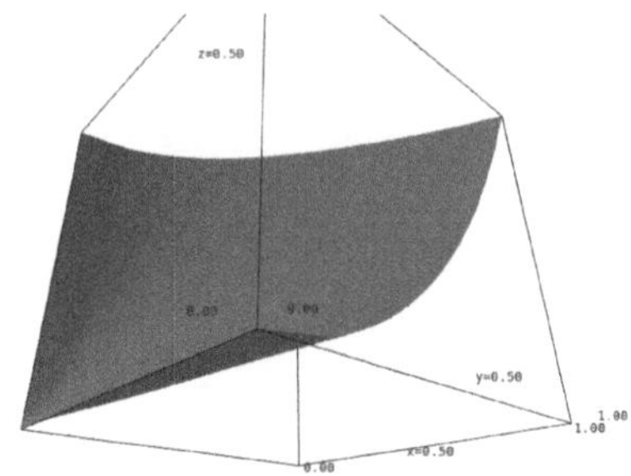

FIGURE 6.2: The surface $xz - y^2 = 0$

If $\mathbf{A}^T\mathbf{A}$ is not invertible, then $\det(\mathbf{A}^T\mathbf{A}) = 0$. By the 2×2 formula for determinants, that gives us the equation

$$xz - y^2 = 0,$$

which defines a surface in 3 dimensions as shown in Figure 6.2. Now imagine selecting (x, y, z) uniformly at random in the box $[0, 1]^3$, which has volume 1. The probability of (x, y, z) landing in a smaller volume inside the box is equal to that volume. For example, the box $[0, 0.5] \times [0, 0.2] \times [0, 0.3]$ has volume

$$0.5 \times 0.2 \times 0.3 = 0.03$$

so there is only a 3% chance of (x, y, z) landing in that box. But a surface has a volume of zero, and so the probability of (x, y, z) landing on the surface shown in Figure 6.2 is zero, even though we can write down examples of (x, y, z) on the surface. The point is, a least squares solution is unique in most real-life situations.

Example 6.4.1. Find a least squares solution $\hat{\mathbf{x}}$ to $\mathbf{Ax} = \mathbf{b}$, where $\mathbf{A}$ and $\mathbf{b}$ are given below:

$$\mathbf{A} = \begin{bmatrix} 1 & 0 \\ 0 & 1 \\ 0 & 0 \end{bmatrix} \quad \text{and} \quad \mathbf{b} = \begin{bmatrix} 1 \\ 2 \\ 3 \end{bmatrix}.$$

Solution

Step 1: Compute $\mathbf{A}^T\mathbf{A}$ and $\mathbf{A}^T\mathbf{b}$:

$$\begin{bmatrix} 1 & 0 & 0 \\ 0 & 1 & 0 \end{bmatrix} \begin{bmatrix} 1 & 0 \\ 0 & 1 \\ 0 & 0 \end{bmatrix} = \begin{bmatrix} 1 & 0 \\ 0 & 1 \end{bmatrix} = \mathbf{A}^T\mathbf{A} \quad \text{and} \quad \begin{bmatrix} 1 & 0 & 0 \\ 0 & 1 & 0 \end{bmatrix} \begin{bmatrix} 1 \\ 2 \\ 3 \end{bmatrix} = \begin{bmatrix} 1 \\ 2 \end{bmatrix} = \mathbf{A}^T\mathbf{b}.$$

Step 2: Row reduce $[\mathbf{A}^T\mathbf{A}|\mathbf{A}^T\mathbf{b}]$:

$$\left[\begin{array}{cc|c} 1 & 0 & 1 \\ 0 & 1 & 2 \end{array}\right]$$

is already reduced. The (unique) solution is

$$\hat{\mathbf{x}} = \begin{bmatrix} 1 \\ 2 \end{bmatrix}.$$

Example 6.4.2. Find a least squares solution $\hat{\mathbf{x}}$ to $\mathbf{Ax} = \mathbf{b}$, where $\mathbf{A}$ and $\mathbf{b}$ are given below:

$$\mathbf{A} = \begin{bmatrix} 1 & 1 \\ 1 & 1 \\ 1 & 1 \end{bmatrix} \quad \text{and} \quad \mathbf{b} = \begin{bmatrix} 1 \\ 2 \\ 3 \end{bmatrix}.$$

Solution

Step 1: Compute $\mathbf{A}^T\mathbf{A}$ and $\mathbf{A}^T\mathbf{b}$:

$$\begin{bmatrix} 1 & 1 & 1 \\ 1 & 1 & 1 \end{bmatrix} \begin{bmatrix} 1 & 1 \\ 1 & 1 \\ 1 & 1 \end{bmatrix} = \begin{bmatrix} 3 & 3 \\ 3 & 3 \end{bmatrix} = \mathbf{A}^T\mathbf{A} \quad \text{and} \quad \begin{bmatrix} 1 & 1 & 1 \\ 1 & 1 & 1 \end{bmatrix} \begin{bmatrix} 1 \\ 2 \\ 3 \end{bmatrix} = \begin{bmatrix} 6 \\ 6 \end{bmatrix} = \mathbf{A}^T\mathbf{b}.$$

Step 2: Row reduce $[\mathbf{A}^T\mathbf{A}|\mathbf{A}^T\mathbf{b}]$:

$$\left[\begin{array}{cc|c} 3 & 3 & 6 \\ 3 & 3 & 6 \end{array}\right] \xrightarrow{(-1)R_1+R_2} \left[\begin{array}{cc|c} 3 & 3 & 6 \\ 0 & 0 & 0 \end{array}\right] \xrightarrow{\frac{1}{3}R_1} \left[\begin{array}{cc|c} 1 & 1 & 2 \\ 0 & 0 & 0 \end{array}\right].$$

The solution is not unique (there is one free variable). We get $x + y = 2$ and $y = y$, so the general solution is

$$\hat{\mathbf{x}} = \begin{bmatrix} 2-y \\ y \end{bmatrix} = \begin{bmatrix} 2 \\ 0 \end{bmatrix} + y \begin{bmatrix} -1 \\ 1 \end{bmatrix}.$$

Example 6.4.3. Find a least squares solution $\hat{\mathbf{x}}$ to $\mathbf{Ax} = \mathbf{b}$, where $\mathbf{A}$ and $\mathbf{b}$ are given below:

$$\mathbf{A} = \begin{bmatrix} 1 & 0 & 0 \\ 0 & 1 & 0 \end{bmatrix} \quad \text{and} \quad \mathbf{b} = \begin{bmatrix} 1 \\ 2 \end{bmatrix}.$$

Solution

First observe:

1. There is already an exact solution:

$$\mathbf{x} = \begin{bmatrix} 1 \\ 2 \\ z \end{bmatrix}.$$

2. There are fewer rows than columns.

Even though $\mathbf{A}$ has a pivot in every row, $\mathbf{A}^T\mathbf{A}$ has no hope of being invertible.

Step 1: Compute $\mathbf{A}^T\mathbf{A}$ and $\mathbf{A}^T\mathbf{b}$:

$$\begin{bmatrix} 1 & 0 \\ 0 & 1 \\ 0 & 0 \end{bmatrix} \begin{bmatrix} 1 & 0 & 0 \\ 0 & 1 & 0 \end{bmatrix} = \begin{bmatrix} 1 & 0 & 0 \\ 0 & 1 & 0 \\ 0 & 0 & 0 \end{bmatrix} = \mathbf{A}^T\mathbf{A} \quad \text{and} \quad \begin{bmatrix} 1 & 0 \\ 0 & 1 \\ 0 & 0 \end{bmatrix} \begin{bmatrix} 1 \\ 2 \end{bmatrix} = \begin{bmatrix} 1 \\ 2 \\ 0 \end{bmatrix} = \mathbf{A}^T\mathbf{b}.$$

Step 2: Row reduce $[\mathbf{A}^T\mathbf{A}|\mathbf{A}^T\mathbf{b}]$:

$$\left[\begin{array}{ccc|c} 1 & 0 & 0 & 1 \\ 0 & 1 & 0 & 2 \\ 0 & 0 & 0 & 0 \end{array}\right]$$

is already reduced. The solution is the same as the exact solution above.

Activity 6.4.4. Find all least squares solutions to $\mathbf{Ax} = \mathbf{b}$ in each case given below:

1.

$$\mathbf{A} = \begin{bmatrix} 1 & 0 \\ 0 & 1 \\ 1 & 1 \end{bmatrix} \quad \text{and} \quad \mathbf{b} = \begin{bmatrix} 3 \\ 1 \\ 1 \end{bmatrix}$$

2.

$$\mathbf{A} = \begin{bmatrix} 1 & 0 & 1 \\ 0 & 1 & 1 \end{bmatrix} \quad \text{and} \quad \mathbf{b} = \begin{bmatrix} 2 \\ 2 \end{bmatrix}$$

Activity 6.4.5. (with Sage) It is natural to look at the least squares problem in terms of calculus. The smallest meaningful example is when $\mathbf{A}$ is 2×1:

$$\mathbf{x} = \begin{bmatrix} x \end{bmatrix}, \quad \mathbf{A} = \begin{bmatrix} a_{11} \\ a_{21} \end{bmatrix}, \quad \text{and} \mathbf{b} = \begin{bmatrix} b_1 \\ b_2 \end{bmatrix}$$

To define $\mathbf{A}$ and $\mathbf{b}$ in Sage, we first must define the variables in them:

```
var('a11,a21,b1,b2')
```

Then $\mathbf{A}$, $\mathbf{b}$ and $\mathbf{x}$ can be defined in sage as follows

```
A = matrix([[a11],[a21]])
b = vector([b1,b2])
X = vector([x])
```

Note that I have used a capital X for the vector in sage to avoid overwriting lowercase x as a variable. We want to minimize the 2-norm of $\mathbf{Ax} - \mathbf{b}$, which by hand is

$$f(x) = \| \mathbf{Ax} - \mathbf{b} \|_2 = \sqrt{(a_{11}x - b_1)^2 + (a_{21}x - b_2)^2}. \tag{6.4}$$

In Sage, the function $f(x)$ can be defined by

```
f = (AX-b).norm()
```

To minimize f, we must compute the first derivative, $f'(x)$. The derivative computation involves chain rule. In Sage, you can compute $f'(x)$ by running the following code:

```
f.diff(x).factor()
```

It should match the result by hand, which is

$$f'(x) = \frac{(a_{11}^2 + a_{21}^2)x - (a_{11}b_1 + a_{21}b_2)}{\sqrt{(a_{11}x - b_1)^2 + (a_{21}x - b_2)^2}}. \tag{6.5}$$

You should observe that the denominator is $f(x)$, which is $\| AX - b \|_2$. To minimize $f(x)$, we must solve $f'(x) = 0$. Since $f'(x)$ can only be zero when the numerator is zero, we get the equation

$$(a_{11}^2 + a_{21}^2)x - (a_{11}b_1 + a_{21}b_2) = 0.$$

But this equation is precisely $(\mathbf{A}^T\mathbf{A})\mathbf{x} - \mathbf{A}^T\mathbf{b} = \mathbf{0}$, which you can check by running the following Sage code:

```
((A.T*A)*X-A.T*b).factor()
```

Of course, to know that it is a minimum, we should check that $f''(x)$ is positive. To get the second derivative in Sage, you can run the following code:

```
f.diff(x,2).factor()
```

Now the denominator is $\| AX - b \|_2^3$ and the numerator is a square, so it is never negative although it is possible for $f''(x) = 0$, which would make the second derivative test inconclusive.

We can still use single variable calculus by increasing the number of rows of $\mathbf{A}$, but if the number of columns is increased, then equation (6.4) generalizes to

$$f(x_1, x_2 \ldots x_n) = \| \mathbf{Ax} - \mathbf{b} \|_2 . \tag{6.6}$$

In multi-variable calculus the first derivative generalizes to the gradient, so (6.5) generalizes to

$$\nabla f(x_1, x_2 \ldots x_n) = \frac{1}{\| \mathbf{Ax} - \mathbf{b} \|_2}(\mathbf{A}^T\mathbf{Ax} - \mathbf{A}^T\mathbf{b}). \tag{6.7}$$

Then setting $\nabla f(x_1, x_2 \ldots x_n)$ equal to the zero vector and clearing the denominator leads to

$$\mathbf{A}^T\mathbf{Ax} - \mathbf{A}^T\mathbf{b} = 0,$$

which agrees with (6.3).

6.4.2 Regression

Suppose we have a data set with N pairs (x_i, y_i) that we want to fit with a line

$$y = mx + b. \tag{6.8}$$

How do we get m and b? We plug in. Plugging in gives us a system of equations

$$\begin{aligned} y_1 &= mx_1 + b \\ y_2 &= mx_2 + b \\ &\vdots \\ y_N &= mx_N + b \end{aligned}$$

with variables m and b. This system can be written as a matrix equation, $\mathbf{Ax} = \mathbf{b}$, with

$$\mathbf{x} = \begin{bmatrix} m \\ b \end{bmatrix}, \quad \mathbf{A} = \begin{bmatrix} x_1 & 1 \\ x_2 & 1 \\ \vdots & \vdots \\ x_N & 1 \end{bmatrix}, \quad \text{and} \quad \mathbf{b} = \begin{bmatrix} y_1 \\ y_2 \\ \vdots \\ y_N \end{bmatrix}.$$

To obtain the least squares solution, we need $\mathbf{A}^T\mathbf{A}$:

$$\begin{array}{cc} & \begin{bmatrix} x_1 & 1 \\ x_2 & 1 \\ \vdots & \vdots \\ x_N & 1 \end{bmatrix} \\ \begin{bmatrix} x_1 & x_2 & \cdots & x_N \\ 1 & 1 & \cdots & 1 \end{bmatrix} & \begin{bmatrix} \sum_{i=1}^N x_i^2 & \sum_{i=1}^N x_i \\ \sum_{i=1}^N x_i & N \end{bmatrix} = N \begin{bmatrix} E[X^2] & E[X] \\ E[X] & 1 \end{bmatrix}. \end{array} \tag{6.9}$$

where

$$E[X] = \frac{1}{N}\sum_{i=1}^{N} x_i \quad \text{and} \quad E[X^2] = \frac{1}{N}\sum_{i=1}^{N} x_i^2;$$

in particular $E[X]$ is just the average or mean of the x coordinates. The least squares solution will be unique if the determinant of $\mathbf{A}^T\mathbf{A}$. By the 2×2 formula, we have

$$\det(\mathbf{A}^T\mathbf{A}) = N^2(E[X^2] - E[X]^2) = N^2\sigma^2$$

where σ is the standard deviation of the x-coordinates. So, we will have a unique least squares solution if and only if the standard deviation isn't zero, which is typical. We also need $\mathbf{A}^T\mathbf{b}$:

$$\begin{bmatrix} x_1 & x_2 & \cdots & x_N \\ 1 & 1 & \cdots & 1 \end{bmatrix} \begin{bmatrix} y_1 \\ y_2 \\ \vdots \\ y_N \end{bmatrix} \begin{bmatrix} \sum_{i=1}^N x_i y_i \\ \sum_{i=1}^N y_i \end{bmatrix} = N \begin{bmatrix} E[XY] \\ E[Y] \end{bmatrix}. \tag{6.10}$$

where

$$E[Y] = \frac{1}{N}\sum_{i=1}^{N} y_i \quad \text{and} \quad E[XY] = \frac{1}{N}\sum_{i=1}^{N} x_i y_i.$$

It follows from equations (6.9) and (6.10) that $[\mathbf{A}^T\mathbf{A}|\mathbf{A}^T\mathbf{b}]$ is

$$N\left[\begin{array}{cc|c} E[X^2] & E[X] & E[XY] \\ E[X] & 1 & E[Y] \end{array}\right], \tag{6.11}$$

and the factor of N in front can be removed, since it is multiplying both sides of a matrix equation. A solution to (6.11) will give us m and b.

This method generalizes to fitting data with any equation with unknown coefficients. For example,

$$y = ax^2 + bx + c \tag{6.12}$$

is not linear, but we still get a system of linear equations, with variables (a, b, c) when we plug in with the pairs (x_i, y_i). Then by doing the same steps as before, it can be shown that

$$\mathbf{A}^T\mathbf{A} = N\begin{bmatrix} E[X^4] & E[X^3] & E[X^2] \\ E[X^3] & E[X^2] & E[X] \\ E[X^2] & E[X] & 1 \end{bmatrix} \quad \text{and} \quad \mathbf{A}^T\mathbf{b} = N\begin{bmatrix} E[X^2Y] \\ E[XY] \\ E[Y] \end{bmatrix}. \tag{6.13}$$

You might also have a data set with more coordinates than two coordinates. Say, for example, that you have N triples (x_i, y_i, z_i) that you want to fit with a plane:

$$z = m_1 x + m_2 y + b. \tag{6.14}$$

Then plugging in with (x_i, y_i, z_i) gives us a system of linear equations in m_1, m_2, b. Then the same steps as before lead to

$$\mathbf{A}^T\mathbf{A} = N\begin{bmatrix} E[X^2] & E[XY] & E[X] \\ E[XY] & E[Y^2] & E[Y] \\ E[X] & E[Y] & 1 \end{bmatrix} \quad \text{and} \quad \mathbf{A}^T\mathbf{b} = N\begin{bmatrix} E[XZ] \\ E[YZ] \\ E[Z] \end{bmatrix}. \tag{6.15}$$

6.5 The Singular Value Decomposition

6.5.1 Introduction to the SVD

Let $\mathbf{A}$ be an $m \times n$ matrix with complex entries. Then both

$$\mathbf{A}^*\mathbf{A} \quad \text{and} \quad \mathbf{A}\mathbf{A}^*$$

are Hermitian. Furthermore, in both cases, their eigenvalues are non-negative. Let

$$\sigma_1 \geq \sigma_2 \geq \cdots \geq \sigma_n \geq 0$$

denote the eigenvalues of $\sqrt{\mathbf{A}^*\mathbf{A}}$ (Explained in Section 5.5) or equivalently the positive square root of the eigenvalues of $\mathbf{A}^*\mathbf{A}$. The numbers σ_i are called the **singular values** of $\mathbf{A}$.

Let r be the largest index such that $\sigma_r > 0$. Then a **Singular Value Decomposition** (SVD) of an $n \times n$ matrix $\mathbf{A}$ has the form

$$\mathbf{A} = \mathbf{U}\boldsymbol{\Sigma}\mathbf{V}^* \tag{6.16}$$

where

$$\mathbf{U}^*\mathbf{U} = \mathbf{I}, \quad \mathbf{V}^*\mathbf{V} = \mathbf{I}, \tag{6.17}$$

and

$$\boldsymbol{\Sigma} = \begin{bmatrix} \sigma_1 & & \\ & \ddots & \\ & & \sigma_r \\ & & \end{bmatrix}, \tag{6.18}$$

is padded with zeros to the right and below if needed to match the dimension of $\mathbf{A}$.

Remark 6.5.1. The matrices $\mathbf{U}$ and $\mathbf{V}$ may not be unique. Example 6.5.1 and Activity 6.5.2 show how this can happen when the singular values are distinct.

Warning 6.5.2. If $\mathbf{A}$ is a square matrix, then it has eigenvalues. While a singular value of $\mathbf{A}$ is an eigenvalue of $\mathbf{D}$, it is possible that it might not be an eigenvalue of $\mathbf{A}$. See Example 6.5.10.

The singular value decomposition is useful for many reasons including:

1. solving a system of equations (Subsection 6.5.2),
2. computing generalized inverses of matrices (Subsection 6.5.3),
3. computing the rank of a matrix, (Subsection 6.5.4), and
4. approximating a matrix by a matrix with lower rank.

Example 6.5.1. Let

$$\mathbf{A} = \begin{bmatrix} 1+i & 0 & 0 \\ 0 & -1 & 0 \end{bmatrix}.$$

The complex conjugate of the transpose of $\mathbf{A}$ is

$$\mathbf{A}^* = \begin{bmatrix} 1-i & 0 \\ 0 & -1 \\ 0 & 0 \end{bmatrix}.$$

Both $\mathbf{A}^*\mathbf{A}$ and $\mathbf{A}\mathbf{A}^*$ are diagonal:

$$\begin{array}{cc} & \begin{bmatrix} 1-i & 0 \\ 0 & -1 \\ 0 & 0 \end{bmatrix} \\ \begin{bmatrix} 1+i & 0 & 0 \\ 0 & -1 & 0 \end{bmatrix} & \begin{bmatrix} 2 & 0 \\ 0 & 1 \end{bmatrix} = \mathbf{A}\mathbf{A}^* \end{array}$$

$$\begin{array}{cc} & \begin{bmatrix} 1+i & 0 & 0 \\ 0 & -1 & 0 \end{bmatrix} \\ \begin{bmatrix} 1+i & 0 \\ 0 & -1 \\ 0 & 0 \end{bmatrix} & \begin{bmatrix} 2 & 0 & 0 \\ 0 & 1 & 0 \\ 0 & 0 & 0 \end{bmatrix} = \mathbf{A}^*\mathbf{A}. \end{array}$$

Since they are diagonal, the eigenvalues are on the diagonal. Notice that the non-zero values are the same. The singular values of $\mathbf{A}$ are the positive square roots of the eigenvalues of $\mathbf{A}^*\mathbf{A}$, hence

$$\sigma_1 = \sqrt{2}, \quad \sigma_2 = 1, \quad \sigma_3 = 0.$$

Since $\sigma_3 = 0$, we have $r = 2$, so

$$\mathbf{\Sigma} = \begin{bmatrix} \sqrt{2} & 0 & 0 \\ 0 & 1 & 0 \end{bmatrix}.$$

What matrices $\mathbf{U}$ and $\mathbf{V}$ give us $\mathbf{A}$? On one hand we can take

$$\mathbf{U} = \begin{bmatrix} \frac{1+i}{\sqrt{2}} & 0 \\ 0 & -1 \end{bmatrix}$$

and $\mathbf{V} = \mathbf{I}_3$. First, we check that $\mathbf{U}\mathbf{U}^* = \mathbf{I}_2$:

$$\begin{array}{cc} & \begin{bmatrix} \frac{1-i}{\sqrt{2}} & 0 \\ 0 & -1 \end{bmatrix} \\ \begin{bmatrix} \frac{1+i}{\sqrt{2}} & 0 \\ 0 & -1 \end{bmatrix} & \begin{bmatrix} \frac{1^2+1^2}{2} & 0 \\ 0 & (-1)^2 \end{bmatrix} = \begin{bmatrix} 1 & 0 \\ 0 & 1 \end{bmatrix}. \end{array}$$

Now we show that $\mathbf{U}\mathbf{\Sigma}\mathbf{V}^* = \mathbf{A}$. Since $\mathbf{V} = \mathbf{I}_3$ then $\mathbf{V}^* = \mathbf{I}_3$, so $\mathbf{U}\mathbf{\Sigma}\mathbf{V}^*$ is

$$\begin{array}{ccc} & \begin{bmatrix} \sqrt{2} & 0 & 0 \\ 0 & 1 & 0 \end{bmatrix} & \begin{bmatrix} 1 & 0 & 0 \\ 0 & 1 & 0 \\ 0 & 0 & 1 \end{bmatrix} \\ \begin{bmatrix} \frac{1+i}{\sqrt{2}} & 0 \\ 0 & -1 \end{bmatrix} & \begin{bmatrix} 1+i & 0 & 0 \\ 0 & -1 & 0 \end{bmatrix} & \begin{bmatrix} 1+i & 0 & 0 \\ 0 & -1 & 0 \end{bmatrix} = \mathbf{A}. \end{array}$$

However Activity 6.5.2 gives another choice of $\mathbf{U}$ and $\mathbf{V}$.

Activity 6.5.2. Let

$$\mathbf{\Sigma} = \begin{bmatrix} \sqrt{2} & 0 & 0 \\ 0 & 1 & 0 \end{bmatrix}, \quad \mathbf{U} = \begin{bmatrix} 1 & 0 \\ 0 & 1 \end{bmatrix}, \quad \text{and} \quad \mathbf{V} = \begin{bmatrix} \frac{1-i}{\sqrt{2}} & 0 & 0 \\ 0 & -1 & 0 \\ 0 & 0 & 1 \end{bmatrix}.$$

1. Compute $\mathbf{V}^*$.

2. Show that $\mathbf{V}^*\mathbf{V} = \mathbf{I}_3$.

3. Show that $\mathbf{U\Sigma V}^* = \mathbf{A}$ from Example 6.5.1.

Example 6.5.3. Let

$$\mathbf{A} = \begin{bmatrix} 2 & 0 \\ 0 & 2 \end{bmatrix} = 2\mathbf{I}.$$

Then we have $\mathbf{A} = \mathbf{\Sigma}$, and we are allowed to take $\mathbf{U} = \mathbf{V} = \mathbf{I}$.

But the identity matrix is not the only choice for $\mathbf{U}$ and $\mathbf{V}$. If $\mathbf{U}$ is any linear isometery of $\mathbb{R}^2$ (so $\mathbf{U}\mathbf{U}^T = \mathbf{I}$), then take $\mathbf{V} = \mathbf{U}$:

$$\mathbf{U\Sigma U}^T = \mathbf{U}(2\mathbf{I})\mathbf{U}^T = 2\mathbf{U}\mathbf{U}^T = 2\mathbf{I} = \mathbf{A}.$$

For instance, taking

$$\mathbf{U} = \mathbf{V} = \begin{bmatrix} 0 & 1 \\ 1 & 0 \end{bmatrix},$$

just reverses the rows and columns. This shows that $\mathbf{U}$ and $\mathbf{V}$ are not unique.

The situation illustrated in Example 6.5.1 generalizes. If

$$\sigma_1, \sigma_2, \ldots \sigma_r$$

are the non-zero singular values of $\mathbf{A}$ then

$$\sigma_1^2, \sigma_2^2, \ldots \sigma_r^2$$

are the non-zero eigenvalues of $\mathbf{A}^*\mathbf{A}$. Since $\mathbf{A}^*\mathbf{A}$ is Hermitian, the Spectral Theorem applies. In particular, by taking the hermitian conjugate of equation (6.16), we have

$$\mathbf{A}^* = (\mathbf{U\Sigma V}^*)^* = (\mathbf{V}^*)^*\mathbf{\Sigma}^*\mathbf{U}^* = \mathbf{V\Sigma}^*\mathbf{U}^*. \tag{6.19}$$

Since $\mathbf{U}^*\mathbf{U} = \mathbf{I}$

$$\begin{aligned} \mathbf{A}^*\mathbf{A} = (\mathbf{V\Sigma}^*\mathbf{U}^*)(\mathbf{U\Sigma V}^*) &= \mathbf{V\Sigma}^*(\mathbf{U}^*\mathbf{U})\mathbf{\Sigma V}^* \\ &= \mathbf{V\Sigma}^*(\mathbf{I})\mathbf{\Sigma V}^* \\ &= \mathbf{V\Sigma}^*\mathbf{\Sigma V}^*. \end{aligned}$$

Since $\mathbf{V}^*\mathbf{V} = \mathbf{I}$, then

$$\begin{aligned} \mathbf{V}^*(\mathbf{A}^*\mathbf{A})\mathbf{V} &= \mathbf{V}^*\mathbf{V\Sigma}^*\mathbf{\Sigma V}^*\mathbf{V} \\ &= \mathbf{I\Sigma}^*\mathbf{\Sigma I} \\ &= \mathbf{\Sigma}^*\mathbf{\Sigma}, \end{aligned}$$

which is diagonal. Our final result is

$$\mathbf{V}^*(\mathbf{A}^*\mathbf{A})\mathbf{V} = \mathbf{\Sigma}^*\mathbf{\Sigma}, \tag{6.20}$$

which shows that conjugation by $\mathbf{V}$ diagonalizes $\mathbf{A}^*\mathbf{A}$. The same calculation with $\mathbf{A}$ and $\mathbf{A}^*$ reversed, shows that

$$\mathbf{U}^*(\mathbf{A}\mathbf{A}^*)\mathbf{U} = \mathbf{\Sigma\Sigma}^*. \tag{6.21}$$

Example 6.5.4. Given the matrix

$$\mathbf{A} = \begin{bmatrix} 1 & 1 & 0 \\ 0 & 0 & 1 \end{bmatrix},$$

do all of the following:

1. Show that $\mathbf{A} = \mathbf{U\Sigma V}^T$, where

$$\mathbf{\Sigma} = \begin{bmatrix} \sqrt{2} & 0 & 0 \\ 0 & 1 & 0 \end{bmatrix}, \ \mathbf{U} = \begin{bmatrix} 1 & 0 \\ 0 & 1 \end{bmatrix}, \text{ and } \mathbf{V} = \begin{bmatrix} \frac{\sqrt{2}}{2} & 0 & \frac{\sqrt{2}}{2} \\ \frac{\sqrt{2}}{2} & 0 & -\frac{\sqrt{2}}{2} \\ 0 & 1 & 0 \end{bmatrix}.$$

2. Compute $\mathbf{AA}^T$ and $\mathbf{A}^T\mathbf{A}$.

3. Show that $\mathbf{U}^T(\mathbf{AA}^T)\mathbf{U}$ is diagonal.

4. Show that $\mathbf{V}^T(\mathbf{A}^T\mathbf{A})\mathbf{V}$ is diagonal.

Solution

We only need transpose here instead of the hermitian conjugate because all of the matrices in this problem are real.
Since $\mathbf{U} = \mathbf{I}$, then $\mathbf{U\Sigma V}^T$ simplifies to just $\mathbf{\Sigma V}^T$ in this case. Thus

$$\begin{bmatrix} \frac{\sqrt{2}}{2} & \frac{\sqrt{2}}{2} & 0 \\ 0 & 0 & 1 \\ \frac{\sqrt{2}}{2} & -\frac{\sqrt{2}}{2} & 0 \end{bmatrix}$$

$$\begin{bmatrix} \sqrt{2} & 0 & 0 \\ 0 & 1 & 0 \end{bmatrix} \begin{bmatrix} 1 & 1 & 0 \\ 0 & 0 & 1 \end{bmatrix} = \mathbf{A}$$

For $\mathbf{AA}^T$ and $\mathbf{A}^T\mathbf{A}$, we have

$$\begin{bmatrix} 1 & 1 & 0 \\ 0 & 0 & 1 \end{bmatrix} \qquad\qquad \begin{bmatrix} 1 & 0 \\ 1 & 0 \\ 0 & 1 \end{bmatrix}$$

$$\begin{bmatrix} 1 & 0 \\ 1 & 0 \\ 0 & 1 \end{bmatrix} \begin{bmatrix} 1 & 1 & 0 \\ 1 & 1 & 0 \\ 0 & 0 & 1 \end{bmatrix} = \mathbf{A}^T\mathbf{A} \qquad \begin{bmatrix} 1 & 1 & 0 \\ 0 & 0 & 1 \end{bmatrix} \begin{bmatrix} 2 & 0 \\ 0 & 1 \end{bmatrix} = \mathbf{AA}^T$$

Since $\mathbf{AA}^T$ is already diagonal and $\mathbf{U}$ is the 2×2 identity matrix, multiplying on the left and right by the identity matrix does not change $\mathbf{AA}^T$
For $\mathbf{V}^T(\mathbf{A}^T\mathbf{A})\mathbf{V}$

$$\begin{bmatrix} 1 & 1 & 0 \\ 1 & 1 & 0 \\ 0 & 0 & 1 \end{bmatrix} \begin{bmatrix} \frac{\sqrt{2}}{2} & 0 & \frac{\sqrt{2}}{2} \\ \frac{\sqrt{2}}{2} & 0 & -\frac{\sqrt{2}}{2} \\ 0 & 1 & 0 \end{bmatrix}$$

$$\begin{bmatrix} \frac{\sqrt{2}}{2} & \frac{\sqrt{2}}{2} & 0 \\ 0 & 0 & 1 \\ \frac{\sqrt{2}}{2} & -\frac{\sqrt{2}}{2} & 0 \end{bmatrix} \begin{bmatrix} \sqrt{2} & \sqrt{2} & 0 \\ 0 & 0 & 1 \\ 0 & 0 & 0 \end{bmatrix} \begin{bmatrix} 2 & 0 & 0 \\ 0 & 1 & 0 \\ 0 & 0 & 0 \end{bmatrix}$$

Activity 6.5.5. Given the matrix

$$\mathbf{A} = \begin{bmatrix} 3 & 4 & 0 \\ 0 & 0 & 1 \end{bmatrix}$$

do all of the following:

1. Show that $\mathbf{A} = \mathbf{U}\boldsymbol{\Sigma}\mathbf{V}^T$, where

$$\boldsymbol{\Sigma} = \begin{bmatrix} 5 & 0 & 0 \\ 0 & 1 & 0 \end{bmatrix}, \quad \mathbf{U} = \begin{bmatrix} 1 & 0 \\ 0 & 1 \end{bmatrix}, \text{ and } \mathbf{V} = \begin{bmatrix} \frac{3}{5} & 0 & \frac{4}{5} \\ \frac{4}{5} & 0 & -\frac{3}{5} \\ 0 & 1 & 0 \end{bmatrix}.$$

2. Compute $\mathbf{A}\mathbf{A}^T$ and $\mathbf{A}^T\mathbf{A}$.
3. Show that $\mathbf{U}^T(\mathbf{A}\mathbf{A}^T)\mathbf{U}$ is diagonal.
4. Show that $\mathbf{V}^T(\mathbf{A}^T\mathbf{A})\mathbf{V}$ is diagonal.

Reflection 6.5.1. How do the numbers on the diagonals of $\mathbf{U}^T(\mathbf{A}\mathbf{A}^T)\mathbf{U}$ and $\mathbf{V}^T(\mathbf{A}^T\mathbf{A})\mathbf{V}$ relate to each other? How do they relate to the values in $\boldsymbol{\Sigma}$?

6.5.2 Solving a Linear System with the SVD

Suppose we want to solve

$$\mathbf{A}\mathbf{x} = \mathbf{b}.$$

Given the SVD for $\mathbf{A}$,

$$\mathbf{A} = \mathbf{U}\boldsymbol{\Sigma}\mathbf{V}^T,$$

we have

$$\mathbf{A}\mathbf{x} = (\mathbf{U}\boldsymbol{\Sigma}\mathbf{V}^T)\mathbf{x} = \mathbf{U}\boldsymbol{\Sigma}(\mathbf{V}^T\mathbf{x}) = \mathbf{b}.$$

Since $\mathbf{U}\mathbf{U}^T = \mathbf{I}$, then after multiplying both sides on the left by $\mathbf{U}^T$,

$$\begin{aligned} \mathbf{U}^T\mathbf{U}\boldsymbol{\Sigma}(\mathbf{V}^T\mathbf{x}) &= \mathbf{U}^T\mathbf{b} \\ \implies \boldsymbol{\Sigma}(\mathbf{V}^T\mathbf{x}) &= \mathbf{U}^T\mathbf{b}. \end{aligned}$$

Converting this equation to augmented form, gives us

$$[\boldsymbol{\Sigma}|\mathbf{U}^T\mathbf{b}],$$

which is almost in reduced form. Solving the system, gives us $\mathbf{y} = \mathbf{V}^T\mathbf{x}$. Then since $\mathbf{V}\mathbf{V}^T = \mathbf{I}$, we get

$$\mathbf{V}\mathbf{y} = \mathbf{V}\mathbf{V}^T\mathbf{x} = \mathbf{x}.$$

Example 6.5.6. Suppose $\mathbf{A} = \mathbf{U}\boldsymbol{\Sigma}\mathbf{V}^T$, where

$$\boldsymbol{\Sigma} = \begin{bmatrix} 5 & 0 & 0 \\ 0 & 3 & 0 \end{bmatrix}, \quad \mathbf{U} = \begin{bmatrix} \frac{5}{13} & -\frac{12}{13} \\ \frac{12}{13} & \frac{5}{13} \end{bmatrix}, \text{ and } \mathbf{V} = \begin{bmatrix} \frac{2}{3} & -\frac{1}{3} & \frac{2}{3} \\ -\frac{1}{3} & \frac{2}{3} & \frac{2}{3} \\ \frac{2}{3} & \frac{2}{3} & -\frac{1}{3} \end{bmatrix}.$$

Solve $\mathbf{A}\mathbf{x} = \mathbf{b}$, where

$$\mathbf{b} = \begin{bmatrix} 13 \\ 0 \end{bmatrix}.$$

Solution

First, we need to compute $\mathbf{U}^T\mathbf{b}$:

$$\begin{array}{cc} & \begin{bmatrix} 13 \\ 0 \end{bmatrix} \\ \begin{bmatrix} \frac{5}{13} & \frac{12}{13} \\ -\frac{12}{13} & \frac{5}{13} \end{bmatrix} & \begin{bmatrix} 5 \\ -12 \end{bmatrix} = \mathbf{U}^T\mathbf{b} \end{array}$$

Next we set up the augmented form and solve. We will use the notation cR_i to mean that R_i is multiplied by c, which amounts to multiplying the corresponding equation by c. See Subsection 2.3.1 for more detail if necessary.

$$\left[\begin{array}{ccc|c} 5 & 0 & 0 & 5 \\ 0 & 3 & 0 & -12 \end{array}\right] \xrightarrow{\frac{1}{5}R_1} \left[\begin{array}{ccc|c} 1 & 0 & 0 & 1 \\ 0 & 3 & 0 & -12 \end{array}\right] \xrightarrow{\frac{1}{3}R_2} \left[\begin{array}{ccc|c} 1 & 0 & 0 & 1 \\ 0 & 1 & 0 & -4 \end{array}\right].$$

Note that this is a solution for $\mathbf{y} = \mathbf{V}^T\mathbf{x}$, not $\mathbf{x}$. The 3rd column does not contain a pivot, so y_3 is free, hence we get

$$\mathbf{y} = \begin{bmatrix} 1 \\ -4 \\ y_3 \end{bmatrix}.$$

To finish the solution, we multiply on the left by $\mathbf{V}$:

$$\begin{array}{cc} & \begin{bmatrix} 1 \\ -4 \\ y_3 \end{bmatrix} \\ \begin{bmatrix} \frac{2}{3} & -\frac{1}{3} & \frac{2}{3} \\ -\frac{1}{3} & \frac{2}{3} & \frac{2}{3} \\ \frac{2}{3} & \frac{2}{3} & -\frac{1}{3} \end{bmatrix} & \begin{bmatrix} 2 + \frac{2}{3}y_3 \\ -3 + \frac{2}{3}y_3 \\ -2 - \frac{1}{3}y_3 \end{bmatrix} = \begin{bmatrix} 2 \\ -3 \\ -2 \end{bmatrix} + \frac{y_3}{3}\begin{bmatrix} 2 \\ 2 \\ -1 \end{bmatrix} = \mathbf{x}. \end{array}$$

Activity 6.5.7. Suppose $\mathbf{A} = \mathbf{U\Sigma V}^T$, where

$$\mathbf{\Sigma} = \begin{bmatrix} 3 & 0 & 0 \\ 0 & 2 & 0 \end{bmatrix} \quad \mathbf{U} = \begin{bmatrix} \frac{3}{5} & \frac{4}{5} \\ -\frac{4}{5} & \frac{3}{5} \end{bmatrix} \quad \text{and} \quad \mathbf{V} = \begin{bmatrix} \frac{2}{3} & \frac{2}{3} & -\frac{1}{3} \\ -\frac{1}{3} & \frac{2}{3} & \frac{2}{3} \\ \frac{2}{3} & -\frac{1}{3} & \frac{2}{3} \end{bmatrix}.$$

Solve $\mathbf{Ax} = \mathbf{b}$, where

$$\mathbf{b} = \begin{bmatrix} 5 \\ 0 \end{bmatrix}$$

6.5.3 Generalized Inverse

The concept of a generalized inverse is fairly natural. Even if $T : \mathbb{R}^n \to \mathbb{R}^m$ does not have an inverse it is still possible to get an inverse by restricting the domain to the row space, and the codomain to the column space. If we are using $\mathbf{\Sigma}$ as the matrix for T, then the matrix of the restricted transformation is the square submatrix of $\mathbf{D}$ with only the non-zero singular values σ_i on the diagonal, say $\sigma_1, \sigma_2, \ldots \sigma_r$. The inverse of that matrix is

$$\begin{bmatrix} \sigma_1 & 0 & \cdots & 0 \\ 0 & \sigma_2 & \cdots & 0 \\ \vdots & \vdots & \ddots & \vdots \\ 0 & 0 & \cdots & \sigma_r \end{bmatrix}^{-1} = \begin{bmatrix} \sigma_1^{-1} & 0 & \cdots & 0 \\ 0 & \sigma_2^{-1} & \cdots & 0 \\ \vdots & \vdots & \ddots & \vdots \\ 0 & 0 & \cdots & \sigma_r^{-1} \end{bmatrix}.$$

But we want the generalized inverse to be a linear transformation $T' : \mathbb{R}^m \to \mathbb{R}^n$ (reversing domain and codomain). So we must also take the transpose, which justifies the following definition.

Definition 6.5.1. If $\mathbf{A} = \mathbf{U\Sigma V}^*$ is a singular value decomposition of $\mathbf{A}$, we define the generalized inverse of $\mathbf{A}$ to be

$$\mathbf{A}' = \mathbf{V\Sigma}'\mathbf{U}^* \tag{6.22}$$

where $\mathbf{\Sigma}'$ is constructed by replacing each non-zero σ_i in Σ with σ_i^{-1} and taking the transpose.

Remark 6.5.3. Equation (6.22) is used to *define* A', but it is not usually the SVD of A'. That is because

$$\sigma_1^{-1}, \sigma_2^{-1}, \ldots \sigma_r^{-1}$$

occur in increasing order, not decreasing order.

The following theorem and its corollary makes the sense in which $\mathbf{A}'$ generalizes $\mathbf{A}^{-1}$ concrete.

Theorem 6.5.2. *If* $\mathbf{A}'$ *is the generalized inverse of* $\mathbf{A}$*, then*

1. $\mathbf{A}'\mathbf{A}$ *is the identity on* $\mathrm{Row}(\mathbf{A})$*, and zero on* $\mathrm{Nul}(\mathbf{A})$*, and*
2. $\mathbf{AA}'$ *is the identity on* $\mathrm{Col}(\mathbf{A})$*, and zero on* $\mathrm{Nul}(\mathbf{A}^T)$*.*

See [2] for proof.

Corollary 6.5.3. *If* $\mathbf{A}$ *is an invertible square matrix, then* $\mathbf{A}' = \mathbf{A}^{-1}$*.*

Remark 6.5.4. By Theorem 6.2.6 we can say that $\mathbf{A}'\mathbf{A}$ is zero on the perpendicular subspace of $\mathrm{Row}(\mathbf{A})$ and $\mathbf{AA}'$ is zero on the perpendicular subspace of $\mathrm{Col}(\mathbf{A})$.

Both $\mathbf{A}'\mathbf{A}$ and $\mathbf{AA}'$ are symmetric.

If $\mathbf{A} \neq \mathbf{0}$, then there will be at least one non-zero singular value. If $\mathbf{A} = \mathbf{0}$, then $\mathbf{\Sigma} = \mathbf{0}$ and $\mathbf{\Sigma}'$ is the transpose of $\mathbf{\Sigma}$.

Example 6.5.8. Compute the generalized inverse of $\mathbf{A}$ from Example 6.5.4, then verify Theorem 6.5.2.

Solution

We can just write down $\mathbf{\Sigma}'$:

$$\begin{bmatrix} \frac{1}{\sqrt{2}} & 0 \\ 0 & 1 \\ 0 & 0 \end{bmatrix}.$$

Next compute $\mathbf{V\Sigma}'\mathbf{U}^T$:

$$\begin{bmatrix} \frac{1}{\sqrt{2}} & 0 \\ 0 & 1 \\ 0 & 0 \end{bmatrix} \begin{bmatrix} 1 & 0 \\ 0 & 1 \end{bmatrix}$$

$$\begin{bmatrix} \frac{\sqrt{2}}{2} & 0 & \frac{\sqrt{2}}{2} \\ \frac{\sqrt{2}}{2} & 0 & -\frac{\sqrt{2}}{2} \\ 0 & 1 & 0 \end{bmatrix} \begin{bmatrix} \frac{1}{2} & 0 \\ \frac{1}{2} & 0 \\ 0 & 1 \end{bmatrix} \begin{bmatrix} \frac{1}{2} & 0 \\ \frac{1}{2} & 0 \\ 0 & 1 \end{bmatrix} = \mathbf{A}'$$

To verify the first part of Theorem 6.5.2, we need $\mathbf{A}'\mathbf{A}$:

$$\begin{array}{cc} & \begin{bmatrix} 1 & 1 & 0 \\ 0 & 0 & 1 \end{bmatrix} \\ \begin{bmatrix} \frac{1}{2} & 0 \\ \frac{1}{2} & 0 \\ 0 & 1 \end{bmatrix} & \begin{bmatrix} \frac{1}{2} & \frac{1}{2} & 0 \\ \frac{1}{2} & \frac{1}{2} & 0 \\ 0 & 0 & 1 \end{bmatrix} \end{array}$$

Since $\mathbf{A}$ has two pivots and one free variable. We get a parameterization of the null space in the usual way: $x + y = 0$, $z = 0$ and $y = y$, give us

$$\begin{bmatrix} x \\ y \\ z \end{bmatrix} = \begin{bmatrix} -y \\ y \\ 0 \end{bmatrix} = y \begin{bmatrix} -1 \\ 1 \\ 0 \end{bmatrix}.$$

If we multiply on the left by $\mathbf{A}'\mathbf{A}$, we should get zero

$$\begin{array}{cc} & \begin{bmatrix} -1 \\ 1 \\ 0 \end{bmatrix} \\ \begin{bmatrix} \frac{1}{2} & \frac{1}{2} & 0 \\ \frac{1}{2} & \frac{1}{2} & 0 \\ 0 & 0 & 1 \end{bmatrix} & \begin{bmatrix} 0 \\ 0 \\ 0 \end{bmatrix} \end{array}$$

and we do. We can check that $\mathbf{A}'\mathbf{A}$ acts as the identity, by checking on $\text{Row}(\mathbf{A})$, which span the row space by definition. But technically we should take the transpose of the rows, since $\mathbf{A}$ is constructed with the idea that it is acting on column vectors in $\mathbb{R}^3$ and producing column vectors in $\mathbb{R}^2$:

$$\begin{array}{cc} & \begin{bmatrix} 1 & 0 \\ 1 & 0 \\ 0 & 1 \end{bmatrix} \\ \begin{bmatrix} \frac{1}{2} & \frac{1}{2} & 0 \\ \frac{1}{2} & \frac{1}{2} & 0 \\ 0 & 0 & 1 \end{bmatrix} & \begin{bmatrix} 1 & 0 \\ 1 & 0 \\ 0 & 1 \end{bmatrix}. \end{array}$$

To verify the second part of Theorem 6.5.2, we need $\mathbf{A}\mathbf{A}'$:

$$\begin{array}{cc} & \begin{bmatrix} \frac{1}{2} & 0 \\ \frac{1}{2} & 0 \\ 0 & 1 \end{bmatrix} \\ \begin{bmatrix} 1 & 1 & 0 \\ 0 & 0 & 1 \end{bmatrix} & \begin{bmatrix} 1 & 0 \\ 0 & 1 \end{bmatrix}. \end{array}$$

Since, this is the identity matrix there is nothing more to check.

Activity 6.5.9. Compute the generalized inverse of $\mathbf{A}$ from Activity 6.5.5, then verify Theorem 6.5.2.

6.5.4 Computing Rank by the SVD

The singular value decomposition provides a theoretical framework for a way of computing the rank of a matrix $\mathbf{A}$. The following example illustrates the idea behind this method.

Example 6.5.10. Consider the matrix

$$\mathbf{A} = \begin{bmatrix} \boxed{2} & 0 & 0 \\ 0 & 0 & \boxed{1} \\ 0 & 0 & 0 \end{bmatrix}.$$

There are a few observations that can be made about $\mathbf{A}$:

1. $\mathbf{A}$ is already in echelon form with the pivots in boxes, so the rank is 2.
2. $\mathbf{A}$ is upper-triangular, so the eigenvalues of $\mathbf{A}$ are 0 and 2.
3. The number of non-zero values on the diagonal is not equal to the rank.

However, both $\mathbf{A}\mathbf{A}^T$ and $\mathbf{A}^T\mathbf{A}$ are symmetric, so they can both be diagonalized. In general, we only need to compute the smaller one, but in this case they are the same size:

$$\mathbf{A}\mathbf{A}^T: \quad \begin{bmatrix} 2 & 0 & 0 \\ 0 & 0 & 1 \\ 0 & 0 & 0 \end{bmatrix} \begin{bmatrix} 2 & 0 & 0 \\ 0 & 0 & 0 \\ 0 & 1 & 0 \end{bmatrix} = \begin{bmatrix} 4 & 0 & 0 \\ 0 & 1 & 0 \\ 0 & 0 & 0 \end{bmatrix} \qquad \mathbf{A}^T\mathbf{A}: \quad \begin{bmatrix} 2 & 0 & 0 \\ 0 & 0 & 0 \\ 0 & 1 & 0 \end{bmatrix} \begin{bmatrix} 2 & 0 & 0 \\ 0 & 0 & 1 \\ 0 & 0 & 0 \end{bmatrix} = \begin{bmatrix} 4 & 0 & 0 \\ 0 & 0 & 0 \\ 0 & 0 & 1 \end{bmatrix}, \tag{6.23}$$

In this case, both $\mathbf{A}\mathbf{A}^T$ and $\mathbf{A}^T\mathbf{A}$ have the characteristic polynomial $(x-4)(x-1)x$. The eigenvalues of $\mathbf{A}\mathbf{A}^T$ and $\mathbf{A}^T\mathbf{A}$ are 4, 1, and 0.

Activity 6.5.11. Given the matrix

$$\mathbf{A} = \begin{bmatrix} 0 & 1 & 0 \\ 0 & 0 & 1 \\ 0 & 0 & 0 \end{bmatrix},$$

compute $\mathbf{A}^T\mathbf{A}$ and $\mathbf{A}\mathbf{A}^T$, then verify that the rank of $\mathbf{A}$ is 2, by:

1. counting the pivots of $\mathbf{A}$, and
2. counting the number of non-zero eigenvalues of $\mathbf{A}^T\mathbf{A}$ or $\mathbf{A}\mathbf{A}^T$.

6.6 Principal Component Analysis

Data is often complex and messy. There are many strategies for cleaning up data and trying to simplify it without loosing important information. Principal Component Analysis (PCA) is one of these strategies. The basic idea is that the most important information is captured by the largest eigenvalues and eigen vectors of the covariance matrix. If a data table has many columns, the covariance matrix could be very big. If only the largest 2 or 3 eigenvalues are kept, then even high dimensional data can be visualized with a scatter plot. But if too much information is lost, it may be useful to keep a few more eigenvalues even though plotting becomes impractical.

Activity 6.6.1. (With Sage and Numpy)

In this activity we will explore PCA by constructing a normal distribution in two variables with a given covariance matrix and plotting it.

First, we create a standard normal distribution with two variables

```
import numpy as np
import matplotlib.pyplot as plt
from matplotlib.patches import Ellipse

n = 1000
x = np.random.randn(n)
y = np.random.randn(n)
SN = np.vstack((x, y))
```

Since it is standard normal, the covariance matrix should be the identity matrix. Since randomness is involved, there will be some error. You can check by running the following code

```
print(np.dot(SN,SN.T)/n - np.eye(2))
```

If you want the error to be smaller, just increase n. Suppose that we want to construct a normal distribution with covariance matrix $\mathbf{C}$ and mean $\mathbf{b}$, where

$$\mathbf{C} = \begin{bmatrix} 9 & 3 \\ 3 & 5 \end{bmatrix} \quad \text{and} \quad \mathbf{b} = \begin{bmatrix} 3 \\ -2 \end{bmatrix}.$$

This can be done using the Cholesky decomposition for $\mathbf{C}$. Since $\mathbf{C} = \mathbf{A}\mathbf{A}^T$, where

$$\mathbf{A} = \begin{bmatrix} 3 & 0 \\ 1 & 2 \end{bmatrix}$$

then the following code constructs the data that we need

```
A = np.array([[3.0, 0.0],
              [1.0, 2.0]])
b = np.array([[3.0],
             [-2.0]])
data = np.dot(A,SN) + b
```

The `data` array is a sideways data table, as can be seen from checking the shape:

```
data.shape
```

You can check that the data has the prescribed mean as follows:

```
mu = np.mean(data, axis=1, keepdims=True)
print(mu)
print(mu-b)
```

To get the covariance matrix of the data, it is necessary to first subtract the mean:

```
X_centered = data - mu
```

Then you can check that the data has the prescribed covariance matrix:

```
print(np.dot(A,A.T))
print(np.dot(A,A.T) - np.dot(X_centered,X_centered.T)/n)
```

What are the eigenvalues and eigenvectors of the covariance matrix?

```
cov = np.dot(X_centered, X_centered.T)/n
eigvals, eigvecs = np.linalg.eigh(cov)

idx = np.argsort(eigvals)[::-1]
eigvals = eigvals[idx]
eigvecs = eigvecs[:,idx]

print("Eigenvalues")
print(eigvals)
print("Eigenvectors")
print(eigvecs)
```

If the data is plotted, then the eigenvectors lie give the directions of the major and minor axis of an ellipse centered at the mean. The lengths of the major and minor axis are the square roots of the eigenvectors of the covariance matrices.

```
fig, ax = plt.subplots(figsize=(6, 6))
ax.scatter(data[0, :], data[1, :], alpha=0.4, s=15, label='Data')
ax.scatter(mu[0], mu[1], color='red', marker='x', s=100, label='Mean')

for i in range(2):
    start = mu.flatten()
    vec = eigvecs[:, i] * np.sqrt(eigvals[i])
    ax.plot([start[0], start[0] + vec[0]],
            [start[1], start[1] + vec[1]],
            linewidth=3,
            label=f'PC{i+1}')

angle = np.degrees(np.arctan2(eigvecs[1, 0], eigvecs[0, 0]))
width, height = 2 * np.sqrt(eigvals)
ellipse = Ellipse(xy=mu.flatten(), width=width, height=height,
    angle=angle, edgecolor='black', fc='none', lw=2, ls='--')
ax.add_patch(ellipse)

ax.set_title("PCA Decomposition")
ax.set_xlabel("x")
ax.set_ylabel("y")
ax.axis('equal')
ax.legend()
ax.grid(True)
plt.show()
```

6.7 Fourier Series

The material in this section uses quite a bit of calculus. If you pick up ten different books on Fourier series, you will likely find ten different approaches. The material in this section follows [13] fairly closely. One of our goals is to discuss why orthogonality occurs for Fourier series, and why it can be used to compute Fourier coefficients. Another goal is to actually discuss how to compute the Fourier coefficients.

Suppose that $f(x)$ is a periodic function on $\mathbb{R}$ with period 2π, in other words

$$f(x) = f(x + 2\pi)$$

for all x. If f and g are real valued periodic functions, and c is a real number, then

$$(f + g)(x) = f(x) + g(x) = f(x + 2\pi) + g(x + 2\pi) = (f + g)(x + 2\pi)$$

and

$$(cf)(x) = cf(x) = cf(x + 2\pi) = (cf)(x + 2\pi).$$

The periodic functions with period 2π form a subspace of the vector space of functions on $\mathbb{R}$. If n is an integer, then $\sin(nx)$ and $\cos(nx)$ are examples of periodic functions with period 2π. Constant functions, like 1, are also periodic. These functions together generate a subspace of the periodic functions:

$$f_n(x) = \frac{a_0}{2} + \sum_{k=1}^{n} a_k \cos(kx) + b_k \sin(kx). \tag{6.24}$$

The functions f_n are linear combinations of periodic functions, and so they are still periodic. We can then take the limit as n goes to infinity:

$$\lim_{n\to\infty} f_n(x) = \frac{a_0}{2} + \sum_{k=1}^{\infty} a_k \cos(kx) + b_k \sin(kx). \tag{6.25}$$

In general, we no longer have a linear combination in the limit because there could be an infinite number of non-zero values among the coefficients a_k, b_k. A linear combination, by definition, must have a finite number of non-zero coefficients. We also have not addressed the convergence of this limit.

We have an inner product

$$\langle f, g\rangle = \frac{1}{\pi}\int_{-\pi}^{\pi} f(x)\overline{g(x)}\, dx, \tag{6.26}$$

where the bar over $g(x)$ is complex conjugation. The functions $\sin(nx)$, $\cos(nx)$, and 1, are orthogonal with respect to this inner product, but they are not quite normalized. Before we show this, it will be helpful to review some important facts.

Definition 6.7.1. A function $f(x)$ is **even** if

$$f(-x) = f(x).$$

A function $f(x)$ is **odd** if

$$f(-x) = -f(x)$$

If $f(x)$ is a polynomial, then it is even if all powers of x are even and odd if all powers of x are odd. For constant functions, it is useful to imagine them as coefficients of x^0. Since the power of x is zero, and zero is even, then all constant functions are even functions. We can also see that constants are even functions because the graph of an even function is symmetric across the y-axis. So, 1 is an even function even though it is an odd number. Again, it is the power of x that matters, not the number itself. If the powers are mixed, then the function is neither even nor odd. For trig functions $\cos(x)$ is even and $\sin(x)$ is odd. This can be understood from their Taylor expansions at zero:

$$\cos(x) = 1 - \frac{1}{2!}x^2 + \frac{1}{4!}x^4 \pm \cdots + \frac{(-1)^n}{(2n)!}x^{2n}, \quad \text{and}$$
$$\sin(x) = x - \frac{1}{3!}x^3 + \frac{1}{5!}x^5 \pm \cdots + \frac{(-1)^n}{(2n+1)!}x^{2n+1};$$

cosine has only even powers and sine has only odd powers.

Proposition 6.7.2. *Even and odd functions have the following arithmetic properties:*

1. *If f and g are odd, then $f+g$ is odd.*
2. *If f and g are even, then $f+g$ is even.*
3. *If f and g are odd, then fg is even.*
4. *If f and g are even, then fg is even.*
5. *If f is odd and g is even, then fg is odd.*
6. *If f is odd and g is odd, then $f \circ g$ is odd.*
7. *If f is even and g is odd, then $f \circ g$ and $g \circ f$ are even.*
8. *If f is even and g is even, then $f \circ g$ is even.*

We generally have avoided proofs, but it is worth seeing how to do this because it builds algebra skills.

Proof. As an example, if f is odd and g is odd, then

$$\begin{aligned}(fg)(-x) = f(-x)g(-x) &= (-f(x)(-g(x)) \\ &= (-1)^2 f(x)g(x) = f(x)g(x) = (fg)(x)\end{aligned}$$

□

We can also illustrate this same property by checking it with familiar odd functions.

Example 6.7.1. Take $f(x) = x^3$ and $g(x) = x^5$, both of which are odd functions. Then

$$f(x)g(x) = x^3x^5 = x^{3+5} = x^8$$

which is an even function.

Activity 6.7.2. Verify that if f is odd and g is even then fg is odd, by mimicking either the proof or Example 6.7.1.

The importance of even and odd functions in this context is that if f is an odd function defined on the interval $[-a, a]$, then

$$\int_{-a}^{a} f(x)\,dx = 0. \tag{6.27}$$

Also if f is an even function defined on $[-a, a]$, then

$$\int_{-a}^{a} f(x)\,dx = 2\int_{0}^{a} f(x)\,dx. \tag{6.28}$$

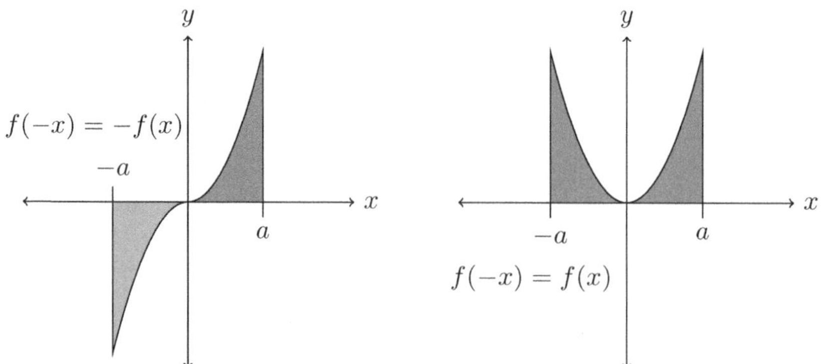

FIGURE 6.3: The graph on the left illustrates why $\int_{-a}^{a} f(x)\,dx = 0$ for an odd function. The integrals to the left and right of the y-axis have opposite sign, but the areas are equal, so they cancel exactly. The graph on the right illustrates why $\int_{-a}^{a} f(x)\,dx = 2\int_0^a f(x)\,dx$. The integrals to the left and right of the y-axis have the same sign are equal, so the right half can be evaluated and doubled

Both equation (6.27) and (6.28) are useful, but the first of these already gives orthogonality in some cases. Since $\cos(nx)$ is even and $\sin(mx)$ is odd, for all positive integers n and m, then $\cos(nx)\sin(mx)$ is odd, hence

$$\langle \cos(nx), \sin(mx) \rangle = \frac{1}{\pi}\int_{-\pi}^{\pi} \cos(nx)\sin(mx)\,dx = 0.$$

We do not have to bother with the complex conjugate since sine and cosine are real valued on $[-\pi, \pi]$. The trouble is that $\cos(nx)\cos(mx)$ and $\sin(nx)\sin(mx)$ are both even functions, so this trick does not work for $\langle \cos(nx), \cos(mx) \rangle$ and $\langle \sin(nx), \sin(mx) \rangle$. Instead, we must use trig identities. Starting with

$$\cos(A + B) = \cos(A)\cos(B) - \sin(A)\sin(B), \tag{6.29}$$

$$\sin(A + B) = \sin(A)\cos(B) + \cos(A)\sin(B), \tag{6.30}$$

we replace B by $-B$ in the equation (6.29) to obtain

$$\begin{aligned}\cos(A - B) &= \cos(A)\cos(-B) - \sin(A)\sin(-B) \\ &= \cos(A)\cos(B) + \sin(A)\sin(B)\end{aligned} \tag{6.31}$$

where we have used the fact that $\cos(-B) = \cos(B)$ and $\sin(-B) = -\sin(B)$; cosine is even and sine is odd. Then by adding equations (6.29) and (6.31), we obtain

$$\cos(A + B) + \cos(A - B) = 2\cos(A)\cos(B), \tag{6.32}$$

a formula which allows us to turn a product of cosines into a sum of cosines. Similarly, by subtraction we obtain

$$\cos(A - B) - \cos(A + B) = 2\sin(A)\sin(B). \tag{6.33}$$

Next, we apply this to $A = nx$ and $B = mx$, so $A + B = (n + m)x$ and $(n - m)x$, which means that

$$\begin{aligned}\langle\cos(nx),\cos(mx)\rangle &= \frac{1}{\pi}\int_{-\pi}^{\pi}\cos(nx)\cos(mx)\,dx\\ &= \frac{1}{\pi}\int_{0}^{\pi}2\cos(nx)\cos(mx)\,dx\\ &= \frac{1}{\pi}\int_{0}^{\pi}\cos((n+m)x)\,dx + \frac{1}{\pi}\int_{0}^{\pi}\cos((n-m)x)\,dx, \qquad (6.34)\end{aligned}$$

where we have used the fact that $\cos(nx)\cos(mx)$ is even to get the first equality, and the identity (6.32) to get the second equality. Now m and n are both positive integers, so $n+m$ is not zero. If $n \neq m$, then $n-m$ is also not zero. In that case we can complete both integrals by u-substitution. For the first integral,

$$\begin{aligned} &u = (n+m)x && x = 0 \longrightarrow u = 0\\ &du = (n+m)\,dx \implies dx = \frac{1}{n+m}\,du && x = \pi \longrightarrow u = (n+m)\pi.\end{aligned}$$

It follows that

$$\begin{aligned}\frac{1}{\pi}\int_0^{\pi}\cos((n+m)x)\,dx &= \frac{1}{(n+m)\pi}\int_0^{(n+m)\pi}\cos(u)\,du\\ &= \frac{1}{(n+m)\pi}\sin(u)\Big|_0^{(n+m)\pi}\\ &= \frac{1}{(n+m)\pi}(\sin((n+m)\pi)-\sin(0)) = 0 \qquad (6.35)\end{aligned}$$

since sine is zero at any integer multiple of π.

Activity 6.7.3. Complete the u-substitution to show that

$$\frac{1}{\pi}\int_0^{\pi}\cos((n-m)x)\,dx = 0. \qquad (6.36)$$

Equations (6.34), (6.35), and (6.36) now show that if $n \neq m$, then

$$\langle\cos(nx),\cos(mx)\rangle = 0.$$

The analogous result for sine is left to exercise 6.11. To complete the verification of orthogonality, it remains to show that $\langle 1,\cos(nx)\rangle = 0$ and $\langle 1,\sin(nx)\rangle = 0$. For sine, this follows from the fact that $\sin(nx)$ is odd, for cosine it follows from the same type of u-substitution calculation that we just did.

We now compute the squared norms. The calculation of $\langle\cos(nx),\cos(mx)\rangle$ still works to the end of equation (6.34), but with $n = m$ the second integral simplifies and we must use power rule instead of u-substitution. Continuing from there, we have

$$\begin{aligned}\langle\cos(nx),\cos(nx)\rangle &= \frac{1}{\pi}\int_0^{\pi}\cos(2nx)\,dx + \frac{1}{\pi}\int_0^{\pi}1\,dx\\ &= \frac{1}{2n\pi}\sin(2nx)\Big|_0^{\pi} + \frac{1}{\pi}x\Big|_0^{\pi}\\ &= \frac{1}{2n\pi}(\sin(2n\pi)-\sin(0)) + \frac{1}{\pi}(\pi-0) = 1\end{aligned}$$

where we applied u-substitution to the first integral. In fact, it is the same u-substitution as above with $m = n$. The squared norm of 1 is easier to compute, but in a way it is an exception

$$\langle 1, 1\rangle = \frac{1}{\pi}\int_{-\pi}^{\pi} 1\,dx = \frac{2}{\pi}\int_{0}^{\pi} 1\,dx = \frac{2}{\pi}x\Big|_0^{\pi} = \frac{2}{\pi}(\pi - 0) = 2.$$

So, 1 isn't normalized! Even so, it is easier to leave it that way. That is one reason that equations (6.25) and (6.26) have $\frac{a_0}{2}$ rather than a_0. For a second reason, see exercise 6.12.

Suppose now that

$$f(x) = \frac{a_0}{2} + \lim_{n\to\infty}\sum_{k=1}^{n} a_k\cos(kx) + b_k\sin(kx).$$

We would like to use orthogonality to compute the coefficients, in a way something like this:

$$\begin{aligned}
\langle f(x), 1\rangle &= \frac{1}{\pi}\int_{-\pi}^{\pi} f(x)\,dx \\
&= \frac{1}{\pi}\int_{-\pi}^{\pi}\left(\frac{a_0}{2} + \lim_{n\to\infty}\sum_{k=1}^{n} a_k\cos(kx) + b_k\sin(kx)\right)dx \\
&\overset{?}{=} \frac{1}{\pi}\int_{-\pi}^{\pi}\frac{a_0}{2}\,dx + \lim_{n\to\infty}\sum_{k=1}^{n}\frac{1}{\pi}\int_{-\pi}^{\pi} a_k\cos(kx)\,dx + \frac{1}{\pi}\int_{-\pi}^{\pi} b_k\sin(kx)\,dx \\
&= \frac{a_0}{2}\langle 1, 1\rangle + \lim_{n\to\infty}\sum_{k=1}^{n} a_k\langle 1, \cos(kx)\rangle + a_k\langle 1, \sin(kx)\rangle = a_0.
\end{aligned}$$

The question mark above the equal sign is justified. We have tried to pull a limit out of an integral, and this is not something you can just do. The integral itself is defined by a limit, and it is generally not possible to switch the order of limits without extra assumptions. In this case the extra assumption is something called "uniform convergence." We will address this point concretely in Example 6.7.4 below. Nonetheless, it motivates the definition

$$a_n = \langle f(x), \cos(nx)\rangle = \frac{1}{\pi}\int_{-\pi}^{\pi} f(x)\cos(nx)\,dx \quad \text{for } n = 0, 1, 2, \ldots, \tag{6.37}$$

$$b_n = \langle f(x), \sin(nx)\rangle = \frac{1}{\pi}\int_{-\pi}^{\pi} f(x)\sin(nx)\,dx \quad \text{for } n = 1, 2, \ldots. \tag{6.38}$$

Notice that $\cos(nx) = 1$ when $n = 0$.

Example 6.7.4. Let $f(x) = x$. Since x is an odd function and $\cos(nx)$ is an even function for all n, then $x\cos(nx)$ is an odd function for all n. It follows that

$$a_n = \langle f(x), \cos(nx)\rangle = 0$$

for all n. What about b_n? Since $\sin(nx)$ is an odd function, then $x\sin(nx)$ is even. So,

$$b_n = \langle f(x), \sin(nx)\rangle = \frac{1}{\pi}\int_{-\pi}^{\pi} x\sin(nx)\,dx = \frac{2}{\pi}\int_{0}^{\pi} x\sin(nx)\,dx.$$

We now apply integration by parts with $x = u$ and $\sin(nx)\,dx = dv$:

$$\begin{aligned}
u &= x & -\frac{1}{n}\cos(nx) &= v \\
du &= dx & \sin(nx)\,dx &= dv.
\end{aligned}$$

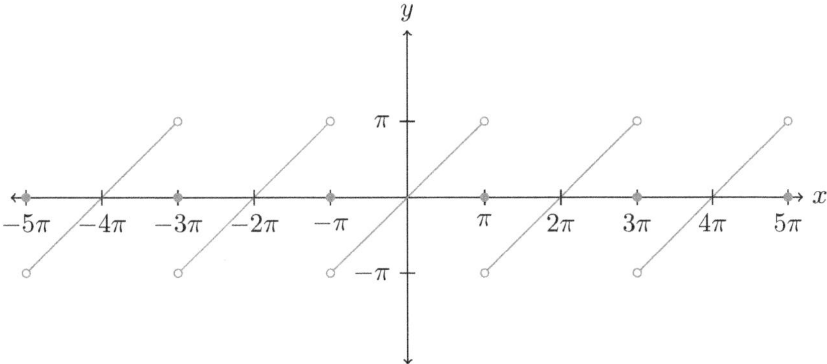

FIGURE 6.4: The saw tooth function

It follows that

$$\begin{aligned}\frac{2}{\pi}\int_0^{\pi} x\sin(nx)\,dx &= \frac{2}{\pi}\left[-\frac{x}{n}\cos(nx)\Big|_0^{\pi} + \int_0^{\pi}\frac{1}{n}\cos(nx)\,dx\right]\\ &= \frac{2}{\pi}\left[-\frac{\pi}{n}(-1)^n + \frac{1}{n}\int_0^{\pi}\cos(nx)\,dx\right].\end{aligned}$$

The n in the denominator is never a problem because b_0 does not occur. When evaluating $\frac{x}{n}\cos(nx)$ at $x = 0$ we get zero because of the factor of x; when evaluating it at $x = \pi$ we use $\cos(n\pi) = (-1)^n$. Continuing from here, use u-substitution with $u = nx$:

$$\begin{aligned}&u = nx && x = 0 \longrightarrow u = 0\\ &du = n\,dx \implies dx = \frac{1}{n}\,du && x = \pi \longrightarrow u = n\pi.\end{aligned}$$

It follows that

$$\begin{aligned}\frac{2}{\pi}\left[-\frac{\pi}{n}(-1)^n + \frac{1}{n}\int_0^{\pi}\cos(nx)\,dx\right] &= \frac{2}{\pi}\left[-\frac{\pi}{n}(-1)^n + \frac{1}{n^2}\int_0^{n\pi}\cos(u)\,du\right]\\ &= \frac{2}{\pi}\left[-\frac{\pi}{n}(-1)^n + \frac{1}{n^2}\sin(u)\Big|_0^{n\pi}\right]\\ &= \frac{2}{\pi}\left[-\frac{\pi}{n}(-1)^n\right] = -\frac{2}{n}(-1)^n.\end{aligned}$$

So the Fourier series that we get for $f(x) = x$ is

$$\sum_{n=1}^{\infty} -\frac{2}{n}(-1)^n \sin(nx).$$

It converges for all x, but matches $f(x) = x$ only on $(-\pi, \pi)$. This should make sense because the Fourier series is periodic, while $f(x) = x$ is not. What it actually converges to is the "saw tooth function."

As Figure 6.4 shows, the saw tooth function is zero at any even multiple of π, but jump discontinuities occur at odd multiples of π. The Fourier series is uniformly convergent away from these jump discontinuities.

Activity 6.7.5. Compute a_n and b_n for $f(x) = |x|$. You should treat a_0 as a special case. Hint: $|x|$ is an even function.

Activity 6.7.6. Compute a_n and b_n for

$$f(x) = \begin{cases} 1 & \text{for } x > 0, \\ 0 & \text{otherwise.} \end{cases}$$

You should treat a_0 as a special case.

Exercises

Problem 6.1. Given the quadratic form

$$Q(\mathbf{x}) = \begin{bmatrix} x_1 & x_2 \end{bmatrix} \begin{bmatrix} 3 & 1 \\ 1 & 5 \end{bmatrix} \begin{bmatrix} x_1 \\ x_2 \end{bmatrix} = 3x_1^2 + 2x_1x_2 + 5x_2^2,$$

do the following steps,

1. Complete the square with $2x_1x_2 + 5x_2^2$.
2. Use your result from the previous step to define variables u_1, u_2 diagonalizing the symmetric matrix.

Problem 6.2. Let

$$\mathbf{x} = \begin{bmatrix} x_1 \\ x_2 \end{bmatrix}, \quad \mathbf{A} = \begin{bmatrix} 3 & 2 \\ 2 & 0 \end{bmatrix}, \quad \text{and} \quad Q(\mathbf{x}) = \mathbf{x}^T\mathbf{A}\mathbf{x} = 3x_1^2 + 4x_1x_2,$$

do the following steps:

1. Compute the eigenvalues for the matrix $\mathbf{A}$.
2. Compute basis vectors for each eigenspace.
3. Show that the eigenspaces are orthogonal.
4. Normalize the eigenvectors and write down a matrix $\mathbf{S}$ such that $\mathbf{S}^{-1} = \mathbf{S}^T$.
5. Calculate $\mathbf{B} = \mathbf{S}^T\mathbf{A}\mathbf{S}$, which will be diagonal with the eigenvalues on the diagonal (if your previous steps are correct).
6. Define $\mathbf{u} = \mathbf{S}^T\mathbf{x}$, and compute $Q(\mathbf{u}) = \mathbf{u}^T\mathbf{B}\mathbf{u}$.
7. Classify the quadratic form Q as to whether it is positive definite, negative definite, or indefinite.

Problem 6.3. Given

$$\mathbf{u} = \begin{bmatrix} 1 \\ 1 \end{bmatrix} \quad \text{and} \quad \mathbf{v} = \begin{bmatrix} 2 \\ 0 \end{bmatrix},$$

1. compute $\mathbf{u} \cdot \mathbf{v}$ using components,
2. compute $\mathbf{u} \cdot \mathbf{v}$ as $\| u \| \cdot \| v \| \cos(\theta)$,
3. compute $\mathbf{w} = \mathbf{u} - \mathbf{v}$ and verify the law of cosines:

$$\| \mathbf{w} \|^2 = \| \mathbf{u} \|^2 + \| \mathbf{v} \|^2 - 2 \| u \| \cdot \| v \| \cos(\theta).$$

Problem 6.4. Let $\mathbf{e}_1$, $\mathbf{e}_2$, and $\mathbf{e}_3$ be the standard basis in $\mathbb{R}^3$.

1. Show that $\mathbf{e}_3$ is orthogonal to $\mathbf{e}_1$ and $\mathbf{e}_2$
2. Let $\mathbf{v}$ be in the span of $\mathbf{e}_1$ and $\mathbf{e}_2$. Express $\mathbf{v}$ as a linear combination of $\mathbf{e}_1$ and $\mathbf{e}_2$.
3. Show that $\mathbf{e}_3$ is also orthogonal to $\mathbf{v}$.

Problem 6.5. Let V be the subspeace of $\mathbb{R}^3$ spanned by

$$\begin{bmatrix} 1 \\ 0 \\ -1 \end{bmatrix} \quad \text{and} \quad \begin{bmatrix} 1 \\ 1 \\ 0 \end{bmatrix}$$

1. Compute $V^{\perp}$
2. Compute $\dim V$ and $\dim V^{\perp}$, then show that the sum of the dimensions is equal to the dimension of $\mathbb{R}^3$.

Problem 6.6. Suppose that a linear transformation $T : \mathbb{R}^5 \to \mathbb{R}^4$ has matrix

$$\mathbf{A} \begin{bmatrix} 1 & 0 & 0 & 0 & 0 \\ 0 & 1 & -3 & 0 & 0 \\ 0 & 0 & 0 & 0 & 1 \\ 0 & 0 & 0 & 0 & 0 \end{bmatrix}.$$

1. Compute a parameterization for $\text{Nul}(A)$
2. Show that $\text{Row}(A)^{\perp} = \text{Nul}(A)$
3. Compute a parameterization for $\text{Nul}(A)^T$
4. Show that $\text{Col}(A)^{\perp} = \text{Nul}(A)^T$

Problem 6.7. Apply Gram-Schmidt to the vectors

$$\begin{bmatrix} 1 \\ 1 \\ 0 \end{bmatrix}, \begin{bmatrix} 1 \\ 0 \\ 1 \end{bmatrix}, \begin{bmatrix} 0 \\ 1 \\ 1 \end{bmatrix}$$

Problem 6.8. Use least squares to find an approximate solution to the system:

$$\begin{aligned} x + y &= 1 \\ x + 2y &= -1 \\ x + 3y &= -3 \end{aligned}$$

Problem 6.9. Compute the singular value decomposition of

$$\mathbf{A} = \begin{bmatrix} 1 & 1 & 1 & 1 \\ 1 & 1 & -1 & -1 \end{bmatrix}$$

Problem 6.10. Show that each part of theorem 6.2.7 is false for the matrix

$$\mathbf{A} = \begin{bmatrix} 1 & 1 \\ 0 & 0 \end{bmatrix}$$

Problem 6.11. Show that if $n \neq m$, then

$$\langle \sin(nx), \sin(mx) \rangle = 0.$$

Problem 6.12. Some people prefer to do Fourier series with exponentials instead of sine and cosine. One advantage is that the functions e^{inx} are actually ortho-normal with respect to the inner product

$$\langle f, g \rangle = \frac{1}{2\pi} \int_{-\pi}^{\pi} f(x)\overline{g(x)}\, dx. \tag{6.39}$$

One disadvantage is that the functions e^{inx} are not real valued.

1. Show that $\langle e^{inx}, e^{inx} \rangle = 1$ for all n with $\langle f, g \rangle$ as defined by 6.39
2. Show that $\langle e^{inx}, e^{imx} \rangle = 0$ for $m \neq n$. You will need to do the cases where $n - m$ is even and $n - m$ is odd separately.
3. The connection with sines and cosines can be made with the help of

   $$e^{inx} = \cos(nx) + i \sin(nx).$$

   Substitute this identity into

   $$\sum_{-\infty}^{\infty} c_n e^{inx}$$

   and show that

   $$a_n = c_n + c_{-n} \quad b_n = c_n - c_{-n}$$

   for all non-negative integers n. This calculation provides a second explanation for the $\frac{a_0}{2}$ term in equations 5.2 and 5.4, at the same time, it shows why b_0 does not occur.

$$\sum_{-\infty}^{\infty} c_n e^{inx}$$

Problem 6.13 (with AI)**.** Use the following AI prompt to apply PCA to an open data set:

"Can you write some code in python that will load the Iris dataset, compute its covariance matrix, perform PCA to reduce dimensions to 2D, and produce a 2D scatter plot with the points representing each species of iris differ by both shape and color?"

How much separation or overlap between the species with PCA? To compare the result to what you would get without PCA, you should also generate at least one scatter plot using two of the raw variables in the dataset. If necessary, ask the AI to give you code for that too.

7

Representation Theory

7.1 Introduction to Linear Representations

As it turns out, we have already seen an example of a linear representation in the context of the symmetries of a square. At the time we did not define what a group is, so we will do that now.

Definition 7.1.1. A **group** G is a set, together with an operation $*$ on elements of the set, satisfying the following properties:

1. There exists an **identity element** e in G satisfying

$$e * g = g * e = g$$

for all g in G.

2. For every element g in G, there exists an **inverse element** g^{-1} in G, satisfying

$$g * g^{-1} = g^{-1} * g = e$$

3. G is **associative**. That is, for all a, b, and c in G,

$$a * (b * c) = (a * b) * c.$$

G is called an **abelian group** if the following additional property holds:

4. G is **commutative**. That is, for all a and b in G,

$$a * b = b * a.$$

There are many important examples, and it is common to name them.

1. $\mathbb{R}$ is an abelian group with addition as the operation. The identity element is 0. The inverse of any number x is $-x$. The same goes for $\mathbb{Z}$, $\mathbb{Q}$, and $\mathbb{C}$.

2. $\mathbb{R}^\times = (-\infty, 0) \cup (0, \infty)$ is an abelian group with multiplication as the operation. The identity element is 1. The inverse of any number x is its reciprocal $1/x$, which is why 0 must be removed to get a group. The same goes for $\mathbb{Q}^\times$ and $\mathbb{C}^\times$. The notation of raising a set to the power $\times$ means that all elements that don't have a multiplicative inverse are removed. So, for example, $\mathbb{Z}^\times$ is just ± 1 with multiplication.

3. Any vector space is an abelian group with vector addition as the operation. The identity is the zero vector, and the inverse of any vector v is $-v$.

DOI: 10.1201/9781003737490-7

4. There are multiple groups of invertible matrices, with matrix multiplication as the operation. The identity matrix is the identity element. Given any matrix $\mathbf{A}$, the inverse is $\mathbf{A}^{-1}$. The zero matrix is never in a multiplicative group because it is not invertible. If F is the field of scalars (e.g. $\mathbb{R}$ for real matrices), then there are many interesting groups of $n \times n$ matrices, including

   (a) $\mathrm{GL}_n(F)$, the "General linear group," is the group of all $n \times n$ invertible matrices

   (b) $\mathrm{SL}_n(F)$, the "Special linear group," is the group of all $n \times n$ matrices with determinant 1,

   (c) $\mathrm{O}_n(F)$, the "Orthogonal group," is the group all $n \times n$ matrices satisfying $\mathbf{A}^T\mathbf{A} = \mathbf{I}$.

   (d) $\mathrm{SO}_n(F)$, the "Special orthogonal group," is the group of all $n \times n$ matrices with determinant 1 satisfying $\mathbf{A}^T\mathbf{A} = \mathbf{I}$.

   Since a 1×1 matrix is just a number, then $\mathrm{GL}_1(F) = F^\times$.

5. There are many symmetry groups, such as the symmetries of the square. The operation is applying symmetries one after another (function composition). The identity is "doing nothing." Many symmetry groups have names. The symmetry group of a regular n-gon is called the "Dihedral group" and written as either D_{2n} or D_n. In this book we will use D_n, so D_4 is the symmetry group of the square. The group $\mathrm{SO}_3(\mathbb{R})$ is the group of rotations in 3 dimensions, which is a subgroup of the symmetries of the sphere; mirror reflections are not included because they have determinant -1.

6. Lastly there are permutation groups. Given the set $S = \{1, 2, \dots n\}$, a permutation is a function $f : S \to S$ that is both one-to-one and onto. Two permutations f and g can be combined by function composition $f \circ g$, and so all of the permutations of $\{1, 2, \dots n\}$ taken together form a group, which is called S_n, the "Symmetric group" on n symbols. The identity element ι is the function sending each number to itself, that is 1 goes to 1, 2 goes to 2, etc. The inverse of any permutation f is given by f^{-1}. It is natural to represent permutations with function tables. However, "cycle notation" is more concise and caries important information that we will need, so in this book, it will be the standard way of representing permutations. For this reason, we take the time to explain cycle notation in the next example.

Example 7.1.1. A cycle is a permutation with exactly one loop of numbers, each number being sent to the next number in the loop. For example the permutation σ, given by

1 2
3 4

contains a cycle sending 1 to 2, 2 to 3, 3 to 1, while 4 fixed. As a sideways function table

$$\sigma = \begin{pmatrix} 1 & 2 & 3 & 4 \\ 2 & 3 & 1 & 4 \end{pmatrix},$$

in which the first row is the input and the second row is the output. We can abbreviate the cycle (1 goes to 2, 2 goes to 3, and 3 goes back to 1) as

$$(1\ 2\ 3),$$

where it is implied that the last number gets sent back to the first. It does not matter what number you start with in a cycle, so

$$(1\ 2\ 3) = (2\ 3\ 1) = (3\ 1\ 2).$$

The number of elements in a cycle is called the cycle length, so $(1\ 2\ 3)$ has length 3. A fixed point is a cycle with length 1, so the cycle notation for σ is

$$\sigma = (1\ 2\ 3)(4).$$

The order of the cycles does not matter if they share no number is common. On the other hand, if two cycles share a number in common, then we apply them in the order of function composition (right to left). For example, let's look at the action of $(1\ 2\ 3)(3\ 4)$ on each number. Anytime a number is not in the cycle, it remains fixed. Anytime a number is in a cycle it gets sent to the next number, or loops back to the first number in the cycle. So,

$$(1\ 2\ 3)(3\ 4)1 = (1\ 2\ 3)1 = 2$$

because 1 is not in $(3\ 4)1$, so that cycle does nothing, and then $(1\ 2\ 3)$ sends 1 to 2. Similarly,

$$\begin{aligned}(1\ 2\ 3)(3\ 4)2 &= (1\ 2\ 3)2 = 3,\\ (1\ 2\ 3)(3\ 4)3 &= (1\ 2\ 3)4 = 4, \text{and}\\ (1\ 2\ 3)(3\ 4)4 &= (1\ 2\ 3)3 = 1.\end{aligned}$$

So 1 goes to 2, 2 goes to 3, 3 goes to 4, and 4 goes to 1, meaning that

$$(1\ 2\ 3)(3\ 4) = (1\ 2\ 3\ 4).$$

With practice, computations with cycles can be done quite fast.

To get the inverse of a permutation from a sideways function table, we can just reverse the rows:

$$\sigma^{-1} = \begin{pmatrix} 2 & 3 & 1 & 4 \\ 1 & 2 & 3 & 4 \end{pmatrix} = \begin{pmatrix} 1 & 2 & 3 & 4 \\ 3 & 1 & 2 & 4 \end{pmatrix}.$$

To get the inverse of a permutation from cycle notation, we can reverse the order of the cycles and the order of the numbers within each cycle

$$\sigma^{-1} = (4)(3\ 2\ 1).$$

Activity 7.1.2. Suppose that σ is the permutation given by

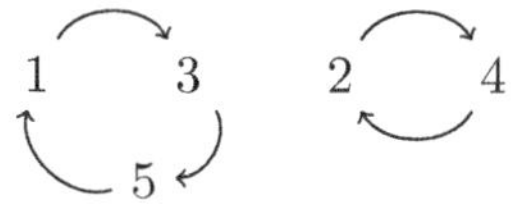

1 3 2 4
5

Do the following:

1. Express σ as a sideways function table.

2. Express σ in cycle notation.

3. Compute σ^{-1} both ways.

4. Compute $\sigma \circ \tau$ where $\tau = (3\ 4)$. Remember that function composition is read right to left, so $(3\ 4)$ gets applied before any of the cycles in σ.

Sage represents permutations in cycle notation too, but there are two things to keep in mind:

1. Sage omits fixed points from the cycle notation,
2. Sage multiplies cycles left to right.

The first is not a problem if it is understood what numbers are being used, e.g.1 through 4. The second is also not a huge problem, since in one sense it is just a convention. So Example 7.1.1 should be done in Sage like this:

```
S4=SymmetricGroup(4)
S4[11]          #(1 2 3)
S4[12]          #(3 4)
S4[12]*S4[11]   #(1 2 3 4)
```

So, for Sage

$$(3\ 4)(1\ 2\ 3) = (1\ 2\ 3\ 4),$$

which is the exact opposite order of multiplication as in this book. The convention used in this book is based on the viewpoint of permutations as functions, and function composition works right to left naturally.

Definition 7.1.2. Given a group G and an F-vector space V with dimension n, a **linear representation** of G is given by a function

$$\rho : G \to \mathrm{GL}_n(F)$$

satisfying the condition that $\rho(gh) = \rho(g)\rho(h)$ for all g and h in G. The **dimension** or **degree** of the representation is the dimension of the vector space V.

Note that ρ, pronounced "rho," is a Greek letter making the sound of an r, and r is the first letter in "representation."

Example 7.1.3. We now have a name for the symmetries of the square, D_4. We will use r for rotation by 90 degrees, and m for mirror reflection. Previously, we obtained a 2 dimensional representation by placing the corners of a square at ends of the vectors

$$\mathbf{A} = \begin{bmatrix} 1 \\ 0 \end{bmatrix} \quad \mathbf{B} = \begin{bmatrix} 0 \\ 1 \end{bmatrix} \quad \mathbf{C} = \begin{bmatrix} -1 \\ 0 \end{bmatrix} \quad \mathbf{D} = \begin{bmatrix} 0 \\ -1 \end{bmatrix}.$$

By looking at the action of m and r on these vectors, we obtained

$$\rho(r) = \mathbf{R} = \begin{bmatrix} 0 & 1 \\ -1 & 0 \end{bmatrix} \quad \text{and} \quad \rho(m) = \mathbf{M} = \begin{bmatrix} 0 & 1 \\ 1 & 0 \end{bmatrix}$$

We will illustrate the property $\rho(gh) = \rho(g)\rho(h)$ by using $g = r$ and $h = r^2$ then $gh = r^3$. Since r is rotation by 90 degrees then h is rotation by 180 degrees, thus we have

$$\rho(r) = \mathbf{R} = \begin{bmatrix} 0 & -1 \\ 1 & 0 \end{bmatrix}, \quad \rho(r^2) = \mathbf{R}^2 = \begin{bmatrix} -1 & 0 \\ 0 & -1 \end{bmatrix}, \quad \text{and} \quad \mathbf{R} \cdot \mathbf{R}^2 = \begin{bmatrix} 0 & 1 \\ -1 & 0 \end{bmatrix}$$

which is the matrix for rotation by 270 degrees as expected.

Example 7.1.4 (Permutation matrices.). We get a natural action of S_n on an n-dimensional vector space V by acting on the *indices* of the vectors $\mathbf{e}_i$ in a basis for V. For example, for $n = 4$, the standard basis is

$$\mathbf{e}_1 = \begin{bmatrix}1\\0\\0\\0\end{bmatrix}, \mathbf{e}_2 = \begin{bmatrix}0\\1\\0\\0\end{bmatrix}, \mathbf{e}_3 = \begin{bmatrix}0\\0\\1\\0\end{bmatrix}, \mathbf{e}_4 = \begin{bmatrix}0\\0\\0\\1\end{bmatrix}.$$

Applying permutation $g = (1\ 2\ 3\ 4)$ to the indices leads to the following action:

$$\begin{aligned} g \cdot \mathbf{e}_1 &= \mathbf{e}_2 = 0\mathbf{e}_1 + 1\mathbf{e}_2 + 0\mathbf{e}_3 + 0\mathbf{e}_4, \\ g \cdot \mathbf{e}_2 &= \mathbf{e}_3 = 0\mathbf{e}_1 + 0\mathbf{e}_2 + 1\mathbf{e}_3 + 0\mathbf{e}_4, \\ g \cdot \mathbf{e}_3 &= \mathbf{e}_4 = 0\mathbf{e}_1 + 0\mathbf{e}_2 + 0\mathbf{e}_3 + 1\mathbf{e}_4, \\ g \cdot \mathbf{e}_4 &= \mathbf{e}_1 = 1\mathbf{e}_1 + 0\mathbf{e}_2 + 0\mathbf{e}_3 + 0\mathbf{e}_4. \end{aligned}$$

Then, as usual, we get a matrix representing σ by taking the transpose of the matrix formed by the coefficients

$$\rho(g) = \begin{bmatrix}0 & 0 & 0 & 1\\1 & 0 & 0 & 0\\0 & 1 & 0 & 0\\0 & 0 & 1 & 0\end{bmatrix}.$$

This construction works irrespective of the choice of scalars or the choice of basis.

Activity 7.1.5. Given g as in the previous example and $h = (1)(2\ 3\ 4)$, do the following:

1. Compute $\rho(h)$.
2. Compute the function $h \circ g$, with g as given in the example above (see Example 7.1.1).
3. Show that
$$\rho(h \circ g) = \rho(h)\rho(g).$$
In other words, compute the matrix for $h \circ g$, and show that it is equal to the product of the matrices for h and g.

When Sage constructs a matrix for a permutation, it does not take the transpose.

```
S4=SymmetricGroup(4)
S4[23].matrix()
```

This is fine if the matrix is understood to act on *row vectors.* Since we acted on column vectors, the transpose is required.

```
S4[23].matrix().T
```

Why not act on row vectors all the time then? Well, then you might have to do column reduction instead of row reduction. There are always trade-offs.

7.2 Characters

Definition 7.2.1. If ρ is a representation of G on an n-dimensional F-vector space, then the trace gives a function $\chi : G \to F$ called a **character**; thus $\chi = \operatorname{tr} \circ \rho$.

As we have seen from Proposition 4.6.2, the trace remains the same after conjugation. Therefore, when making a function table for a character, it is common to only specify the value for each conjugacy class. There are various ways of labeling conjugacy classes, but for concreteness, and so as not to muddy the waters, in this book we will use representatives for each class.

Example 7.2.1. In D_4, the conjugacy classes are as follows:

1. The identity e is only conjugate to itself, so $|\mathcal{C}_e| = 1$.
2. The rotations r and r^3 are conjugate to each other, so $|\mathcal{C}_r| = 2$.
3. The rotation r^2 is only conjugate to itself, so $|\mathcal{C}_{r^2}| = 1$.
4. The mirrors m and mr^2 are conjugate, so $|\mathcal{C}_m| = 2$.
5. The mirrors mr and mr^3 are conjugate, so $|\mathcal{C}_{mr}| = 2$.

For the 2 dimensional representation ρ in Example 7.1.3, we get the matrices

$$\rho(e) = \begin{bmatrix} 1 & 0 \\ 0 & 1 \end{bmatrix}, \quad \rho(r) = \begin{bmatrix} 0 & -1 \\ 1 & 0 \end{bmatrix}, \quad \rho(r^2) = \begin{bmatrix} -1 & 0 \\ 0 & -1 \end{bmatrix},$$
$$\rho(m) = \begin{bmatrix} 0 & 1 \\ 1 & 0 \end{bmatrix}, \quad \text{and} \quad \rho(mr) = \begin{bmatrix} 1 & 0 \\ 0 & -1 \end{bmatrix}.$$

By taking the trace of each matrix, we obtain the following function table for $\chi = \operatorname{tr} \circ \rho$:

g	e	r	r^2	m	mr
$\lvert\mathcal{C}_g\rvert$	1	2	1	2	2
$\chi(g)$	2	0	-2	0	0

In the future it will also be convenient to keep track of the number of conjugates, so we list that in row labeled $|\mathcal{C}_g|$.

Activity 7.2.2. Complete the following function table for $\chi = \operatorname{tr} \circ \rho$, where ρ is the representation of S_3 given by acting on the indices of the standard basis $\mathbf{e}_1$, $\mathbf{e}_2$, $\mathbf{e}_3$.

g	(1)	(1 2)	(1 2 3)
$\lvert\mathcal{C}_g\rvert$	1	3	2
$\chi(g)$			

7.3 Irreducible Representations

Given a representation ρ of a group G on a vector space V, it is possible that there may be an invariant subspace V_1, much as eigenspaces are invariant subspaces for a single matrix, except that here we are talking about invariance under all matrices representing the group G. If this happens, then a basis can be found bringing the representation into block upper-triangular form:

$$\rho(g) = \begin{bmatrix} \rho_1(g) & \beta(g) \\ \mathbf{0} & \rho_2(g) \end{bmatrix}, \tag{7.1}$$

where ρ_1 and ρ_2 are representations on V_1 and V_2 respectively and $V = V_1 \oplus V_2$. It is critical to understand that the basis must work for all g.

Definition 7.3.1. If it is possible to bring a representation ρ into the form given by equation (7.1), then the representation ρ is called **reducible** and the representations ρ_1 and ρ_2 are called **sub representations**. If this is not possible, then ρ is **irreducible**.

It may be possible to go even further and obtain a form that is block-diagonal:

$$\rho(g) = \begin{bmatrix} \rho_1(g) & \mathbf{0} \\ \mathbf{0} & \rho_2(g) \end{bmatrix}, \tag{7.2}$$

Definition 7.3.2. If it is possible to bring a representation ρ into the form given by equation (7.2), then the representation ρ is said to be **decomposable**. If this is not possible, then ρ is **indecomposable**.

If G is an infinite group, then it may be possible to obtain block upper triangular form, but not block diagonal form, meaning that it is possible for ρ to be indecomposable but not irreducible. However, when G is finite, Maschke's theorem shows that these concepts are equivalent.

Intuitively, we are trying to understand larger representations in terms of smaller ones. The power of Maschke's theorem is that, for a finite group G, it is possible to treat the irreducible representations as building blocks of larger representations. How do we know when we have found all of the irreducible representations of G?

Proposition 7.3.3. *For a finite group G the number of irreducible representations is equal to the number of conjugacy classes.*

The task of constructing the irreducible representations is another story, but there are some useful facts to keep in mind. But first, we need new terminology.

Definition 7.3.4. The **order** of a finite group G, is the number of elements in the group. It is sometimes written as $|G|$ or as $\#G$.[1]

Here are a few properties that are sometimes useful for computing the dimensions of irreducible representations.

1. The dimension of an irreducible representation must be a factor of the order of a group.

2. The sum of the squares of the dimensions must be equal to the order of a group.

3. The one dimensional representations form a group by multiplication.

Example 7.3.1 (Irreducible representations of D_4 (Symmetries of the square)). D_4 has order 8. There are 5 conjugacy classes total. The 2-dimensional representation of D_4 in Example 7.1.4 is irreducible and will henceforth be called ρ_5. There must be four remaining irreducible representations, all of which must have dimension 1 for the sum of the squares to be 8:

$$1^2 + 1^2 + 1^2 + 1^2 + 2^2 = 8.$$

There are two easy irreducible representations of dimension 1. First,

$$\rho_1(g) = 1$$

for all g. Second,

$$\rho_2(g) = \begin{cases} 1 & \text{if } g \text{ is orientation preserving (i.e. a rotation)} \\ -1 & \text{if } g \text{ is orientation reversing (i.e. a mirror),} \end{cases}$$

[1]The advantage of $|G|$ is that it can be compared with absolute value as having something to do with size. The disadvantage of $|G|$ is that we also use $|$ for divisibility. If you want to talk about a group of order divisible by 3 how friendly is $3 \mid |G|$?

which is the determinant of the 2-dimensional irreducible representation. Generally, taking the determinant of a higher dimensional representation leads to a 1-dimensional representation, so this is not a surprise.

We get a third irreducible representation, as follows.

$$\rho_3(g) = \begin{cases} 1 & \text{if } g \text{ preserves the diagonals of the square} \\ -1 & \text{if } g \text{ reverses the diagonals of the square} \end{cases}.$$

To get the fourth irreducible representation of dimension 1, we multiply the second and third: $\rho_4(g) = \rho_2(g)\rho_3(g)$. Thus we obtain the following table of characters, in which χ_i is the character for ρ_i.

g	e	r	r^2	m	mr
$\lvert\mathcal{C}_g\rvert$	1	2	1	2	2
$\chi_1(g)$	1	1	1	1	1
$\chi_2(g)$	1	1	1	-1	-1
$\chi_3(g)$	1	-1	1	1	-1
$\chi_4(g)$	1	-1	1	-1	1
$\chi_5(g)$	2	0	-2	0	0

We can check our work with Sage using the following code:

```
D4 = DihedralGroup(4)
D4.character_table()
D4.conjugacy_classes_representatives()
```

The first line defines the group D_4. The second line creates the character table, but does not label the columns. Each column corresponds to a conjugacy class, so the third line asks Sage to list a representative of each conjugacy class. Sage lists them in cycle notation in the same order. See Example 7.3.4 for more.

Let G be a finite group, and let α and β be two complex valued functions that are well-defined on the conjugacy classes of G. Then $\alpha + \beta$ and $c\alpha$ are as well, for any $c \in \mathbb{C}$. Hence, these functions are a vector space. The dimension of the vector space is equal to the number of conjugacy classes, and

$$\begin{aligned} \langle \alpha, \beta \rangle &= \frac{1}{|G|} \sum_{g \in G} \alpha(g)\overline{\beta(g)} \\ &= \frac{1}{|G|} \sum_{\mathcal{C}_g} |\mathcal{C}_g| \cdot \alpha(g)\overline{\beta(g)} \end{aligned} \tag{7.3}$$

defines an inner product, where $\overline{a + bi} = a - bi$ is complex conjugation. As this equation shows, on one hand we can sum over all g, on the other hand, we can collect terms together belonging to the same conjugacy class, since the values of α an β remain the same.

Proposition 7.3.5 (Schur orthogonality). *The characters of the irreducible representations of G form an orthonormal basis with respect to the inner product* (7.3). *That is*

$$\langle \chi_i, \chi_j \rangle = \begin{cases} 1 & \text{if } i = j \\ 0 & \text{if } i \neq j. \end{cases}$$

Example 7.3.2. We look at two examples using the character table for D_4. It is important to either

1. sum over all elements of g, or

2. if we sum just over the representatives, then we must count the number of conjugates.

All of the important details are listed in Examples 7.2.1 and 7.3.1. As an illustration of normality:

$$\begin{aligned}\langle \chi_5, \chi_5 \rangle &= \frac{1}{8}\left(|\chi_5(e)|^2 + |\chi_5(r)|^2 + |\chi_5(r^2)|^2 + |\chi_5(r^3)|^2\right. \\ &\quad \left. + |\chi_5(m)|^2 + |\chi_5(mr)|^2 + |\chi_5(mr^2)|^2 + |\chi_5(mr^3)|^2\right) \\ &= \frac{1}{8}\left(1 \cdot |\chi_5(e)|^2 + 2 \cdot |\chi_5(r)|^2 + 1 \cdot |\chi_5(r^2)|^2\right. \\ &+2 \cdot |\chi_5(m)|^2 + 2 \cdot |\chi_5(mr)|^2\left.\right) \\ &= \frac{1}{8}\left(1 \cdot 2^2 + 2 \cdot 0 + 1 \cdot (-2)^2 + 2 \cdot 0 + 2 \cdot 0\right) \\ &= \frac{1}{8}(4 + 0 + 4 + 0 + 0) = \frac{1}{8} \cdot 8 = 1.\end{aligned}$$

The simplification between the first and second equal sign occurs because r and r^3 are conjugates, m and mr^2 are conjugates, and mr and mr^3 are conjugates. For convenience, the number of conjugates are listed in the character table. Then we look up the values in the character table. In this case they are all integers, so the absolute value is overkill; the absolute value would be necessary with complex numbers.

As an illustration of orthogonality:

$$\begin{aligned}\langle \chi_3, \chi_5 \rangle &= \frac{1}{8}\Big(1 \cdot \chi_3(e)\overline{\chi_5(e)} + 2 \cdot \chi_3(r)\overline{\chi_5(r)} + 1 \cdot \chi_3(r^2)\overline{\chi_5(r^2)} \\ &\quad +2 \cdot \chi_3(m)\overline{\chi_5(m)} + 2 \cdot \chi_3(mr)\overline{\chi_5(mr)}\Big) \\ &= \frac{1}{8}(1 \cdot 1 \cdot 2 + 2 \cdot (-1) \cdot 0 + 1 \cdot 1 \cdot (-2) + 2 \cdot 1 \cdot 0 + 2 \cdot (-1) \cdot 0) \\ &= \frac{1}{8}(2 + 0 - 2 + 0 + 0) = \frac{1}{8} \cdot 0\end{aligned}$$

Activity 7.3.3. Using the character table for D_4, compute $\langle \chi_2, \chi_2 \rangle$ and $\langle \chi_1, \chi_2 \rangle$.

The point of Schur orthogonality is that it enables us to efficiently determine which irreducible representations are present in a larger representation. Let's say that we have a representation ρ of finite group G, and that χ is the character of ρ. Then χ is a linear combination of the characters χ_i of the irreducible representations ρ_i of G:

$$\chi = c_1\chi_1 + c_2\chi_2 + \cdots c_m\chi_m$$

Then Schur orthogonality gives us

$$\langle \chi, \chi_i \rangle = c_i,$$

which is the number of times that the irreducible representation ρ_i occurs when ρ is brought into block diagonal form.

Example 7.3.4. Suppose we label the vertices of a square as follows:

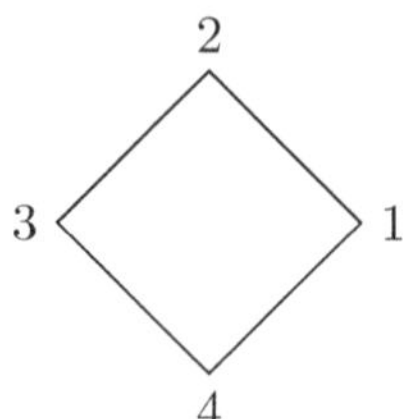

Then the rotation r acts on the vertices as the permutation (1 2 3 4) and the mirror m acts on the vertices as the permutation (1 4)(2 3). This, in fact, is how Sage represents the elements of D_4. Therefore, we can get a 4-dimensional representation of D_4 with permutation matrices:

$$\rho(r) = \begin{bmatrix} 0 & 0 & 0 & 1 \\ 1 & 0 & 0 & 0 \\ 0 & 1 & 0 & 0 \\ 0 & 0 & 1 & 0 \end{bmatrix} \quad \text{and} \quad \rho(m) = \begin{bmatrix} 0 & 0 & 0 & 1 \\ 0 & 0 & 1 & 0 \\ 0 & 1 & 0 & 0 \\ 1 & 0 & 0 & 0 \end{bmatrix},$$

the first of which should be familiar from Example 7.1.3. Both of these matrices have trace 0. We still need the trace of e,r^2, and mr. For r^2 in cycle notation,

$$(1\ 2\ 3\ 4)(1\ 2\ 3\ 4) = (1\ 3)(2\ 4) \quad \text{so} \quad \rho(r^2) = \begin{bmatrix} 0 & 0 & 1 & 0 \\ 0 & 0 & 0 & 1 \\ 1 & 0 & 0 & 0 \\ 0 & 1 & 0 & 0 \end{bmatrix}, \tag{7.4}$$

which has trace 0. For mr in cycle notation,

$$(1\ 4)(2\ 3)(1\ 2\ 3\ 4) = (1\ 3) \quad \text{so} \quad \rho(mr) = \begin{bmatrix} 0 & 0 & 1 & 0 \\ 0 & 1 & 0 & 0 \\ 1 & 0 & 0 & 0 \\ 0 & 0 & 0 & 1 \end{bmatrix}, \tag{7.5}$$

which has trace 2. And $\rho(e)$ is the 4×4 identity matrix which has trace 4. Thus we get the following values for the character χ of ρ:

g	e	r	r^2	m	mr
$\lvert C_g \rvert$	1	2	1	2	2
$\chi(g)$	4	0	0	0	2

The representation ρ is reducible. It can be broken into irreducible pieces with the help of Schur orthogonality. Let's see if the 2-dimensional representation shows up. Because of the zeros of $\chi(g)$ we only need to sum over the conjugacy classes of e and mr,

$$\langle \chi, \chi_5 \rangle = \frac{1}{8}(1 \cdot 4 \cdot 2 + 2 \cdot 2 \cdot 0) = \frac{1}{8}8 = 1,$$

so we one 2×2 block. Since the rest of the irreducible representations are 1-dimensional, then there should also be two 1 by 1 blocks, but what are they? They come from the representations ρ_1 and ρ_4. Looking at the corresponding characters, we have

$$\langle \chi, \chi_1 \rangle = \langle \chi, \chi_4 \rangle = \frac{1}{8}(1 \cdot 4 \cdot 1 + 2 \cdot 2 \cdot 1) = \frac{1}{8}8 = 1.$$

The other irreducible representations do not occur at all because

$$\langle \chi, \chi_2 \rangle = \langle \chi, \chi_3 \rangle = \frac{1}{8}(1 \cdot 4 \cdot 1 + 2 \cdot 2 \cdot (-1)) = \frac{1}{8}0 = 0.$$

This means that there must be a change of basis for which

$$\rho(r) = \begin{bmatrix} 0 & -1 & & \\ 1 & 0 & & \\ & & 1 & \\ & & & -1 \end{bmatrix} \quad \text{and} \quad \rho(m) = \begin{bmatrix} 0 & 1 & & \\ 1 & 0 & & \\ & & 1 & \\ & & & -1 \end{bmatrix},$$

with ρ_5 occurring in the top left corner, ρ_1 occurring next, and ρ_4 occurring in the bottom right corner. We will discuss how to find this change of basis shortly.

Remark 7.3.1. The representation in Example 7.3.4 is a permutation representation. For such a representation, the trace is equal to the number of fixed points. The fixed points are the numbers that don't appear in the disjoint cycle notation. For instance, equation (7.5) shows that $\rho(mr)$ has trace 2, because there are 2 ones on the diagonal.

Activity 7.3.5. In Activity 7.2.2, you looked at the representation of S_3 obtained from the natural action on the indices of $\mathbf{e}_1, \mathbf{e}_2, \mathbf{e}_3$. That representation is reducible. The character table for S_3 is

g	(1)	(1 2)	(1 2 3)
$\lvert C_g \rvert$	1	3	2
$\chi_1(g)$	1	1	1
$\chi_2(g)$	1	−1	1
$\chi_3(g)$	2	0	−1

Use this table and your result from Activity 7.2.2 to compute $\langle \chi, \chi_1 \rangle$, $\langle \chi, \chi_2 \rangle$ and $\langle \chi, \chi_3 \rangle$.

In Example 7.3.4, no irreducible representation occurs more than once. In such a case, we can obtain the block decomposition with the help of the following theorem:

Theorem 7.3.6. *Let χ_i be an irreducible character of G of dimension n_i, and let ρ be a representation of G on a $\mathbb{C}$-vector space V. Then*

$$\mathbf{P}_i = \frac{n_i}{|G|} \sum_{g \in G} \overline{\chi_i}(g) \rho(g) \tag{7.6}$$

defines a projection onto the subspace of V, containing all copies of the irreducible representation ρ_i.

This theorem and its proof can be found in [23].

Before diving into an example, it is worth making a couple observations:

1. For each $g \in G$, $\overline{\chi_i}(g)$ is a scalar, and
2. the value of n_i is equal to $\chi_i(e)$.

Example 7.3.6. We continue with the representation from Example 7.3.4, and show how to construct a change of basis bringing it into block form. The irreducible representations of D_4 that occur are ρ_1, ρ_4 and ρ_5. Their characters are

g	e	r	r^2	m	mr
$\lvert C_g \rvert$	1	2	1	2	2
$\chi_1(g)$	1	1	1	1	1
$\chi_4(g)$	1	−1	1	−1	1
$\chi_5(g)$	2	0	−2	0	0

the easiest of which is χ_5 because of the zeros. The conjugacy classes are

$$e, \quad r, r^3, \quad , r^2, \quad m, mr^2, \quad mr, mr^3.$$

Since $|D_4| = 8$ and $n_5 = \chi_5(e) = 2$, then we have

$$\mathbf{P}_5 = \frac{n_5}{|G|} \sum_{g \in G} \overline{\chi_5}(g)\rho(g) = \frac{2}{8}(2\rho(e) - 2\rho(r^2)) = \frac{1}{2}(\rho(e) - \rho(r^2))$$

$$= \frac{1}{2}\left(\begin{bmatrix} 1 & 0 & 0 & 0 \\ 0 & 1 & 0 & 0 \\ 0 & 0 & 1 & 0 \\ 0 & 0 & 0 & 1 \end{bmatrix} - \begin{bmatrix} 0 & 0 & 1 & 0 \\ 0 & 0 & 0 & 1 \\ 1 & 0 & 0 & 0 \\ 0 & 1 & 0 & 0 \end{bmatrix}\right) = \frac{1}{2}\begin{bmatrix} 1 & 0 & -1 & 0 \\ 0 & 1 & 0 & -1 \\ -1 & 0 & 1 & 0 \\ 0 & -1 & 0 & 1 \end{bmatrix}.$$

The rank of $\mathbf{P}_5$ is 2, and the first two columns alone generate the column space. This is the subspace that $\mathbf{P}_5$ projects onto.

Next we look at ρ_1. Since $n_1 = 1$ and $\chi_1(g) = 1$ for all g, then we basically get the average of all matrices in the representation ρ:

$$\mathbf{P}_1 = \frac{n_1}{|G|} \sum_{g \in G} \overline{\chi_1}(g)\rho(g) = \frac{1}{8} \sum_{g \in G} \rho(g)$$

$$= \frac{1}{8}\left(\begin{bmatrix} 1 & 0 & 0 & 0 \\ 0 & 1 & 0 & 0 \\ 0 & 0 & 1 & 0 \\ 0 & 0 & 0 & 1 \end{bmatrix} + \begin{bmatrix} 0 & 0 & 0 & 1 \\ 1 & 0 & 0 & 0 \\ 0 & 1 & 0 & 0 \\ 0 & 0 & 1 & 0 \end{bmatrix} + \begin{bmatrix} 0 & 1 & 0 & 0 \\ 0 & 0 & 1 & 0 \\ 0 & 0 & 0 & 1 \\ 1 & 0 & 0 & 0 \end{bmatrix} + \begin{bmatrix} 0 & 0 & 1 & 0 \\ 0 & 0 & 0 & 1 \\ 1 & 0 & 0 & 0 \\ 0 & 1 & 0 & 0 \end{bmatrix}\right.$$

$$\left. + \begin{bmatrix} 0 & 0 & 0 & 1 \\ 0 & 0 & 1 & 0 \\ 0 & 1 & 0 & 0 \\ 1 & 0 & 0 & 0 \end{bmatrix} + \begin{bmatrix} 0 & 1 & 0 & 0 \\ 1 & 0 & 0 & 0 \\ 0 & 0 & 0 & 1 \\ 0 & 0 & 1 & 0 \end{bmatrix} + \begin{bmatrix} 1 & 0 & 0 & 0 \\ 0 & 0 & 0 & 1 \\ 0 & 0 & 1 & 0 \\ 0 & 1 & 0 & 0 \end{bmatrix} + \begin{bmatrix} 0 & 0 & 1 & 0 \\ 0 & 1 & 0 & 0 \\ 1 & 0 & 0 & 0 \\ 0 & 0 & 0 & 1 \end{bmatrix}\right)$$

$$= \frac{1}{8}\begin{bmatrix} 2 & 2 & 2 & 2 \\ 2 & 2 & 2 & 2 \\ 2 & 2 & 2 & 2 \\ 2 & 2 & 2 & 2 \end{bmatrix} = \frac{1}{4}\begin{bmatrix} 1 & 1 & 1 & 1 \\ 1 & 1 & 1 & 1 \\ 1 & 1 & 1 & 1 \\ 1 & 1 & 1 & 1 \end{bmatrix}.$$

Finally, we look at $\mathbf{P}_4$.

$$\mathbf{P}_4 = \frac{n_4}{|G|} \sum_{g \in G} \overline{\chi_4}(g)\rho(g)$$

$$= \frac{1}{8}(\rho(e) + \rho(r^2) + \rho(mr) + \rho(mr^3) - \rho(m) - \rho(mr^2) - \rho(r) - \rho(r^3))$$

$$= \frac{1}{8}\left(\begin{bmatrix} 1 & 0 & 0 & 0 \\ 0 & 1 & 0 & 0 \\ 0 & 0 & 1 & 0 \\ 0 & 0 & 0 & 1 \end{bmatrix} + \begin{bmatrix} 0 & 0 & 1 & 0 \\ 0 & 0 & 0 & 1 \\ 1 & 0 & 0 & 0 \\ 0 & 1 & 0 & 0 \end{bmatrix} + \begin{bmatrix} 1 & 0 & 0 & 0 \\ 0 & 0 & 0 & 1 \\ 0 & 0 & 1 & 0 \\ 0 & 1 & 0 & 0 \end{bmatrix} + \begin{bmatrix} 0 & 0 & 1 & 0 \\ 0 & 1 & 0 & 0 \\ 1 & 0 & 0 & 0 \\ 0 & 0 & 0 & 1 \end{bmatrix}\right.$$

$$\left. - \begin{bmatrix} 0 & 0 & 0 & 1 \\ 0 & 0 & 1 & 0 \\ 0 & 1 & 0 & 0 \\ 1 & 0 & 0 & 0 \end{bmatrix} - \begin{bmatrix} 0 & 1 & 0 & 0 \\ 1 & 0 & 0 & 0 \\ 0 & 0 & 0 & 1 \\ 0 & 0 & 1 & 0 \end{bmatrix} - \begin{bmatrix} 0 & 0 & 0 & 1 \\ 1 & 0 & 0 & 0 \\ 0 & 1 & 0 & 0 \\ 0 & 0 & 1 & 0 \end{bmatrix} - \begin{bmatrix} 0 & 1 & 0 & 0 \\ 0 & 0 & 1 & 0 \\ 0 & 0 & 0 & 1 \\ 1 & 0 & 0 & 0 \end{bmatrix}\right)$$

$$= \frac{1}{8}\begin{bmatrix} 2 & -2 & 2 & -2 \\ -2 & 2 & -2 & 2 \\ 2 & -2 & 2 & -2 \\ -2 & 2 & -2 & 2 \end{bmatrix} = \frac{1}{4}\begin{bmatrix} 1 & -1 & 1 & -1 \\ -1 & 1 & -1 & 1 \\ 1 & -1 & 1 & -1 \\ -1 & 1 & -1 & 1 \end{bmatrix}.$$

We get a change of basis by constructing a matrix consisting of a basis for the column space of each of these projections. For $\mathbf{P}_5$, we take the first two columns, and for $\mathbf{P}_1$ and $\mathbf{P}_4$, we take the first column. Luckily, they are all orthogonal.[2] So, by normalizing them, we get a matrix satisfying $\mathbf{S}^{-1} = \mathbf{S}^T$:

$$\mathbf{S} = \begin{bmatrix} \frac{1}{\sqrt{2}} & 0 & \frac{1}{2} & \frac{1}{2} \\ 0 & \frac{1}{\sqrt{2}} & \frac{1}{2} & -\frac{1}{2} \\ -\frac{1}{\sqrt{2}} & 0 & \frac{1}{2} & \frac{1}{2} \\ 0 & -\frac{1}{\sqrt{2}} & \frac{1}{2} & -\frac{1}{2} \end{bmatrix}.$$

The change of basis $\mathbf{S}^{-1}\rho(g)\mathbf{S}$ then gives us a matrix in block form for any $g \in D_4$. For $\mathbf{S}^{-1}\rho(r)\mathbf{S}$ we find

$$\mathbf{S}^{-1}\rho(r)\mathbf{S} = \begin{bmatrix} 0 & -1 & & \\ 1 & 0 & & \\ & & 1 & \\ & & & -1 \end{bmatrix}$$

and for $\mathbf{S}^{-1}\rho(m)\mathbf{S}$

$$\mathbf{S}^{-1}\rho(m)\mathbf{S} = \begin{bmatrix} 0 & -1 & & \\ -1 & 0 & & \\ & & 1 & \\ & & & -1 \end{bmatrix}$$

Since D_4 is generated by r and m, then we can get the rest of the matrices in this new basis by multiplying these two together.

While the elements of a matrix have two indices, the calculations we have just done are best understood by introducing an array with a third index. In order to use such an array in Sage, we begin by importing NumPy. Each index corresponds to what NumPy calls an "axis."

```
import numpy as np

def matrix_sum(L):
    return sum([g.matrix() for g in L])

def IrredProj(G):
    C=G.conjugacy_classes()
    char=G.character_table()

    A=np.array([matrix_sum(L) for L in C])
    R=np.dot(A.T,char.T).T/len(G)
    return np.ascontiguousarray(R)

def IrredBasis(G):
    P = IrredProj(G)
    L = [matrix(M).row_space().matrix() for M in P]
    return matrix(np.vstack(L))
```

[2]The column spaces of different $\mathbf{P}_i$ and $\mathbf{P}_j$ will generally be orthogonal, but to obtain an orthogonal basis for the column space of one of the projections $\mathbf{P}_i$ it may be required to use Gram-Schmidt.

```
def BlockForm(G):
    S=IrredBasis(G)
    return [S*g.matrix()*S^-1 for g in G]

D4 = DihedralGroup(4)

IrredProj(D4)
IrredBasis(D4)
BlockForm(D4)
```

The array `A` in the definition of `IrredProj` is created by first adding the matrices within each conjugacy class, and then stacking them. To use the dot product in NumPy, it is critical to pay attention to how it is defined. As of NumPy version 1.23, the specification for `np.dot(A,B)` says that the dot product is applied to the last axis of `A` and the *second to last* axis of `B`. We need to compute the dot product over conjugacy classes. When we computed `A` by stacking, this gave us conjugacy classes on the first axis. Meanwhile the character table has the conjugacy classes in columns (the last axis). Therefore, we need the transpose of both before the dot product can make sense. The final result of the function `IrredProj` is an array with the projections $\mathbf{P}_i$ stacked in layers.
The rest of the functions have simpler definitions. Since the matrices produced by Sage act on row vectors, then we need a basis for the row space instead of the column space. `IrredBasis` finds them, and stacks them into a matrix $\mathbf{S}$ that can be used in a change of basis. The `BlockForm` function then applies the change of basis. Note that $\mathbf{S}^{-1}$ is applied on the right, because of the action on rows instead of columns. Taking the transpose, switches from columns to rows, and switches the order of multiplication.

Activity 7.3.7. In Activity 7.3.5, ρ was the representation of S_3 by permutation matrices. You computed the irreducible representations occurring in ρ. Now you will compute the projections $\mathbf{P}_i$. You will need $|S_3| = 6$, the character table for S_3,

g	(1)	(1 2)	(1 2 3)
$\lvert\mathcal{C}_g\rvert$	1	3	2
$\chi_1(g)$	1	1	1
$\chi_2(g)$	1	-1	1
$\chi_3(g)$	2	0	-1

and of course, the permutation matrices themselves.

1. For each irreducible representation ρ_i found in Activity 7.3.5, compute

$$\mathbf{P}_i = \frac{n_i}{|G|} \sum_{g \in G} \overline{\chi_i(g)} \rho(g).$$

2. For each $\mathbf{P}_i$ determine a basis for the column space (this can be done by stare down).

3. Write down a matrix $\mathbf{S}$ giving a change of basis to block form.

   If you want the matrix $\mathbf{S}^{-1} = \mathbf{S}^T$, *then you will need Gram-Schmidt. Without this property, you will need row reduction to find the inverse.*

4. Compute $\mathbf{S}^{-1}\rho((1\ 2))\mathbf{S}$ and $\mathbf{S}^{-1}\rho((1\ 2\ 3))\mathbf{S}$.

If an irreducible representation occurs more than once, it is still possible to obtain the block decomposition, but we need something a little more fine grained than what Theorem 7.3.6 allows. Suppose ρ_i is one of the irreducible representations that occurs, where n_i is its dimension. Then we need an explicit realization of ρ_i, that is we need matrices for $\rho_i(g)$. We will refer to the component in row α and column β as $[\rho_i(g)]_{\alpha\beta}$.

Theorem 7.3.7. *Let*

$$\mathbf{P}_{i,\alpha\beta} = \frac{n_i}{|G|} \sum_{g\in G} [\rho_i(g)]_{\alpha\beta}\rho(g)^{-1}, \tag{7.7}$$

and let $\mathbf{v}_1, \mathbf{v}_2, \ldots, \mathbf{v}_m$ *be a basis for the column space of* $\mathbf{P}_{i,11}$. *Then for each* j

$$\mathbf{P}_{i,11}\mathbf{v}_j, \mathbf{P}_{i,12}\mathbf{v}_j, \ldots, \mathbf{P}_{i,1n_i}\mathbf{v}_j,$$

is a basis for exactly one copy of the irreducible representation ρ_i. *Furthermore,*

$$\mathbf{P}_i = \sum_{\alpha,\beta} \mathbf{P}_{i,\alpha\beta}.$$

This theorem and its proof can be found in [23].

Example 7.3.8. Consider the representation of D_4 defined by

$$\rho(e) = \begin{bmatrix} \mathbf{I} & \mathbf{0} \\ \mathbf{0} & \mathbf{I} \end{bmatrix}, \quad \rho(r) = \begin{bmatrix} \mathbf{0} & -\mathbf{I} \\ \mathbf{I} & \mathbf{0} \end{bmatrix}, \quad \rho(r^2) = \begin{bmatrix} -\mathbf{I} & \mathbf{0} \\ \mathbf{0} & -\mathbf{I} \end{bmatrix},$$

$$\rho(r^3) = \begin{bmatrix} \mathbf{0} & \mathbf{I} \\ -\mathbf{I} & \mathbf{0} \end{bmatrix}, \quad \rho(m) = \begin{bmatrix} \mathbf{0} & \mathbf{I} \\ \mathbf{I} & \mathbf{0} \end{bmatrix}, \quad \rho(mr) = \begin{bmatrix} \mathbf{I} & \mathbf{0} \\ \mathbf{0} & -\mathbf{I} \end{bmatrix}, \tag{7.8}$$

$$\rho(mr^2) = \begin{bmatrix} \mathbf{0} & -\mathbf{I} \\ -\mathbf{I} & \mathbf{0} \end{bmatrix}, \quad \rho(mr^3) = \begin{bmatrix} -\mathbf{I} & \mathbf{0} \\ \mathbf{0} & \mathbf{I} \end{bmatrix},$$

where $\mathbf{I}$ is the 2×2 identity matrix and $\mathbf{0}$ is the 2×2 zero matrix. Then

g	e	r	r^2	m	mr
$\lvert C_g \rvert$	1	2	1	2	2
$\chi(g)$	4	0	-4	0	0

and

$$\langle \chi, \chi_5 \rangle = \frac{1}{8}(1 \cdot 4 \cdot 2 + 1 \cdot (-4) \cdot (-2)) = \frac{1}{8}(8 + 8) = 2.$$

Since ρ_5 has dimension 2 and we have two copies of it, then we get a total dimension of 4, which is equal to the dimension of ρ. So, ρ_5 is the only irreducible representation that occurs. The matrices in (4.1) give us an explicit realization of ρ_5. For example,

$$\rho_5(r) = \begin{bmatrix} 0 & -1 \\ 1 & 0 \end{bmatrix},$$

so

$$[\rho_5(r)]_{11} = 0, \quad [\rho_5(r)]_{12} = -1, \quad [\rho_5(r)]_{21} = 1, \quad [\rho_5(r)]_{22} = 0$$

is just the usual way that we identify the components of a matrix. It is important to understand that in equation (7.7), α and β are fixed while we are summing over g. That means we are getting exactly one component of each matrix in the irreducible representation of interest in the same spot. For clarity, we will collect these components in a table:

$\alpha\beta$	e	r	r^2	r^3	m	mr	mr^2	mr^3
11	1	0	−1	0	0	1	0	−1
12	0	−1	0	1	1	0	−1	0
21	0	1	0	−1	1	0	−1	0
22	1	0	−1	0	0	−1	0	1

(7.9)

It is also important to observe the inverse in equation (7.7). It is certainly true that $\rho(g^{-1}) = \rho(g)^{-1}$, but since inverses of matrices are probably more familiar than inverses in groups, we put the inverse outside. In the case of D_4, each element is its own inverse except for r and r^3 which are inverse of each other. So

$$[\rho_5(r)]_{\alpha\beta}\rho(r)^-1 = [\rho_5(r)]_{\alpha\beta}\rho(r^3),$$

and the same is true with r and r^3 reversed. We can omit terms for which $[\rho_5(g)]_{\alpha\beta} = 0$, so let's look at $\mathbf{P}_{5,12}$ first. Using the coefficients in the second row of the table (7.9) and the matrices $\rho(g)$ listed in (7.8), with $|G| = 8$ and $n_5 = 2$, we have

$$\begin{aligned}\mathbf{P}_{5,12} &= \frac{2}{8}\left(-\begin{bmatrix}\mathbf{0} & \mathbf{I}\\ -\mathbf{I} & \mathbf{0}\end{bmatrix} + \begin{bmatrix}\mathbf{0} & -\mathbf{I}\\ \mathbf{I} & \mathbf{0}\end{bmatrix} + \begin{bmatrix}\mathbf{0} & \mathbf{I}\\ \mathbf{I} & \mathbf{0}\end{bmatrix} - \begin{bmatrix}\mathbf{0} & -\mathbf{I}\\ -\mathbf{I} & \mathbf{0}\end{bmatrix}\right)\\ &= \frac{1}{4}\left(\begin{bmatrix}\mathbf{0} & -2\mathbf{I}\\ 2\mathbf{I} & \mathbf{0}\end{bmatrix} + \begin{bmatrix}\mathbf{0} & 2\mathbf{I}\\ 2\mathbf{I} & \mathbf{0}\end{bmatrix}\right) = \begin{bmatrix}\mathbf{0} & \mathbf{0}\\ \mathbf{I} & \mathbf{0}\end{bmatrix}.\end{aligned}$$

As for $\mathbf{P}_{5,11}$, we have $[\rho_5(r)]_{11} = [\rho_5(r^3)]_{11} = 0$, so it is a little more straightforward:

$$\begin{aligned}\mathbf{P}_{5,11} &= \frac{2}{8}\left(\begin{bmatrix}\mathbf{I} & \mathbf{0}\\ \mathbf{0} & \mathbf{I}\end{bmatrix} - \begin{bmatrix}-\mathbf{I} & \mathbf{0}\\ \mathbf{0} & -\mathbf{I}\end{bmatrix} + \begin{bmatrix}\mathbf{I} & \mathbf{0}\\ \mathbf{0} & -\mathbf{I}\end{bmatrix} - \begin{bmatrix}-\mathbf{I} & \mathbf{0}\\ \mathbf{0} & +\mathbf{I}\end{bmatrix}\right)\\ &= \frac{1}{4}\left(\begin{bmatrix}2\mathbf{I} & \mathbf{0}\\ \mathbf{0} & 2\mathbf{I}\end{bmatrix} + \begin{bmatrix}2\mathbf{I} & \mathbf{0}\\ \mathbf{0} & -2\mathbf{I}\end{bmatrix}\right) = \begin{bmatrix}\mathbf{I} & \mathbf{0}\\ \mathbf{0} & \mathbf{0}\end{bmatrix}.\end{aligned}$$

Similarly it can be shown that

$$\mathbf{P}_{5,21} = \begin{bmatrix}\mathbf{0} & \mathbf{I}\\ \mathbf{0} & \mathbf{0}\end{bmatrix} \quad \text{and} \quad \mathbf{P}_{5,22} = \begin{bmatrix}\mathbf{0} & \mathbf{0}\\ \mathbf{0} & \mathbf{I}\end{bmatrix}, \tag{7.10}$$

although they are not actually needed. The column space of $\mathbf{P}_{5,11}$ has basis

$$\mathbf{v}_1 = \begin{bmatrix}1\\0\\0\\0\end{bmatrix}, \quad \text{and} \quad \mathbf{v}_2 = \begin{bmatrix}0\\1\\0\\0\end{bmatrix}.$$

We now apply the transformations $\mathbf{P}_{5,11}$ and $\mathbf{P}_{5,12}$ to each of the vectors $\mathbf{v}_j$ in order:

$$\mathbf{P}_{5,11}\mathbf{v}_1 = \begin{bmatrix}1\\0\\0\\0\end{bmatrix}, \quad \mathbf{P}_{5,12}\mathbf{v}_1 = \begin{bmatrix}0\\0\\1\\0\end{bmatrix}, \quad \mathbf{P}_{5,11}\mathbf{v}_2 = \begin{bmatrix}0\\1\\0\\0\end{bmatrix}, \quad \text{and} \quad \mathbf{P}_{5,12}\mathbf{v}_2 = \begin{bmatrix}0\\0\\0\\1\end{bmatrix}. \tag{7.11}$$

The first two give a basis for one block, and the second two give a basis for the other block. We now take these vectors in order as the columns of a matrix

$$\mathbf{Q} = \begin{bmatrix}1 & 0 & 0 & 0\\ 0 & 0 & 1 & 0\\ 0 & 1 & 0 & 0\\ 0 & 0 & 0 & 1\end{bmatrix}.$$

Luckily $\mathbf{Q}^{-1} = \mathbf{Q}$, and it can be checked that

$$\mathbf{Q}^{-1}\rho(r)\mathbf{Q} = \left[\begin{array}{cc|cc} 0 & -1 & 0 & 0 \\ 1 & 0 & 0 & 0 \\ \hline 0 & 0 & 0 & -1 \\ 0 & 0 & 1 & 0 \end{array}\right] \quad \text{and} \quad \mathbf{Q}^{-1}\rho(m)\mathbf{Q} = \left[\begin{array}{cc|cc} 0 & 1 & 0 & 0 \\ 1 & 0 & 0 & 0 \\ \hline 0 & 0 & 0 & 1 \\ 0 & 0 & 1 & 0 \end{array}\right]. \tag{7.12}$$

Activity 7.3.9. Verify equations (7.10), (7.11), and (7.12).

Exercises

Problem 7.1. Compute the character of each representation below

1. The group S_4 has conjugacy classes

$$(1),\ (1,2),\ (1,2)(3,4),\ (1,2,3),\ (1,2,3,4)$$

   which have sizes 1, 6, 3, 8, 6 respectively. Let ρ be the representation by permutation matrices.

2. The group D_5 is generated by r and m, where $r^5 = e$, $m^2 = e$, and $mr = r^4m$. The conjugacy class representatives are

$$e, m, r, r^2,$$

   which have sizes 1, 5, 2, 2 respectively. Let the vertices of the pentagon be labeled by 1,2,3,4,5, and let ρ be the 5 dimensional representation arising in the same way as example 7.1.3.

8

Applications

8.1 Chemical Equations

The importance of balancing chemical equations will be familiar to most introductory chemistry students and can be found in texts such as [14]. The mathematical viewpoint is often not found in chemistry books but can be found in other Linear Algebra books, including [15] and [8].

When considering a chemical reaction, we may want the proportions of the reactants and of the products. Ultimately, this boils down to solving a system of linear equations. Even if sometimes the solution is obvious enough that linear algebra is not needed, we would still like to understand what is going on from the viewpoint of the theory that linear algebra has to offer.

First, how do we get a system of linear equations?

1. We get one variable for each reactant and each product.

2. We get one equation for each atom type.

3. If ions are involved, then we get an additional equation for the charge.

The equations for each atom type occur because of conservation of atomic mass. If ions are involved, then we also need to pay attention to conservation of charge, which is why we get an additional equation in that case.

Example 8.1.1. Consider the combustion of benzene

$$x_1\,\mathrm{C_6H_6} + x_2\,\mathrm{O_2} \longrightarrow x_3\,\mathrm{CO_2} + x_4\,\mathrm{H_2O}$$

There are two reactants and two products for a total of four variables. Charge is not involved, but there are 3 atom types: carbon, hydrogen, and oxygen. We get the equations by counting the number of each atom type in each molecule and multiplying by the corresponding variable, then adding up on the left and right separately and setting the results equal to each other.

For carbon, we get $6x_1$ on the left and x_3 on the right, so

$$6x_1 = x_3 \quad \text{or} \quad 6x_1 - x_3 = 0\ .$$

For hydrogen, we get $6x_1$ on the left and $2x_4$ on the right, so

$$6x_1 = 2x_4 \quad \text{or} \quad 6x_1 - 2x_4 = 0\ .$$

For oxygen, we get $2x_2$ on the left and $2x_3 + x_4$ on the right, so

$$2x_2 = 2x_3 + x_4 \quad \text{or} \quad 2x_2 - 2x_3 - x_4 = 0\ .$$

DOI: 10.1201/9781003737490-8

Together this leads to the following augmented form

$$\left[\begin{array}{cccc|c} 6 & 0 & -1 & 0 & 0 \\ 6 & 0 & 0 & -2 & 0 \\ 0 & 2 & -2 & -1 & 0 \end{array}\right].$$

As usual we solve by row reduction, however, we are not concerned with getting leading ones necessarily. Instead we proceed as follows:

$$\left[\begin{array}{cccc|c} 6 & 0 & -1 & 0 & 0 \\ 6 & 0 & 0 & -2 & 0 \\ 0 & 2 & -2 & -1 & 0 \end{array}\right] \xrightarrow{(-1)R_1} \left[\begin{array}{cccc|c} -6 & 0 & 1 & 0 & 0 \\ 6 & 0 & 0 & -2 & 0 \\ 0 & 2 & -2 & -1 & 0 \end{array}\right]$$

$$\xrightarrow{(-\frac{1}{2})R_2} \left[\begin{array}{cccc|c} -6 & 0 & 1 & 0 & 0 \\ -3 & 0 & 0 & 1 & 0 \\ 0 & 2 & -2 & -1 & 0 \end{array}\right] \xrightarrow{2R_1+R_3} \left[\begin{array}{cccc|c} -6 & 0 & 1 & 0 & 0 \\ -3 & 0 & 0 & 1 & 0 \\ -12 & 2 & 0 & -1 & 0 \end{array}\right]$$

$$\xrightarrow{R_2+R_3} \left[\begin{array}{cccc|c} -6 & 0 & 1 & 0 & 0 \\ -3 & 0 & 0 & 1 & 0 \\ -15 & 2 & 0 & 0 & 0 \end{array}\right] \xrightarrow{\frac{1}{2}R_3} \left[\begin{array}{cccc|c} -6 & 0 & \boxed{1} & 0 & 0 \\ -3 & 0 & 0 & \boxed{1} & 0 \\ -\frac{15}{2} & \boxed{1} & 0 & 0 & 0 \end{array}\right].$$

When we first introduced row reduction, we always chose a leading non-zero entry in a row to be a pivot. Sometimes the leading entry in a row is the best choice, but not always. Here we saw that 6 is divisible by both -1 and -2, which made the location of the -1 in row 1 and the -2 in row 2 good choices for pivot positions. That is why we multiplied row 1 by -1 and row 2 by $-\frac{1}{2}$. Once we had ones in those positions, it was an easy matter to clear the remaining numbers in those columns. After that, then the 2 in row 3 jumped out because it was already the lone non-zero entry in its column. Hence, we multiplied row 3 by $\frac{1}{2}$. The pivot positions in the final matrix are boxed in for clarity. The first column does not contain a pivot, so x_1 is a free variable. We then convert back into equations to get a parameterization of the solution set:

$$\begin{aligned} x_1 &= x_1 \\ -6x_1 + x_3 &= 0 \\ -3x_1 + x_4 &= 0 \\ -\frac{5}{2}x_1 + x_2 &= 0 \end{aligned} \longrightarrow \begin{aligned} x_1 &= x_1 \\ x_3 &= 6x_1 \\ x_4 &= 3x_1 \\ x_2 &= \frac{15}{2}x_1 \end{aligned} \longrightarrow \begin{bmatrix} x_1 \\ x_2 \\ x_3 \\ x_4 \end{bmatrix} = x_1 \begin{bmatrix} 1 \\ \frac{15}{2} \\ 6 \\ 3 \end{bmatrix}.$$

There is only one denominator here, which is 2, so the smallest positive integer solution is obtained by taking $x_1 = 2$. This leads to the balanced equation

$$2C_6H_6 + 15O_2 \longrightarrow 12CO_2 + 6H_2O.$$

Remark 8.1.1. If more than one denominator occurs, then the smallest value of x_1 would be the least common multiple of the denominators when each fraction is already in reduced form. It is also worth mentioning that in Example 8.1.1, it is not possible to do row reduction in any way without eventually ending up with fractions. That is because none of the values of the variables in the smallest positive integer solution is equal to one.

Activity 8.1.2. Try balancing the following chemical reaction by row reduction

$$x_1\, C_6H_6 + x_2\, O_2 \longrightarrow x_3\, CO + x_4\, H_2O$$

Now that we have seen how the process of balancing equations works, it is worthwhile to consider what the theory of linear algebra implies. First of all, the augmented column will always be zero, so it is not possible for the system to be inconsistent. There are two reasons that it may not be possible to balance a chemical equation:

1. There may be no free variables.
2. There may be free variables but no solution in which all variables are positive, like a line in two dimensions through the origin with negative slope.

The first of these definitely occurs as can be seen in Example 8.1.3. It is also possible for there to be more than one free variable as can be seen in Example 8.1.4.

Example 8.1.3. Suppose that we try to remove carbon dioxide from the products of the combustion of benzene

$$x_1\,\mathrm{C_6H_6} + x_2\,\mathrm{O_2} \longrightarrow x_3\,\mathrm{H_2O}$$

That amounts to removing the 3rd column from the initial augmented matrix in Example 8.1.1:

$$\left[\begin{array}{ccc|c} 6 & 0 & 0 & 0 \\ 6 & 0 & -2 & 0 \\ 0 & 2 & -1 & 0 \end{array}\right].$$

This time we can do row reduction all the way until there is a pivot in every column to the left of the augmented column.

$$\left[\begin{array}{ccc|c} 6 & 0 & 0 & 0 \\ 6 & 0 & -2 & 0 \\ 0 & 2 & -1 & 0 \end{array}\right] \xrightarrow{\frac{1}{6}R_1} \left[\begin{array}{ccc|c} 1 & 0 & 0 & 0 \\ 6 & 0 & -2 & 0 \\ 0 & 2 & -1 & 0 \end{array}\right] \xrightarrow{(-6)R_1+R_2} \left[\begin{array}{ccc|c} 1 & 0 & 0 & 0 \\ 0 & 0 & -2 & 0 \\ 0 & 2 & -1 & 0 \end{array}\right]$$

$$\xrightarrow{\frac{1}{-2}R_2} \left[\begin{array}{ccc|c} 1 & 0 & 0 & 0 \\ 0 & 0 & 1 & 0 \\ 0 & 2 & -1 & 0 \end{array}\right] \xrightarrow{R_2+R_3} \left[\begin{array}{ccc|c} 1 & 0 & 0 & 0 \\ 0 & 0 & 1 & 0 \\ 0 & 2 & 0 & 0 \end{array}\right] \xrightarrow{\frac{1}{2}R_3} \left[\begin{array}{ccc|c} 1 & 0 & 0 & 0 \\ 0 & 0 & 1 & 0 \\ 0 & 1 & 0 & 0 \end{array}\right]$$

$$\xrightarrow{R_2\leftrightarrow R_3} \left[\begin{array}{ccc|c} 1 & 0 & 0 & 0 \\ 0 & 1 & 0 & 0 \\ 0 & 0 & 1 & 0 \end{array}\right]$$

So there are no free variables; $x_1 = x_2 = x_3 = 0$ is the unique solution. In terms of chemistry, this means that the reaction does not occur.

Example 8.1.4. Suppose we want to balance the reaction

$$x_1\,\mathrm{XeF_4} + x_2\,\mathrm{H_2O_4} \longrightarrow x_3\,\mathrm{Xe} + x_4\,\mathrm{XeO_3} + x_5\,\mathrm{HF} + x_6\,\mathrm{O_2}$$

(see [27]). There are 4 atoms, 6 variables, and charge is not involved. This means that there will be at least 2 free variables at the end of row reduction, because $6 - 4 = 2$. By counting the atoms on each side, we obtain the following equations

$$\begin{array}{ll} \mathrm{H} & 2x_2 = x_5 \\ \mathrm{O} & x_2 = 3x_4 + 2x_6 \\ \mathrm{F} & 4x_1 = x_5 \\ \mathrm{Xe} & x_1 = x_3 + x_4 \end{array} \longrightarrow \begin{array}{r} 2x_2 - x_5 = 0 \\ x_2 - 3x_4 - 2x_6 = 0 \\ 4x_1 - x_5 = 0 \\ x_1 - x_3 - x_4 = 0 \end{array}$$

Next we convert to augmented form and row reduce. Row operations can be done in parallel so long as there is no conflict between the rows.

$$
\left[\begin{array}{cccccc|c}
0 & 2 & 0 & 0 & -1 & 0 & 0\\
0 & 1 & 0 & -3 & 0 & -2 & 0\\
4 & 0 & 0 & 0 & -1 & 0 & 0\\
1 & 0 & -1 & -1 & 0 & 0 & 0
\end{array}\right]
$$

$$
\xrightarrow{\substack{(-2)R_2+R_1\\(-4)R_4+R_3}}
\left[\begin{array}{cccccc|c}
0 & 0 & 0 & 6 & -1 & 4 & 0\\
0 & 1 & 0 & -3 & 0 & -2 & 0\\
0 & 0 & 4 & 4 & -1 & 0 & 0\\
1 & 0 & -1 & -1 & 0 & 0 & 0
\end{array}\right]
$$

$$
\xrightarrow{(-1)R_3+R_1}
\left[\begin{array}{cccccc|c}
0 & 0 & -4 & 2 & 0 & 4 & 0\\
0 & 1 & 0 & -3 & 0 & -2 & 0\\
0 & 0 & 4 & 4 & -1 & 0 & 0\\
1 & 0 & -1 & -1 & 0 & 0 & 0
\end{array}\right]
$$

$$
\xrightarrow{\substack{\frac{1}{2}R_1\\(-4)R_3}}
\left[\begin{array}{cccccc|c}
0 & 0 & -2 & 1 & 0 & 2 & 0\\
0 & 1 & 0 & -3 & 0 & -2 & 0\\
0 & 0 & -4 & -4 & 1 & 0 & 0\\
1 & 0 & -1 & -1 & 0 & 0 & 0
\end{array}\right]
$$

$$
\xrightarrow{\substack{3R_1+R_2\\4R_1+R_3\\R_1+R_4}}
\left[\begin{array}{cccccc|c}
0 & 0 & -2 & \boxed{1} & 0 & 2 & 0\\
0 & \boxed{1} & -6 & 0 & 0 & 4 & 0\\
0 & 0 & -12 & 0 & \boxed{1} & 8 & 0\\
\boxed{1} & 0 & -3 & 0 & 0 & 2 & 0
\end{array}\right]
$$

The pivots are boxed in the final matrix for clarity. The free variables correspond to any column to the left of the augmented column not containing a pivot, so x_3 and x_6 are the free variables. Converting back to equations, we have

$$
\begin{aligned}
x_1 - 3x_3 + 2x_6 &= 0\\
x_2 - 6x_3 + 4x_6 &= 0\\
x_3 &= x_3\\
x_4 - 2x_3 + 2x_6 &= 0\\
x_5 - 12x_3 + 8x_6 &= 0\\
x_6 &= x_6
\end{aligned}
\longrightarrow
\begin{bmatrix} x_1\\ x_2\\ x_3\\ x_4\\ x_5\\ x_6 \end{bmatrix}
=
\begin{bmatrix} 3x_3 - 2x_6\\ 6x_3 - 4x_6\\ x_3\\ 2x_3 - 2x_6\\ 12x_3 - 8x_6\\ x_6 \end{bmatrix}.
$$

Each component of this vector must be positive, so we get a system of inequalities:

$$
\begin{aligned}
3x_3 - 2x_6 > 0, \quad 6x_3 - 4x_6 > 0, \quad x_3 > 0,\\
2x_3 - 2x_6 > 0, \quad 12x_3 - 8x_6 > 0, \quad x_6 > 0.
\end{aligned}
$$

The middle inequalities in each line above are multiples of the first, so both can be discarded. The inequality $2x_3 - 2x_6 > 0$ simplifies to $x_3 > x_6$. Since $x_3 > x_6$ and $x_6 > 0$ imply $x_3 > 0$ by the transitive property, then $x_3 > 0$ can be discarded as well. The first inequality can be dealt with as follows:

$$
\begin{aligned}
&x_3 > x_6 \implies 2x_3 > 2x_6\\
&2x_3 > 2x_6 \quad \text{and} \quad x_3 > 0 \implies 3x_3 > 2x_6.
\end{aligned}
$$

As a result, the system of inequalities simplifies to just two:

$$
x_6 > 0 \quad \text{and} \quad x_3 > x_6,
$$

and of course, x_3 and x_6 must both be integers. But there are no other restrictions, and so not only can the equation be balanced, it can be balanced in infinitely many different ways that are not multiples of each other. For example

$$\begin{aligned} 4\mathrm{XeF}_4 + 8\mathrm{H}_2\mathrm{O}_4 &\longrightarrow 2\mathrm{Xe} + 2\mathrm{XeO}_3 + 16\mathrm{HF} + \mathrm{O}_2, \quad \text{and} \\ 5\mathrm{XeF}_4 + 10\mathrm{H}_2\mathrm{O}_4 &\longrightarrow 3\mathrm{Xe} + 2\mathrm{XeO}_3 + 20\mathrm{HF} + 2\mathrm{O}_2, \end{aligned}$$

the first of which is obtained with $x_6 = 1$ and $x_3 = 2$ and the second of which is obtained with $x_6 = 2$ and $x_3 = 3$.

Activity 8.1.5. Try balancing

$$x_1\,\mathrm{C_6H_6} + x_2\,\mathrm{O_2} \longrightarrow x_3\,\mathrm{CO_2} + x_4\,\mathrm{CO} + x_5\,\mathrm{H_2O}$$

using row reduction.

Finally we look at an example where charge is involved.

Example 8.1.6. Consider the reaction

$$x_1\,\mathrm{MnO_4^-} + x_2\,\mathrm{Fe^{2+}} + x_3\,\mathrm{H^+} \longrightarrow x_4\,\mathrm{Mn^{2+}} + x_5\,\mathrm{Fe^{3+}} + x_6\,\mathrm{H_2O}\,.$$

We take the usual approach of getting an equation for each type of atom involved, and for the charge, we determine the total charge on both sides.[1] Pay close attention to the sign coming from the charge!

$$\begin{array}{rlcr} \text{Mn:} & x_1 = x_4 & & x_1 - x_4 = 0 \\ \text{Fe:} & x_2 = x_5 & & x_2 - x_5 = 0 \\ \text{H:} & x_3 = 2x_6 & \longrightarrow & x_3 - 2x_6 = 0 \\ \text{O:} & 4x_1 = x_6 & & 4x_1 - x_6 = 0 \\ +\text{:} & -x_1 + 2x_2 + x_3 = 2x_4 + 3x_5 & & -x_1 + 2x_2 + x_3 - 2x_4 - 3x_5 = 0 \end{array}$$

These equations lead to the following augmented form

$$\left[\begin{array}{cccccc|c} 1 & 0 & 0 & -1 & 0 & 0 & 0 \\ 0 & 1 & 0 & 0 & -1 & 0 & 0 \\ 0 & 0 & 1 & 0 & 0 & -2 & 0 \\ 4 & 0 & 0 & 0 & 0 & -1 & 0 \\ -1 & 2 & 1 & -2 & -3 & 0 & 0 \end{array}\right]$$

The solution is given by the following reduced form:

$$\left[\begin{array}{cccccc|c} 1 & 0 & 0 & 0 & 0 & -\frac{1}{4} & 0 \\ 0 & 1 & 0 & 0 & 0 & -\frac{5}{4} & 0 \\ 0 & 0 & 1 & 0 & 0 & -2 & 0 \\ 0 & 0 & 0 & 1 & 0 & -\frac{1}{4} & 0 \\ 0 & 0 & 0 & 0 & 1 & -\frac{5}{4} & 0 \end{array}\right]$$

Since the 6th column does not have a pivot, x_6 is a free variable. It follows that the parameterization of the solution set is

$$\begin{bmatrix} x_1 \\ x_2 \\ x_3 \\ x_4 \\ x_5 \\ x_6 \end{bmatrix} = x_6 \begin{bmatrix} \frac{1}{4} \\ \frac{5}{4} \\ 2 \\ \frac{1}{4} \\ \frac{5}{4} \\ 1 \end{bmatrix}.$$

[1]We could count electrons, but I believe there is a risk of dyslexia.

Because of the denominators, in order for the solution to have integers, x_6 must be a multiple of 4. Therefore, taking $x_6 = 4$ give the smallest such solution:

$$MnO_4^- + 5Fe^{2+} + 8H^+ \longrightarrow Mn^{2+} + 5Fe^{3+} + 4H_2O\ .$$

Activity 8.1.7. Use row reduction to balance

$$x_1\, Al(OH)_3 + x_2\, H_3O^+ \longrightarrow x_3\, Al^{3+} + x_4\, H_2O\ .$$

8.2 Structural Engineering of Trusses

I began my college studies as a major in architecture. Eventually I changed my mind, but there are many things I learned that I remember to this day, one of which is the approach that we were taught to the theory of trusses. The approach we learned is sometimes called the method of joints or the force method as it is in [21]. In this section, we will look at the force method from the viewpoint of linear algebra. But first we must begin with some basic concepts and definitions.

Trusses can be 2 or 3 dimensional. The 2 dimensional trusses are called planar trusses, and the 3 dimensional trusses are called space trusses. Planar trusses are commonly found in houses or bridges, and can be abstracted by a 2 dimensional diagram such as the one in Example 8.2.1. Trusses may be made of wood or steel members. The places at which these members are connected to each other or to the rest of the structure are called joints.

Forces are vectors. A vector in 2 dimensions can be broken into horizontal and vertical components, and a vector in 3 dimensions can be broken into 3 components. However, in most cases it is also necessary to consider rotational forces that cause bending or twisting. Structural joints are classified into the types of forces that they resist, and different symbols are used in diagrams to specify the joint type. Here is the classification in 2 dimensions:

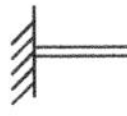

A fixed joint is rigid and provides resistance to rotational forces, horizontal forces, and vertical forces.

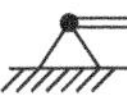

A pin joint provides resistance only to horizontal and vertical force. The members connected at a pin are allowed to rotate freely, so there is no resistance to rotation.

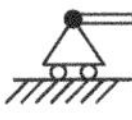

A roller joint provides resistance only to vertical forces. Free rotation and free horizontal motion are allowed.

We would like to solve the problem of computing the forces in a truss by analyzing free body diagrams at the joints, breaking the forces into components, and solving the system of equations that arise. Free body diagrams are very common in physics and can be found in many introductory texts, such as [22].

Our ability to solve the problem in this way requires us to assume that the forces are transmitted axially through the members of the truss, and that no rotational forces occur. Therefore, we must assume that:

1. the members of the truss are connected at only pin joints,
2. the forces applied to a truss occur only at the pin joints,
3. the truss is supported only with pin or roller joints, and
4. the length of the members of the truss does not change.

FIGURE 8.1: Possible axial loads on a truss member

In practice, these assumptions are almost never true. Wood trusses are commonly made using bolts or nails, and in metal trusses you may see rivets, any of which would make a fixed joint, not a pin. Trusses are often connected to the rest of the structure with fixed joints as well.

Suppose, however, that we manage to build a truss using only pin joints, and that it is only supported with pins or rollers. How do we ensure that the forces on the truss only occur at the joints? If the truss is open air, the wind would produce a force on the members. So, the truss has to be enclosed. In the case of a truss used for a roof, it would have to be separated from the roof deck by purlins placed directly at the joints. The roof deck may bend, but it must be prevented from bending so much as to touch the members of the trusses.

If a force is applied axially to a member: it can be in compression or it can be in tension (see Figure 8.1). Hooke's law can be used as a model for how a material changes its lengths under tension or compression. Changes in the length of the members of a truss will cause the angles in the truss to change by the law of sines, and so the problem will no longer be linear. So we only have a linear approximation, but that approximation may be accurate enough if the constant in Hooke's law is very small, and that is generally the case in practice. Even so, it is wise to be aware of the assumptions made by theory, and to be aware of when those assumptions are no longer accurate enough.

We will number the joints of a truss. The forces applied to a truss may come from the weight of a vehicle, a strong wind, snow accumulation, or the weight of the individual members making it up (although this is sometimes ignored for an introduction). A force applied to a truss will be called a load, and denoted with $\mathbf{L}_i$ if the load is applied to joint i. The loads are transmitted through the members to the supports of the truss and are resisted with reaction forces. A reaction force at joint i will be denoted with $\mathbf{R}_i$. The members are always connected to two joints. We will label the axial forces along the member connecting joint i and joint j with $\mathbf{F}_{ij}$. The forces $\mathbf{F}_{ij}$ are internal to the truss, while the forces $\mathbf{L}_i$ and $\mathbf{R}_i$ are external.

A choice has to be made about the direction that the forces $\mathbf{F}_{ij}$ are drawn along the members. In this book we use the convention of drawing the forces toward the joints. This is equivalent to assuming that the members of the truss are in compression. Imagine having pin joints at each end of the member in Figure 8.1a. For equilibrium to be maintained at the joint, the member in compression must react with a force toward the joint. If this guess is wrong for a member and it is actually in tension, then the value computed for the magnitude of $\mathbf{F}_{ij}$ will be negative.

In the force method, we obtain a system of equations for a planar truss as follows:

1. For each member we get one variable, the common length (magnitude) of the forces $\mathbf{F}_{ij}$ and $\mathbf{F}_{ji}$. These forces are equal in magnitude and opposite in direction. Their components are determined by the geometry of the truss.

2. For the joints supporting the truss we get one variable for each roller joint, and two for each pin joint because the ratio between the components is not determined by the geometry of the truss.

3. The forces in a free body diagram sum to zero. Each joint gives us one free body diagram. Each free body diagram provides two equations, one for the horizontal components, and the other for the vertical components.

Once the equations are set up, they can be solved in the usual way.

Example 8.2.1. If the magnitude of $\mathbf{L}_1$ is 6 newtons in the diagram below, solve for the remaining forces. Then determine which members are in tension, and which are in compression.

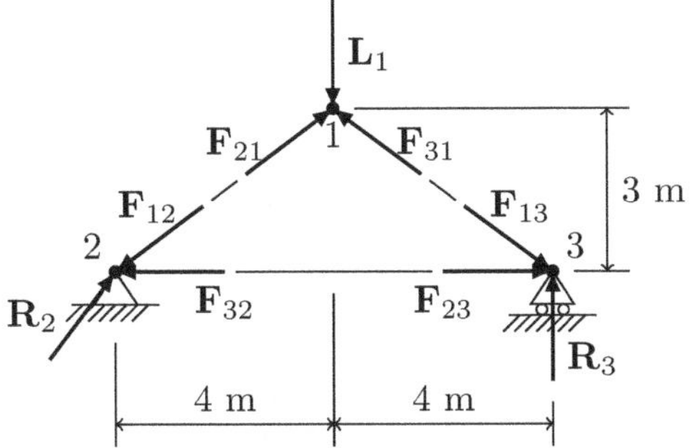

Solution

The truss is formed by pasting two 3-4-5 right triangles together. Since $\mathbf{F}_{21}$ points in the direction of the 1st quadrant and $\mathbf{F}_{31}$ points in the direction of the 2nd quadrant, the free body diagram at joint 1 gives us

$$\mathbf{F}_{21} + \mathbf{F}_{31} + \mathbf{L}_1 = f_{21}\begin{bmatrix} \frac{4}{5} \\ \frac{3}{5} \end{bmatrix} + f_{31}\begin{bmatrix} -\frac{4}{5} \\ \frac{3}{5} \end{bmatrix} + \begin{bmatrix} 0 \\ -6 \end{bmatrix} = \mathbf{0}.$$

Then the horizontal and vertical components give us the equations

$$\frac{4}{5}f_{21} - \frac{4}{5}f_{31} = 0 \quad \text{and} \quad \frac{3}{5}f_{21} + \frac{3}{5}f_{31} - 6 = 0. \tag{8.1}$$

At joint 2, we need to remember that $\mathbf{F}_{12} = -\mathbf{F}_{21}$, while $\mathbf{F}_{32}$ has a vertical component of zero. So the free body diagram at joint 2 gives us

$$\mathbf{R}_2 + \mathbf{F}_{12} + \mathbf{F}_{32} = \begin{bmatrix} R_{2,x} \\ R_{2,y} \end{bmatrix} - f_{21}\begin{bmatrix} \frac{4}{5} \\ \frac{3}{5} \end{bmatrix} + f_{32}\begin{bmatrix} -1 \\ 0 \end{bmatrix} = \mathbf{0},$$

and the horizontal and vertical components yield the equations

$$R_{2,x} - \frac{4}{5}f_{21} - f_{32} = 0 \quad \text{and} \quad R_{2,y} - \frac{3}{5}f_{21} = 0. \tag{8.2}$$

Finally, at joint 3, we need $\mathbf{F}_{13} = -\mathbf{F}_{31}$ and $\mathbf{F}_{23} = -\mathbf{F}_{32}$. Since joint 3 is supported on a roller joint, the horizontal component of $\mathbf{R}_3$ is zero. Thus the free body diagram at joint 3 gives us

$$\mathbf{R}_3 + \mathbf{F}_{13} + \mathbf{F}_{23} = \begin{bmatrix} 0 \\ R_{3,y} \end{bmatrix} + f_{31}\begin{bmatrix} \frac{4}{5} \\ -\frac{3}{5} \end{bmatrix} + f_{32}\begin{bmatrix} 1 \\ 0 \end{bmatrix} = \mathbf{0},$$

and the horizontal and vertical components lead to the equations

$$\frac{4}{5}f_{31} + f_{32} = 0 \quad \text{and} \quad R_{3,y} - \frac{3}{5}f_{31} = 0. \tag{8.3}$$

By moving the constants in equations (8.1), (8.2), and (8.3) over to the right-hand side, we obtain the following augmented form

$$\left[\begin{array}{cccccc|c} 0 & 0 & 0 & \frac{4}{5} & -\frac{4}{5} & 0 & 0 \\ 0 & 0 & 0 & \frac{3}{5} & \frac{3}{5} & 0 & 6 \\ 1 & 0 & 0 & -\frac{4}{5} & 0 & -1 & 0 \\ 0 & 1 & 0 & -\frac{3}{5} & 0 & 0 & 0 \\ 0 & 0 & 0 & 0 & \frac{4}{5} & 1 & 0 \\ 0 & 0 & 1 & 0 & -\frac{3}{5} & 0 & 0 \end{array}\right], \tag{8.4}$$

where the order of the columns is $R_{3,y}$, $R_{2,x}$, $R_{2,y}$, f_{21}, f_{31}, f_{32}. Now we row reduce to find the solution. There are already friendly ones in the last 4 rows, so we work in such a way as to avoid ruining them. Therefore, we begin by multiplying row 1 by $\frac{5}{4}$ and row 2 by $\frac{5}{3}$ to get ones there:

$$\left[\begin{array}{cccccc|c} 0 & 0 & 0 & 1 & -1 & 0 & 0 \\ 0 & 0 & 0 & 1 & 1 & 0 & 10 \\ 1 & 0 & 0 & -\frac{4}{5} & 0 & -1 & 0 \\ 0 & 1 & 0 & -\frac{3}{5} & 0 & 0 & 0 \\ 0 & 0 & 0 & 0 & \frac{4}{5} & 1 & 0 \\ 0 & 0 & 1 & 0 & -\frac{3}{5} & 0 & 0 \end{array}\right]$$

$$\xrightarrow{R_2+R_1} \left[\begin{array}{cccccc|c} 0 & 0 & 0 & 2 & 0 & 0 & 10 \\ 0 & 0 & 0 & 1 & 1 & 0 & 10 \\ 1 & 0 & 0 & -\frac{4}{5} & 0 & -1 & 0 \\ 0 & 1 & 0 & -\frac{3}{5} & 0 & 0 & 0 \\ 0 & 0 & 0 & 0 & \frac{4}{5} & 1 & 0 \\ 0 & 0 & 1 & 0 & -\frac{3}{5} & 0 & 0 \end{array}\right]$$

$$\xrightarrow{\frac{1}{2}R_1} \left[\begin{array}{cccccc|c} 0 & 0 & 0 & 1 & 0 & 0 & 5 \\ 0 & 0 & 0 & 1 & 1 & 0 & 10 \\ 1 & 0 & 0 & -\frac{4}{5} & 0 & -1 & 0 \\ 0 & 1 & 0 & -\frac{3}{5} & 0 & 0 & 0 \\ 0 & 0 & 0 & 0 & \frac{4}{5} & 1 & 0 \\ 0 & 0 & 1 & 0 & -\frac{3}{5} & 0 & 0 \end{array}\right]$$

$$\xrightarrow{\substack{(-1)R_1 + R_2 \\ (\frac{4}{5})R_1 + R_3 \\ (\frac{3}{5})R_1 + R_4}} \left[\begin{array}{cccccc|c} 0 & 0 & 0 & 1 & 0 & 0 & 5 \\ 0 & 0 & 0 & 0 & 1 & 0 & 5 \\ 1 & 0 & 0 & 0 & 0 & -1 & 4 \\ 0 & 1 & 0 & 0 & 0 & 0 & 3 \\ 0 & 0 & 0 & 0 & \frac{4}{5} & 1 & 0 \\ 0 & 0 & 1 & 0 & -\frac{3}{5} & 0 & 0 \end{array}\right]$$

$$\xrightarrow{\substack{(-\frac{4}{5})R_2 + R_5 \\ (\frac{3}{5})R_2 + R_6}} \left[\begin{array}{cccccc|c} 0 & 0 & 0 & 1 & 0 & 0 & 5 \\ 0 & 0 & 0 & 0 & 1 & 0 & 5 \\ 1 & 0 & 0 & 0 & 0 & -1 & 4 \\ 0 & 1 & 0 & 0 & 0 & 0 & 3 \\ 0 & 0 & 0 & 0 & 0 & 1 & -4 \\ 0 & 0 & 1 & 0 & 0 & 0 & 3 \end{array}\right]$$

$$\xrightarrow{R_5+R_3} \left[\begin{array}{cccccc|c} 0 & 0 & 0 & 1 & 0 & 0 & 5 \\ 0 & 0 & 0 & 0 & 1 & 0 & 5 \\ 1 & 0 & 0 & 0 & 0 & 0 & 0 \\ 0 & 1 & 0 & 0 & 0 & 0 & 3 \\ 0 & 0 & 0 & 0 & 0 & 1 & -4 \\ 0 & 0 & 1 & 0 & 0 & 0 & 3 \end{array}\right].$$

The solution is unique, and gives the following values in newtons:

$$R_{2,x} = 0, R_{2,y} = 3, R_{3,y} = 3, f_{21} = 5, f_{31} = 5, f_{32} = -3.$$

Since f_{32} is negative, the horizontal member connecting joint 2 and 3 is in tension, while the others are in compression.

Activity 8.2.2. If the magnitude of $\mathbf{L}_1$ is 8 newtons in the diagram below, solve for the remaining forces. Then determine which members are in tension, and which are in compression.

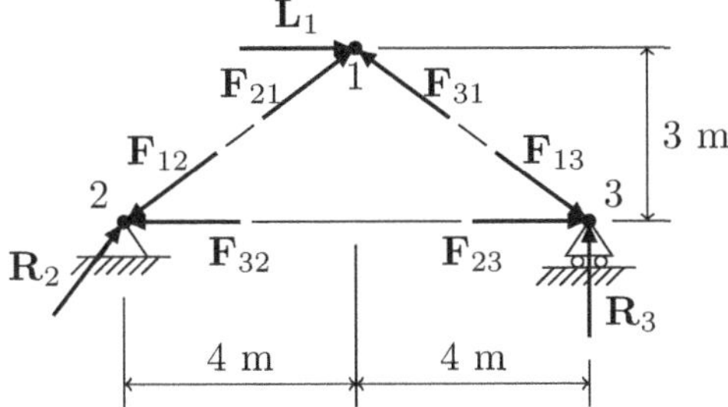

Hint: Only the free body diagram at joint 1 has changed. Now the vertical component of $\mathbf{L}_1$ is zero, while the horizontal component is not. The equations from joints 2 and 3 can be reused.

Remark 8.2.1. It is fairly common to set up equations for the external forces as well, which also must sum to zero. In the case of Example 8.2.1, we would have

$$\mathbf{L}_1 + \mathbf{R}_2 + \mathbf{R}_3 = \begin{bmatrix} 0 \\ -6 \end{bmatrix} + \begin{bmatrix} R_{2,x} \\ R_{2,y} \end{bmatrix} + \begin{bmatrix} 0 \\ R_{3,y} \end{bmatrix} = \mathbf{0},$$

so

$$R_{2,x} = 0 \quad \text{and} \quad -6 + R_{2,y} + R_{3,y} = 0. \tag{8.5}$$

Using the same order for the variables as in (8.4) would give us two extra rows:

$$\left[\begin{array}{cccccc|c} 1 & 0 & 0 & 0 & 0 & 0 & 0 \\ 0 & 1 & 1 & 0 & 0 & 0 & 6 \end{array}\right]. \tag{8.6}$$

However, there are two natural dependence relations that will always occur: the top row of the matrix in (8.6) is the sum of the odd numbered rows of the matrix in (8.4) (a dependence relation involving only the horizontal components) and the bottom row of the matrix in (8.6) is the sum of the even numbered rows of the matrix in (8.4) (a dependence relation involving only the vertical components). Since any vector occurring in a dependence relation can be removed, we are allowed to omit one equation from the horizontal components and one equation from the vertical components, whichever ones we like the least. But notice how there is exactly one 1 in the first three columns of the matrix in (8.4), with zeros above and below. That happens naturally if we use the equations coming from the free body diagrams at the joints because each reaction occurs only at one joint. However, when we sum the equations, then we may get several ones in the same row as in the 2nd row of the matrix in (8.6). If you are solving the problem without matrices, then having the extra equations is helpful. The top row of (8.6) tells us immediately that $R_{2,x} = 0$. However, with matrices it is more helpful to have zeros above and below ones. Therefore, the most sensible thing is to use only the equations coming from the free body diagrams at each joint as in Example 8.2.1.

The system of equations in Example 8.2.1 has a unique solution. When that happens the system is called **statically determinate**. If we end up with one or more free variable, then the system is called **statically indeterminate**. If there is no solution to the system, it is an indication that the structure is unstable as shown in Example 8.2.5. How can we ensure that a truss is statically determinate? If a coefficient matrix has more columns than rows, then it is impossible to get a unique solution there are either infinitely many or no solutions. For matrices with more rows than columns it is possible to get a unique solution if dependence relations occur between rows, but by far the easiest case is when the coefficient matrix is a square matrix. Then uniqueness of a solution can be determined by any part of the invertible matrix theorem, such as not having zero as an eigenvalue. It must be emphasized that eigenvalues do not apply to matrices that are not square. In that case, the singular value decomposition should be used instead.

Given a planar truss with j joints and m members, we get $2j$ equations and m variables. We also get one variable per non-zero component of the reactions, so let this number be r. Then the coefficient matrix will be square if and only if

$$2j = m + r. \tag{8.7}$$

For a space truss the 2 is changed to a 3. Given a truss diagram without reactions indicated, we can ask whether it is possible to introduce reactions at certain places in such a way that the resulting square matrix has a non-zero determinant. Reordering the rows changes only the sign of the determinant.

Example 8.2.3. For each truss diagram below, determine if it is possible to add reactions in such a way that the truss is statically determinate.

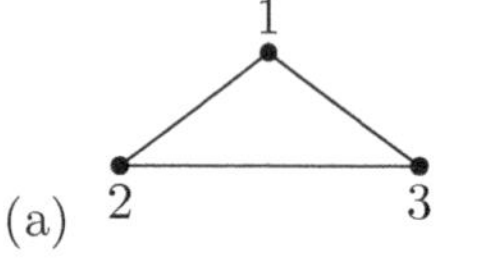

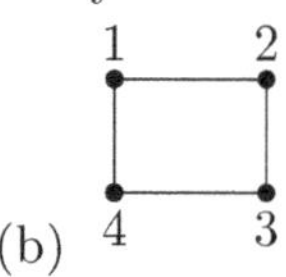

Solution

The truss in (a) is the same as in Example 8.2.1, just without supports. We have $j = 3$ and $m = 3$, so we must have $r = 2j - m = 6 - 3 = 3$, which is why we used a pin and a roller to get 3 reaction components. Reordering the rows of a square matrix changes only the sign of its determinant. We can reorder the rows of the coefficient matrix in (8.4) so that it becomes

$$\left[\begin{array}{ccc|ccc} 1 & 0 & 0 & -\frac{4}{5} & 0 & -1 \\ 0 & 1 & 0 & -\frac{3}{5} & 0 & 0 \\ 0 & 0 & 1 & 0 & -\frac{3}{5} & 0 \\ \hline 0 & 0 & 0 & \frac{4}{5} & -\frac{4}{5} & 0 \\ 0 & 0 & 0 & \frac{3}{5} & \frac{3}{5} & 0 \\ 0 & 0 & 0 & 0 & \frac{4}{5} & 1 \end{array}\right] \tag{8.8}$$

which is in block upper-triangular form. Thanks to Theorem 3.1.6 we know that the determinant is equal to the product of the determinants of the blocks on the diagonal.

The block in the upper left corner is the 3×3 identity matrix, and we will always get such a matrix from the reaction components that are introduced. So the determinant is equal to the determinant in the lower left corner, which is in block lower-triangular form. The transpose in is block upper-triangular form, and thus has determinant

$$\left(\frac{4}{5} \cdot \frac{3}{5} - -\frac{4}{5} \cdot \frac{3}{5}\right)(1) = \frac{24}{25}$$

which is why Example 8.2.1 is statically determinate. Once we have seen this, it become clear how to approach the problem in general. Before reaction forces are introduced we have only the right 3 columns of the matrix in (8.8). The trick is to eliminate rows in such a way that the remaining square matrix has non-zero determinant. The rows that are eliminated correspond to either the horizontal or vertical component of a specific joint, which is where a reaction must be added.

For (b), we have $j = 4$ and $m = 4$, so $r = 2j - m = 8 - 4$ reactions must be added. But where? We first need to get equations from the free body diagrams, with the members assumed to be in compression. At joint 1

$$\mathbf{F}_{21} + \mathbf{F}_{41} = f_{21}\begin{bmatrix} -1 \\ 0 \end{bmatrix} + f_{41}\begin{bmatrix} 0 \\ 1 \end{bmatrix},$$

at joint 2

$$\mathbf{F}_{12} + \mathbf{F}_{32} = f_{21}\begin{bmatrix} 1 \\ 0 \end{bmatrix} + f_{32}\begin{bmatrix} 0 \\ 1 \end{bmatrix},$$

at joint 3

$$\mathbf{F}_{43} + \mathbf{F}_{23} = f_{43}\begin{bmatrix} 1 \\ 0 \end{bmatrix} + f_{32}\begin{bmatrix} 0 \\ -1 \end{bmatrix},$$

and at joint 4

$$\mathbf{F}_{34} + \mathbf{F}_{14} = f_{43}\begin{bmatrix} -1 \\ 0 \end{bmatrix} + f_{41}\begin{bmatrix} 0 \\ -1 \end{bmatrix}.$$

If the variables are taken in the order $f_{21}, f_{41}, f_{32}, f_{43}$

$$\begin{bmatrix} -1 & 0 & 0 & 0 \\ 0 & 1 & 0 & 0 \\ 1 & 0 & 0 & 0 \\ 0 & 0 & 1 & 0 \\ 0 & 0 & 0 & 1 \\ 0 & 0 & -1 & 0 \\ 0 & 0 & 0 & -1 \\ 0 & -1 & 0 & 0 \end{bmatrix}. \tag{8.9}$$

We must eliminate 4 rows in such a way as to get an invertible matrix. This can be done in more than one way. We could keep only the rows with positive ones. That corresponds to putting a pin at joint 4, a roller at joint 3, and a vertical roller at joint 1 (which must provide horizontal resistance). On the other hand, we could eliminate rows 3,4,7 and 8, which corresponds to putting pin supports at joints 2 and 4 (opposite corners of the square). But beware! While it is true that both of these approaches give us a statically determinate situation, and while it may be possible to actually build such a thing, it does not follow that it is wise. There are other structural methods that need to be considered, and experience tells us that triangles provide the best structural stability.

Activity 8.2.4. For each truss diagram below, determine if it is possible to add reactions in such a way that the truss is statically determinate.

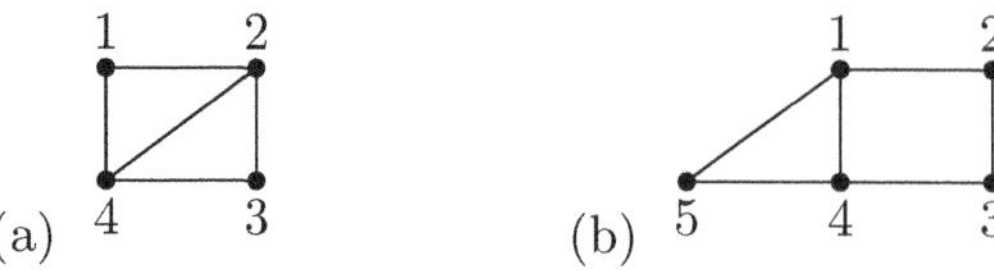

You may assume that the triangles are 3-4-5 right triangles.

In Example 8.2.3, we saw that a rectangle needs 4 properly placed reaction components to be statically determinate. If one of them is removed, then the coefficient matrix has more rows than columns. Depending on the load that is applied, it is possible that there would be no-solution.

Example 8.2.5. If the magnitude of $\mathbf{L}_1$ is 2 newtons in the diagram below, show that there is no solution. What what would happen physically?

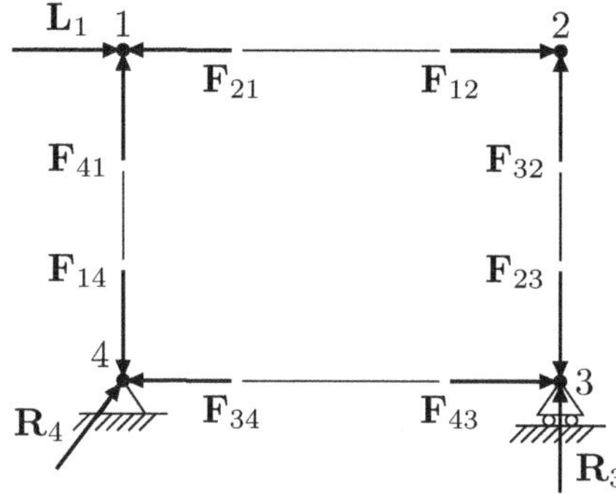

Solution

The placement of the reactions $\mathbf{R}_3$ and $\mathbf{R}_4$ amounts to appending the 3×3 identity matrix on the left of the bottom 3 rows of the matrix in (8.9), with zeros above. The load $\mathbf{L}_1$ has only a horizontal component, and so the augmented column will have a 2 in the top entry and zeros below it. By just one step of row reduction

$$
\left[\begin{array}{ccccccc|c}
0 & 0 & 0 & -1 & 0 & 0 & 0 & 2 \\
0 & 0 & 0 & 0 & 1 & 0 & 0 & 0 \\
0 & 0 & 0 & 1 & 0 & 0 & 0 & 0 \\
0 & 0 & 0 & 0 & 0 & 1 & 0 & 0 \\
0 & 0 & 0 & 0 & 0 & 0 & 1 & 0 \\
1 & 0 & 0 & 0 & 0 & -1 & 0 & 0 \\
0 & 1 & 0 & 0 & 0 & 0 & -1 & 0 \\
0 & 0 & 1 & 0 & -1 & 0 & 0 & 0
\end{array}\right]
\xrightarrow{R_3+R_1}
\left[\begin{array}{ccccccc|c}
0 & 0 & 0 & 0 & 0 & 0 & 0 & 2 \\
0 & 0 & 0 & 0 & 1 & 0 & 0 & 0 \\
0 & 0 & 0 & 1 & 0 & 0 & 0 & 0 \\
0 & 0 & 0 & 0 & 0 & 1 & 0 & 0 \\
0 & 0 & 0 & 0 & 0 & 0 & 1 & 0 \\
1 & 0 & 0 & 0 & 0 & -1 & 0 & 0 \\
0 & 1 & 0 & 0 & 0 & 0 & -1 & 0 \\
0 & 0 & 1 & 0 & -1 & 0 & 0 & 0
\end{array}\right]
$$

we obtain a matrix with zeros in the top row to the left of the augmented column, while

2 remains on the right. Thus there is no solution to the system.
Physically, the pin support at joint 4 holds that corner in place. But all of the members are free to rotate. Without any horizontal resistance along the top, even a load much smaller than 2 Newtons would push the square over so that it falls flat. That is a dynamic solution, not a static solution.

Activity 8.2.6. If the magnitude of $\mathbf{L}_1$ is 6 newtons in the diagram below, show that there is no solution. What what would happen physically?

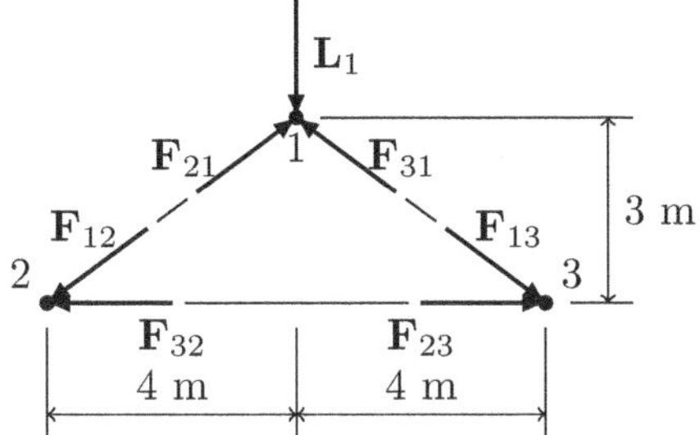

Hint: this truss is the same as in Example 8.2.1, except that there is no support. So the first 3 columns of the matrix in (8.4) need to be removed. Think logically about what would happen when a downward force is applied to a truss with no support.

For other applications of linear algebra to structures including the displacement method, see [21].

8.3 Linear Programming Problems

The material in this section is based mostly on [28]. A linear programming problem consists of the following features:

1. A linear function of several variables called an **objective function**
2. A system of linear inequalities ($\leq$ or $\geq$) and equations restricting the domain of the objective function. These are called the **constraints** and set of points satisfying the constraints is called the **feasible region**.

The goal is to find the maximum and minimum values of the objective function when restricted to the feasible region (if they exist). The linear inequalities always include equality, which means that the feasible region is closed. If it is bounded as well, then both a maximum and minimum exist by the extreme value theorem on $\mathbb{R}^n$. Of course in real life we often want integer solutions, and it is not necessarily the case that the maximum and minimum on the feasible region will occur at a point with integer coordinates. Whole books have been devoted to the subject of finding solutions with various algorithms. Here we will only consider finding the maximum on a bounded region and we will look at only two algorithms: the graphical approach in 2D, and the simplex method.

In practical contexts, the objective function might be a profit function, which we would aim to maximize. The constraints typically come from practical restrictions such as a limited amount of each material on hand, or meeting a quota.

8.3.1 The Graphical Solution in 2D

If only two variables are involved, it is possible to solve a system graphically. To do so, it is first necessary to graph the feasible region. Graphing inequalities can be done in three steps:

1. Graph the boundary, i.e. the points for which equality holds.
2. Pick a test point away from the boundary and determine if it is in the true region or the false region.
3. Shade the true region.

Example 8.3.1. To graph the inequality $3x + 2y \geq 6$, we first need to graph $3x + 2y = 6$ using the method from Section 2.1. For $x = 0$ we get $y = 3$ and for $y = 0$ we get $x = 2$. We plot the points $(0, 3)$ and $(2, 0)$ and the line through them. The point $(0, 0)$ is clearly not on the line, so we can use it as a test point. Is

$$3 \cdot 0 + 2 \cdot 0 = 0 \geq 6?$$

No. So $(0, 0)$ is on the false side and we shade away from it.

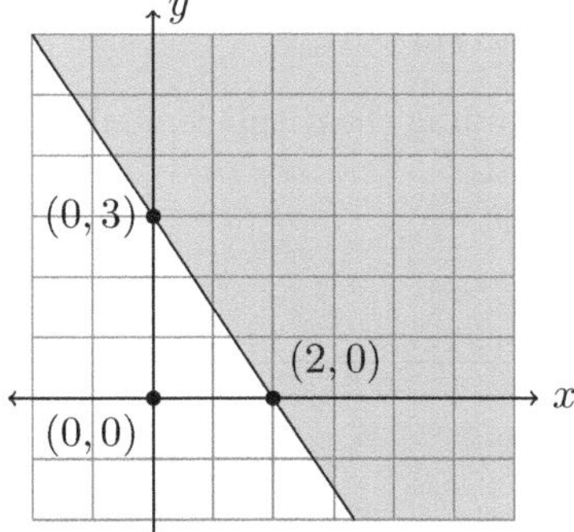

On the other hand, if we have a line through $(0, 0)$, then to plot it we need to take x and y different from zero to get a second point, and we should pick either $(1, 0)$ or $(0, 1)$ as a test point (at least one of which will not be on the line). For example for $2x - 3y \geq 0$, we get $x = 3$ and $y = 2$. We plot the line through $(0, 0)$ and $(3, 2)$. The point $(1, 0)$ is not on the line, and

$$2 \cdot 1 - 3 \cdot 0 = 2 \geq 0$$

is true, so we share in the direction of $(1, 0)$.

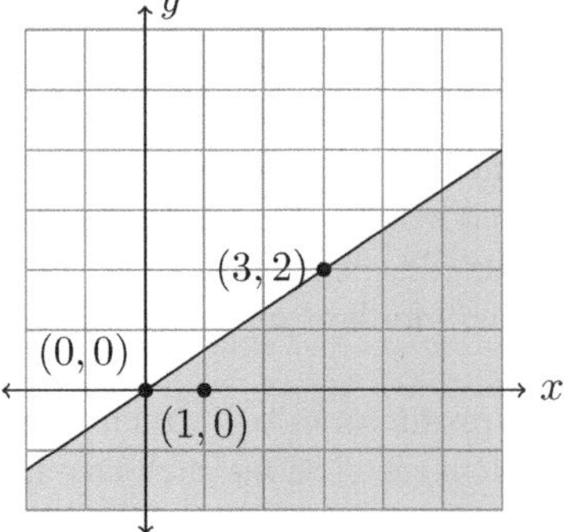

Activity 8.3.2. Graph each inequality below

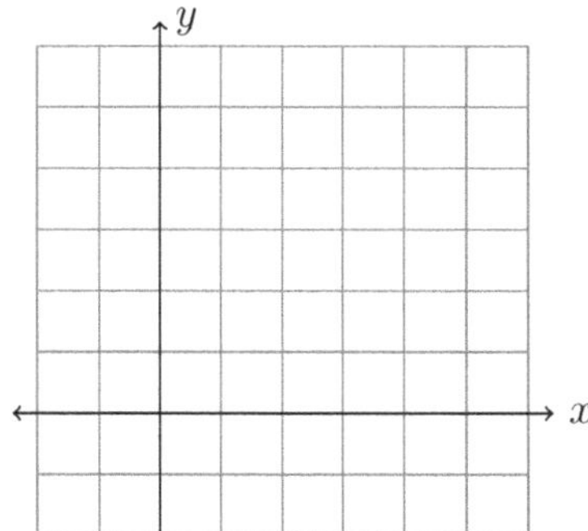

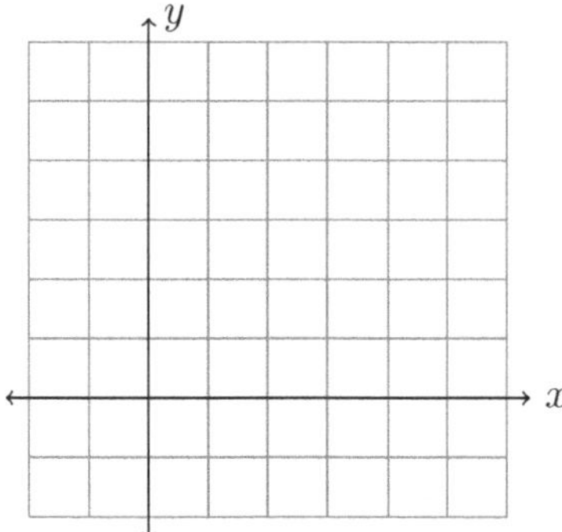

To graph a feasible region, it is usually necessary to graph a system of inequalities. The shaded region consists of the for which all of the inequalities are true. It may be helpful to draw arrows in the true direction for all of the lines before shading them. Then shade only the region that all of the arrows point toward.

Example 8.3.3. Given the inequalities

$$4x + 3y \leq 24$$
$$2x + 3y \leq 18$$
$$x \geq 0, y \geq 0$$

the last two define the first quadrant. Looking at just the first two, we must graph the lines

$$4x + 3y = 24$$
$$2x + 3y = 18$$

The points $(0, 8)$ and $(6, 0)$ are on the first line, and the points $(9, 0)$ and $(6, 0)$ are on the second line. The point $(0, 0)$ is not on either of the lines, so it can be used as a test point for both of them.

$$4 \cdot 0 + 3 \cdot 0 = 0 \leq 24$$
$$2 \cdot 0 + 3 \cdot 0 = 0 \leq 18$$

is true in both cases, hence we shade toward $(0, 0)$ while remaining in the 1st quadrant (because of $x \geq 0$ and $y \geq 0$.

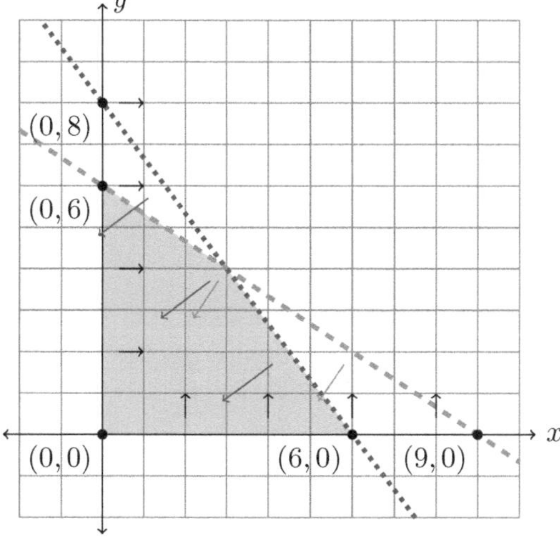

Activity 8.3.4. Graph the system of inequalities and find the coordinates of all corner points.

$$y + 2x \geq 6$$
$$x + y \leq 4$$
$$x \geq 0, y \geq 0$$

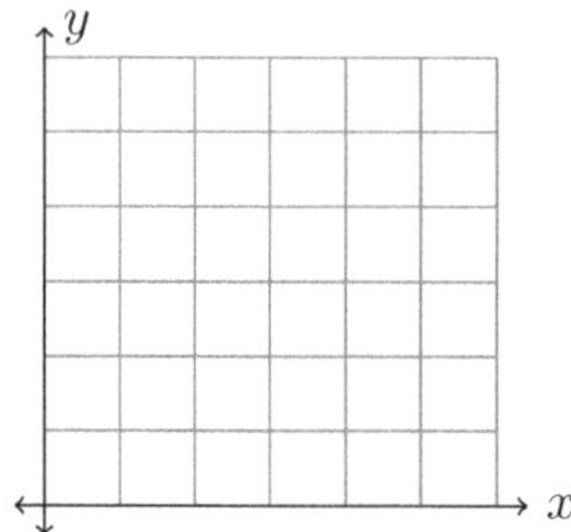

Once we have a graph of the feasible region, the solution is found by determining the direction in which the objective function is increasing. Constant values of the objective function are parallel lines, some of which will cut through the feasible region. We can find the direction in which the objective function increases in two steps

1. Set the objective function equal to a constant for which it is easy to plot the line.

2. Pick a test point not on the line to see if the value increases or decreases in that direction.

Once we know the direction of increasing, we can determine the maximum by checking the value at the most likely corner points.

Example 8.3.5. Suppose now that we want to maximize the objective function $3x + 4y$ on the feasible region given in Example 8.3.3. We will call the value of this function z. In the graph below we have plotted the lines

$$3x + 4y = 12 \quad \text{and} \quad 3x + 4y = 24$$

The choice of $z = 12$ is relatively easy since 12 is the product of 3 and 4. The point $(0, 0)$ is not on the line and

$$3 \cdot 0 + 4 \cdot 0 = 0 \leq 12$$

so the direction of increasing is away from the origin. The direction of increasing is indicated by arrows. The only corner point not found in Example 8.3.3 is $(3, 4)$. If the coordinates are not clear from a graph, then they can be found by row reduction. As we let z increase, the line

$$3x + 4y = z$$

sweeps through the feasible region and $(3, 4)$ is the last point that it cuts through; the slope of $3x + 4y = 12$ is in between that of the dashed line and of the dotted line. However, it is not always so easy to tell visually. While it may be clear that the point $(0, 0)$ is in the wrong direction, and perhaps $(6, 0)$ can be discarded as well, the difference between $(0, 6)$ and $(3, 4)$ may challenge visual acuity.

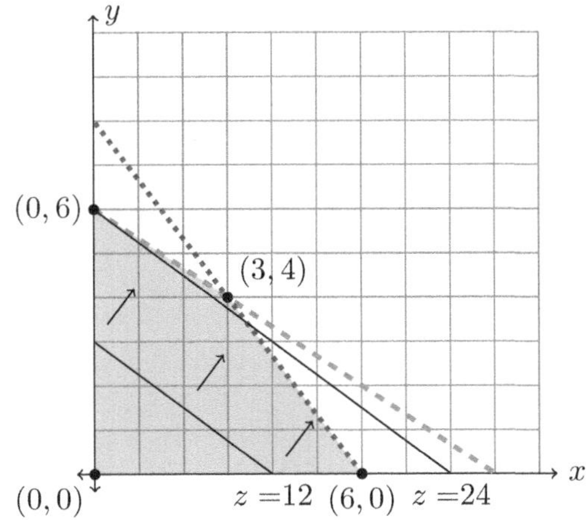

(x, y)	$z = 3x + 4y$
$(0, 0)$	0
$(6, 0)$	18
$(0, 6)$	24
$(3, 4)$	25 max.

Activity 8.3.6. Given the objective function $2y - x$ and the feasible region

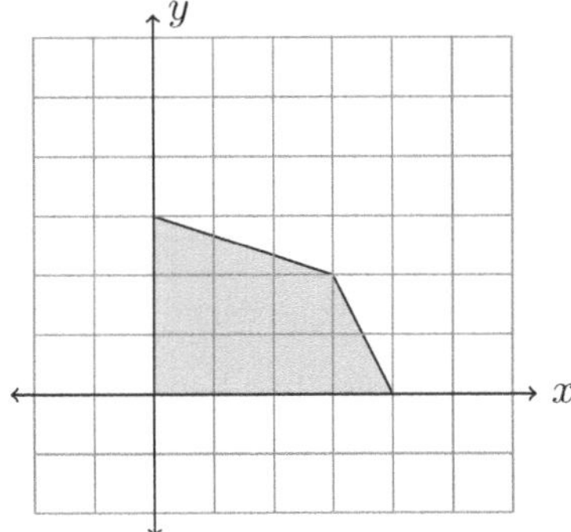

1. Draw the line $2y - x = -2$ and indicate the direction of increasing.
2. make a table listing all corner points, and the values of the objective function,
3. indicate the maximum value.

8.3.2 The Simplex Method

The first step of the simplex method is to convert the constraints into a system of equations in which all variables must be non-negative. The system of equations must also include an equation for the objective function. When represented by an augmented matrix, the top row should represent the equation for the objective function. This is called the **standard form** of a linear programming problem. Inequalities can be turned into equations by introducing new variables, but we may introduce new variables even when one of the constraints is given by an equation. We also aim to obtain an augmented form in which each row already contains a pivot, in which case it is automatically true that the rows are linearly independent. These pivots may not be leading, but the telltale sign of a pivot is that zeros occur everywhere else in the same column.

Definition 8.3.1. A variable corresponding to a pivot is called a **basic variable**. The remaining variables – that is the free variables – are called **non-basic variables**.

At each stage of the algorithm, a solution to the system of equations can be obtained by setting the free variables (non-basic variables) equal to zero, which will yield the coordinates of a corner point of the feasible region. The simplex method uses a strategy to change the

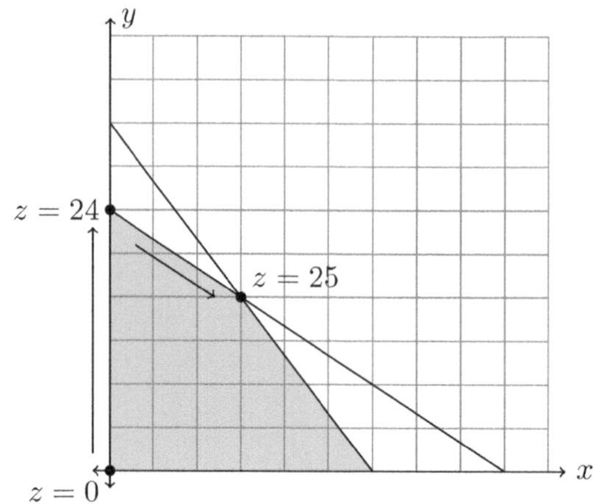

FIGURE 8.2: Pivoting between adjacent corner points

choice of pivot to improve the value of the objective function. We call this "pivoting." To determine how to pivot, we must decide on two things:

1. a non-basic variable to turn into a basic variable (called an "entering variable"),
2. a basic variable to turn into a non-basic variable (called an "exiting variable").

This terminology should make sense since an entering variable is one that becomes a pivot, and an exiting variable is one that will no longer be a pivot. For a maximization problem on a bounded feasible region, we use the following procedure for determining these variables.

1. Look in top row (the row representing the objective function) for the column with the most negative coefficient. The variable corresponding to that column will be the entering variable.
2. For all remaining rows, compute the quotients

$$\frac{\text{value in the augmented column}}{\text{value in the column of the entering variable}},$$

   disregarding those where the denominator is zero. Find the row with the smallest quotient. The variable corresponding to the current pivot in that row is the exiting variable.

These rules have two notable effects:

1. The row reduction operations to change to the new pivot are prescribed and can actually be done in parallel.
2. The new solution is an adjacent corner point. See Figure 8.2 for an illustration.

Example 8.3.7. Suppose that we want to maximize the objective function

$$3x + 4y$$

subject to the constraints

$$\begin{aligned} 4x + 3y &\le 24 \\ 2x + 3y &\le 18 \\ x \ge 0, y &\ge 0 \end{aligned} \tag{8.10}$$

The inequalities $x \geq 0$ and $y \geq 0$ already say that x and y must be non-negative, so to obtain standard form we only need to deal with the remaining two constraints in (8.10). We solve them in such a way that we get zero less than or equal to something, which will become a new variable:

$$\begin{aligned} 4x + 3y \leq 24 &\implies 0 \leq 24 - 4x - 3y, \\ 2x + 3y \leq 18 &\implies 0 \leq 18 - 2x - 3y. \end{aligned}$$

If we define $s_1 = 24 - 4x - 3y$ and $s_2 = 18 - 2x - 3y$, then these two inequalities become $s_1 \geq 0$ and $s_2 \geq 0$. We also need to define a variable for the value of the objective function, which we will call z, so $z = 3x + 4y$. Since x and y can't be negative then z also will not be negative. Thus we have obtained the following standard form:

$$\begin{aligned} -3x - 4y + z &= 0 \\ 2x + 3y + s_1 &= 18 \\ 4x + 3y + s_2 &= 24 \\ x, y, z, s_1, s_2 &\geq 0 \end{aligned} \tag{8.11}$$

The variables s_1 and s_2 are called "slack variables," because inequalities such as $4x+3y \leq 24$ are often caused by resource limitations. If a solution has a positive value of one of the slack variables, then not all of the resources have been used up. Other inequalities may limit the downside, in which case it makes sense to refer to them as "excess variables" instead, and they are named e_i. With the variables in the order x, y, z, s_1, s_2 the augmented form is

$$\left[\begin{array}{ccccc|c} -3 & -4 & 1 & 0 & 0 & 0 \\ 4 & 3 & 0 & 1 & 0 & 24 \\ 2 & 3 & 0 & 0 & 1 & 18 \end{array}\right] \tag{8.12}$$

The three pivots jump out immediately: columns 3, 4, and 5 each contain exactly one 1 and zeros everywhere else. So s_1, s_2 and z are the basic variables, and x and y are the non-basic variables (the free variables). By setting x and y equal to zero, we obtain $s_1 = 18$ and $s_2 = 24$ (no resources have been used). And indeed, $x = 0$ and $y = 0$ corresponds to a corner point in the feasible region (see Figure 8.2). The value of the objective function here is $z = 0$. Can we do better? Yes.

We look in the top row for the most negative coefficient, which is -4. Since -4 is in the 2nd column, that means that y will be the entering variable. We now go down that column and compute the quotients:

$$\frac{24}{3} = 8 \quad \text{and} \quad \frac{18}{3} = 6.$$

The smaller of these and is 6, which is in the 3rd row. The new pivot will be at the intersection of the 2nd column and the 3rd row, and it will replace the pivot currently in the 3rd row. That pivot is in column 5, which corresponds to s_2, so s_2 will be the departing variable. To get a one in the 2nd column and 3rd row, we begin by multiplying row 2 by the reciprocal of the current value in that spot.

$$\left[\begin{array}{ccccc|c} -3 & -4 & 1 & 0 & 0 & 0 \\ 4 & 3 & 0 & 1 & 0 & 24 \\ 2 & 3 & 0 & 0 & 1 & 18 \end{array}\right] \xrightarrow{\frac{1}{3}R_3} \left[\begin{array}{ccccc|c} -3 & -4 & 1 & 0 & 0 & 0 \\ 4 & 3 & 0 & 1 & 0 & 24 \\ \frac{2}{3} & 1 & 0 & 0 & \frac{1}{3} & 6 \end{array}\right]$$

Next we clear the values above and below the new one by row reduction (in this case only above).

$$\left[\begin{array}{ccccc|c} -3 & -4 & 1 & 0 & 0 & 0 \\ 4 & 3 & 0 & 1 & 0 & 24 \\ \frac{2}{3} & 1 & 0 & 0 & \frac{1}{3} & 6 \end{array}\right] \xrightarrow[(-3)R_3 + R_2]{4R_3 + R_1} \left[\begin{array}{ccccc|c} -\frac{1}{3} & 0 & 1 & 0 & \frac{4}{3} & 24 \\ 2 & 0 & 0 & 1 & -1 & 6 \\ \frac{2}{3} & 1 & 0 & 0 & \frac{1}{3} & 6 \end{array}\right].$$

Now the basic variables are s_1, y, z and the non-basic variables are x and s_2. When we set $x = 0$ and $s_2 = 0$ then we have $y = 6$, $z = 24$, and $s_1 = 6$. This means that the objective function has the value 24 at the point $x = 0$ and $y = 6$, which is another corner point of the feasible region. Certainly 24 is an improvement over 0, but we still have a negative value in the first row. That value is $-\frac{1}{3}$ and it is in the 1st column. We look down the first column and compute

$$\frac{6}{2} = 3 \quad \text{and} \quad \frac{6}{\frac{2}{3}} \cdot \frac{3}{3} = \frac{6 \cdot 3}{2} = 9.$$

Since 3 is smaller, we use row 2. We begin by multiplying row two by the value at the intersection of row 2 and column 1:

$$\left[\begin{array}{cccccc|c} -\frac{1}{3} & 0 & 1 & 0 & \frac{4}{3} & 24 \\ 2 & 0 & 0 & 1 & -1 & 6 \\ \frac{2}{3} & 1 & 0 & 0 & \frac{1}{3} & 6 \end{array}\right] \xrightarrow{\frac{1}{2}R_2} \left[\begin{array}{cccccc|c} -\frac{1}{3} & 0 & 1 & 0 & \frac{4}{3} & 24 \\ 1 & 0 & 0 & \frac{1}{2} & -\frac{1}{2} & 3 \\ \frac{2}{3} & 1 & 0 & 0 & \frac{1}{3} & 6 \end{array}\right].$$

Then we clear the values above and below the new 1 by row operations:

$$\left[\begin{array}{ccccc|c} -\frac{1}{3} & 0 & 1 & 0 & \frac{4}{3} & 24 \\ 1 & 0 & 0 & \frac{1}{2} & -\frac{1}{2} & 3 \\ \frac{2}{3} & 1 & 0 & 0 & \frac{1}{3} & 6 \end{array}\right] \xrightarrow[(-\frac{2}{3})R_2 + R_2]{\frac{1}{3}R_2 + R_1} \left[\begin{array}{ccccc|c} 0 & 0 & 1 & \frac{1}{6} & \frac{7}{6} & 25 \\ 1 & 0 & 0 & \frac{1}{2} & -\frac{1}{2} & 3 \\ 0 & 1 & 0 & -\frac{1}{3} & \frac{2}{3} & 4 \end{array}\right].$$

The new basic variables are x, y, z and the non-basic variables are s_1, s_2. If we set s_1 and s_2 to zero, then we obtain $x = 3$, $y = 4$, and $z = 25$. The point $(3, 4)$ is a corner point of the feasible region. Since there is no longer a negative number in the first row, we have the optimal solution: 25 is the maximum value of the objective function $3x + 4y$ on the feasible region.

Remark 8.3.1. For a constraint that is an equation, it does not make sense to speak of slack or excess. Instead we call it a dummy variable and we use the notation d_i. Ultimately we need $d_i = 0$ for a solution to make sense, however, introducing dummy variables has the benefit of producing a pivot in each row automatically.

8.4 Markov Chains

Markov chains are a type of mathematical model involving probabilities. There are two important facts about about probabilities that you should bear in mind at all times:

1. Probabilities are values between 0 and 1.

2. The sum of the probabilities of all possible things that can happen must add up to 1.

A probability of 0 means that a particular possibility never happens, and a probability of 1 means that it always happens.

Definition 8.4.1. A **Markov chain** is a mathematical model in discrete time consisting of a system that satisfies two conditions:

1. the system can only be in a fixed number of states.

2. the probability of a transition from any given state to any other state depends only on the two states.

The assumption of discrete time is important. For our purposes, it will also be useful to think about observing the system at different instances in time, separated by regular intervals. So we may observe a system at the beginning of one day, then leave, come back a day later and look at it again, leave, come back the next day, etc. The idea is that the state (condition) the system is in at each time we look at it is governed by the rules of probability. Those probabilities should not depend on previous states or conditions that the system has been in, only on where it is now and where it is going.

Example 8.4.1. Consider flipping a fair coin twice.

- The system is the coin.
- The states are heads and tails.

The probability to go from any state to any other state on one flip is 1/2.

We can represent this process with a **transition diagram.**

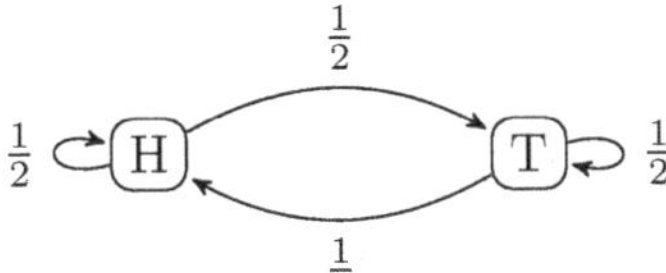

The arrows in a transition diagram point from the current state to the new state, and are labeled with the probability for that transition. Since something must happen, the probabilities on the arrows leaving a given state must add up to one.

8.4.1 The Transition Matrix

Consider a Markov chain with n states. Let p_{ij} be the probability of making a direct transition from state i to state j, $1 \leq i \leq n$, $1 \leq j \leq n$. The matrix

$$\mathbf{P} = [p_{ij}]$$

is called the **one-step transition matrix**, or simply the **transition matrix** for the Markov chain.

Example 8.4.2. Write down the transition matrix for the fair coin.

Solution

The probabilities in the transition diagram in Example 8.4.1 are all 1/2 so the matrix we get is

$$\begin{bmatrix} \frac{1}{2} & \frac{1}{2} \\ \frac{1}{2} & \frac{1}{2} \end{bmatrix}.$$

Example 8.4.3. On an average morning, when the alarm goes off, I have a 75% chance of hitting the snooze button. The states involve being in bed (I), and being out of bed (O). In that problem I can only go from being in bed to being out of bed. Draw the transition diagram and the corresponding transition matrix.

Solution

In the transition diagram the arrows leaving a state must add up to 1. For the transition matrix, this means that the rows must add up to 1. The probability of staying in state

I is $\frac{3}{4}$, so the probability of going from I to O must be $\frac{1}{4}$. Once I am in state O, I can't leave, so the probability of going from O back to I is 0 and the probability of going from O to O is 1. The resulting transition diagram and transition matrix are given below:

$$\begin{bmatrix} \frac{3}{4} & \frac{1}{4} \\ 0 & 1 \end{bmatrix}$$

Activity 8.4.4. A math student deep in thought is pacing back and forth along a line going East and West. There is a 1/3 probability that the student will start off walking west (W) and 2/3 probability that the student will start off walking east (E). The system is the pacing math student. The states are walking east and walking west. Draw the transition diagram, and write down the corresponding transition matrix.

Activity 8.4.5. A biologist observes a family of voles at regular intervals during the day. The biologist notes that if a vole is eating on one observation, there is a .2 probability that it will be eating on the next observation, and .8 probability that it will be underground in its burrow. If the vole is originally in its burrow, there is a .4 probability that it will still be in its burrow on the next observation, and .6 probability that it will be eating. Draw the transition diagram, and write down the corresponding transition matrix.

Activity 8.4.6. The daily weather forecast for a town can be sunny(S), rainy(R), or cloudy (C). If it is sunny today, there is a 1/3 chance it will be sunny tomorrow, and 2/3 chance it will be cloudy tomorrow. If it is cloudy today then there is 1/6 chance that it will be sunny tomorrow, 1/2 chance that it will be cloudy, and 1/3 chance that it will be rainy. If it is rainy today then there is 1/6 chance that it will be rainy tomorrow and 5/6 chance that it will be sunny tomorrow. Suppose it is sunny today. Draw the transition diagram, and write down the corresponding transition matrix.

Activity 8.4.7. Arnold Schoenberg, the serialist composer, plans to write a piano sonata in C-minor. When he plays the tonic triad in the C-minor scale, there is a 5/6 probability that he will play the 1st inversion of this chord next, and a 1/6 probability that he will play the tonic triad again. When he plays the 1st inversion of the tonic triad, there is a 1/2 probability that he will go back to the tonic triad, 1/6 probability that he will repeat the 1st inversion, and 1/3 probability that he will play the dominant triad (in C-major). When he plays the dominant triad there is a 1/3 probability that he will play it again, and 2/3 probability that he will follow with the tonic triad instead. Draw the transition diagram and write down the corresponding transition matrix.

Definition 8.4.2. A **probability vector**, is a vector

$$\mathbf{x} = [x_1, x_2, \ldots x_N],$$

with coordinates that are non-negative, i.e. $x_i \geq 0$ for all i, and which add to 1, i.e.

$$x_1 + x_2 + \cdots + x_N = 1.$$

Definition 8.4.3. Consider a Markov chain with N states. A **state vector** for the Markov chain is a probability vector, where x_i is the probability of being in the i-th state.

Example 8.4.8. Suppose the weather system from Activity 8.4.6 started out as sunny.
Then the state vector is

$$\begin{bmatrix} 1 & 0 & 0 \end{bmatrix}.$$

When we multiply a state vector by a one-step transition matrix, we get a new state vector representing the probabilities of being in different states the next time the system is observed.

Example 8.4.9. Given the transition matrix

$$\mathbf{P} = \begin{bmatrix} .1 & .5 & 0 & .4 \\ 0 & .7 & .3 & 0 \\ 0 & .6 & .1 & .3 \\ 0 & .8 & 0 & .2 \end{bmatrix},$$

answer the following questions:

1. If the system starts out in state 2, what is the probability that we will be in state 3 on the next observation?

2. If the system starts out in state 1 what is the probability that it successively occupies states 1,4, and 2 on the next 3 observations?

3. If the system starts out in state 2 on the 1st observation, what is the probability that it will be in state 2 on the 3rd observation.

Solution

1. If the system starts in state 2 and goes to state 3, this is just the probability in the 2nd row and 3rd column: .3.

2. If we are interested in the system occupying a specific sequence of states, then we multiply the probabilities for the transitions between those states. Below each transition is labelled with its probability

$$1 \xrightarrow{.4} 4 \xrightarrow{.8} 2 \xrightarrow{.3} 3$$

so the probability of this sequence of transitions is

$$.4 \times .8 \times .3 = .32 \times .3 = .096$$

3. If we are interested in an initial and final state, and a certain number of transitions in between, then there may be multiple paths. So, we have a "sum over histories." If the system starts out in state 2 in the 1st observation, then the state vector for that observation is

$$\begin{bmatrix} 0 & 1 & 0 & 0 \end{bmatrix}.$$

For it to end up in state 2 on the 3rd observation, there have to be 2 transitions $(3 - 1 = 2)$. In this case, since we cannot go from state 2 to state 1 or 4, there are only two histories

$$2 \xrightarrow{.7} 2 \xrightarrow{.7} 2$$
$$2 \xrightarrow{.3} 3 \xrightarrow{.6} 2$$

with probabilities .49 and .18 respectively. Summing over the histories gives us a final result of $.49 + .18 = .67$. However, matrix multiplication keeps track of all of this for us. Since there are two transition we must multiply the initial state vector by the transition matrix twice (once for each transition).

$$\begin{array}{c} \\ \\ \\ \\ \begin{bmatrix} 0 & 1 & 0 & 0 \end{bmatrix} \end{array} \begin{array}{c} \begin{bmatrix} .1 & .5 & 0 & .4 \\ 0 & .7 & .3 & 0 \\ 0 & .6 & .1 & .3 \\ 0 & .8 & 0 & .2 \end{bmatrix} \\ \begin{bmatrix} 0 & .7 & .3 & 0 \end{bmatrix} \end{array} \begin{array}{c} \begin{bmatrix} .1 & .5 & 0 & .4 \\ 0 & .7 & .3 & 0 \\ 0 & .6 & .1 & .3 \\ 0 & .8 & 0 & .2 \end{bmatrix} \\ \begin{bmatrix} 0 & .67 & .24 & .09 \end{bmatrix} \end{array}$$

We can now see that the probability for state 2 is .67. In this example, computing the sum over individual histories isn't too hard, but in general matrix multiplication is computationally expedient.

Activity 8.4.10. Markov chains can be used to generate semi-coherent text. For a computer implementation of this go to `http://joshmillard.com/garkov/`. For a modest example, let the states be: (1) "to be," (2) "or not," and (3) "gazorninplat". If the system is in state 1, then state 1 will not be repeated and there is an equal probability of going to state 2 or state 3. If the system is in state 2, then it is guaranteed to be in state 1 next. If the system is in state 3, then it will not go to state 2 next, and is 3 times as likely to repeat state 3, as it is to go to state 1. Use this data to do the following problems:

1. Find the transition diagram and transition matrix for this Markov chain.
2. Given that the system is in state 1 on the 1st observation, find the probability that it successively occupies the states 2, 1, and 3 on the next 3 observations.

Activity 8.4.11. Find the probability that it is sunny three days from now, given that it is Sunny today. Use the results from Activity 8.4.6 and Example 8.4.8.

Activity 8.4.12. Recall the Schoenberg problem from Activity 8.4.7 Suppose he starts off playing the tonic triad. Find the state vectors for the next two chords he plays.

Activity 8.4.13. Suppose that a vole is eating on the 1st observation. Find the probability that a vole is also eating on the 4th observation. Use the results from Activity 8.4.5.

8.4.2 Classification and Long Term Behavior

The k-step transition matrix for large values of k is an indication of long term behavior of a system. For many Markov chains the probabilities will stabilize for large values of k.

Definition 8.4.4. A Markov chain with transition matrix $\mathbf{P}$ is called **regular** if there is a positive integer k such that $\mathbf{P}^k$ has all positive entries.

Example 8.4.14. Consider 3 Markov chains, where the corresponding one-step transition matrices are given by

$$\mathbf{A} = \begin{bmatrix} \frac{3}{4} & \frac{1}{4} \\ 0 & 1 \end{bmatrix} \quad \mathbf{B} = \begin{bmatrix} \frac{3}{4} & \frac{1}{4} \\ 1 & 0 \end{bmatrix} \quad \mathbf{C} = \begin{bmatrix} 0 & 1 \\ 1 & 0 \end{bmatrix}.$$

Determine which of these Markov chains is regular by looking at the k-step transition matrices for larger values of k.

Solution

To check when a Markov chain is regular, we do not need the exact values of the entries. Instead, we just need to observer when something is positive or zero. In matrix multiplication, whenever at least one pair of numbers multiplied together is positive, the result is positive.
So for **A** we have

$$\begin{array}{cc} & \begin{bmatrix} + & + \\ 0 & + \end{bmatrix} \\ \begin{bmatrix} + & + \\ 0 & + \end{bmatrix} & \begin{bmatrix} + & + \\ 0 & + \end{bmatrix}. \end{array}$$

We get the same matrix back, and so we will get the same matrix no matter how many times we multiply. There will always be a zero in the bottom left corner, and so the Markov chain is not regular.
For **B** we have

$$\begin{array}{cc} & \begin{bmatrix} + & + \\ + & 0 \end{bmatrix} \\ \begin{bmatrix} + & + \\ + & 0 \end{bmatrix} & \begin{bmatrix} + & + \\ + & + \end{bmatrix}, \end{array}$$

so the Markov chain is regular. Of course, for regular Markov chains, it is not always the case that we get all positive entries in one step of multiplication. We may need to multiply more than once.
For **C** we have

$$\begin{array}{ccc} & \begin{bmatrix} 0 & + \\ + & 0 \end{bmatrix} & \begin{bmatrix} 0 & + \\ + & 0 \end{bmatrix} \\ \begin{bmatrix} 0 & + \\ + & 0 \end{bmatrix} & \begin{bmatrix} + & 0 \\ 0 & + \end{bmatrix} & \begin{bmatrix} 0 & + \\ + & 0 \end{bmatrix}. \end{array}$$

Here we find ourselves going in loops, back and forth between the matrices

$$\begin{bmatrix} 0 & + \\ + & 0 \end{bmatrix} \quad \text{and} \quad \begin{bmatrix} + & 0 \\ 0 & + \end{bmatrix},$$

so the Markov chain is not regular.

Activity 8.4.15. Given the transition matrices

$$\mathbf{A} = \begin{bmatrix} \frac{1}{2} & 0 & \frac{1}{2} \\ 0 & 1 & 0 \\ \frac{1}{2} & 0 & \frac{1}{2} \end{bmatrix} \quad \mathbf{B} = \begin{bmatrix} 0 & 0 & 1 \\ 1 & 0 & 0 \\ 0 & 1 & 0 \end{bmatrix} \quad \mathbf{C} = \begin{bmatrix} 0 & \frac{1}{2} & \frac{1}{2} \\ 1 & 0 & 0 \\ 0 & 1 & 0 \end{bmatrix}$$

determine whether or not the corresponding Markov chains are regular.

As pointed out in Example 8.4.14, for a regular Markov chain, it may be necessary to multiply more than once before getting + in every entry. But if a Markov chain is not regular, we will never get + in every entry no matter how many times we multiply. How do we know when to stop? One of two things will happen. After enough steps of multiplication, either

1. the matrix will stabilize as it did for **A** and **B** in Example 8.4.14, or

2. we will end up going in a loop.

So, we can stop once we get + everywhere or once we have encountered the same matrix twice. Encountering the same matrix twice in a row, means that it has stabilized: we will get the same matrix for each step thereafter. On the other hand, in some cases it can be seen that a Markov chain is not regular without going through any multiplication steps at all. The matrices **B** and **C** in Example 8.4.14 both illustrate this point.

Definition 8.4.5. A Markov chain is called **absorbing** if the one-step transition matrix has a 1 on the main diagonal. The state with that 1 is called an **absorbing state**.

Remark 8.4.1. The intuition behind an absorbing state, is that once you are in an absorbing state, you cannot leave it. If there is a non-zero probability of going from another state to the absorbing state, then we are pulled into it. In this way, the use of the term "absorbing" should make sense.

Example 8.4.16. The Markov chain with matrix **A** in Example 8.4.14 is absorbing, while **B** and **C** are not. In fact **A** should be familiar from the alarm clock problem in Example 8.4.3. Being out of bed is an "absorbing" state: even with a high probability of hitting the snooze, I will eventually get out of bed.

A Markov chain is not regular if any of the following is true:

1. the Markov chain has at least one absorbing state,
2. the Markov chain has a cycle, or
3. the Markov chain has more than one connected component.

By a "cycle" we mean a loop, in which the probability of going from one state to the next in the loop is always 1. Arbitrary permutations are made up of one or more cycles, so a Markov chain with a permutation matrix as its transition matrix is not regular, such as **C** in Example 8.4.14.

Remark 8.4.2. If we can make observations, about cycles, absorbing states, and connected components, it is not necessary to go through all the multiplication steps in Example 8.4.14. On the other hand, for a Markov chain with a large number of states – while it is easy to check for rows and columns with only zeros and ones – matrix multiplication is a practical way for a computer to identify the connected components, especially if implemented in binary.

Example 8.4.17. Determine the connected components for each Markov chain below.

$$\mathbf{A} = \begin{bmatrix} 0 & 1 & 0 & 0 \\ 1 & 0 & 0 & 0 \\ 0 & 0 & 1 & 0 \\ 0 & 0 & \frac{1}{2} & \frac{1}{2} \end{bmatrix} \quad \text{and} \quad \mathbf{B} = \begin{bmatrix} 0 & 0 & 1 & 0 \\ 0 & 0 & \frac{1}{2} & \frac{1}{2} \\ 0 & 1 & 0 & 0 \\ 1 & 0 & 0 & 0 \end{bmatrix}$$

Solution

For a human, one way to find the connected components is to draw the transition diagram in such a way that arrows between two states are only drawn, if there is a non-zero probability for the transition.
For **A** the transition diagram is

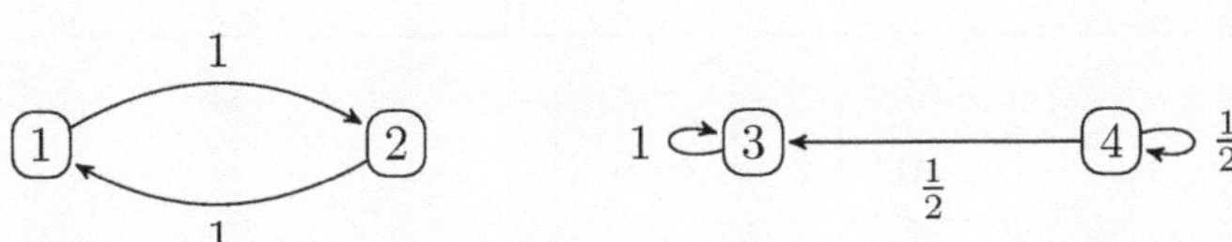

so there are two connected components. There is a cycle between states 1 and 2, and while it possible to go from state 4 to state 3, state 3 is absorbing.
For **B**, so there is one connected component.

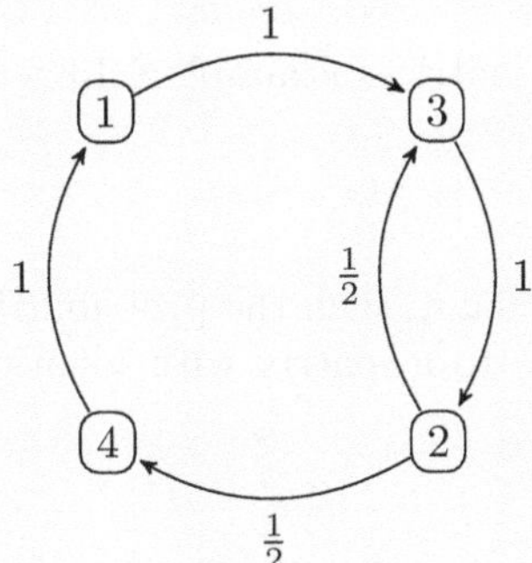

In matrix form, the connected components can be seen in a different way. Both **A** and **B** can be divided into 2×2 blocks. For **A**, the blocks off the diagonal are zero, which means that there is no mixing between the blocks. So, the connected components correspond to the blocks on the diagonal. On the other hand, for **B** the blocks on the diagonal are zero, and the blocks off the diagonal are not zero. So, there is mixing between the blocks. We have to be a little more careful here because there could be a different way of dividing up the matrix. Simply replacing the first row of **B** by

$$\begin{bmatrix} 0 & 0 & 0 & 1 \end{bmatrix}$$

yields a Markov chain with two connected components again because then there is a cycle between 1 and 4 that we never get out of. But if we can easily spot a way of dividing up the matrix with zeros off the diagonal, then we know there is more than one connected component.

Activity 8.4.18. Draw the transition diagram corresponding to the transition matrices in Activity 8.4.15. Which are connected?

Definition 8.4.6. If $\mathbf{x}$ is the current state vector, the k-step transition matrix $\mathbf{P}(k)$ is defined to be the matrix such that $\mathbf{xP}(k)$ will be the state vector k stages in the future.

Proposition 8.4.7. *For any one-step transition matrix* $\mathbf{P}$*, the* k *step transition matrix is simply*

$$\mathbf{P}(k) = \mathbf{P}^k.$$

This is illustrated by the following example.

Example 8.4.19. One of the transition matrices in Example 8.4.14 was

$$\mathbf{B} = \begin{bmatrix} \frac{3}{4} & \frac{1}{4} \\ 1 & 0 \end{bmatrix}$$

Compute the 2 and 3 step transition matrices.

Solution

$$
\begin{array}{ccc}
 & \begin{bmatrix} \frac{3}{4} & \frac{1}{4} \\ 1 & 0 \end{bmatrix} & \begin{bmatrix} \frac{3}{4} & \frac{1}{4} \\ 1 & 0 \end{bmatrix} \\
\begin{bmatrix} \frac{3}{4} & \frac{1}{4} \\ 1 & 0 \end{bmatrix} & \begin{bmatrix} \frac{13}{16} & \frac{3}{16} \\ \frac{3}{4} & \frac{1}{4} \end{bmatrix} & \begin{bmatrix} \frac{51}{64} & \frac{13}{64} \\ \frac{7}{16} & \frac{9}{16} \end{bmatrix}
\end{array}
$$

Activity 8.4.20. Find the 2-step transition matrix for the weather system in Activity 8.4.6.

Activity 8.4.21. Find the 2-step, 3-step and 4-step transition matrices Example 8.4.3 with the alarm clock.

Hint: $4^2 = 16$, $4^3 = 64$, $4^4 = 256$.

If $\mathbf{P}$ is the one-step transition matrix for a regular Markov chain, then the probabilities in $\mathbf{P}(k)$ stabilize for large values of k, and becomes very close to a matrix with identical rows

$$
\begin{bmatrix} \mathbf{w} \\ \mathbf{w} \\ \vdots \\ \mathbf{w} \end{bmatrix} = \begin{bmatrix} w_1 & w_2 & \cdots & w_N \\ w_1 & w_2 & \cdots & w_N \\ & \cdot\ \cdot\ \cdot\ \cdot\ \cdot & & \\ w_1 & w_2 & \cdots & w_N \end{bmatrix} \tag{8.13}
$$

where

$$
\mathbf{w} = \begin{bmatrix} w_1 & w_2 & \cdots w_N \end{bmatrix}.
$$

Example 8.4.22. The transition matrix from the alarm clock problem in Example 8.4.3,

$$
\mathbf{P} = \begin{bmatrix} \frac{3}{4} & \frac{1}{4} \\ 0 & 1 \end{bmatrix},
$$

isn't regular, but it still has this property. The k-step transition matrix is

$$
\mathbf{P}(k) = \begin{bmatrix} \left(\frac{3}{4}\right)^k & 1 - \left(\frac{3}{4}\right)^k \\ 0 & 1 \end{bmatrix}.
$$

Since $\left(\frac{3}{4}\right)^k \to 0$ as $k \to \infty$, then

$$
\lim_{k \to \infty} \begin{bmatrix} \left(\frac{3}{4}\right)^k & 1 - \left(\frac{3}{4}\right)^k \\ 0 & 1 \end{bmatrix} = \begin{bmatrix} 0 & 1 \\ 0 & 1 \end{bmatrix},
$$

so $\mathbf{w} = \begin{bmatrix} 0 & 1 \end{bmatrix}$.

8.4.3 Stable State Vector

For a regular Markov chain with transition matrix $\mathbf{P}$ we have seen that

$$
\lim_{k \to \infty} \mathbf{P}^k
$$

is a matrix in which the rows are all equal to a single vector $\mathbf{w}$. This vector $\mathbf{w}$ has the very interesting property that

$$
\mathbf{wP} = \mathbf{w}.
$$

That means that **w** is an eigenvector with eigenvalue 1. The only difference with what we did in Chapter 5 is that here the matrix **P** is applied on the right-hand side of **w** and **w** is a row vector, whereas in Chapter 5 we acted on the left of a column vector. However, all we have to do is apply the transpose to both sides of the equation.

Since **w** is unchanged by applying **P**, the following definition should seem natural:

Definition 8.4.8. A state vector **w** with the property that

$$\mathbf{wP} = \mathbf{w}, \tag{8.14}$$

is called a **stable state vector**.

8.4.4 Computing the Stable State Vector

The stable state vector of a transition matrix **P** is a probability vector **w** satisfying the equation

$$\mathbf{wP} = \mathbf{w}.$$

It is an example of an eigenvector with eigenvalue 1, except that it is on the left. We can find it by the same method that we usually use for eigenvectors if we begin by taking the transpose. We also are not interested in just any eigenvector. We need probabilities. This extra feature can be dealt with in one of two ways

1. we can choose the value of the free variable at the end, or

2. we can include an extra row corresponding to the equation

$$w_1 + w_2 + \cdots w_N = 1.$$

Although both of these are possible, the second approach is better in practice, since it gives a leading one automatically.

It is also helpful to use some facts that we know about dependence relations. The rows of **P** add up to 1. When we subtract the identity matrix, the rows add up to zero. When we take the transpose, that becomes true of the columns. Since the columns of $(\mathbf{P} - \mathbf{I})^T$ add up to 1, it means that there is a dependence relation between the rows, and therefore one of the rows can be thrown out.

Finally, we can avoid fractions until we get to echelon form by clearing denominators. As a result of these observations, we get the following approach:

1. Compute $(\mathbf{P} - \mathbf{I})^T = \mathbf{P}^T - \mathbf{I}$, and augment by a column of zeros.

2. Cross off your least favorite row (the one with the fewest zeros and most annoying denominators). Add a new row of ones at the top.

3. Multiply each row by the least common multiple of the denominators in that row to get integers.

4. Bring the matrix into reduced form by row reduction.

Remark 8.4.3. Once row reduction is complete, you can check your work by applying the rule of probability to the augmented column. The final values in the augmented column

1. must be between zero and one and

2. must add up to one.

If either of these is not true, then a mistake must have been made somewhere either during row reduction, or in the original setup of the problem. Check your work.

Example 8.4.23. Find the stable state vector for the matrix

$$\mathbf{A} = \begin{bmatrix} \frac{3}{4} & \frac{1}{4} \\ 0 & 1 \end{bmatrix}$$

from Example 8.4.14 above, which is actually the transition matrix for the Markov chain in Example 8.4.3. It was already mentioned above that this has a stable state vector even though it is absorbing, not regular. Using the initial state vector $[1, 0]$ compare the state vectors 3-steps in the future with the stable state vector.

Solution

Step 1: Compute $\mathbf{A}^T - \mathbf{I}$, and augment by a column of zeros.

$$\mathbf{A}^T - \mathbf{I} = \begin{bmatrix} \frac{3}{4} & 0 \\ \frac{1}{4} & 1 \end{bmatrix} - \begin{bmatrix} 1 & 0 \\ 0 & 1 \end{bmatrix} = \begin{bmatrix} -\frac{1}{4} & 0 \\ \frac{1}{4} & 0 \end{bmatrix} \to \left[\begin{array}{cc|c} -\frac{1}{4} & 0 & 0 \\ \frac{1}{4} & 0 & 0 \end{array}\right]$$

Steps 2: Cross off your least favorite row and add a new row of ones at the top.

$$\left[\begin{array}{cc|c} -\frac{1}{4} & 0 & 0 \\ \frac{1}{4} & 0 & 0 \end{array}\right] \to \left[\begin{array}{cc|c} 1 & 1 & 1 \\ -\frac{1}{4} & 0 & 0 \end{array}\right]$$

Steps 3: Multiply each row by the least common multiple of the denominators in that row.

$$\left[\begin{array}{cc|c} 1 & 1 & 1 \\ -\frac{1}{4} & 0 & 0 \end{array}\right] \xrightarrow{4R_2} \left[\begin{array}{cc|c} 1 & 1 & 1 \\ -1 & 0 & 0 \end{array}\right]$$

Steps 4: Bring the matrix into reduced form by row reduction.

$$\left[\begin{array}{cc|c} 1 & 1 & 1 \\ -1 & 0 & 0 \end{array}\right] \xrightarrow{R_1+R_2} \left[\begin{array}{cc|c} 1 & 1 & 1 \\ 0 & 1 & 1 \end{array}\right] \xrightarrow{(-1)R_2+R_1} \left[\begin{array}{cc|c} 1 & 0 & 0 \\ 0 & 1 & 1 \end{array}\right]$$

The stable state vector is the transpose of the augmented column:

$$\mathbf{w} = \begin{bmatrix} 1 & 0 \end{bmatrix}$$

Remark 8.4.4. We see from this example that the stable state vector represents long term probabilities of the system described by the Markov chain. It may not be clear from this example, but the long term behavior doesn't depend on which state vector we start from.

Activity 8.4.24. Find the stable state vector for the Schoenberg piano sonata in Example 8.4.7.

$$\mathbf{P} = \begin{bmatrix} \frac{1}{6} & 0 & \frac{5}{6} \\ \frac{2}{3} & \frac{1}{3} & 0 \\ \frac{1}{2} & \frac{1}{3} & \frac{1}{6} \end{bmatrix}$$

Interpret this in terms of which chords are favored over the whole course of the sonata.

Activity 8.4.25. Find the stable state vector for the weather system in Activity 8.4.6.

$$\mathbf{P} = \begin{bmatrix} \frac{1}{3} & \frac{2}{3} & 0 \\ \frac{1}{6} & \frac{1}{2} & \frac{1}{3} \\ \frac{2}{3} & 0 & \frac{1}{3} \end{bmatrix}$$

Interpret this in terms of weather patterns in the long term.

8.4.5 Existence and Uniqueness

A stable state vector always exists. If $\mathbf{P}$ is a transition matrix, then the rows must add up to 1. This means that a column vector of ones is a right eigenvector with eigenvalue 1. Since 1 occurs as an eigenvalue, then there is also a left eigenvector with that eigenvalue.

Example 8.4.26. Using the matrix $\mathbf{P}$ from Example 8.4.22

$$\begin{bmatrix} 1 \\ 1 \end{bmatrix}$$

$$\begin{bmatrix} \frac{3}{4} & \frac{1}{4} \\ 0 & 1 \end{bmatrix} \begin{bmatrix} 1 \\ 1 \end{bmatrix},$$

which shows that 1 occurs as an eigenvalue of $\mathbf{P}$. We have already seen in Example 8.4.23, that $\mathbf{w} = \begin{bmatrix} 0 & 1 \end{bmatrix}$ occurs as a left eigenvector with eigenvalue 1.

Remark 8.4.5. The maximum eigenvalue of $\mathbf{P}$ is 1. First, since the rows of $\mathbf{P}$ add up to 1, then by Theorem 5.4.3 $\rho(\mathbf{P}) \leq 1$. Since 1 occurs as an eigenvalue, then $\rho(\mathbf{P}) \geq 1$ as well so $\rho(\mathbf{P}) = 1$.

The stable state vector is also unique if and only if the 1-eigenspace has dimension 1. The following theorem gives a partial answer to the question of when we have dimension 1 and therefore uniqueness:

Theorem 8.4.9. *Let* $\mathbf{P}$ *be a finite Markov chain. Then there is a unique stable state vector* $\mathbf{w}$ *if*

1. *the Markov chain is regular,*
2. *the Markov chain is cyclic, or*
3. *the transition diagram is connected and there is one absorbing state.*

But in general the stable state vector is not unique.

Example 8.4.27. Consider the transition matrix

$$\mathbf{A} = \begin{bmatrix} 0 & 1 & 0 & 0 \\ 1 & 0 & 0 & 0 \\ 0 & 0 & 1 & 0 \\ 0 & 0 & \frac{1}{2} & \frac{1}{2} \end{bmatrix}.$$

Find a parameterization of the left eigenvectors, then use it to find all of the stable state vectors.

Solution

State 3 is an absorbing state for $\mathbf{A}$ and there is only one. However, we saw in Example 8.4.17 that the transition diagram for $\mathbf{A}$ has two connected components. The general approach to finding a left eigenvector with eigenvalue 1 is to take the transpose, subtract

1 from the diagonal and row reduce:

$$[(\mathbf{A}-\mathbf{I})^T|\mathbf{0}] = \left[\begin{array}{cccc|c} -1 & 1 & 0 & 0 & 0 \\ 1 & -1 & 0 & 0 & 0 \\ 0 & 0 & 0 & \frac{1}{2} & 0 \\ 0 & 0 & 0 & -\frac{1}{2} & 0 \end{array}\right]$$

$$\xrightarrow[R_3+R_4]{R_2+R_1} \left[\begin{array}{cccc|c} 0 & 0 & 0 & 0 & 0 \\ 1 & -1 & 0 & 0 & 0 \\ 0 & 0 & 0 & \frac{1}{2} & 0 \\ 0 & 0 & 0 & 0 & 0 \end{array}\right].$$

Even though this is not in reduced form, we essentially have a pivot in columns 1 and 4, while columns 2 and 3 are free variables, the dimension is 2. Row 2 gives us the equation $w_1 - w_2 = 0$, so $w_1 = w_2$, and Row 3 gives us the equation $\frac{1}{2}w_4 = 0$, so $w_4 = 0$. So far, this gives us the parameterization

$$\mathbf{w} = \left[\begin{array}{cccc} w_2 & w_2 & w_3 & 0 \end{array}\right].$$

But since $\mathbf{w}$ is a probability vector, the components must add up to 1, so $2w_2 + w_3 = 1$, which we can use to eliminate w_3:

$$\mathbf{w} = \left[\begin{array}{cccc} w_2 & w_2 & 1 - 2w_2 & 0 \end{array}\right].$$

The components also have to be between 0 and 1, so $0 \le w_2 \le 1$ and $0 \le 1 - 2w_2 \le 1$. The second of these is stronger: solving for w_2 yields $0 \le w_2 \le \frac{1}{2}$. A value of w_2 in this range will yield a valid probability vector. At the extremes, we have

$$\left[\begin{array}{cccc} 0 & 0 & 1 & 0 \end{array}\right] \text{ and } \left[\begin{array}{cccc} \frac{1}{2} & \frac{1}{2} & 0 & 0 \end{array}\right]$$

both of which are valid stable state vectors for $\mathbf{A}$.

Activity 8.4.28. Consider the transition matrix

$$\mathbf{A} = \left[\begin{array}{ccc} 1 & 0 & 0 \\ \frac{1}{4} & \frac{1}{2} & \frac{1}{4} \\ 0 & 0 & 1 \end{array}\right].$$

Find a parameterization of the left eigenvectors, then use it to find all of the stable state vectors.

In the cyclic case, it is easy to write down a stable state vector.

Example 8.4.29. The transition matrix

$$\mathbf{C} = \left[\begin{array}{ccc} 0 & 1 & 0 \\ 0 & 0 & 1 \\ 1 & 0 & 0 \end{array}\right]$$

has $\mathbf{w} = \left[\begin{array}{ccc} \frac{1}{3} & \frac{1}{3} & \frac{1}{3} \end{array}\right]$ as a stable state vector.

Even though powers of $\mathbf{C}$ in Example 8.4.29 do not converge on a matrix with $\mathbf{w}$ in each row, we achieve convergence by averaging the powers:

$$\lim_{n\to\infty} \frac{1}{n}\sum_{k=0}^{n-1} \mathbf{C}^k = \begin{bmatrix} \frac{1}{3} & \frac{1}{3} & \frac{1}{3} \\ \frac{1}{3} & \frac{1}{3} & \frac{1}{3} \\ \frac{1}{3} & \frac{1}{3} & \frac{1}{3} \end{bmatrix}$$

This can be illustrated with the following Sage code:

```
C = matrix.circulant([0,1,0])
for m in range(1,16):
    print(2^m)
    print(N(sum([C^k for k in range(2^m)])/2^m))
```

8.5 Leontief Input-Output Model

Wassily Leontief won the Nobel prize in economics in 1973 for his development of the Input-Output Model. The idea of the model is as follows. An economy is divided into sectors. Because of interdependence between the sectors, each sector may require input from other sectors to produce some output. There may also be demand for each sector from consumers. It is assumed that

1. the demand vector $\mathbf{d}$ and
2. the consumption matrix $\mathbf{C}$, which describes the interdependence between sectors,

are both known. The problem is then to determine $\mathbf{x}$ such that

$$\mathbf{x} - \mathbf{C}\mathbf{x} = \mathbf{d}. \tag{8.15}$$

In this equation, $\mathbf{x}$ is the amount produced by each sector. Then $\mathbf{C}\mathbf{x}$ describes how much is used up, $\mathbf{x} - \mathbf{C}\mathbf{x}$ is the amount remaining. If the amount remaining is equal to the demand, then we have equilibrium. Since $\mathbf{x} = \mathbf{I}\mathbf{x}$, then we can factor equation (8.15)

$$(\mathbf{I} - \mathbf{C})\mathbf{x} = \mathbf{d}. \tag{8.16}$$

In this form it becomes clear that we can solve for $\mathbf{x}$ by either

1. row reducing $[\mathbf{I} - \mathbf{C}|\mathbf{d}]$, or
2. computing $(\mathbf{I} - \mathbf{C})^{-1}$ (if it exists) and then

$$\mathbf{x} = (\mathbf{I} - \mathbf{C})^{-1}\mathbf{d}.$$

Example 8.5.1. Suppose we have an economy with two sectors: manufacturing and agriculture. The consumption matrix is given by

	manufacturing	agriculture
manufacturing	0.7	0.5
agriculture	0.1	0.3

Suppose, for example, that manufacturing produces 10 units. Then manufacturing would use $0.7 \times 10 = 7$ of those and agriculture would use $0.1 \times 10 = 1$ of those. Since $7+1 = 8$ and $10 - 8 = 2$ then only 2 units would be left to meet consumer demand for manufacturing.

But if demand for manufacturing is 240 units and demand for agriculture is 80 units, then how much actually needs to be produced? In both methods we need

$$\mathbf{I} - \mathbf{C} = \begin{bmatrix} 0.3 & -0.5 \\ -0.1 & 0.7 \end{bmatrix}.$$

Using row reduction

$$\left[\begin{array}{cc|c} 0.3 & -0.5 & 240 \\ -0.1 & 0.7 & 80 \end{array}\right] \xrightarrow[10R_2]{10R_1} \left[\begin{array}{cc|c} 3 & -5 & 2400 \\ -1 & 7 & 800 \end{array}\right] \xrightarrow{3R_2+R_1} \left[\begin{array}{cc|c} 0 & 16 & 4800 \\ -1 & 7 & 800 \end{array}\right]$$

$$\xrightarrow{\frac{1}{16}R_2+R_1} \left[\begin{array}{cc|c} 0 & 1 & 300 \\ -1 & 7 & 800 \end{array}\right] \xrightarrow{R_1 \leftrightarrow R_2} \left[\begin{array}{cc|c} -1 & 7 & 800 \\ 0 & 1 & 300 \end{array}\right]$$

$$\xrightarrow{(-1)R_1} \left[\begin{array}{cc|c} 1 & -7 & -800 \\ 0 & 1 & 300 \end{array}\right] \xrightarrow{7R_2+R_1} \left[\begin{array}{cc|c} 1 & 0 & 1300 \\ 0 & 1 & 300 \end{array}\right].$$

Since the solution is unique, then the inverse of $\mathbf{I}-C$ exists by the invertible matrix theorem. For the determinant, we have

$$\det(\mathbf{I} - C) = 0.3 \times 0.7 - (-0.1) \times (-0.5) = 0.21 - 0.05 = 0.16 = \frac{16}{100}$$

so

$$(\mathbf{I} - C)^{-1} = \frac{100}{16} \begin{bmatrix} 0.7 & 0.5 \\ 0.1 & 0.3 \end{bmatrix} = \frac{5}{8} \begin{bmatrix} 7 & 5 \\ 1 & 3 \end{bmatrix}.$$

Then we multiply by $\mathbf{d}$ to get $\mathbf{x}$

$$\mathbf{x} = (\mathbf{I} - C)^{-1}\mathbf{d} = \frac{5}{8} \begin{bmatrix} 7 & 5 \\ 1 & 3 \end{bmatrix} \begin{bmatrix} 240 \\ 80 \end{bmatrix} = \frac{5}{8} \begin{bmatrix} 2080 \\ 480 \end{bmatrix} = \frac{5}{2} \begin{bmatrix} 520 \\ 120 \end{bmatrix} = 10 \begin{bmatrix} 130 \\ 30 \end{bmatrix} = \begin{bmatrix} 1300 \\ 300 \end{bmatrix},$$

so in both cases, we get the same result.

Activity 8.5.2. Suppose we have an economy with two sectors: manufacturing and agriculture. The consumption matrix is given by

	manufacturing	agriculture
manufacturing	0.4	0.2
agriculture	0.1	0.3

If demand for manufacturing is 240 units and demand for agriculture is 80 units, then find $\mathbf{x}$ satisfying

$$(\mathbf{I} - \mathbf{C})\mathbf{x} = \mathbf{d}$$

using

1. row reduction, and
2. the inverse of $\mathbf{I} - \mathbf{C}$.

Your results should match each other.

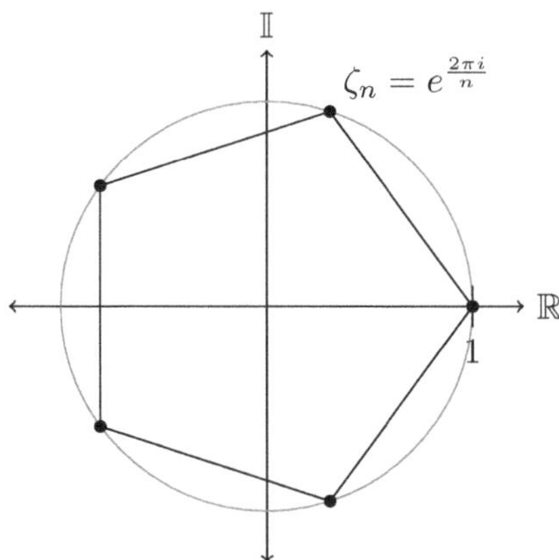

FIGURE 8.3: The roots of unity in the complex plane for $n = 5$

8.6 The Fast Fourier Transform

Although the main goal of this section is to describe the Fast Fourier Transform (FFT) and its application to fast multiplication, it is wise to build up in steps. So, we begin by describing the Discrete Fourier Transform (DFT) without concern for speed. We then describe the extra idea needed to improve the speed. We conclude by illustrating how multiplication can be done with the help of either the DFT or the FFT, bearing in mind that it is only the FFT that is worth using in practice.

8.6.1 The Discrete Fourier Transform

Suppose we have a vector $\mathbf{x}$ in $\mathbb{C}^n$. We will index the components starting from 0:

$$\mathbf{x} = \begin{bmatrix} x_0 & x_1 & \cdots & x_{n-1} \end{bmatrix}. \tag{8.17}$$

We will need the n-th roots of unity. In the complex plane, the n-th roots of unity lie on the unit circle and form the vertices of a regular n-gon with one vertex at 1 (see Figure 8.3). They are all powers of $\zeta_n = e^{\frac{2\pi i}{n}}$. If we denote the Discrete Fourier Transform of $\mathbf{x}$ by $\mathcal{F}(\mathbf{x})$, then the k-th component is defined to be

$$\mathcal{F}(\mathbf{x})_k = \sum_{j=0}^{n-1} x_j \zeta_n^{-jk}. \tag{8.18}$$

Equation (8.18) can be rewritten as

$$\mathcal{F}(\mathbf{x}) = \mathbf{x}\mathbf{F}. \tag{8.19}$$

where $\mathbf{F}$ is an $n \times n$ matrix with ζ_n^{-jk} in row j and column k, keeping in mind that in this section rows and columns are numbered starting from zero.

Example 8.6.1. Let $n = 4$ and $\mathbf{x} = \begin{bmatrix} 0 & 1 & 2 & 3 \end{bmatrix}$. Then $\zeta_n = i$, so by matrix multiplication

$$\begin{bmatrix} 0 & 1 & 2 & 3 \end{bmatrix} \begin{bmatrix} 1 & 1 & 1 & 1 \\ 1 & -i & -1 & i \\ 1 & -1 & 1 & -1 \\ 1 & i & -1 & -i \end{bmatrix} = \begin{bmatrix} 6 & -2+2i & -2 & -2-2i \end{bmatrix} = \mathcal{F}(\mathbf{x}).$$

To be clear about how the matrix $\mathbf{F}$ was computed, consider the last row. The last row has index 3, and the columns have indices 0, 1, 2, and 3. The products $-jk$ are therefore

$$-3 \cdot 0 = 0, \quad -3 \cdot 1 = -3, \quad -3 \cdot 2 = -6, \quad \text{and} \quad -3 \cdot 3 = -9.$$

Since $\zeta_n = i$, then

$$\zeta_n^0 = 1, \quad \zeta_n^{-3} = i, \quad \zeta_n^{-6} = -1, \quad \text{and} \quad \zeta_n^{-9} = -i.$$

Activity 8.6.2. Let $n = 4$ and $\mathbf{x} = \begin{bmatrix} 4 & 3 & 2 & 1 \end{bmatrix}$. Compute $\mathcal{F}(\mathbf{X})$

If we want to get back to $\mathbf{x}$ from $\mathcal{F}(\mathbf{x})$, we need the inverse Fourier transform, which we will write as $\mathcal{F}^{-1}$. The k-th component is

$$\mathcal{F}^{-1}(\mathbf{x})_k = \frac{1}{n} \sum_{j=0}^{n-1} \mathbf{x}_j \zeta_n^{jk}. \tag{8.20}$$

We can also use the inverse of $\mathbf{F}$:

$$\mathcal{F}^{-1}(\mathbf{x}) = \mathbf{x}\mathbf{F}^{-1}. \tag{8.21}$$

Thankfully, equation (8.20) tells us exactly what $\mathbf{F}^{-1}$ is: $\frac{1}{n}\zeta_n^{jk}$ is the component in row j and column k.

Example 8.6.3. Suppose $n = 4$ and $\mathbf{x} = \begin{bmatrix} 6 & -2+2i & -2 & -2-2i \end{bmatrix}$, which was the result in Example 8.6.1. Then using equation (8.21)

$$\begin{bmatrix} 6 & -2+2i & -2 & -2-2i \end{bmatrix} \begin{bmatrix} 1 & 1 & 1 & 1 \\ 1 & i & -1 & -i \\ 1 & -1 & 1 & -1 \\ 1 & -i & -1 & i \end{bmatrix} = \begin{bmatrix} 0 & 4 & 8 & 12 \end{bmatrix} \cdot \frac{1}{4}$$
$$= \begin{bmatrix} 0 & 1 & 2 & 3 \end{bmatrix}.$$

So, $\mathcal{F}^{-1}$ brings us back to the original vector in Example 8.6.1.

Activity 8.6.4. Apply $\mathcal{F}^{-1}$ to your result from Activity 8.6.2. You should get $\begin{bmatrix} 4 & 3 & 2 & 1 \end{bmatrix}$.

8.6.2 Improving the Speed

Since the Discrete Fourier Transform can be defined as multiplication by equations (8.19) and (8.21), then the algorithm efficiency cannot be worse than that of matrix multiplication.

In this case $\mathbf{F}$ is an $n \times n$ matrix, and $\mathbf{x}$ is a vector with n components, so by Figure 1.8 the sequential approach can be no worse than $O(n^2)$ in time and the parallel approach can be no worse than $O(\log(n))$. Any implementation that is faster than $O(n^2)$ sequentially may be called a **Fast Fourier Transform** (FFT).

How is it possible to do better than $O(n^2)$ sequentially? First suppose $\mathbf{x}$ is in $\mathbb{C}^n$ and that $n = n_1 n_2$, where n_1 and n_2. Then

$$\zeta_n^{n_1} = \zeta_{n_2}, \quad \zeta_n^{n_2} = \zeta_{n_1}, \quad \text{and} \quad \zeta_n^{n_1 n_2} = 1.$$

Let

$$j = an_1 + b \quad \text{and} \quad k = cn_2 + d$$

where $0 \leq b, c < n_1$ and $0 \leq a, d < n_2$. Then

$$jk = (an_1 + b)(cn_2 + d) = acn_1 n_2 + adn_1 + bcn_2 + bd,$$

so by exponent rules

$$\zeta_n^{-jk} = \zeta_{n_2}^{-ad} \zeta_{n_1}^{-bc} \zeta_n^{-bd}.$$

By applying this to equation (8.18)

$$\begin{aligned} \mathcal{F}(\mathbf{x})_{cn_2+d} = \sum_{j=0}^{n-1} \mathbf{x}_j \zeta_n^{-jk} &= \sum_{a=0}^{n_2-1} \sum_{b=0}^{n_1-1} \mathbf{x}_{an_1+b} \zeta_{n_2}^{-ad} \zeta_{n_1}^{-bc} \zeta_n^{-bd} \\ &= \sum_{b=0}^{n_1-1} \left(\left(\sum_{a=0}^{n_2-1} \mathbf{x}_{an_1+b} \zeta_{n_2}^{-ad} \right) \zeta_n^{-bd} \right) \zeta_{n_1}^{-bc}. \end{aligned} \tag{8.22}$$

Equation (8.22) can described as a five step process.

Step 1. Reshape the vector $\mathbf{x}$ into an $n_2 \times n_1$ matrix; its indices are (a, b).

Step 2. The $\zeta_{n_2}^{-ad}$ factor can be thought of as a matrix with (a, d) indices. The inner sum is over a, and can be thought of as multiplying this matrix by the matrix from Step 1 resulting in a matrix with (b, d) indices.

Step 3. The ζ_n^{-bd} factor can be thought of as a matrix with (b, d) indices. Since the matrix from Step 2 also has (b, d) indices, multiplication is component-wise.

Step 4. The factor $\zeta_{n_1}^{-bc}$ can be thought of as a matrix with (b, c) indices. The outer sum is over b, and can be thought of as multiplying this matrix by the matrix from Step 3, resulting in a matrix with (c, d) indices.

Step 5. Take the transpose of the matrix from Step 4 and reshape back into a vector, which is $\mathcal{F}(\mathbf{x})$.

Example 8.6.5. Let $n = 6$, $n_1 = 3$, and $n_2 = 2$, then $\zeta_2 = -1$, $\zeta_3 = \frac{-1+i\sqrt{3}}{2}$, and $\zeta_6 = \zeta_3 + 1$. It also helps to know that $\zeta_3^2 = -1 - \zeta_3$ Let $\mathbf{x} = \begin{bmatrix} 0 & 1 & 2 & 3 & 4 & 5 \end{bmatrix}$. For Steps 1 and 2, we reshape $\mathbf{x}$ into a 2×3 matrix and multiply on the left by

$$\begin{bmatrix} \zeta_2^0 & \zeta_2^0 \\ \zeta_2^0 & \zeta_2^{-1} \end{bmatrix} = \begin{bmatrix} 1 & 1 \\ 1 & -1 \end{bmatrix},$$

which is the usual matrix $\mathbf{F}$ when $n_2 = 2$. So far that gives us

$$\begin{array}{cc} & \begin{bmatrix} 0 & 1 & 2 \\ 3 & 4 & 5 \end{bmatrix} \\ \begin{bmatrix} 1 & 1 \\ 1 & -1 \end{bmatrix} & \begin{bmatrix} 3 & 5 & 7 \\ -3 & -3 & -3 \end{bmatrix} \end{array}.$$

Next we do component-wise multiplication with

$$\begin{bmatrix} \zeta_6^0 & \zeta_6^0 & \zeta_6^0 \\ \zeta_6^0 & \zeta_6^{-1} & \zeta_6^{-2} \end{bmatrix} = \begin{bmatrix} 1 & 1 & 1 \\ 1 & -\zeta_3 & -1-\zeta_3 \end{bmatrix}$$

which makes

$$\begin{bmatrix} 3 & 5 & 7 \\ -3 & 3\zeta_3 & 3+3\zeta_3 \end{bmatrix}$$

the result after Step 3. For Step 4, we multiply on the right by the usual matrix $\mathbf{F}$ when $n = 3$,

$$\begin{array}{cc} & \begin{bmatrix} 1 & 1 & 1 \\ 1 & -1-\zeta_3 & \zeta_3 \\ 1 & \zeta_3 & -1-\zeta_3 \end{bmatrix} \\ \begin{bmatrix} 3 & 5 & 7 \\ -3 & 3\zeta_3 & 3+3\zeta_3 \end{bmatrix} & \begin{bmatrix} 15 & -2+2\zeta_3 & -4-2\zeta_3 \\ 6\zeta_3 & -3 & -6-6\zeta_3 \end{bmatrix}. \end{array}$$

Finally, we take the transpose and reshape to get a vector:

$$\mathcal{F}(\mathbf{x}) = \begin{bmatrix} 15 & 6\zeta_3 & -2+2\zeta_3 & -3 & -4-2\zeta_3 & -6-6\zeta_3 \end{bmatrix}.$$

Using the results from Figure 1.8 we may conclude that it takes only $O(n(n_1 + n_2))$ in time. If we continue by factoring n_1 and n_2 etc., then eventually we can go all the way down to the prime factors. If

$$n = p_1^{e_1} p_2^{e_2} \cdots p_r^{e_r},$$

then we can get $O(n(e_1p_1 + e_2p_2 + \cdots e_rp_r))$. The simplest case is when n is a prime power, in which case we get $O(n \log(n))$. We will now look at the case where n is a power of 2, but it is not too hard to generalize to other prime powers.

I know of three different ways of implementing the FFT when n is a power of 2: recursively, iteratively, or with arrays. The recursive and iterative approaches are described in [29], on which much of the material in this section is based. With arrays, we proceed as follows:

Step 0. Given an input vector $\mathbf{x}$ with length 2^k, reshape to a $2^k \times 1$ matrix, which we will call $\mathbf{A}_0$.

Step j+1. This step is done for $0 \leq j < k$. In general, $\mathbf{A}_j$ will have shape $2^{k-j} \times 2^j$. Given $\mathbf{A}_j$ as an input, $\mathbf{A}_{j+1}$ is constructed as follows:

1. Construct the vector $\mathbf{u}_j$ consisting of half of the 2^{j+1} roots of unity:

$$\mathbf{u}_j = \begin{bmatrix} 1 & \zeta_{2^{j+1}} & \zeta_{2^{j+1}}^2 & \cdots & \zeta_{2^{j+1}}^{2^j-1} \end{bmatrix}.$$

2. Multiply the rows of the bottom half of $\mathbf{A}_j$ component-wise by $\mathbf{u}_j$ to get $\mathbf{B}_j$.

3. The left half of $\mathbf{A}_{j+1}$ is the top half of $\mathbf{A}_j$ plus $\mathbf{B}_j$.

4. The right half of $\mathbf{A}_{j+1}$ is the top half of $\mathbf{A}_j$ minus $\mathbf{B}_j$.

Step k+1. Reshape $\mathbf{A}_k$ to match the shape of $\mathbf{x}$ to get $\mathcal{F}(\mathbf{x})$.

Remark 8.6.1. If $\mathbf{x}$ is an array with shape $(2^k, \ldots)$, this algorithm generalizes by taking $\mathbf{A}_j$ to have shape $(2^{k-j}, 2^j, \ldots)$. In this way, the FFT can be computed for several vectors in parallel. Since the operations in Step $j+1$ are all component-wise, then most of it can be done in parallel as well.

Example 8.6.6. Suppose $\mathbf{x} = \begin{bmatrix} 0 & 1 & 2 & 3 & 4 & 5 & 6 & 7 \end{bmatrix}$, which has length $8 = 2^3$, so $k = 3$.

Step 0. We write $\mathbf{x}$ as a column vector and call it $\mathbf{A}_0$.

Step 1. The 2^1 roots of unity are 1 and -1, so $\mathbf{u}_0 = \begin{bmatrix} 1 \end{bmatrix}$. Since everything is multiplied by 1, $\mathbf{B}_0$ is just the bottom half of $\mathbf{A}_0$, so it consists of the numbers 4 to 7. The top half of $\mathbf{A}_0$ consists of the numbers 0 to 3, so

$$\mathbf{A}_1 = \begin{bmatrix} 0+4 & 0-4 \\ 1+5 & 1-5 \\ 2+6 & 2-6 \\ 3+7 & 3-7 \end{bmatrix} = \begin{bmatrix} 4 & -4 \\ 6 & -4 \\ 8 & -4 \\ 10 & -4 \end{bmatrix}.$$

Step 2. The 2^2-roots of unity are 1, i, -1, and $-i$, so

$$\mathbf{u}_1 = \begin{bmatrix} 1 & i \end{bmatrix}.$$

When we multiply each row of $\mathbf{A}_1$ component-wise by $\mathbf{u}_1$, it has the effect of multiplying the only second column of $\mathbf{A}_1$ by i:

$$\mathbf{B}_1 = \begin{bmatrix} 8 & -4i \\ 10 & -4i \end{bmatrix}.$$

By adding and subtracting this from the top half of $\mathbf{A}_1$, we get

$$\mathbf{A}_2 = \begin{bmatrix} 4+8 & -4-4i & 4-8 & -4+4i \\ 6+10 & -4-4i & 6-10 & -4+4i \end{bmatrix} = \begin{bmatrix} 12 & -4-4i & -4 & -4+4i \\ 16 & -4-4i & -4 & -4+4i \end{bmatrix}$$

Step 3. The 2^3 roots of unity give us

$$\mathbf{u}_2 = \begin{bmatrix} 1 & \frac{\sqrt{2}}{2} + \frac{\sqrt{2}}{2}i & i & -\frac{\sqrt{2}}{2} + \frac{\sqrt{2}}{2}i \end{bmatrix}.$$

The bottom half of $\mathbf{A}_2$ is now just the bottom row, so by multiplying component-wise by $\mathbf{u}_2$, we get

$$\begin{aligned} \mathbf{B}_2 &= \begin{bmatrix} 16 & -4\sqrt{2}i & -4i & -4\sqrt{2}i \end{bmatrix} \\ &\approx \begin{bmatrix} 16 & -5.7i & -4i & -5.7i \end{bmatrix}. \end{aligned}$$

We get $\mathbf{A}_2$ by adding and subtracting this from the top row of $\mathbf{A}_2$:

$$\begin{bmatrix} 28 & -4-9.7i & -4-4i & -4-1.7i & -4 & -4+1.7i & -4+4i & -4+9.7i \end{bmatrix}$$

Step 4. Reshaping from 1×8 to 8 is necessary with NumPy and with other programming languages as well.

8.6.3 Multiplication

Suppose we have two numbers with N places in base B

$$x = \sum_{i=0}^{N-1} x_i B^i \quad \text{and} \quad y = \sum_{i=0}^{N-1} y_i B^i. \tag{8.23}$$

Their product is

$$xy = \left(\sum_{i=0}^{N-1} x_i B^i \right) \left(\sum_{j=0}^{N-1} y_j B^j \right) = \sum_{i=0}^{N-1} \sum_{j=0}^{N-1} x_i y_j B^{i+j}.$$

Let $k = i + j$, so that $j = k - i$. We want to collect common powers of B together, which means we want an outer sum over k, and an inner sum over i:

$$xy = \sum_k \left(\sum_i x_i y_{k-i} \right) B^k. \tag{8.24}$$

Since $0 \leq i, j \leq N - 1$, it should be clear that

$$0 \leq k \leq 2N - 2,$$

so these are the values of k that must be summed over. As for the inner sum, it looks like a convolution:

Definition 8.6.1. Given two vectors

$$\mathbf{x} = \begin{bmatrix} x_0 & x_1 & \cdots x_{n-1} \end{bmatrix} \quad \text{and} \quad \mathbf{y} = \begin{bmatrix} y_0 & y_1 & \cdots y_{n-1} \end{bmatrix},$$

the **convolution** of $\mathbf{x}$ and $\mathbf{y}$ is written as $\mathbf{x} * \mathbf{y}$ and the k-th component is

$$(\mathbf{x} * \mathbf{y})_k = \sum_{i=0}^{n-1} x_i y_{k-i}.$$

If $k - i$ is negative, then add n to get a valid index for y.

Example 8.6.7. Compute the convolution of

$$\mathbf{x} = \begin{bmatrix} 1 & 1 & 0 & 0 \end{bmatrix} \quad \text{and} \quad \mathbf{y} = \begin{bmatrix} 0 & 1 & 2 & 3 \end{bmatrix}.$$

Solution

We can interpret this in terms of matrix multiplication by multiplying $\mathbf{x}$ by the **circulant matrix** constructed by taking $\mathbf{y}$ as the first row of a square matrix, and shifting it one unit to the right for each subsequent row:

$$\begin{array}{cc} & \begin{bmatrix} 0 & 1 & 2 & 3 \\ 3 & 0 & 1 & 2 \\ 2 & 3 & 0 & 1 \\ 1 & 2 & 3 & 0 \end{bmatrix} \\ \begin{bmatrix} 1 & 1 & 0 & 0 \end{bmatrix} & \begin{bmatrix} 3 & 1 & 3 & 5 \end{bmatrix} \end{array}$$

Activity 8.6.8. Compute the convolution of

$$\mathbf{x} = \begin{bmatrix} 1 & 1 & 0 & 0 \end{bmatrix} \quad \text{and} \quad \mathbf{y} = \begin{bmatrix} 4 & 3 & 2 & 1 \end{bmatrix}.$$

Unfortunately, the sum over i in (8.24) is not exactly a convolution as it stands, because when $k - i$ is negative it doesn't just wrap around to a valid index of $\mathbf{y}$. Instead, that term drops out completely. There is also a problem when $k - i \geq N$; again, those terms of the sum drop out. The trick is to extend both x and y by padding with zeros. Then we allow the indices to wrap around as in the definition of convolution, and when $x_i = 0$ or $y_{k-i} = 0$, then $x_i y_{k-i} = 0$ and those terms drop out naturally.

Theorem 8.6.2. *If* $\mathbf{x}$ *and* $\mathbf{y}$ *are in* $\mathbb{C}^n$ *and* c *is a scalar, then*

1. $\mathcal{F}(\mathbf{x} + \mathbf{y}) = \mathcal{F}(\mathbf{x}) + \mathcal{F}(\mathbf{y})$,

2. $\mathcal{F}(c\mathbf{x}) = c\mathcal{F}(\mathbf{x})$,

3. $\mathcal{F}(\mathbf{x} * \mathbf{y}) = \mathcal{F}(\mathbf{x})\mathcal{F}(\mathbf{y})$,

4. $\mathcal{F}(\mathbf{xy}) = \mathcal{F}(\mathbf{x}) * \mathcal{F}(\mathbf{y})$,

where $\mathbf{xy}$ *and* $\mathcal{F}(\mathbf{x})\mathcal{F}(\mathbf{y})$ *are component-wise multiplication.*

The first two properties mean that $\mathcal{F} : \mathbb{C}^n \to \mathbb{C}^n$ is a linear transformation, which is why we are able to get a matrix $\mathbf{F}$ for it. It is the third and fourth property that make it possible to do fast multiplication. Here is the strategy:

Step 1. The numbers to be multiplied are expressed as vectors $\mathbf{x}$ and $\mathbf{y}$. Pad $\mathbf{x}$ and $\mathbf{y}$ with zeros so that their length is the smallest power of 2 greater than the number of places in the product.

Step 2. Apply the FFT to both vectors from Step 1.

Step 3. Multiply the vectors from Step 2 component-wise.

Step 4. Apply the inverse FFT to the result from Step 3.

Step 5. Carry, then remove the padding.

How the padding is removed depends on:

1. whether the digits are written left to right or right to left, and

2. where the zeros are appended to the vectors $\mathbf{x}$ and $\mathbf{y}$.

In this book we will write the digits left to right, and pad with zeros on the left (in higher places). In this case, it is necessary to remove one zero on the right, and any trailing zeros on the left. [2]

Example 8.6.9. In this example we will use the slow version of the DFT to avoid making things too confusing. This is going to look like more work than ordinary multiplication, because the numbers are small. With larger numbers and with the FFT, it becomes quite fast.

Suppose we want to multiply 20 and 45. As vectors, we start with

$$\mathbf{x} = \begin{bmatrix} 2 & 0 \end{bmatrix} \text{ and } \quad \mathbf{y} = \begin{bmatrix} 4 & 5 \end{bmatrix},$$

with one component for each place. The number of places in the product is 1 less than the sum of the number of places of the two numbers being multiplied. Since $2 + 2 - 1 = 3$, and 4 is the smallest power of two bigger than 3, then we pad them to length 4, which gives us:

$$\begin{bmatrix} 0 & 0 & 2 & 0 \end{bmatrix} \quad \text{and} \quad \begin{bmatrix} 0 & 0 & 4 & 5 \end{bmatrix}$$

For Step 2 we multiply by the same matrix $\mathbf{F}$ from Example 8.6.1:

$$\begin{array}{cc} & \begin{bmatrix} 1 & 1 & 1 & 1 \\ 1 & -i & -1 & i \\ 1 & -1 & 1 & -1 \\ 1 & i & -1 & -i \end{bmatrix} \\ \begin{bmatrix} 0 & 0 & 2 & 0 \\ 0 & 0 & 4 & 5 \end{bmatrix} & \begin{bmatrix} 2 & -2 & 2 & -2 \\ 9 & -4+5i & -1 & -4-5i \end{bmatrix} \end{array}.$$

[2] You may notice that the indices in equation (8.23) would lead to the places being written right to left. That is fine for a computer, but this book is written for humans.

Now we multiply component-wise, which gives

$$\begin{bmatrix} 18 & 8-10i & -2 & 8+10i \end{bmatrix}$$

For Step 4 we multiply by $\mathbf{F}^{-1}$ from Example 8.6.3:

$$\begin{aligned} & \begin{bmatrix} 1 & 1 & 1 & 1 \\ 1 & i & -1 & -i \\ 1 & -1 & 1 & -1 \\ 1 & -i & -1 & i \end{bmatrix} \\ \begin{bmatrix} 18 & 8-10i & -2 & 8+10i \end{bmatrix} & \begin{bmatrix} 32 & 40 & 0 & 0 \end{bmatrix} \cdot \frac{1}{4} \\ & = \begin{bmatrix} 8 & 10 & 0 & 0 \end{bmatrix}. \end{aligned}$$

There are no extra zeros to the left that need to be dropped, but there will always be an extra zero on the right that needs to be dropped. So, finally, we carry with

$$\begin{bmatrix} 8 & 10 & 0 \end{bmatrix}$$

to get $\begin{bmatrix} 9 & 0 & 0 \end{bmatrix}$. This checks out with basic mental math since

$$45 \times 20 = 90 \times 10 = 900.$$

Activity 8.6.10. Use the FFT to multiply 2 and 36.

The FFT can be done in $\log(N)$ steps. Each step requires a number of operations proportional to N, which can be done in parallel. As a consequence, we get the values in the following table:

	Space	Time
FFT (sequential)	$O(N)$	$O(N\log(N))$
FFT (parallel)	$O(N)$	$O(\log(N))$

FIGURE 8.4: Asymptotic efficiency of the FFT

If the FFT is applied to multiplication of big numbers, we must not forget carrying. Full carrying cannot be done in parallel and is $O(N)$ in time. So in the sequential case, it is the FFT that dominates, but in the parallel case it is carrying that dominates, leading to the times in Figure 1.6. On the other hand multiplication of polynomials does not involve carrying, and so if the FFT is used for multiplying polynomials with small number coefficients, the efficiency is given by Figure 8.4.

8.7 Molecular Symmetries

This section combines concepts from chemistry, physics, and representation theory. The main reference on chemical symmetries themselves is by Albert Cotton [5]. Cotton's book does include background on the necessary math and physics, but the necessary background on physics is also in [10]. For the representation theory on representation theory, see Chapter 7 in this book.

Atoms have different energy levels that can be occupied by electrons. Each energy level may consist of various types of orbitals: s, p, d, f and so on.[3] These orbitals correspond to different eigenspaces of a linear operator called the Hamiltonian, which we will denote by $\mathbf{H}$. In quantum mechanics any observable quantity is represented by a hermitian operator. By the complex case of the spectral theorem, a hermitian operator can be diagonalized and therefore the eigenvalues are all real, which makes sense for observable quantities. The Hamiltonian is a hermitian operator, and its eigenvalues are energies. In fact, it is probably not an exaggeration to say that the Hamiltonian in atomic theory provides the quintessential example of the spectral theorem. After all, what is a spectrum? When we burn a molecule and observe the spectrum, we are recording wavelengths of light. Each wavelength λ of light corresponds to a different energy E by the formulas

$$c = \lambda\nu \quad \text{and} \quad E = h\nu$$

where c is the speed of light and h is Planck's constant. What does this Hamiltonian operator act on? It acts on a space of functions that we call wave functions. These wave functions can be thought of as representing a particle in the following way. Given the inner product

$$\langle f, g\rangle = \int_{\mathbb{R}^3} f(x,y,z)\overline{g(x,y,z)}\, dxdydz \tag{8.25}$$

if we normalize a wave function ϕ, so that

$$\| \phi \|^2 = \langle \phi, \phi\rangle = 1,$$

then the function

$$\phi\overline{\phi} = |\phi|^2$$

is a probability density function for the position of the particle. For example,

$$\langle \phi, x\phi\rangle = \int_{\mathbb{R}^3} \phi(x,y,z)\overline{x\phi(x,y,z)}\, dxdydz$$

gives the expected value of the position in the x-direction.[4] If E is an eigenvalue of $\mathbf{H}$, then we are interested in the eigenfunctions ϕ:

$$\mathbf{H}\phi = E\phi. \tag{8.26}$$

For an atom, the wave functions represent electrons, and the dimension of each eigenspace depends the orbital type. The dimensions are odd numbers: type s has dimension 1, type p has dimension 3, type d has dimension 5, and so on, which is why we speak of one s orbital, 3 p orbitals, etc. Each orbital can be filled with up to two electrons so long as those electrons have opposite spin. The order in which these orbitals are filled with electrons leads to the structure of the periodic table.

Each quantum system – whether an atom, a molecule, or something else – has its own Hamiltonian operator. For a molecule with n atoms, we may assign a number to each atom in the molecule. Each atom will have its own Hamiltonian operator $\mathbf{H}_i$, and its own atomic orbitals. Suppose that for each atom we take a wave function ϕ_i. Generally, it is not reasonable to assume that the eigenfunctions of the Hamiltonians $\mathbf{H}_i$ of the individual atoms

[3]This strange collection of letters corresponds to actual names: sharp (s), principle (p), diffuse (d), and fundamental (f). After f though, we just continue with the alphabet: g, h, etc.

[4]Multiplication by x is also a hermitian operator.

will also be eigenfunctions of the Hamiltonian $\mathbf{H}$ for the whole molecule. Nevertheless, it is common to use this idea as a first approximation. So, let

$$\mathbf{H}\phi_i = E\phi_i$$

for all i. Then, by applying the inner product (8.25), we have

$$H_{ij} = \langle \phi_i, \mathbf{H}\phi_j \rangle = \langle \phi_i, E\phi_j \rangle = E\langle \phi_i, \phi_j \rangle. \tag{8.27}$$

It is reasonable to expect that the functions ϕ_i are linearly independent, which means that the values H_{ij} give us a matrix for $\mathbf{H}$ on the vector space consisting of all linear combinations

$$\psi = \sum_{i=1}^{n} c_1 \phi_i.$$

The Hükel approximation assumes further that

1. the functions ϕ_i are orthonormal[5]
2. $H_{ij} = 0$ unless $i = j$ or the i-th and j-th atom are adjacent.

Under these assumptions, equation (8.27) now implies that E is the eigenvalue of the matrix with coefficients H_{ij}.

If all of the atoms are of the same type, such as the carbon atoms in a hydrocarbon, then we may assume further that there exists α and β such that

$$\alpha = H_{11} = H_{22} = \cdots = H_{nn}, \text{ and}$$
$$\beta = H_{ij}, \text{ for all adjacent atoms } i \text{ and } j.$$

If we subtract α from the diagonal of a matrix, then all of the eigenvalues get shifted by α. If we then rescale the energies so that $\beta = 1$, then the problem reduces to that of finding the eigenvalues for the adjacency matrix of the molecule.

Example 8.7.1. Benzene has the chemical formula C_6H_6. If we look just at the carbon atoms, then it has the form of a hexagon.

With the atoms labeled in cyclic order, the adjacency matrix takes the following form

$$\mathbf{A} = \begin{bmatrix} 0 & 1 & 0 & 0 & 0 & 1 \\ 1 & 0 & 1 & 0 & 0 & 0 \\ 0 & 1 & 0 & 1 & 0 & 0 \\ 0 & 0 & 1 & 0 & 1 & 0 \\ 0 & 0 & 0 & 1 & 0 & 1 \\ 1 & 0 & 0 & 0 & 1 & 0 \end{bmatrix} \tag{8.28}$$

The matrix in (8.28) is a special type of a matrix called a circulant. Since Sage has a special function for constructing circulant matrices, the matrix can be defined by using just the first row.

```
A = matrix.circulant([0,1,0,0,0,1])
A.eigenvalues()
```

[5] While different eigenspaces of the same operator are automatically orthogonal, this need not be true of eigenspaces of different operators such as the Hamiltonians of different atoms in a molecule.

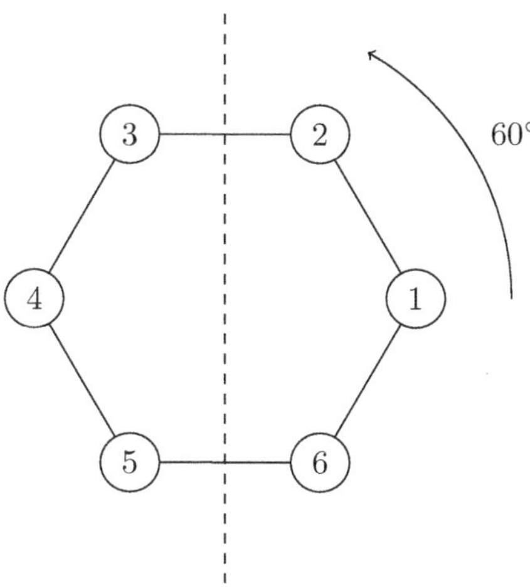

FIGURE 8.5: Benzene

This shows that the eigenvalues are ± 1 and ± 2. The fact that the eigenvalues are all integers is a bit of luck.

Molecules are 3 dimensional objects, so it is necessary to consider all symmetries arising in a 3 dimensional context. Some of the symmetries lead to permutation matrices coming from the numbering of the atoms, but other symmetries are best understood by their action on the wave functions. For example, if we think of the atoms of a molecule as lying the xy-plane, then there is also a symmetry coming from

$$\phi(x, y, z) \mapsto \phi(x, y, -z).$$

If ϕ is an odd function in z, then we get $\phi(x, y, z) = -\phi(x, y, z)$, and if if it is an even function, then we will get $\phi(x, y, z) = \phi(x, y, z)$.

Example 8.7.2. Benzene has all of the symmetries coming from D_6, with the usual generators r and m. Additionally it has a symmetry coming from $z \mapsto -z$ as just described, and we will call σ_z. The full group of symmetries is called D_{6h} and is generated by r, m, σ_z, with the relations

$$r^6 = e,\ m^2 = e,\ \sigma_z^2 = m,\ r^5 m = mr,\ \sigma_z r = r\sigma_z,\ \sigma_z m = m\sigma_z.$$

There are 12 conjugacy classes. For the permutation matrices that arise, the trace is equal to the number of fixed points (see Remark 7.3.1). So for example if m acts on the numbering of the atoms as $(3\ 5)(2\ 6)$, then it has two fixed points (atoms 1 and 4), so

$$\chi(m) = 2,$$

but if r has no fixed points, so $\chi(r) = 0$. If the functions ϕ_i are odd in the z-coordinate, then $\chi(\sigma_z)$ is the negative of the identity matrix, so $\chi(\sigma_z) = -6$, while $\chi(e) = 6$. In general, when σ_z occurs as a factor in an element of D_{6h}, this will change the sign of the trace. By these considerations, we arrive at the following result for the character

g	e	σ_z	m	$\sigma_z m$	mr	$\sigma_z mr$	r	$\sigma_z r$	r^2	$\sigma_z r^2$	r^3	$\sigma_z r^3$
$\lvert\mathcal{C}_g\rvert$	1	1	3	3	3	3	2	2	2	2	1	1
$\chi(g)$	6	-6	2	-2	0	0	0	0	0	0	0	0

Using Sage, we can get the character table.

g	e	σ_z	m	$\sigma_z m$	mr	$\sigma_z mr$	r	$\sigma_z r$	r^2	$\sigma_z r^2$	r^3	$\sigma_z r^3$
$\lvert\mathcal{C}_g\rvert$	1	1	3	3	3	3	2	2	2	2	1	1
$\chi_1(g)$	1	1	1	1	1	1	1	1	1	1	1	1
$\chi_2(g)$	1	−1	−1	1	−1	1	1	−1	1	−1	1	−1
$\chi_3(g)$	1	−1	−1	1	1	−1	−1	1	1	−1	−1	1
$\chi_4(g)$	1	−1	1	−1	−1	1	−1	1	1	−1	−1	1
$\chi_5(g)$	1	−1	1	−1	1	−1	1	−1	1	−1	1	−1
$\chi_6(g)$	1	1	−1	−1	−1	−1	1	1	1	1	1	1
$\chi_7(g)$	1	1	−1	−1	1	1	−1	−1	1	1	−1	−1
$\chi_8(g)$	1	1	1	1	−1	−1	−1	−1	1	1	−1	−1
$\chi_9(g)$	2	−2	0	0	0	0	−1	1	−1	1	2	−2
$\chi_{10}(g)$	2	−2	0	0	0	0	1	−1	−1	1	−2	2
$\chi_{11}(g)$	2	2	0	0	0	0	−1	−1	−1	−1	2	2
$\chi_{12}(g)$	2	2	0	0	0	0	1	1	−1	−1	−2	−2

It can then be checked that

$$\langle \chi, \chi_i \rangle = \begin{cases} 1 & \text{if } i = 4, 5, 9, \text{ or } 10, \\ 0 & \text{otherwise.} \end{cases} \tag{8.29}$$

Here is the Sage code uses to get the character table above:

```
D6 = DihedralGroup(6)
C2 = groups.permutation.Cyclic(2)
D6h = direct_product_permgroups([D6,C2])
D6h.character_table()
D6h.conjugacy_classes_representatives()

sz,r,m = D6h.gens()
[sz,m,sz*m,m*r,sz*m*r,r,sz*r,r^2,sz*r^2,r^3,sz*r^3]

def rho(G): return [g.matrix() for g in G]
def chi(G):
    L = G.conjugacy_classes_representatives()
    return [g.matrix().trace() for g in L]

rho(D6)
chi(D6)
```

In Sage I used `sz` to stand for σ_z.I renamed the conjugacy classes in terms of the generators σ_z, r, and m. Those without σ_z are from D_6. The new conjugacy classes get σ_z as an extra factor, and occur in alternating order.
The easiest way to get the appropriate representation is to start with the matrices from D_6, and remember that $\rho(\sigma_z) = -\mathbf{I}$. So, any time that σ_z occurs as a factor, you will get the negative of the matrix, which will have the negative trace. This is why the functions `rho` and `chi` are applied to `D6`.

Example 8.7.3. Since equation (8.29) shows that no irreducible representation occurs more than once in the representation ρ in Examples 8.7.1 and 8.7.2, then Theorem 7.3.6 is sufficient to obtain block form. For other molecules it may be necessary to use Theorem 7.3.7 For the particular representations involved here, we have $\rho(\sigma_z g) = -\rho(g)$ and $\chi_i(\sigma_z g) = -\chi(g)$ for all g and $i = 4$, 5, 9 or 10. As a result

$$\mathbf{P}_i = \frac{n_1}{|D_{6h}|} \sum_{g \in D_{6h}} \overline{\chi_i(g)}\rho(g) = \frac{2n_1}{|D_{6h}|} \sum_{g \in D_6} \overline{\chi_i(g)}\rho(g),$$

where the second sum is only over g not having σ_z as a factor. Since $|D_{6h}| = 24$ and $n_9 = n_{10} = 2$, then

$$\begin{aligned}\mathbf{P}_9 =& \frac{4}{24}\left(2\rho(e) - \rho(r) - \rho(r^5) - \rho(r^2) - \rho(r^4) + 2\rho(r^3)\right)\\ =& \frac{1}{6}\begin{bmatrix} 2 & -1 & -1 & 2 & -1 & -1 \\ -1 & 2 & -1 & -1 & 2 & -1 \\ -1 & -1 & 2 & -1 & -1 & 2 \\ 2 & -1 & -1 & 2 & -1 & -1 \\ -1 & 2 & -1 & -1 & 2 & -1 \\ -1 & -1 & 2 & -1 & -1 & 2 \end{bmatrix}\end{aligned}$$

and

$$\begin{aligned}\mathbf{P}_{10} =& \frac{4}{24}\left(2\rho(e) + \rho(r) + \rho(r^5) - \rho(r^2) - \rho(r^4) - 2\rho(r^3)\right)\\ =& \frac{1}{6}\begin{bmatrix} 2 & 1 & -1 & -2 & -1 & 1 \\ 1 & 2 & 1 & -1 & -2 & -1 \\ -1 & 1 & 2 & 1 & -1 & -2 \\ -2 & -1 & 1 & 2 & 1 & -1 \\ -1 & -2 & -1 & 1 & 2 & 1 \\ 1 & -1 & -2 & -1 & 1 & 2 \end{bmatrix}\end{aligned}$$

For $\mathbf{P}_4$ and $\mathbf{P}_5$ we have $n_4 = n_5 = 1$. so

$$\begin{aligned}\mathbf{P}_4 =& \frac{2}{24}\big(\rho(e) + \rho(m) + \rho(mr^2) + \rho(mr^4) - \rho(mr) - \rho(mr^3) - \rho(mr^5)\\ &-\rho(r) - \rho(r^5) + \rho(r^2) + \rho(r^4) - \rho(r^3)\big)\\ =& \frac{1}{12}\begin{bmatrix} 2 & -2 & 2 & -2 & 2 & -2 \\ -2 & 2 & -2 & 2 & -2 & 2 \\ 2 & -2 & 2 & -2 & 2 & -2 \\ -2 & 2 & -2 & 2 & -2 & 2 \\ 2 & -2 & 2 & -2 & 2 & -2 \\ -2 & 2 & -2 & 2 & -2 & 2 \end{bmatrix}\end{aligned}$$

and

$$\begin{aligned}\mathbf{P}_5 =& \frac{2}{24}\big(\rho(e) + \rho(m) + \rho(mr^2) + \rho(mr^4) + \rho(mr) + \rho(mr^3) + \rho(mr^5)\\ &+\rho(r) + \rho(r^5) + \rho(r^2) + \rho(r^4) + \rho(r^3)\big)\\ =& \frac{1}{12}\begin{bmatrix} 2 & 2 & 2 & 2 & 2 & 2 \\ 2 & 2 & 2 & 2 & 2 & 2 \\ 2 & 2 & 2 & 2 & 2 & 2 \\ 2 & 2 & 2 & 2 & 2 & 2 \\ 2 & 2 & 2 & 2 & 2 & 2 \\ 2 & 2 & 2 & 2 & 2 & 2 \end{bmatrix}\end{aligned}$$

The first two columns of $\mathbf{P}_9$ and $\mathbf{P}_{10}$ generate the column spaces of those projections, while for $\mathbf{P}_4$ and $\mathbf{P}_5$ just the first column is needed. Thus the matrix

$$\mathbf{S} = \begin{bmatrix} 2 & -1 & 2 & 1 & 1 & 1 \\ -1 & 2 & 1 & 2 & -1 & 1 \\ -1 & -1 & -1 & 1 & 1 & 1 \\ 2 & -1 & -2 & -1 & -1 & 1 \\ -1 & 2 & -1 & -2 & 1 & 1 \\ -1 & -1 & 1 & -1 & -1 & 1 \end{bmatrix}$$

gives a change of basis with the property that the representation ρ is put in block form. But remember that the vector space here is the space of wave functions generated by ϕ_1 through ϕ_6, and we used the same basis for the adjacency matrix $\mathbf{A}$. Under the change of basis the adjacency matrix becomes

$$\mathbf{S}^{-1}\mathbf{A}\mathbf{S} = \begin{bmatrix} -1 & 0 & 0 & 0 & 0 & 0 \\ 0 & -1 & 0 & 0 & 0 & 0 \\ 0 & 0 & 1 & 0 & 0 & 0 \\ 0 & 0 & 0 & 1 & 0 & 0 \\ 0 & 0 & 0 & 0 & -2 & 0 \\ 0 & 0 & 0 & 0 & 0 & 2 \end{bmatrix}$$

8.8 Ray Tracing

Ray tracing is a method of optics that has become very popular in computer graphics. Some modern graphics cards even have ray tracing built-in, making the computations much faster than could ever be achieved by software. In this section we will look at some of the physics and computer science concepts involved.

Physically, we think of rays of light as originating from a source, such as the sun, a candle, or a light bulb. The rays move in one direction away from the source until they encounter an object. Once the rays arrive at the object they may be reflected, transmitted, or absorbed. Light also scatters from surfaces and from small particles in the air, going in random directions. In any case, the rays continue on in their new direction until they encounter another object and the process repeats. In computer graphics, each one of these changes in direction is called a bounce. The vast majority of these rays never reach our eye, but in computer graphics we care only about the ones that do. For this reason, it tends to be more practical to think of tracing the rays backwards from the eye.

If the goal of ray tracing is to render a scene as close to real life as possible, then we must admit that there are some challenges and shortcomings. The most serious shortcoming is the fact that light is both a particle and a wave, but rays are not naturally able to account for some wave phenomena. For example, waves will naturally bend around a barrier, which is know as diffraction. One consequence of diffraction is that shadow edges can be soft, whereas basic ray tracing would produce a hard edge. There are ways to deal this and other problems, and in most practical cases the results are convincing enough.

The position of the eye relative to the screen makes a big difference. The viewing angle as illustrated in Figure 8.6 is determined by the formula

$$\alpha = 2\tan^{-1}\left(\frac{d}{2f}\right) \tag{8.30}$$

where f is the focal length and d is the length of the screen or image.

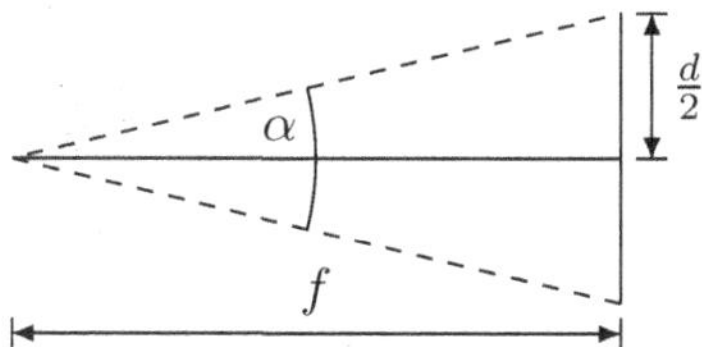

FIGURE 8.6: Viewing angle

A lens with a focal length f of 50mm is considered normal for a 35mm film camera. An image on 35mm film is 36mm wide by 24mm high, so the vertical viewing angle is

$$2\tan^{-1}\left(\frac{24}{2\cdot 50}\right)\approx 27\text{degrees}.$$

An HD monitor is 1920 × 1080 pixels (px). In order to keep the same viewing angle, we must keep the slope the same, and use the new value of $d = 1080$ to solve for f:

$$\frac{24}{2\cdot 50}=\frac{1080}{2f}$$

so $f = 2250$ in pixels. Now we have a problem because the units are in pixels but not all monitors have the same physical size. The monitor of a laptop may be 12 inches wide and a desktop monitor may be 20 inches wide, while both monitors may display 1920 px horizontally. For the desktop we have $1920/20 = 96$ pixels per inch. How far would the eye be from the screen? To get px to cancel, we need in/px:

$$2250\text{ px}\cdot\frac{1\text{ in}}{96\text{ px}}\approx 23.4\text{ in}.$$

That is a comfortable viewing distance, so that is the conversion between pixels and physical distance that I will use. Even though you are likely sitting at your computer, the scene displayed on the screen may assume that a person is standing. Eye height varies as well, but for computational convenience I will use 62.5 inches, because the conversion gives us 6000 px.

With that being said, all spatial coordinates will now be in pixels. The xy-plane will be the floor and the positive z direction will be up. To allow for a person to move, we need a vector $\mathbf{E}$ for the position of the eye, and a basis $\mathbf{v}_1, \mathbf{v}_2, \mathbf{v}_3$ describing the viewing direction and angle, which will be combined as the columns of a single matrix $\mathbf{V}$. The vector $\mathbf{v}_1$ will point to the center of the screen, and originally, we will take it to be in the positive y direction. Screen coordinates are measured from the top left corner of the screen , so it is natural to take $\mathbf{v}_2$ to point right by one unit, and $\mathbf{v}_3$ to point down by one unit. So, far this give us

$$\mathbf{E}=\begin{bmatrix}0\\0\\6000\end{bmatrix}\quad \mathbf{V}=[\mathbf{v}_1|\mathbf{v}_2|\mathbf{v}_3]=\begin{bmatrix}0&1&0\\1&0&0\\0&0&-1\end{bmatrix}.$$

Screen coordinates are measured from the top left corner. If the monitor is $w\times h$, then the center is at $(\frac{w}{2},\frac{h}{2})$. To get from the eye to the pixel with screen coordinates (i, j), we must go f pixels in the direction of $\mathbf{v}_1$, $i-\frac{w}{2}$, in the direction of $\mathbf{v}_2$, and $j-\frac{h}{2}$ in the direction of $\mathbf{v}_3$, which gives us the vector

$$\mathbf{T}=\begin{bmatrix}f\\i-\frac{w}{2}\\j-\frac{h}{2}\end{bmatrix}.$$

We can now convert from screen coordinates (i, j) to x, y, z coordinates by

$$\mathbf{E} + \mathbf{VT}, \tag{8.31}$$

which is valid even when the eye position and viewing direction change. More importantly, **VT** is a tangent vector to the ray stating from the eye in x, y, z coordinates. The ray from a point **P** in the direction **u**, where is given by

$$\mathbf{P} + s\mathbf{u}. \tag{8.32}$$

This is similar to equation (2.24), except that we want **u** to be a unit vector. Then s is the distance from **P** to the point on the ray. When not using a unit vector it is better to use t instead of s to make it clear that it is not a distance.

Example 8.8.1. Compute the formula for the ray from the eye to the pixel with screen coordinates $(210, 415)$ for an HD screen.

Solution

An HD screen is 1920×1080, so

$$i - \frac{w}{2} = 210 - 960 = -750, j - \frac{h}{2} = 415 - 540 = -125.$$

Since $f = 2250$, then

$$\begin{bmatrix} 0 & 1 & 0 \\ 1 & 0 & 0 \\ 0 & 0 & -1 \end{bmatrix} \begin{bmatrix} 2250 \\ -750 \\ -125 \end{bmatrix} = \begin{bmatrix} -750 \\ 2250 \\ 125 \end{bmatrix} = \mathbf{VT}$$

The length of **VT**

$$\| \mathbf{VT} \| = \sqrt{(-750)^2 + 2250^2 + 125^2} = 2375,$$

so

$$\mathbf{u} = \frac{1}{\| \mathbf{Q} - \mathbf{E} \|}(\mathbf{Q} - \mathbf{E}) = \frac{1}{2375} \begin{bmatrix} -750 \\ 2250 \\ 125 \end{bmatrix} = \begin{bmatrix} -\frac{6}{19} \\ \frac{18}{19} \\ \frac{1}{19} \end{bmatrix}.$$

Since the ray is starting at the eye, then $\mathbf{P} = \mathbf{E}$, so the formula for the ray is

$$\mathbf{E} + s\mathbf{u} = \begin{bmatrix} 0 \\ 0 \\ 6000 \end{bmatrix} + s \begin{bmatrix} -\frac{6}{19} \\ \frac{18}{19} \\ \frac{1}{19} \end{bmatrix}.$$

For example, taking $s = 2375$ gives us the point on the screen in x, y, z coordinates.

Activity 8.8.2. Compute the formula for the ray from the eye to the pixel with screen coordinates $(60, 360)$ for an HD screen.

Now that we have a ray, we need to consider how to detect if it intersects with a surface, which is called a hit. While it may be possible to describe surfaces with calculus, it is common to represent surfaces with wire frame models. Imagine approximating a surface by small polygons, the edges of the polygons form a wire frame. An edge can be described by two vectors, one for each endpoint. However, to detect a hit, we want not the edges, but the faces. How can we get a list of faces from a list of edges? We can go through a process of joining edges one after another until we get a closed loop, forming the perimeter of a face. While it is possible to handle polygons with an arbitrary number of sides, its seems the simplest and probably the fastest to use only triangles. Then each face can be represented by three vectors, and it is known that any pair of them gives us a valid edge.

Example 8.8.3. Consider an octahedron with vertices

$$\begin{bmatrix} 1 \\ 0 \\ 0 \end{bmatrix}, \begin{bmatrix} 0 \\ 1 \\ 0 \end{bmatrix}, \begin{bmatrix} 0 \\ 0 \\ 1 \end{bmatrix}, \begin{bmatrix} -1 \\ 0 \\ 0 \end{bmatrix}, \begin{bmatrix} 0 \\ -1 \\ 0 \end{bmatrix}, \begin{bmatrix} 0 \\ 0 \\ -1 \end{bmatrix}. \tag{8.33}$$

The edges are formed by joining each vector with every other vector except its negative. By joining only vectors that occur to the right, we avoid duplicate edges:

$$\begin{gathered} \begin{bmatrix} 1 & 0 \\ 0 & 1 \\ 0 & 0 \end{bmatrix}, \begin{bmatrix} 1 & 0 \\ 0 & 0 \\ 0 & 1 \end{bmatrix}, \begin{bmatrix} 1 & 0 \\ 0 & -1 \\ 0 & 0 \end{bmatrix}, \begin{bmatrix} 1 & 0 \\ 0 & 0 \\ 0 & -1 \end{bmatrix}, \\ \begin{bmatrix} 0 & 0 \\ 1 & 0 \\ 0 & 1 \end{bmatrix}, \begin{bmatrix} 0 & -1 \\ 1 & 0 \\ 0 & 0 \end{bmatrix}, \begin{bmatrix} 0 & 0 \\ 1 & 0 \\ 0 & -1 \end{bmatrix}, \begin{bmatrix} 0 & -1 \\ 0 & 0 \\ 1 & 0 \end{bmatrix}, \\ \begin{bmatrix} 0 & 0 \\ 0 & -1 \\ 1 & 0 \end{bmatrix}, \begin{bmatrix} -1 & 0 \\ 0 & -1 \\ 0 & 0 \end{bmatrix}, \begin{bmatrix} -1 & 0 \\ 0 & 0 \\ 0 & -1 \end{bmatrix}, \begin{bmatrix} 0 & 0 \\ -1 & 0 \\ 0 & -1 \end{bmatrix}. \end{gathered} \tag{8.34}$$

Together, these matrices are the wire frame model for the octahedron with vertices (8.33). To get faces, we look for matrices that share the same first column. We then must check to see if the second columns, which don't match, form another edge. If they do, then we have a face, if they don't then we move on. For example, the first and third matrix in (8.34) have the same first column, but the second columns are

$$\begin{bmatrix} 0 \\ 1 \\ 0 \end{bmatrix} \quad \text{and} \quad \begin{bmatrix} 0 \\ -1 \\ 0 \end{bmatrix},$$

which are not the columns of a single matrix in (8.34), and thus do not form an edge. In this way, proceeding left to right, we find the following faces:

$$\begin{gathered} \begin{bmatrix} 1 & 0 & 0 \\ 0 & 1 & 0 \\ 0 & 0 & 1 \end{bmatrix}, \begin{bmatrix} 1 & 0 & 0 \\ 0 & 1 & 0 \\ 0 & 0 & -1 \end{bmatrix}, \begin{bmatrix} 1 & 0 & 0 \\ 0 & 0 & -1 \\ 0 & 1 & 0 \end{bmatrix}, \begin{bmatrix} 1 & 0 & 0 \\ 0 & -1 & 0 \\ 0 & 0 & -1 \end{bmatrix}, \\ \begin{bmatrix} 0 & 0 & -1 \\ 1 & 0 & 0 \\ 0 & 1 & 0 \end{bmatrix}, \begin{bmatrix} 0 & -1 & 0 \\ 1 & 0 & 0 \\ 0 & 0 & -1 \end{bmatrix}, \begin{bmatrix} 0 & -1 & 0 \\ 0 & 0 & -1 \\ 1 & 0 & 0 \end{bmatrix}, \begin{bmatrix} -1 & 0 & 0 \\ 0 & -1 & 0 \\ 0 & 0 & -1 \end{bmatrix}. \end{gathered} \tag{8.35}$$

Activity 8.8.4. A pyramid is given by the vertices

$$\mathbf{A} = \begin{bmatrix} 1 \\ 0 \\ 0 \end{bmatrix}, \mathbf{B} = \begin{bmatrix} 0 \\ 1 \\ 0 \end{bmatrix}, \mathbf{C} = \begin{bmatrix} -1 \\ 0 \\ 0 \end{bmatrix}, \mathbf{D} = \begin{bmatrix} 0 \\ -1 \\ 0 \end{bmatrix}, \mathbf{E} = \begin{bmatrix} 0 \\ 0 \\ 1 \end{bmatrix},$$

with edges between any pair of vectors that are not negatives of each other, plus one extra edge between **A** and **C**. Find

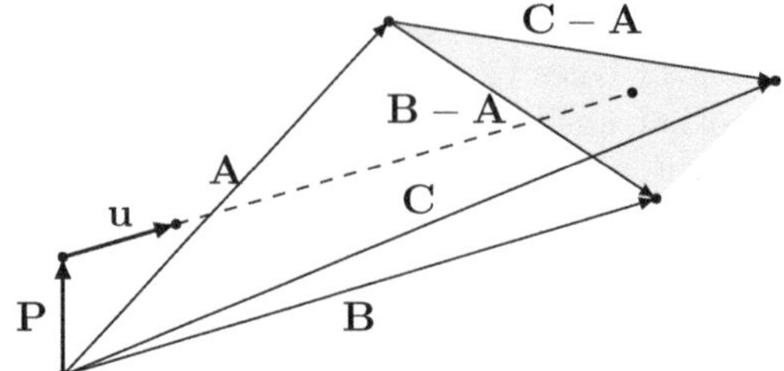

FIGURE 8.7: Determining where a triangular face is hit

1. matrices for the edges defining the wire frame model, and
2. matrices for the faces.

There should be 9 edges and 6 faces.

How do we determine when there is a hit? Suppose that a given face has vertices **A**, **B**, and **C** as illustrated by Figure 8.7. The face lies in the plane determined by **A**, **B**, and **C**, which can be described in vector notation by

$$\mathbf{A} + t_1(\mathbf{B} - \mathbf{A}) + t_2(\mathbf{C} - \mathbf{A}), \tag{8.36}$$

where t_1 and t_2 are arbitrary parameters, similar to (8.32) for rays. The face consists only of the points satisfying

$$0 \le t_1, \quad 0 \le t_2, \quad t_1 + t_2 \le 1. \tag{8.37}$$

So a hit can be detected by the following steps:

1. compute the intersection of the ray and the plane,
2. check if the inequalities (8.37) are satisfied.

The first step can be done by equating (8.32) and (8.37),

$$\mathbf{A} + t_1(\mathbf{B} - \mathbf{A}) + t_2(\mathbf{C} - \mathbf{A}) = \mathbf{P} + s\mathbf{u}, \tag{8.38}$$

which can then be simplified and put into augmented form:

$$\left[\begin{array}{ccc|c} -\mathbf{u} & \mathbf{B} - \mathbf{A} & \mathbf{C} - \mathbf{A} & \mathbf{P} - \mathbf{A} \end{array}\right]$$

where the variables are in the order s, t_1, t_2.

Example 8.8.5. Suppose that we stretch the octahedron in Example 8.8.3 by 100 and move it so that it is centered at $(-1500, 4500, 6200)$. Then two of the faces are

$$\mathbf{F}_1 = \begin{bmatrix} -1400 & -1500 & -1500 \\ 4500 & 4500 & 4400 \\ 6200 & 6300 & 6200 \end{bmatrix}, \quad \mathbf{F}_2 = \begin{bmatrix} -1400 & -1500 & -1500 \\ 4500 & 4400 & 4500 \\ 6200 & 6200 & 6100 \end{bmatrix}.$$

Show that $\mathbf{F}_1$ is hit by the ray computed in Example 8.8.1, while $\mathbf{F}_2$ is not. Also find the vector for the location of the hit.

Solution

For $\mathbf{F}_1$, if we take the columns in the order $\mathbf{A}$, $\mathbf{B}$, and $\mathbf{C}$, then

$$\mathbf{B}-\mathbf{A}=\begin{bmatrix}-100\\0\\100\end{bmatrix},\quad \mathbf{C}-\mathbf{A}=\begin{bmatrix}-100\\-100\\0\end{bmatrix},\quad \mathbf{P}-\mathbf{A}=\begin{bmatrix}1400\\-4500\\-200\end{bmatrix},$$

since $\mathbf{P}=\mathbf{E}$ in this case. Together with $-\mathbf{u}$, this gives us

$$\left[\begin{array}{ccc|c}\frac{6}{19} & -100 & -100 & 1400\\ -\frac{18}{19} & 0 & -100 & -4500\\ -\frac{1}{19} & 100 & 0 & -200\end{array}\right].$$

Using Sage to get reduced form, we obtain

$$\left[\begin{array}{ccc|c}1 & 0 & 0 & \frac{19\times 5700}{23}\\ 0 & 1 & 0 & \frac{11}{23}\\ 0 & 0 & 1 & \frac{9}{23}\end{array}\right],$$

so $t_1=\frac{11}{23}$ and $t_2=\frac{9}{23}$. Clearly t_1 and t_2 are positive, and

$$t_1+t_2=\frac{22}{23}$$

is less than 1, so the inequalities (8.37) are satisfied. We have a hit! It is also important to note that s is positive, which means that the object is in front of us. To get a vector for the location of the hit, we plug back into either side of equation (8.38). Using the right side, we have with the help of Sage

$$\mathbf{P}+s\mathbf{u}=\begin{bmatrix}0\\0\\6000\end{bmatrix}+\frac{19\times 5700}{23}\begin{bmatrix}\frac{6}{19}\\-\frac{18}{19}\\-\frac{1}{19}\end{bmatrix}=\frac{1}{23}\begin{bmatrix}-34200\\102600\\143700\end{bmatrix}.$$

But for $\mathbf{F}_2$, the augmented matrix that we get is

$$\left[\begin{array}{ccc|c}\frac{6}{19} & -100 & -100 & 1400\\ -\frac{18}{19} & -100 & 0 & -4500\\ -\frac{1}{19} & 0 & -100 & -200\end{array}\right].$$

This time, solving with Sage gives us

$$\left[\begin{array}{ccc|c}1 & 0 & 0 & 4636\\ 0 & 1 & 0 & \frac{27}{25}\\ 0 & 0 & 1 & -\frac{11}{25}\end{array}\right].$$

Since $t_2=-\frac{11}{25}$ is negative, we do not even need to check t_1+t_2. The inequalities are not satisfied, so $\mathbf{F}_2$ is not hit.

It is possible that a ray hits more than one face. We care only about the face that it hits first, which we can find by computing the minimum positive value of s among all hits. Once that face has been determined, then we can compute the exact point of intersection, which becomes the new value of $\mathbf{P}$ for any new ray. The new value of $\mathbf{u}$ depends on whether the ray is the result of reflection, refraction, or scattering. In all cases we will need a unit vector $\mathbf{N}$ that is normal to the surface, which can be computed by the cross product

$$\mathbf{N} = \frac{(\mathbf{B}-\mathbf{A}) \times (\mathbf{C}-\mathbf{A})}{\|(\mathbf{B}-\mathbf{A}) \times (\mathbf{C}-\mathbf{A})\|}. \tag{8.39}$$

Example 8.8.6. Compute $\mathbf{N}$ for the face $\mathbf{F}_1$ from Example 8.8.5.

Solution

We compute $(\mathbf{B}-\mathbf{A}) \times (\mathbf{C}-\mathbf{A})$ from the determinant with the standard basis vectors in the components of the first column, $\mathbf{B}-\mathbf{A}$ in the second column and $\mathbf{C}-\mathbf{A}$ in the third column. Borrowing the results for $\mathbf{B}-\mathbf{A}$ and $\mathbf{C}-\mathbf{A}$, we have

$$\begin{vmatrix} \mathbf{e}_1 & -100 & -100 \\ \mathbf{e}_2 & 0 & -100 \\ \mathbf{e}_3 & 100 & 0 \end{vmatrix} = \begin{vmatrix} 0 & -100 \\ 100 & 0 \end{vmatrix} \mathbf{e}_1 - \begin{vmatrix} -100 & -100 \\ 100 & 0 \end{vmatrix} \mathbf{e}_2 + \begin{vmatrix} -100 & -100 \\ 0 & -100 \end{vmatrix} \mathbf{e}_3$$

$$= -10000\mathbf{e}_1 - 10000\mathbf{e}_2 + 10000\mathbf{e}_3 = 10000 \begin{bmatrix} -1 \\ -1 \\ 1 \end{bmatrix}.$$

The length of this vector is $10000\sqrt{(-1)^2 + (-1)^2 + 1^2} = 10000\sqrt{3}$, so

$$\mathbf{N} = \frac{1}{\sqrt{3}} \begin{bmatrix} -1 \\ -1 \\ 1 \end{bmatrix}$$

For reflection, the angle between the incoming ray and $\mathbf{N}$ must be the same as the angle of the outgoing ray and $\mathbf{N}$. But the direction changes: in Figure 8.8a, the vector $\mathbf{u}_1$ is toward the point of reflection and $\mathbf{u}_2$ is away. By equation (1.14) we have

$$\mathbf{u}_1 \cdot \mathbf{N} = \cos(\theta) \quad \text{and} \quad \mathbf{u}_2 \cdot \mathbf{N} = \cos(\pi - \theta) = -\cos(\theta),$$

thus $\mathbf{u}_1 \cdot \mathbf{N} = -\mathbf{u}_2 \cdot \mathbf{N}$. Then by vector projection, we obtain

$$\mathbf{u}_2 = \mathbf{u}_1 - 2(\mathbf{u}_1 \cdot \mathbf{N})\mathbf{N} \tag{8.40}$$
$$= (\mathbf{I} - 2(\mathbf{N} \otimes \mathbf{N}))\mathbf{u}_1. \tag{8.41}$$

The second equality is due to associativity of matrix multiplication:

$$(\mathbf{N} \otimes \mathbf{N})\mathbf{u}_1 = (\mathbf{N}\mathbf{N}^T)\mathbf{u}_1 = \mathbf{N}(\mathbf{N}^T\mathbf{u}_1) = \mathbf{N}(\mathbf{N} \cdot \mathbf{u}_1),$$

but $\mathbf{N} \cdot \mathbf{u}_1$ is a scalar, and it can be written on either side of $\mathbf{N}$. This explains why (4.25) is a reflection.

(a) Reflection (b) Refraction

FIGURE 8.8: Reflected and refracted rays

Example 8.8.7. Compute the reflection of the ray computed in Example 8.8.1, when it hits the face $\mathbf{F}_1$ in Example 8.8.5.

Solution

Using $\mathbf{N}$ from Example 8.8.6 and $\mathbf{u}_1 = \mathbf{u}$ from Example 8.8.1 we find

$$\mathbf{N} \cdot \mathbf{u}_1 = \frac{1}{19\sqrt{3}}((-1)(-6) + (-1)(18) + (1)(1)) = \frac{-11}{19\sqrt{3}}.$$

Then by equation (8.40)

$$\mathbf{u}_2 = \begin{bmatrix} -\frac{6}{19} \\ \frac{18}{19} \\ \frac{1}{19} \end{bmatrix} - 2 \cdot \frac{-11}{19\sqrt{3}} \begin{bmatrix} -\frac{1}{\sqrt{3}} \\ -\frac{1}{\sqrt{3}} \\ \frac{1}{\sqrt{3}} \end{bmatrix} = \begin{bmatrix} -\frac{18}{57} \\ \frac{54}{57} \\ \frac{3}{57} \end{bmatrix} + \begin{bmatrix} -\frac{22}{57} \\ -\frac{22}{57} \\ \frac{22}{57} \end{bmatrix} = \begin{bmatrix} -\frac{40}{57} \\ \frac{32}{57} \\ \frac{25}{57} \end{bmatrix}.$$

This gives us the value of $\mathbf{u}$ needed for the reflected ray. The new value of $\mathbf{P}$ is the location of the hit, which was computed in Example 8.8.5. Thus the general formula for the reflected ray is

$$\mathbf{P} + s\mathbf{u} = \frac{1}{23} \begin{bmatrix} -34200 \\ 102600 \\ 143700 \end{bmatrix} + s \begin{bmatrix} -\frac{40}{57} \\ \frac{32}{57} \\ \frac{25}{57} \end{bmatrix}.$$

For refraction, the incoming and outgoing rays are related by Snell's law

$$n_1 \sin(\theta_1) = n_2 \sin(\theta_2) \tag{8.42}$$

where n_1 is the index of refraction of the medium that the incoming ray is in, and n_2 is the index of refraction of the medium that the outgoing ray is in. Some common indices of refraction are:

Vacuum	1
Air	1.0003
Water	1.333
Window glass	1.5

The index of reflection depends on the wavelength of light (the color), which leads to rainbows. We now show how to get $\mathbf{u}_2$ using vector projection. Let $\mathbf{e}_1$ and $\mathbf{e}_2$ be an orthonormal basis in the plane that $\mathbf{u}_1$ lies in, with $\mathbf{e}_1 = \mathbf{N}$. Then

$$\mathbf{u}_1 = \cos(\theta_1)\mathbf{e}_1 + \sin(\theta_1)\mathbf{e}_2.$$

Since $\mathbf{u}_1$ and $\mathbf{N}$ are both unit vectors, then $\mathbf{u}_1 \cdot \mathbf{N} = \cos(\theta_1)$, so

$$\cos(\theta_1)\mathbf{e}_1 = (\mathbf{u}_1 \cdot \mathbf{N})\mathbf{N}, \quad \text{and} \quad \sin(\theta_1)\mathbf{e}_2 = \mathbf{u}_1 - (\mathbf{u}_1 \cdot \mathbf{N})\mathbf{N}. \tag{8.43}$$

We want to obtain

$$\mathbf{u}_2 = \cos(\theta_2)\mathbf{e}_1 + \sin(\theta_2)\mathbf{e}_2.$$

By equation (8.42) and (8.43),

$$\sin(\theta_2)\mathbf{e}_2 = \frac{\sin(\theta_2)}{\sin(\theta_1)} \sin(\theta_1)\mathbf{e}_2 = \frac{n_1}{n_2} (\mathbf{u}_1 - (\mathbf{u}_1 \cdot \mathbf{N})\mathbf{N}). \tag{8.44}$$

The squared length of this vector is

$$\frac{n_1^2}{n_2^2} \sin^2(\theta_1) = \frac{n_1^2}{n_2^2} \left(1 - \cos^2(\theta_1)\right), \quad \text{where} \quad \cos^2(\theta_1) = (\mathbf{u}_1 \cdot \mathbf{N})^2. \tag{8.45}$$

If the length of (8.44) is greater than 1, then any component we add will only make it longer so it will be impossible to obtain $\mathbf{u}_2$. When light travels from a high index of refraction to a low index of refraction, such as from water to air, it can only escape if θ_1 is less than some critical angle. The critical angle occurs when θ_2 is 90 degrees. Then $\sin(\theta_2) = 1$, and solving equation (8.42) for θ_1 gives

$$\theta_1 = \sin^{-1}\left(\frac{n_2}{n_1}\right). \tag{8.46}$$

For example, if the ray goes from water to air, then $n_1 = 1.333$ and $n_2 = 1.0003$, so the critical angle is

$$\sin^{-1}\left(\frac{1.0003}{1.333}\right) \approx 48.6 \text{ degrees.}$$

If the value of (8.45) is less than 1, then θ_1 is less than the critical angle, and we can compute $\cos(\theta_2)$. Using the Pythagorean trig identities, we find:

$$\begin{aligned}\frac{\cos^2(\theta_2)}{\cos^2(\theta_1)} &= \frac{1-\sin^2(\theta_2)}{\cos^2(\theta_1)} = \frac{1}{\cos^2(\theta_1)} - \frac{\sin^2(\theta_1)}{\cos^2(\theta_1)} \cdot \frac{\sin^2(\theta_2)}{\sin^2(\theta_1)} \\ &= \sec^2(\theta_1) - \tan^2(\theta_1) \cdot \frac{n_1^2}{n_2^2} = \sec^2(\theta_1) - (\sec^2(\theta_1) - 1) \cdot \frac{n_1^2}{n_2^2} \\ &= \sec^2(\theta_1)\left(1 - \frac{n_1^2}{n_2^2}\right) + \frac{n_1^2}{n_2^2}.\end{aligned}$$

Then $\mathbf{u}_2$ can be computed as follows.

Step 1: Compute the length of (8.44) by using (8.45). If the length is less than one, proceed to the next step.

Step 2: Compute the vector (8.44).

Step 3: Compute $\sec^2(\theta_1) = \frac{1}{\cos^2(\theta_1)}$ using the value of cosine computed in step 1, then compute

$$\cos(\theta_2)\mathbf{e}_2 = \frac{\cos(\theta_2)}{\cos(\theta_1)}\cos(\theta_1)\mathbf{e}_2 = \sqrt{\sec^2(\theta_1)\left(1 - \frac{n_1^2}{n_2^2}\right) + \frac{n_1^2}{n_2^2}}\left(\frac{\mathbf{u}_1 \cdot \mathbf{N}}{\| \mathbf{N} \|^2}\mathbf{N}\right) \tag{8.47}$$

Step 4: Add the vectors (8.44) and (8.47) to obtain $\mathbf{u}_2$.

```
import numpy as np

Screen = np.zeros((1920,1080,3),dytpe=ushort)

E = np.array((0,0,6000))
V = np.array([[0,1,0],[1,0,0],[0,0,-1]])
W = np.arange(w)-w//2
H = np.arange(h)-h//2

def ViewVectors(V, f = 2250, w = 1920, h = 1080):
    A = np.array(np.meshgrid(f,W,H))
    return np.dot(np.transpose(A.reshape((3,w,h)),(1,2,0)),V)

def unit(A,axis=-1): return A/np.linalg.norm(A,axis=axis)

#make initial ray with E and ViewVectors(V)
```

```
#def HitMask(A):
# M = np.logical_and.reduce(A>=0,axis=-1)
# M &= A[...:1]+A[...:2]<=1
# return M
#X0 = np.vstack([np.identity(3),-np.identity(3)]).astype(int)
```

Exercises

Problem 8.1. Use row reduction to balance each of the following chemical equations

1. $x_1\,H_2S + x_2\,O_2 \longrightarrow x_3\,H_2O + x_4\,SO_2$
2. $x_1\,KO_2 + x_2\,CO_2 \longrightarrow x_3\,K_2CO_3 + x_4\,O_2$
3. $x_1\,NH_3 + x_2\,O_2 \longrightarrow x_3\,NO + x_4\,H_2O$
4. $x_1\,NH_3 + x_2\,O_2 \longrightarrow x_3\,NO_2 + x_4\,H_2O$
5. $x_1\,NH_3 + x_2\,O_2 \longrightarrow x_3\,NO + x_4\,NO_2 + x_5\,H_2O$
6. $x_1\,C_2H_5OH + x_2\,O_2 \longrightarrow x_3\,CO_2 + x_4\,H_2O$
7. $x_1\,C_3H_8 + x_2\,O_2 \longrightarrow x_3\,CO_2 + x_4\,H_2O$
8. $x_1\,C_3H_8 + x_2\,O_2 \longrightarrow x_3\,CO + x_4\,H_2O$
9. $x_1\,C_3H_8 + x_2\,O_2 \longrightarrow x_3\,CO + x_4\,CO_2 + x_5\,H_2O$
10. $x_1\,C_8H_{18} + x_2\,O_2 \longrightarrow x_3\,CO_2 + x_4\,H_2O$
11. $x_1\,C_8H_{18} + x_2\,O_2 \longrightarrow x_3\,CO + x_4\,H_2O$
12. $x_1\,C_8H_{18} + x_2\,O_2 \longrightarrow x_3\,CO + x_4\,CO_2 + x_5\,H_2O$
13. $x_1\,Zn(OH)_2 + x_2\,H_3O^+ \longrightarrow x_3\,Zn^{2+} + x_4\,H_2O$
14. $x_1\,Sn(OH)_4 + x_2\,H_3O^+ \longrightarrow x_3\,Sn^{4+} + x_4\,H_2O$
15. $x_1\,MnO_4^- + x_2\,Sn^{2+} + x_3\,H^+ \longrightarrow x_4\,Mn^{2+} + x_5\,Sn^{4+} + x_6\,H_2O$
16. $x_1\,CrO_7^{-2} + x_2\,Fe^{2+} + x_3\,H^+ \longrightarrow x_4\,Cr^{3+} + x_5\,Fe^{3+} + x_6\,H_2O$

Problem 8.2. A cat has 3 activities: eating, sleeping and, going outside. If the cat is eating on one observation, it will not be eating on the next observation, and is equally likely to be asleep or outside. If the cat is sleeping on one observation it will not be outside on the next observation, and is twice as likely to be asleep as it is to be eating. If the cat is outside on one observation, then it will not be eating on the next observation, and the probability that it will be sleeping is $\frac{1}{4}$.

1. Draw the transition diagram, and find the corresponding transition matrix.
2. Suppose the cat is sleeping on the 1st observation.
   (a) Find the probability that it is sleeping on the 3rd observation.
   (b) Find the probability that it alternates between sleeping and eating for the first 4 observations.
3. Find the stable state vector for the transition matrix.

Problem 8.3. A taxi cab driver in New York City is driving around aimlessly looking for work. If the driver turned left at the last intersection, then he will not turn right at this intersection, and there is a $\frac{1}{6}$ probability that he will go straight. If the driver went straight through the last intersection, then he will not go straight through this one, and he is equally likely to turn left or right. If he turned right at the last intersection, then he will not turn left at this one, and he is 3 times more likely to turn right again than he is to go straight.

1. Draw the transition diagram, and find the corresponding transition matrix.
2. Suppose the taxi cab driver goes straight through the 1st intersection.
   (a) Find the probability that he goes straight through the 3rd intersection.
   (b) Find the probability that he turns left at the next 3 intersections.
3. Find the stable state vector for the transition matrix.

Problem 8.4. Pick one of the following 3 matrices, and show that it corresponds to a regular Markov chain.

$$1.\ \begin{bmatrix} \frac{1}{2} & \frac{1}{2} & 0 \\ \frac{2}{5} & 0 & \frac{3}{5} \\ 0 & 1 & 0 \end{bmatrix} \qquad 2.\ \begin{bmatrix} 0 & \frac{2}{5} & \frac{3}{5} \\ \frac{1}{2} & \frac{1}{2} & 0 \\ 1 & 0 & 0 \end{bmatrix} \qquad 3.\ \begin{bmatrix} 0 & \frac{3}{5} & \frac{2}{5} \\ 1 & 0 & 0 \\ \frac{1}{2} & 0 & \frac{1}{2} \end{bmatrix}$$

Problem 8.5. Given the transition matrix

$$\mathbf{P} = \begin{bmatrix} \frac{1}{2} & \frac{1}{2} \\ 1 & 0 \end{bmatrix}$$

the general expression for $\mathbf{P}^k$ is

$$\mathbf{P}^k = \frac{1}{3} \begin{bmatrix} 2 + \left(-\frac{1}{2}\right)^k & 1 - \left(-\frac{1}{2}\right)^k \\ 2 + \left(-\frac{1}{2}\right)^{k-1} & 1 - \left(-\frac{1}{2}\right)^{k-1} \end{bmatrix}$$

(you do not need to show this). Do the following steps:

1. Show that the formula for $\mathbf{P}^k$ is valid for $k = 1$ and $k = 2$ by direct calculation. Conclude that $\mathbf{P}$ is regular.
2. Calculate the stable state vector $\mathbf{w}$ for $\mathbf{P}$.
3. Show that

$$\lim_{k\to\infty} \frac{1}{3} \left[2 + \left(-\tfrac{1}{2}\right)^k \quad 1 - \left(-\tfrac{1}{2}\right)^k\right]$$

   is equal to the stable state vector for $\mathbf{P}$, hence $\lim_{k\to\infty} \mathbf{P}^k = \mathbf{M}$ where each row of $\mathbf{M}$ is equal to the stable state vector $\mathbf{w}$.

4. Keeping the same definitions as in the previous part, let $\mathbf{w} = [w_1, w_2]$ be an arbitrary state vector. Show by direct calculation that

$$\mathbf{wM} = \mathbf{w}$$

regardless of the choice of $\mathbf{w}$, hence together with $\lim_{k\to\infty} \mathbf{P}^k = \mathbf{M}$ we can conclude that

$$\lim_{k\to\infty} \mathbf{wP}^k = \mathbf{w}.$$

Problem 8.6 (With AI). Give the following prompt to an AI:

"Construct a Markov chain model for a simplified Monopoly game with only 12 squares in the following order: Go, Properties 1-2, Chance, Properties 3-4, Jail, Properties 5-6, Railroad, Properties 7-8. Assume a deck of 10 chance cards, each of which sends the player to one of the squares other than Go or Chance. Also assume that a player must wait 1 turn in jail, so that one extra state is needed. Construct a 13 by 13 transition matrix compatible with Sage in which the rows sum to 1. Then use the built-in functions in Sage to find the steady-state vector."

Possible extensions of this problem

(a) Ask the AI to explain an optimal strategy for buying properties based on the assumption that the property value increases linearly.

(b) Ask the AI to write an actual version of this game in JavaScript that can be used in any web browser.

(c) Ask the AI to write an AI opponent with different hardness settings, the hardest of which should use the optimal strategy.

Problem 8.7. Use the FFT to compute each product below:

a 12×21 b 12×42 c 34×43

Hints

Chapter 1

Problem 1.5. Take the matrices

$$\begin{bmatrix} 0 & 1 \\ 0 & 1 \end{bmatrix} \begin{bmatrix} 1 & 1 \\ 0 & 0 \end{bmatrix}$$

as **A** and **B** respectively. Design **C** in such a way that **AC** is the same as **AB**.

Chapter 2

Problem 2.1 parts c and d. Try to think in terms of the classification given in section 2.1. A line in two dimensions is given by an equation with two variables. Using 3 equations gives us an example for part c. A plane in three dimensions is given by an equation with 3 variables. Using 2 equations gives us an example for part d.

Problem 2.6. This can be done a couple ways. Row reduction is not really necessary if it is approached following the examples and classification in section 2.1. For instance, for infinitely many solutions, the second equation must be a multiple of the first, which determines h and k. For no solution, the second equation must be parallel to the first, which can be done by changing k. On the other hand, by row reduction, we can get

$$\left[\begin{array}{cc|c} 1 & -3 & 1 \\ 0 & f_1(h) & f_2(k) \end{array}\right]$$

where $f_1(h)$ is an expression involving h and $f_2(k)$ is an expression involving k. For infinitely many solutions, the second row must have all zeros, so set the expressions in h and k equation to zero and solve them. For no solutions, we must have zeros to the left of the vertical bar, and a non-zero value to the right.

Chapter 3

Problem 3.1 parts 1 and 5. Problem 3.5 provides a relevant example.

Problem 3.1 parts 2, 3 and 4. See the definition, and example 3.3.5.

Problem 3.6. This is supposed to be "dependence relations by stare down." If only two vectors are involved, they must be scalar multiples of each other.

Problem 3.8. See example 3.4.1

Chapter 4

Problem 4.1. There are many relevant details in section 4.2. First, and foremost is the definition of one-to-one and onto (Definition 4.3.6), and example 4.3.7 used to illustrate it. Co-domain and range are defined in definitions 4.3.2 and 4.3.1 respectively. Definition 4.3.3 introduces the column space as another name for the range, and theorem 4.3.5 provides useful information relating the column space with the null space. Taking all of these facts into account, the question can be understood in terms of pivots and free variables.

Problem 4.4. For parts 1 and 2, start by writing down an arbitrary linear combination with a, b, c, and d as coefficients. Then ask yourself the following two questions: is it possible to get any 2×2 matrices, and how many ways are there of getting the zero matrix. For part 3, the matrix for T is constructed by the same method used in examples 4.2.1, 4.2.4, 4.2.7, and 4.2.9. Take the transpose of each of the matrices $\mathbf{e}_{11}$, $\mathbf{e}_{12}$, $\mathbf{e}_{21}$, $\mathbf{e}_{22}$, then write the results as a linear combinations in the same basis. The coefficients will give you a 4×4 matrix that naturally acts on the column vectors $\mathbf{f}_1$, $\mathbf{f}_2$, $\mathbf{f}_3$, and $\mathbf{f}_4$.

Problem 4.8. Example 4.2.7 shows how to compute a matrix for a linear transformation, involving the evaluation of a polynomial.

Problem 4.11. Examples 4.2.1, 4.2.4, 4.2.7, and 4.2.9 show how to construct matrices from a transformation. Example 4.3.7 illustrates the concepts of one-to-one and onto.

Chapter 5

Problem 5.1 part 1. Consider counter-clockwise rotation by 90 degrees.

Problem 5.1 part 2. Consider example 5.3.3.

Problem 5.1 part 4. Consider the very first example of eigenvectors.

Problem 5.6. See example 5.3.1 for the real case and example 5.3.4 for the complex case.

Chapter 6

Problem 6.3. It may help to graph the vectors $\mathbf{u}$ and $\mathbf{v}$. The $\cos(\theta)$ has an exact value that may be familiar.

Problem 6.6. See example 6.2.7.
Problem 6.7. See example 6.3.1.
Problem 6.7. Start with

$$\mathbf{A} = \begin{bmatrix} a & b \\ c & d \end{bmatrix}.$$

Compute $\mathbf{A}\mathbf{v}_1$, and set it equal to $0\mathbf{v}_1$. The resulting equations should reduce the number of variables. Next compute $\mathbf{A}\mathbf{v}_2$ and set it equal to $4\mathbf{v}_2$.

Chapter 7

Problem 7.1. See example 7.2.1.

Chapter 8

Problems 8.2 and 8.3. For constructing a transition diagram and transition matrix, see example 8.4.3. For computing probabilities of subsequent observations, see example 8.4.9. For computing the stable state vector, see example 8.4.23.

Problem 8.4. You should use + and 0 as in example 8.4.14.

Problem 8.5. See example 8.4.22.

Problem 8.7. See example 8.6.9.

Bibliography

[1] et al. Almarode, John T. *How Learning Works : A Playbook.* Corwin Press, 2021.

[2] Sheldon Axler. *Linear algebra done right.* Undergraduate Texts in Mathematics. Springer, Cham, fourth edition, 2024.

[3] Benjamin Braun, Priscilla Bremser, Art Duval, Elise Lockwood, and Diana White. What does active learning mean for mathematicians? *Notices of the American Mathematical Society*, 64:124–129, 02 2017.

[4] P.C. Brown, H.L. Roediger, and M.A. McDaniel. *Make It Stick: The Science of Successful Learning.* Harvard University Press, 2014.

[5] F. Albert (Frank Albert) Cotton. *Chemical applications of group theory.* Wiley, New York, 3rd ed. edition, 1990.

[6] P. G. L. Dirichlet. *Lectures on number theory*, volume 16 of *History of Mathematics.* American Mathematical Society, Providence, RI; London Mathematical Society, London, 1999. Supplements by R. Dedekind, Translated from the 1863 German original and with an introduction by John Stillwell.

[7] Alan Edelman. The mathematics of the pentium division bug. *SIAM Rev.*, 39:54–67, 1997.

[8] T.J. Fletcher. *Linear Algebra; Through Its Applications.* Van Nostrand Reinhold, 1972.

[9] I. M. Gelfand. *Lectures on linear algebra.* Dover Books on Advanced Mathematics. Dover Publications, Inc., New York, russian edition, 1989. With the collaboration of Z. Ya. Shapiro, Reprint of the 1961 translation.

[10] Griffiths. *Introduction to Quantum Mechanics.* Pearson Prentice Hall, 2nd edition, 2004.

[11] Nicholas J. Higham. *Accuracy and Stability of Numerical Algorithms.* Society for Industrial and Applied Mathematics, second edition, 2002.

[12] Adolf Hurwitz. *Mathematische Werke. Bd. II: Zahlentheorie, Algebra und Geometrie.* Birkhäuser Verlag, Basel-Stuttgart, 1963. Herausgegeben von der Abteilung für Mathematik und Physik der Eidgenössischen Technischen Hochschule in Zürich.

[13] Wilfred Kaplan. *Advanced calculus.* Addison-Wesley Publishing Company, Advanced Book Program, Reading, MA, third edition, 1984.

[14] J. Kotz, P. Treichel, and G. Weaver. *Chemistry and Chemical Reactivity.* Brooks/Cole, 6th edition, 2005.

[15] D.C. Lay, S.R. Lay, and J.J. McDonald. *Linear Algebra and Its Applications.* Addison Wesley, 6th edition, 2005.

[16] et al. Lovett, Marsha C. *How Learning Works : Eight Research-Based Principles for Smart Teaching.* John Wiley & Sons, 2 edition, 2023.

[17] J.T. Moore. *Elementary Linear and Matrix Algebra: The Viewpoint of Geometry.* McGraw-Hill, 1972.

[18] J. Oberg. Why the mars probe went off course [accident investigation]. *IEEE Spectrum*, 36(12):34–39, 1999.

[19] C. D. Olds. *Continued fractions.* Random House, New York, 1963.

[20] G. Polya. *How to solve it.* Princeton Science Library. Princeton University Press, Princeton, NJ, 2004. A new aspect of mathematical method, Expanded version of the 1988 edition, with a new foreword by John H. Conway.

[21] J.S. Przemieniecki. *Theory of Matrix Structural Analysis.* Dover Civil and Mechanical Engineering. Dover, 1985.

[22] R. Resnick, D. Halliday, and K.S. Krane. *Physics.* Number v. 1 in Physics. John Wiley & Sons, Incorporated, 5th edition, 2001.

[23] Jean-Pierre Serre. *Linear representations of finite groups*, volume Vol. 42 of *Graduate Texts in Mathematics.* Springer-Verlag, New York-Heidelberg, 1977.

[24] J. Stoer and R. Bulirsch. *Introduction to numerical analysis*, volume 12 of *Texts in Applied Mathematics.* Springer-Verlag, New York, second edition, 1993. Translated from the German by R. Bartels, W. Gautschi and C. Witzgall.

[25] Richard S. Varga. *Geršgorin and his circles*, volume 36 of *Springer Series in Computational Mathematics.* Springer-Verlag, Berlin, 2004.

[26] Martin H. Weissman. *An illustrated theory of numbers.* American Mathematical Society, Providence, RI, 2017.

[27] Stanley M. Williamson and Charles W. Koch. Xenon Tetrafluoride: Reaction with Aqueous Solutions. *Science*, 139(3559):1046–1047, March 1963.

[28] W.L. Winston, M. Venkataramanan, and Winston. *Introduction to Mathematical Programming: Applications and Algorithms.* Brooks/Cole, 2002.

[29] Stefan Wörner. Fast fourier transform numerical analysis seminar. 2008.

Index